RANGE
Repair
Manual
1970-1985

This manual contains supplements covering Range Rover Five Speed
Manual gearbox, Transfer gearbox, Automatic gearbox and changes and
additions to the Electrical equipment.

Land-Rover Ltd
A Managing Agent for Land Rover UK Limited
PUBLICATION No. AKM 3630 (EDITION 7)

Land Rover Limited
Lode Lane
Solihul
West Midlands B92 8NW
England

INTRODUCTION

The purpose of this cumulative manual is to assist skilled mechanics in the efficient repair and maintenance of Range-Rover vehicles from their inception. The procedures detailed, carried out in the sequence given and using the appropriate service tools, will enable the operations to be completed in the time stated in the Repair Operation Times.

Indexing

For convenience, this Manual is divided into a number of divisions. A contents page listing the titles and reference numbers of the various divisions is shown opposite.

A list of the operations within each of the divisions appears in alphabetical order on the contents page preceding each of the divisions.

Operation Numbering

Each operation is followed by the number allocated to it in a master index. The number consists of six digits arranged in three pairs.

The master index of operations has been compiled for universal application to vehicles manufactured by BL Limited and therefore continuity of the numbering sequence is not maintained throughout the Manual. To assist with locating information, each division of the Manual is preceded by a contents page listing the operations in alphabetical order.

Each instruction within an operation has a sequence number and, to complete the operation in the minimum time it is essential that these instructions are performed in numerical sequence commencing at 1 unless otherwise stated. Where applicable the sequence numbers identify the components in the appropriate illustration.

Where performance of an operation requires the use of service tool, the tool number is quoted under the operation heading and is repeated in, or following the instruction involving its use.

An illustrated list of all service tools necessary to complete the operations described in the Manual is also included.

References

References to the left or right hand side in the Manual are made when viewing the vehicle from the rear. With the engine and gearbox assembly removed, the water pump end of the engine is referred to as the front.

Repairs and Replacements

When service parts are required it is essential that only genuine Land Rover replacements are used.

Attention is particularly drawn to the following points concerning repairs and the fitting of replacement parts and accessories.

Safety features embodied in the vehicle may be impaired if other than genuine parts are fitted.

In certain territories, legislation prohibits the fitting of parts not to the vehicle manufacturers' specification.

Torque wrench setting figures given in the Repair Operation Manual must be strictly adhered to. Locking devices, where specified, must be fitted. If the efficiency of a locking device is impaired during removal it must be renewed.

Owners purchasing accessories while travelling abroad should ensure that the accessory and its fitted location on the car conform to mandatory requirements existing in their country of origin.

The vehicle warranty may be invalidated by the fitting of other than genuine Land Rover parts.

All Land Rover replacements have the full backing of the factory warranty.

Land Rover and Range Rover Distributors and Dealers are obliged to supply only genuine service parts.

Purchasers are advised that the specification details set out in this Manual apply to a range of vehicles and not to any particular vehicle. For the specification of any particular vehicle Purchasers should consult their Distributor or Dealer.

The Manufacturers reserve the right to vary their specifications with or without notice, and at such times and in such manner as they think fit. Major as well as minor changes may be involved in accordance with the manufacturer's policy of constant product improvement.

Whilst every effort is made to ensure the accuracy of the particulars contained in this Manual, neither the Manufacturer nor the Distributor or Dealer, by whom this Manual is supplied, shall in any circumstances be held liable for any inaccuracy or the consequences thereof.

© Land Rover Limited 1985

CONTENTS

SAFETY

POISONOUS SUBSTANCES

WARNING

Many liquids and other substances used in motor vehicles are poisonous and should under no circumstances be consumed and should as far as possible be kept away from open wounds. These substances among others include antifreeze, brake fluid, fuel, windscreen washer additives, lubricants and various adhesives.

FUEL HANDLING PRECAUTIONS

The following information provides basic precautions which must be observed if petrol (gasoline) is to be handled safely. It also outlines the other areas of risk which must not be ignored.

This information is issued for basic guidance only, and in any case of doubt appropriate enquiries should be made of your local Fire Officer.

General

Petrol/gasoline vapour is highly flammable and in confined spaces is also very explosive and toxic.

When petrol/gasoline evaporates it produces 150 times its own volume in vapour, which when diluted with air becomes a readily ignitable mixture. The vapour is heavier than air and will always fall to the lowest level. It can readily be distributed throughout a workshop by air current, consequently, even a small spillage of petrol/gasoline is potentially very dangerous.

Always have a fire extinguisher containing FOAM CO_2 GAS or POWDER close at hand when handling or draining fuel, or when dismantling fuel systems and in areas where fuel containers are stored.

Always disconnect the vehicle battery BEFORE carrying out dismantling or draining work on a fuel system.

Whenever petrol/gasoline is being handled, drained or stored, or when fuel systems are being dismantled all forms of ignition must be extinguished or removed, any headlamps used must be flameproof and kept clear of spillage.

NO ONE SHOULD BE PERMITTED TO REPAIR COMPONENTS ASSOCIATED WITH PETROL/GASOLINE WITHOUT FIRST HAVING HAD SPECIALIST TRAINING.

Fuel Tank Draining

WARNING: PETROL/GASOLINE MUST NOT BE EXTRACTED OR DRAINED FROM ANY VEHICLE WHILST IT IS STANDING OVER A PIT.

Draining or extracting petrol/gasoline from vehicle fuel tank must be carried out in a well ventilated area.

The receptacle used to contain the petrol/gasoline must be more than adequate for the full amount of fuel to be extracted or drained. The receptacle should be clearly marked with its contents, and placed in a safe storage area which meets the requirements of local authority regulations.

WHEN PETROL/GASOLINE HAS BEEN EXTRACTED OR DRAINED FROM A FUEL TANK THE PRECAUTIONS GOVERNING NAKED LIGHTS AND IGNITION SOURCES SHOULD BE MAINTAINED.

Fuel Tank Removal

On vehicle where the fuel line is secured to the fuel tank outlet by a spring steel clip, it is recommended that such clips are released before the fuel line is disconnected or the fuel tank unit is removed. This procedure will avoid the possibility of residual petrol fumes in the fuel tank being ignited when the clips are released.

As an added precaution fuel tanks should have a PETROL/GASOLINE VAPOUR warning label attached to them as soon as they are removed from the vehicle.

Fuel Tank Repair

Under no circumstances should a repair to any tank involving heat treatment be carried out without first rendering the tank SAFE, by using one of the following methods:

(a) STEAMING: With the filler cap and tank unit removed, empty the tank. Steam the tank for at least 2 hours with low pressure steam. Position the tank so that condensation can drain away freely, ensuring that any sediment and sludge not volatised by the steam, is washed out during the steaming process.

(b) BOILING: With the filler cap and tank unit removed, empty the tank. Immerse the tank completely in boiling water containing an effective alkaline degreasing agent or a detergent, with the water filling and also surrounding the tank for at least 2 hours.

After steaming or boiling a signed and dated label to this effect should be attached to the tank.

ABBREVIATIONS AND SYMBOLS USED IN THIS MANUAL

Across flats (bolt size)	AF
After bottom dead centre	ABDC
After top dead centre	ATDC
Alternating current	a.c.
Ampere	amp
Ampere-hour	amp hr
Atmospheres	Atm
Before bottom dead centre	BBDC
Before top dead centre	BTDC
Bottom dead centre	BDC
Brake mean effective pressure	BMEP
Brake horse power	bhp
British Standards	BS
Carbon monoxide	CO
Centimetre	cm
Centigrade (Celcius)	C
Cubic centimetre	cm^3
Cubic inch	in^3
Degree (angle)	deg or °
Degree (temperature)	deg or °
Diameter	dia.
Direct current	d.c.
Fahrenheit	F
Feet	ft
Feet per minute	ft/min
Fifth	5th
Figure (illustration)	Fig.
First	1st
Fourth	4th
Gramme (force)	gf
Gramme (mass)	g
Gallons	gal
Gallons (US)	US gal
High compression	h.c.
High tension (electrical)	H.T.
Hundredweight	cwt
Independent front suspension	i.f.s.
Internal diameter	i.dia.
Inches of mercury	in.Hg
Inches	in
Kilogramme (force)	kgf
Kilogramme (mass)	kg
Kilogramme centimetre (torque)	kgf.cm
Kilogramme per square centimetre	kg/cm^2
Kilogramme metres (torque)	kgf.m
Kilometres	km
Kilometres per hour	km/h
Kilovolts	kV
King pin inclination	k.p.i.
Left-hand steering	LHStg
Left-hand thread	LHThd
Litres	litre
Low compression	l.c.
Low tension	l.t.
Maximum	max.
Metre	m
Microfarad	mfd

ABBREVIATIONS AND SYMBOLS USED IN THIS MANUAL

Midget edison screw	MES
Millimetre	mm
Miles per gallon	mpg
Miles per hour	mph
Minimum	min
minute (angle)	'
Minus (of tolerance)	−
Negative (electrical)	−
Number	No.
Ohms	ohm
Ounces (force)	ozf
Ounces (mass)	oz
Ounce inch (torque)	ozf.in.
Outside diameter	o.dia.
Paragraphs	para.
Part number	Part No.
Percentage	%
Pints	pt
Pints (US)	US pt
Plus (tolerance)	+
Positive (electrical)	+
Pound (force)	lbf
Pounds feet (torque)	lbf.ft.
Pounds inches (torque)	lbf.in.
Pound (mass)	lb
Pounds per square inch	lb/in^2
Radius	r
Rate (frequency)	c/min
Ratio	:
Reference	ref.
Revolution per minute	rev/min
Right-hand	RH
Right-hand steering	RHStg
Second (angle)	"
Second (numerical order)	2nd
Single carburetter	SC
Specific gravity	sp.gr.
Square centimetres	cm^2
Square inches	in^2
Standard	std.
Standard wire gauge	s.w.g.
Synchroniser/synchromesh	synchro.
Third	3rd
Top dead centre	TDC
Twin carburetters	TC
United Kingdom	UK
Vehicle Identification Number	VIN
Volts	V
Watts	W

SCREW THREADS

American Standard Taper Pipe	NPTF
British Association	BA
Standard Standard Fine	BSF
British Standard Pipe	BSP
British Standard Whitworth	Whit.
Unified Coarse	UNC
Unified Fine	UNF

ENGINE

Type	V8
Number of cylinders	Eight, two banks of four
Bore	88,90 mm (3.500 in)
Stroke	71,12 mm (2.800 in)
Capacity	3528 cm³ (215 in³)
Valve operation	Overhead by pushrod

Main bearings

Number and type	5 Vandervell shells
Material	Lead indium

Connecting rods

Type	Horizontally split big end, solid small end

Valve timing

	Inlet	Exhaust
Opens	30° BTDC	68° BBDC
Closes	75° ABDC	37° ATDC
Duration	285°	285°
Valve peak	112.5° ATDC	105.5° BTDC

Valve timing (9.35:1 engine, low lift camshaft)

	Inlet	Exhaust
Opens	36° BTDC	74° BBDC
Closes	64° ABDC	26° ATDC
Duration	280°	280°
Valve peak	99° ATDC	119° BTDC

Big end bearings

Type and material	Vandervell VP lead indium/aluminium tin

Pistons

Type	Aluminium alloy, concave crown.
	Plain skirt (early 8.25:1 engines)
	'W' slot skirt (late 8.25:1, all 8.13:1 and 9.35:1 engines)

Gudgeon pin

Type	Press fit in connecting rod

Camshaft

Drive	Chain and sprocket from crankshaft
Timing chain	9,52 mm (0.375 in) pitch 54 links x 22.22 mm (0.875 in) width

Tappets

Type	Hydraulic, non adjustable

Lubrication

System type	Wet sump, pressure fed
System pressure	2,1 to 2,8 kg/cm² (30 to 40 lb/in²) at 2400 rev/min
Oil pump type	Gear
Oil filter	Full-flow, self contained cartridge

COOLING SYSTEM

Type	Pressurized spill return system with thermostat control, pump and fan assisted
Type of pump	Centrifugal

FUEL SYSTEM

Carburetter

Make/type	Twin Zenith, 175 CD-SE
	Refer to engine tuning data for other details.

Fuel pump

Make/type	AC, mechanical (earlier models) or electrical (later models)

CLUTCH

Make/type	Borg & Beck, diaphragm type
Clutch plate diameter	266,5 mm (10.5 in)
Facing material	Raybestos WR7 or H.K. Porter 11046
Number of damper springs	6
Damper spring colour	Light grey/green
Clutch release bearing	Ball journal

GEARBOX

Main gearbox

Type	Single helical constant mesh
Speeds	4 forward 1 reverse
Synchromesh	All forward speeds
Ratios Fourth (Top)	1:1
Third	1.505:1
Second	2.448:1
First	4.069:1
Reverse	3.664:1

Transfer gearbox

Type — Two speed reduction on main gearbox output. Front and rear drive permanently engaged via a lockable differential

Ratios Gearbox suffix	'A' & 'B'	'C'	*	**
High	1.174:1	1.113:1	1.123:1	0.996:1
Low	3.321:1	3.321:1	3.321:1	3.321:1

Overall ratios (Final drive)

	high transfer				low transfer
Gearbox suffix	'A' & 'B'	'C'	*	**	All models
Fourth (Top)	4.16:1	3.94:1	3.97:1	3.53:1	11.76:1
Third	6.25:1	5.93:1	5.98:1	5.30:1	17.69:1
Second	10.17:1	9.64:1	9.72:1	8.63:1	28.78:1
First	16.91:1	16.03:1	16.16:1	14.34:1	47.84:1
Reverse	15.23:1	14.43:1	14.56:1	12.91:1	43.08:1

* Suffix 'C' from gearbox No. 355 94060 C

** Prefix '12C'

PROPELLER SHAFTS

Type	Open type 50,8 mm (2 in) diameter
Universal joints	1310 type, wide angle variety on front shaft only (Gaiter fitted to sliding coupling of front shaft)

REAR AXLE

Type	Spiral bevel, fully floating shafts
Ratio	3.54:1

FRONT AXLE

Type	Spiral bevel, enclosed universal joints
Angularity of universal joint on full lock	32°
Ratio	3.54:1

ROAD SPRING SPECIFICATION – 1981
(interchangeable only in sets for earlier vehicles)

STANDARD SUSPENSION	† SPECIFICATION	HEAVY DUTY SUSPENSION	† SPECIFICATION
RHD FRONT	A	RHD FRONT	E
LHD FRONT	(right side) A (left side) B	LHD FRONT	(right side) E (left side) F
RHD REAR	C	RHD REAR	G
LHD REAR	(right side) C (left side) D	LHD REAR	(right side) G (left side) H

ROAD SPRING DATA – ALL VEHICLES

†SPECIFICATION	PART NUMBER	COLOUR CODE	RATING	FREE LENGTH	NO. OF COILS
not current	90575625	yellow stripe	2321.5 kg/m (130 lb/in)	414.29mm (16.34 in)	7.11
not current	620101	yellow stripe	3035.86 kg/m (170 lb/in)	430.53mm (16.95 in)	8.85
C & E	NRC 2119	green stripe	2678.7 kg/m (150 lb/in)	409.70mm (16.13 in)	7.63
A	572315	blue stripe	2375.1 kg/m (133 lb/in)	391.16mm (15.4 in)	7.18
G	NRC 4234	green & yellow stripe	3035.86 kg/m (170 lb/in)	411.48mm (16.20 in)	7.00
H	NRC 4304	red & white stripe	3035.86 kg/m (170 lb/in)	450.53mm (16.95 in)	7.00
D & F	NRC 4305	red & yellow stripe	2678.7 kg/m (150 lb/in)	436.4mm (17.18 in)	7.65
B	NRC 4306	blue & white stripe	2375.1 kg/m (133 lb/in)	417.57mm (16.44 in)	7.55

SHOCK ABSORBERS (DAMPERS)
Type (internal valving modified 1981) — Hydraulic, double acting telescopic, 35mm bore (interchangeable only in sets for earlier vehicles)

LEVELLING UNIT
Type — Boge Hydromat, self-energising

STEERING
Manual steering
Make/type — Burman/recirculating ball, worm and nut
Ratio — Variable: straight ahead 18.2:1 (Non Tie Bar Type) 20.55:1 (Tie Bar Type)

Steering wheel turns, lock-to-lock — 4.99 early models, 5.55 later models

Power steering
Make/type — Adwest Varamatic/linkage
Ratio — Variable: straight ahead 17.5:1
Steering wheel turns, lock-to-lock — 3.375
Steering wheel diameter — 431,8 mm (17 in)
Front wheel alignment — 1,2 to 2,4 mm (0.046 to 0.094 in) toe out
Camber angle — 0° Check with vehicle in static unladen condition, that is, vehicle with water,
Castor angle — 3° oil and five gallons of fuel. Rock the vehicle up and down at the front to
Swivel pin inclination — 7° allow it to take up a static position.

HYDRAULIC DAMPERS
Type — Telescopic, double acting non adjustable
Bore diameter — 35,47 mm (1.375 in)

BRAKES
Foot brake
Type — Disc
Operation — Hydraulic, servo assisted, self adjusting

Front brake
Type — Outboard discs with four pistons
Disc diameter — 298,17 mm (11.75 in)

Rear brake
Type   Outboard discs with two pistons
Disc diameter   290,0 mm (11.42 in)
Total pad area   317,34 cm² (49.2 in²)
Total swept area   3199,2 cm² (496 in²)
Pad material   Ferodo F2430, F2431 or Don 230

Transmission brake
Type   Mechanical, hand operated, duo-servo drum brake
                                                  on rear of transfer gearbox output shaft
Diameter   184,05 mm (7.25 in)
Width   76,2 mm (3.0 in)
Lining material   Don Capasco 24

WHEELS
Size/type   6.00 JK x 16

Tyres
Size/type   205 x 16 Radial ply (tubed), Michelin X M + S, Goodyear
                                                  'Wingfoot', or Firestone Town and Country

ELECTRICAL EQUIPMENT
System   12 volt, negative earth
Fuses   35 amp, blow rating

Battery
Make/type – basic   Chloride (6TWL9Z1) 291
           – heavy duty   Chloride (6TWZ13R) 369
                                                  or Lucas (OCP 13/11)389

Starter motor
Make/type – early models   Lucas M45, pre-engaged
           – later models   Lucas 3M100 pre-engaged

Wiper motor
Make/type – Windscreen   Lucas 17W, two speed, self-switching
           – Tailgate and headlamp   Lucas 14W single speed, self-switching

Horns
Make/type   Lucas 6H or Mixo TR89

Distributor   Refer to engine tuning data

Alternator

Manufacturer	Lucas	Lucas	Lucas	Lucas
Type	16 ACR	18 ACR	25 ACR	133/65
Polarity	Negative earth	Negative earth	Negative earth	Negative earth
Brush length				
– New	12,70mm (0.5 in)	12,70mm (0.5 in)	12,70mm (0.5 in)	20mm (0.78 in)
– Worn, minimum free protrusion from brush box	5,00mm (0.2 in)	8,00mm (0.3 in)	8,00mm (0.3 in)	10mm (0.39 in)
Brush spring pressure				
– flush with brush box face	198 to 283 g (7 to 10 oz)	255 to 370 g (9 to 13 oz)	255 to 370 g 9 to 13 oz)	136 to 279 g (5 to 10 oz)
Rectifier pack				
– output rectification	6 diodes (3 live side and 3 earth side)	6 diodes (3 live side and 3 earth side)	6 diodes (3 live side and 3 earth side)	6 diodes (3 live side and 3 earth side)
– field winding supply rectification	3 diodes	3 diodes	3 diodes	3 diodes
Stator windings	3 phase–star connected	3 phase–delta connected	3 phase–delta connected	3 phase–delta connected
Field winding rotor				
– poles	12	12	12	12
– maximum speed	15,000 rev/min	15,000 rev/min	15,000 rev/min	15,000 rev/min
– shaft thread	9/16 in–18 U.N.F.	M16–1.5 g	M16–1.5 g	M15–1.5–6 g
– winding resistance at 20°C	4.3 ± 5% ohms	3.2 ohms	3.0 to 3.5 ohms	3.2 ohms
Control	Dual–battery sensed with machine sensed safety control	Dual–battery sensed with machine sensed safety control	Dual–battery sensed with machine sensed safety control	Dual–battery sensed with machine sensed safety control
Regulation–type	8 TR	8 TR	8 TR	15 TR
–voltage	14.1 to 14.5 volts	14.1 to 14.5 volts	14.1 to 14.5 volts	13.6 to 14.4 volts
Nominal output				
– condition	Hot	Hot	Hot	Hot
– alternator speed	6000 rev/min	6000 rev/min	6000 rev/min	6000 rev/min
– control voltage	14 volt	14 volt	14 volt	14 volt
– amp	34 amp	43 amp	65 amp	65 amp

Replacement Bulbs		Type
Headlamps . . . . ⌐		Lucas No. SP472 60/55/W (Halogen type) . . .
Headlamps – France		Phillips SP467 60/55W (Halogen type) . . . . .
Sidelamps		Lucas No. 233, 12v, 4w . . . .
Stop/tail lamps	Exterior	Lucas No. 380, 12v, 6/21w . . . .
Reverse lamps	lamps	Lucas No. 382, 12v, 21w . . . .
Rear fog guard lamps		Lucas No. 382, 12v, 21w . . . .
Direction indicator lamps		Lucas No. 382, 12v, 21w . . . .
Side repeater lamps		Lucas No. 989, 12v, 6w . . . .
Number plate lamps . . . . ⌋		Lucas No. 233, 12v, 4w . . . .
Instrument panel lamps and warning lamps . ⌐		Smith No. 4062110974, 12v, 2.2w capless . .
Hazard warning switch lamp		Lucas No. 281, 12v, 2w . . . .
Interior roof lamp 'festoon' bulbs	Interior	Lucas No. 254, 12v, 6w . . . .
Differential lock warning lamp	lamps	Lucas No. 987, 12v, 2.2w . . . .
Clock illumination		Lucas No. 281, 12v, 2w . . . .
Under bonnet illumination . . . . ⌋		Wedge base, 12v, 5w capless . . . .

Vehicle Dimensions

Overall length . . . . 4,47m (176 in)
Overall width . . . . 1,78m (70 in)
Overall height . . . . 1,78m (70 in)
Wheelbase . . . . 2,54m (100 in)
Track: front and rear . . . . 1,48m (58.5 in)
Ground clearance: under differential . . . . 190mm (7.5 in)
Turning circle . . . . 11,28m (37 ft)
Loading height . . . . 660mm (26 in)
Maximum cargo height . . . . 1,01m (40 in)
Rear opening height . . . . 1,01m (40 in)
Usable luggage capacity, rear seat folded . . . . 2,00 cu.m (70 cu.ft)
Usable luggage capacity, rear seat in use:
— four door vehicles . . . . 1,02 cu.m (36.18 cu.ft)
— two door vehicles . . . . 1,17 cu.m (41.48 cu.ft)
Maximum roof rack load . . . . 75 kg (165 lb)

Vehicle weights

Vehicle Type	UNLADEN WEIGHT			EEC KERB WEIGHT			GROSS VEHICLE WEIGHT		
	Front Axle	Rear Axle	Total	Front Axle	Rear Axle	Total	Front Axle	Rear Axle	Total
Four Door	909kg 2004lb	884kg 1949lb	1793kg 3953lb	928kg 2046lb	999kg 2202lb	1927kg 4248lb	1000kg 2205lb	1510kg 3329lb	2510kg 5534lb
Two Door	893kg 1969lb	869kg 1916lb	1762kg 3885lb	912kg 2011lb	983kg 2167lb	1895kg 4178lb	1000kg 2205lb	1510kg 3329lb	2510kg 5534lb

Note: UNLADEN WEIGHT is the minimum vehicle specification, excluding fuel and driver.
EEC KERB WEIGHT is the minimum vehicle specification, plus full fuel tank and 75 kg (165 lb) driver.
GROSS VEHICLE WEIGHT is the maximum all-up weight of the vehicle including driver, passengers, payload and equipment. This figure is liable to vary according to legal requirements in certain countries.

Maximum permissible towed weights

	On-road	Off-road
Trailers without brakes	500kg 1100lb	500kg 1100lb
Trailers with overrun brakes	2000kg 4400lb	1000kg 2200lb
4-wheel trailers with continuous or semi-continuous brakes i.e. coupled brakes	4000kg 8800lb	1000kg 2200lb

Note: It is the Owner's responsibility to ensure that all regulations with regard to towing are complied with. This applies also when towing abroad. All relevant information should be obtained from the appropriate motoring organisation.

ENGINE

Type	V8
Capacity	3528 cm³ (215 in³)
Compression ratio (depending on market and model year)	8.25:1, 8.13:1 or 9.35:1
Firing order	1–8–4–3–6–5–7–2
Cylinder numbering system front to rear	
– Left bank	1–3–5–7
– Right bank	2–4–6–8
Compression pressure (minimum)	9,5 kg/cm² (135 lb/in²)
Timing marks	On crankshaft pulley
Valve clearance	Not adjustable

IGNITION

Coil Make/type	Lucas 16 C 6 with ballast resistor

ENGINE TUNING

The following tables show the engine specification changes introduced from the inception of the Range Rover.
One of the table headings is 'Tuning Procedure' with numbers 1 to 4 listed against each carburetter identification tag.
The numbers 1 to 4 indicate which of the tuning procedures listed after the table should be used in each particular case. In the case of Suffix 'F' emission controlled engines use the tuning procedure 3 or 4 depending upon whether the carburetters are non-tamperproofed or tamperproofed type.

ENGINE NUMBER SUFFIX	*	CARBURETTER IDENTIFICATION TAG	CARBURETTER NEEDLES	NEEDLE ADJUSTMENT	TUNING PROCEDURE	DISTRIBUTOR	TIMING	FUEL OCTANE RATING (RON)	SPARK PLUG AND GAP	COMPRESSION RATIO
A & B	NE	3293	2AQ	Bottom	1	Rover No. 611390 Lucas No. 41325	3 Degrees B.T.D.C.	91–93	L87Y	8.25:1
							T.D.C.	85	0.025in (0.635mm)	
A & B	NE	3394	2AQ	Bottom	1	Rover No. 611390 Lucas No. 41325	3 Degrees B.T.D.C.	91–93	L87Y	8.25:1
							T.D.C.	85	0.025in (0.635mm)	
C	E	3318	B2AS	Top	3	Rover No. 611390 Lucas No. 41325	3 Degrees B.T.D.C.	91–93	L87Y	8.25:1
							T.D.C.	85	0.025in (0.635mm)	
C	E	3677	B1DF	Bottom	2	Rover No. 614003 Lucas No. 41382	5 Degrees A.T.D.C.	91–93	L87Y	8.25:1
							8 Degrees A.T.D.C.	85	0.025in (0.635mm)	
D	NE	3394	2AQ	Bottom	1	Rover No. 614179 Lucas No. 41487	3 Degrees B.T.D.C.	91–93	L92Y	8.25:1
							T.D.C.	85	0.025in (0.635mm)	
D	E	3753	B1DW	Top	2	Rover No. 614003 Lucas No. 41382	5 Degrees A.T.D.C.	91–93	L92Y	8.25:1
							8 Degrees A.T.D.C.	85	0.025in (0.635mm)	
E	NE	3394	2AQ	Bottom	1	Rover No. 614179 Lucas No. 41487	3 Degrees B.T.D.C.	91–93	L92Y	8.25:1
							T.D.C.	85	0.025in (0.635mm)	
E	E	3753	B1DW	Top	3	Rover No. 614003 Lucas No. 41382	5 Degrees A.T.D.C.	91–93	L92Y	8.25:1
							8 Degrees A.T.D.C.	85	0.025in (0.625mm)	
F	NE	3881	I.E.L.	Bottom	1	Rover No. ERC3342 Lucas No. 41680A	6 Degrees B.T.D.C.	91–93	N12Y	8.13:1
							3 Degrees B.T.D.C.	85	0.030in (0.80mm)	
F	E	3854	B1EJ	Top	3 or 4	Rover No. ERC3341 Lucas No. 41681A	5 Degrees A.T.D.C.	91–93	N12Y	8.13:1
							8 Degrees A.T.D.C.	85	0.030in (0.80mm)	
Prefix 11D	E	4104	B1FH	Top	4	Rover No. ERC7131 Lucas No. 41873	6 Degrees B.T.D.C.	97	N12Y 0.030in (0.80mm)	9.35:1

* NE – NON EMISSION ENGINES   E – EMISSION CONTROLLED ENGINES

TUNING PROCEDURE 1 (BOTTOM ADJUSTMENT)

Idle Setting
With engine at operating temperature (Warm Air Intake Valve Open, where fitted) the mixture change is effected by adjusting the jet assembly. Idle speed 550 to 650 rev/min.

TUNING PROCEDURE 2 (BOTTOM ADJUSTMENT)

Idle Setting
With engine at operating temperature (Warm Air Intake Valve Open) the mixture change is effected by adjusting the jet assembly. Zenith tool, Part No. B 24667, is required for this operation.

Method
1. Locate the tool in the slots at the base of the jet assembly.
2. Turn in a clockwise direction to richen mixture, and anti-clockwise to weaken the mixture.

Mixture settings
Exhaust C.O. 3.0 to 4.5% Idle Speed 700 to 750 rev/min.
NOTE: On new engines idle speed will be lower for the initial running in period.

Fast idle
Operation of the choke from 'ON' to 'OFF' should result in a fast idle speed of 1200 ± 50 rev/min when the choke control is approximately 12.70 mm (0.5 in) from the fully-in position.
Should adjustment be required, the following procedure must be used:—
1. Remove the L.H. carburetter from the engine.
2. Put the Starter Unit Adjusting Pin into the full choke position (pin in the groove).
3. With the Starter Unit held in the full choke position, it should be just possible to insert a 0.9 mm (No. 65) drill between the top edge of the throttle plate and the throttle barrel wall.
4. To adjust the setting, release the locknut on the dome head screw fitting in the throttle lever that contacts the Starter Unit Cam, adjust the screw to obtain the required setting, tighten the locknut and recheck the gap.
5. Replace the adjusting pin in the reduced choke position.
6. Refit the L.H. carburetter to the engine.
The following components are preset and non-adjustable.
a. Temperature compensator.
b. Poppet valve.
Should these units be suspected of malfunction they must be changed.

Fuel Deflector Plate
This must be fitted between the inlet manifold and the carburetter insulating block, with the saw teeth facing downstream of the carburetter. The plate will foul the carburetter butterfly plate if fitted between the carburetter and insulator.

TUNING PROCEDURE 3 (TOP ADJUSTMENT)

Idle Speed and Mixture Setting
Exhaust C.O. 3.0 to 4.5% at 700 to 750 rev/min.
NOTE: On new engines idle speed will be lower for the running in period.

Fast Idle Setting
1200 ± 50 rev/min with the engine at normal operating temperature.

Idle Mixture Setting
Before proceeding with adjustment the engine should be at normal operating temperature (Warm Air Intake Valve Open) and the Carburetters balanced.

Equipment required:—
    CO Meter
    Calibrated Tachometer
    *Mixture Adjustment Tool — Part No. S353 (Zenith B 20379) 'Allen' key type or Part No. MS 80 (Zenith B 25860) slotted socket type.
    *To meet certain new legislative requirements recent carburetters employ a raised blade instead of an 'Allen' key socket for adjusting the mixture needle. This necessitates the provision of alternative tools.

Mixture adjustment is made by changing the metering needle position relative to the fixed jet orifice using the appropriate adjustment tool, as follows:–

1. Remove the piston damper.
2. Insert the adjustment tool, into the piston guide rod, ensuring that the peg on the outer barrel of the tool engages in the slot in the guide rod, this prevents the piston from rotating and consequently damaging the diaphragm.
3. Push down the adjustment tool until it engages in the metering needle housing.
4. Turn the tool in a clockwise direction to richen the mixture, and anti-clockwise to weaken the mixture.

NOTE: The adjustment tool should be removed from the guide rod after every adjustment, to allow the engine to stabilise following the piston being depressed during adjustment. Run the engine for a few seconds at 2000 rev/min to aid stabilisation.

5. Refit the piston damper.
6. Slight adjustment to idle speed and/or carburetter balance may be required after refitting the piston damper, therefore run the engine at approximately 2000 rev/min for 20 seconds, make idle and balance adjustments, if necessary, before taking final C.O. reading.

NOTE: The metering needle will normally be set with the shoulder of the needle flush with the face of the piston; this is termed the datum position. If difficulty is experienced with carburation the needle should be set to this datum for investigation.

When using the adjustment tool, a positive stop will be felt when the needle reaches the fully rich position. In the anti-clockwise or weakening direction there is no stop and it is possible to disengage the needle from the adjustment screw if more than two turns are made from the datum position. Should disengagement occur, it can be rectified by applying light pressure in an upwards direction, to the shoulder of the needle at the piston face, while turning the adjustment tool in a clockwise direction.

Poppet Valve and Temperature Compensator

These assemblies are preset and non-adjustable. If a malfunction occurs on either component, the carburetter must be replaced.

Fast Idle Setting

Operation of the choke from 'ON' to 'OFF' should result in a fast idle speed of 1200 ± 50 rev/min when the choke control is approximately 12.70 mm (0.5 in) from the fully-in position.

Should adjustment be required, proceed as follows:–

1. Slacken the choke cable clamping screw at the carburetter.
2. Pull the choke control knob out to a distance of approximately 12.70 mm (0.5 in) and turn to lock-in position.
3. Turn the fast idle cam, allowing the choke cable to slide through the trunnion until the punched mark on the cam flank aligns with the centre of the domed fast idle screw and tighten the clamping screw.
4. With the cam held in this position, adjust the fast idle screw to obtain a fast idle speed of 1200 ± 50 rev/min and retighten the locknut.
5. Push the choke control knob fully home and check that normal idle speed is regained.

ADJUSTMENT PROCEDURE 4 (TOP ADJUSTMENT TAMPERPROOFED)

To comply with ECE exhaust emission regulations, all carburetters must be tamperproofed on the idle adjustment screws. Therefore, after mixture and speed tests have been finalised, the carburetter must be tamperproofed by fitting a cap to the nylon shroud on the idle adjusting screw.

Cap   –   Part No. ERC 3429
Cap fitting tool   –   Part No. ERC 3786

Should, for any reason, the cap require removal, this can be effected by piercing the cap with a sharp pointed tool and prising out.

The following tools will be required to adjust idle speed, mixture and tamperproof carburetter:–

Idle speed adjustment tool   –   Part No. MS 86 (Zenith B25243)
*Idle Mixture adjustment tool –   Part No. S353 (Zenith B20374) 'Allen' key type or Part No. MS 80 (Zenith B25860) slotted socket type
Tamperproof Cap fitting tool –   Part No. ERC 3786

A numerical code exists for the tamperproofed cap and must be adhered to:

Cap fitted by Land Rover Service Departments: Part No. ERC 3429

*To meet certain new legislative requirements, recent carburetters employ a raised blade instead of an 'Allen' key socket for adjusting the mixture needle. This necessitates the provision of alternative tools.

The carburetters will be preset by the manufacturers to engineering requirements and should not normally require adjustment.

However, should adjustment be required, the following method must be used.

Idle Setting – with engine to normal operating temperatures

Exhaust C.O. 3.0 to 4.5% (9.35:1 engine 2.0 to 3.5%) Idle Speed 700 to 750 rev/min.

Fast Idle Setting: 1200 ± 50 rev/min

NOTE: On new engines idle speed will be lower for the initial running in period.

Idle Mixture Adjustment

Engine should be at normal operating temperature (Warm Air Intake Valve Open).

Mixture adjustment is effected by changing the metering needle position, relative to the fixed jet orifice, using the appropriate adjustment tool.

1. Remove the piston damper.
2. Insert the adjustment tool into the piston guide rod, ensuring the peg on the outer barrel of the tool locates in the slot in the guide rod, this prevents the piston twisting with consequent damage to the diaphragm.
3. Push down the adjustment tool until it engages in the metering needle housing.
4. Turn the tool in a clockwise direction to rich the mixture and anti-clockwise to weaken the mixture.

NOTE: The adjustment tool should be removed from the guide rod after every adjustment, to allow the engine to stabilise, following the piston being depressed during adjustment. Run the engine for a few seconds at 2000 rev/min to aid stabilisation.

5. Refit the piston damper.
6. Slight adjustment to idle speed and/or carburetter balance may be required after refitting the piston damper, therefore run engine at approximately 2000 rev/min for 20 seconds, make idle and balance adjustments, if necessary before taking final C.O. reading.

NOTE: The metering needle will normally be set with the shoulder of the needle flush with the face of the piston, this is termed datum position. If difficulty is experienced with carburation the needle should be set to this datum for investigation.

When using the adjustment tool, a positive stop will be felt when the needle reaches the full rich position. In the anti-clockwise direction there is no stop and it is possible to disengage the needle from the adjusting screw, if more than two turns are made from the datum position. Should disengagement occur, it can be rectified by applying light pressure, in an upwards direction, to the shoulder of the needle at the piston face, while turning in a clockwise direction.

Idle Speed Adjustment

To adjust the idle speed, the following procedure must be followed:–

1. Slacken the small bolts on the throttle levers to allow independent adjustment.
2. Using tool, Part No. MS86 (Zenith B25243) slacken the locknuts on the idle speed adjustment screws.
3. Adjust the screws to obtain correct idle speed and just nip the locknuts.
4. Check C.O. reading, check carburetter balance, correct if necessary.
5. When 3 and 4 are satisfactory, tighten the locknuts and reset the throttle.
6. Recheck the idle settings.
7. Fit cap, Part No. ERC 3428, using tool, Part No. ERC 3786 to the nylon shroud surrounding the adjusting screw.

Fast Idle Setting

Operation of the choke from 'ON' to 'OFF' should result in a fast idle speed of 1200 ± 50 rev/min when the choke control is approximately 12.70 mm (0.5 in) from the fully-in position.

Should adjustment be required, proceed as follows:–

1. Slacken the choke cable clamping screw at the carburetter.
2. Pull the choke control knob out and push in to a distance of approximately 12.70 mm (0.5 in) and lock in position.
3. Turn the starter cam, allowing the choke cable to slslide through the trunnion until the punched mark on the cam flank aligns with the centre of the domed screw on the starter/throttle lever and tighten the clamping screw.
4. With the cam held in this position, adjust the fast idle screws to obtain a speed of 1200 ± 50 rev/min and retighten the locknut.
5. Push the choke cable fully home and check that normal idle speed is regained.

Vacuum Retard Switch European (Pre 1975) and all Australian Vehicles

Setting: The gap between the cam and switch pad to be 0.900 to 1.02 mm (0.035 to 0.040 in).

Procedure:—

1. Slacken the nuts retaining the switch to the bracket.
2. Depress the switch pad fully.
3. Adjust the switch to correct the gap between the cam and the switch pad.
4. Tighten the switch retaining nuts.
5. Open the throttle fully, and close slowly to ensure the throttle is not propped open by the cam contacting the top edge of the switch pad. If the cam is propped, this will probably be due to misalignment of the switch pad to cam.

Vacuum Capsule Test

To ensure that the vacuum capsule is functioning correctly, it is important that the following procedure is carried out: Remove the vacuum retard pipe from the distributor (this will increase the idle speed). If the vacuum capsule is functioning correctly, an advance to between 6 to 14 degrees B.T.D.C. should be noted. If this advance is not achieved the distributor should be changed.

DISTRIBUTOR

NOTE: Different distributors have been fitted according to model year and market requirements. The information given below is for guidance only and the unit should be identified by the number stamped on the distributor body.

Make/type	Lucas 35D8
Rotation of rotor	Clockwise
Dwell angle	26° to 28°
Contact breaker gap	0,36 mm to 0,40 mm (0.14 in to 0.16 in)
Condenser capacity	0.18 to 0.25 microfarad

Engine suffix identification	A to C (Non-emission)	D & E (Non-emission)	C, D & E (Emission)
Serial number	41325	41487	41382
Idling speed	600 to 650 rev/min	650 to 750 rev/min	750 to 850 rev/min
Fast idle speed	1000 to 1200 rev/min	1100 to 1300 rev/min	1200 to 1250 rev/min
Ignition timing, static and dynamic at idling speed rev/min	3° BTDC for use with 91 to 93 research octane number fuel. TDC for use with 85 minimum research octane number fuel.	3° BTDC for use with 91 to 93 research octane number fuel. TDC for use with 85 minimum research octane number fuel.	5° ATDC for use with 91 to 93 research octane number fuel. 8° ATDC for use with 85 minimum research octane number fuel.
Vacuum Capsule	Advance only (connected)	Advance only (connected)	Advance & Retard (connected)
Decelerating check with vacuum unit disconnected			
engine rev/min	Crankshaft angle	Crankshaft angle	Crankshaft angle
4,800	27 to 31°	23 to 27°	22 to 26°
3,800	23 to 27°	18 to 22°	17 to 21°
1,800	15 to 19°	10 to 16°	2.5 to 6.5°
1,400	11 to 15°	6 to 10°	0 to 4°
1,000	5 to 9°	0 to 5°	no advance
600	3° below	no advance	no advance

Engine suffix identification	F (Non-emission)	F (Emission) – European	F (Emission) – Australian
Serial number	41680A	41681A	41681A
Idling speed	550 to 650 rev/min	700 to 750 rev/min	*850 to 950 rev/min
Fast idle speed	1100 to 1300 rev/min	1100 to 1300 rev/min	1400 to 1500 rev/min
Ignition timing, static and dynamic at idle speed rev/min	6° BTDC for use with 91 to 93 research octane number fuel. 3° BTDC for use with 85 minimum research octane number fuel.	5° ATDC for use with 92 Research octane number fuel.	5° ATDC for use with 92 Research octane number fuel. *750 to 800 rev/min (with diverter valve)
Vacuum Capsule	Advance only (connected)	Advance & retard (connected)	Advance & retard (connected)
Decelerating check with vacuum unit disconnected			
engine rev/min	Crankshaft angle	Crankshaft angle	Crankshaft angle
4,800	24 to 28°	22 to 26°	22 to 26°
3,800	20 to 24°	17 to 21°	17 to 21°
1,800	12 to 16°	2.5 to 6.5°	2.5 to 6.5°
1,400	8 to 12°	0 to 4°	0 to 4°
1,000	1 to 5°	No advance	No advance
600	No advance	No advance	No advance

continued

Distributor 35D8 *(continued)*

Engine prefix identification	11D (Emission)
Serial number	41873 (sliding contact type)
Idle speed	700 to 750 rev/min
Fast idle speed	1150 to 1250 rev/min
Ignition timing, static and dynamic at idling speed rev/min	6° BTDC for use with 97 research octane number fuel.
Vacuum Capsule	Advance & retard, retard disconnected
Decelerating check with vacuum retard pipe disconnected	
engine rev/min	Crankshaft angle
3,600	20 to 24°
3,080	19 to 23°
2,400	14 to 18°
1,200	8 to 12°
400	No advance

ELECTRONIC DISTRIBUTOR
Make/type   Lucas 35DM ELECTRONIC
Rotation of rotor   Clockwise

EUROPE AND AUSTRALIA
9.35:1 Compression ratio
(Emission controlled)

Engine prefix identification	15D, 16D, 17D, & 19D
Distributor serial number	42092 Electronic
Idle speed	700 to 750 rev/min
Fast idle speed	1150 to 1250 rev/min
Ignition timing, static and dynamic at idling speed rev/min	6°BTDC for use with 97 research octane number fuel.
Vacuum Capsule	Vacuum advance pipe disconnected
Decelerating check with vacuum pipe disconnected	
engine rev/min	Crankshaft angle
2900	12 to 16°
2400	8 to 12°
1600	2 to 6°
800	No advance

SAUDI ARABIA AND GULF STATES
8.13:1 Compression ratio
(Emission controlled)

Engine prefix identification	20D & 21D
Distributor serial number	42056 Electronic
Idle speed	650 to 750 rev/min
Fast idle speed	1150 to 1250 rev/min
Ignition timing, static and dynamic at idling speed rev/min	6°BTDC for use with 90 to 93 research octane number fuel.
Vacuum Capsule	Advance & retard, retard disconnected
Decelerating check with vacuum retard pipe disconnected	
engine rev/min	Crankshaft angle
4600	211 to 25°
3600	16 to 20°
3000	12 to 16°
2400	7 to 11°
1600	1 to 5°
200	No advance

ALL OTHER TERRITORIES
8.13:1 Compression ratio
(Non emission)

Engine prefix identification	13D & 18D
Distributor serial number	42092 Electronic
Idle speed	550 to 650 rev/min
Fast idle speed	1150 to 1250 rev/min
Ignition timing, static and Vacuum Capsule	6°BTDC for use with 90 to 93 research octane number fuel. Advance pipe disconnected
Decelerating check with vacuum retard pipe disconnected	
engine rev/min	Crankshaft
4200	23 to 27°
3500	20 to 24°
3000	16 to 20°
2000	8 to 12°
1200	2 to 6°
400	No advance

TORQUE WRENCH SETTINGS

	kgf.m	lbf.ft
ENGINE		
Connecting rod cap nuts	4,0 to 4,9	30 to 35
Main bearing cap bolts, numbers one to four	7,0 to 7,6	50 to 55
Rear main bearing cap bolts	9,0 to 9,6	65 to 70
Cylinder head bolts, numbers 1 to 10 +) see page	9,0 to 9,6	65 to 70
Cylinder head bolts, numbers 11 to 14 +) 12–22	5,6 to 6,2	40 to 45
Rocker shaft bolts	3,5 to 4,0	25 to 30
Flywheel bolts	7,0 to 8,5	50 to 60
Oil pump cover bolts	1,2	9
Oil pressure relief valve	4,0 to 4,9	30 to 35
Timing chain cover bolts	2,8 to 3,5	20 to 25
Crankshaft starter dog	19,3 to 22,3	140 to 160
Distributor drive gear to camshaft bolt	5,5 to 6,2	40 to 45
Engine mounting rubbers	1,8 to 2,2	13 to 16
Water pump fixing bolts – 1/4 in	0,9 to 1,4	7 to 10
Water pump fixing bolts – 5/16 in	2,2 to 2,7	16 to 20
COOLING SYSTEM		
Water pump housing bolts 7/16 in AF	0,8 to 1,0	6 to 8
Water pump housing bolts 1/2 in AF	2,8 to 3,5	20 to 25
MANIFOLDS AND EXHAUST SYSTEM		
Induction manifold bolts	3,5 to 4,0	25 to 30
Induction manifold gasket clamp bolts	1,4 to 2,0	10 to 15
Exhaust manifold bolts	1,4 to 2,0	10 to 15
CLUTCH		
Clutch cover bolts	4,9 to 5,2	35 to 38
GEARBOX		
Main gear lever retainer bolts	1,5	11
Front output flange nut	11,75	85
Rear output flange nut	11,75	85
Transmission brake shoe pivot bolts	5,9	43
Transmission brake back plate bolts	3,5	25
Gearbox casing to bell housing studs/bolts – Larger diameter	16,6	120
– Smaller diameter	9,6	70
Speedometer drive housing	3,1	22
PROPELLER SHAFTS		
Coupling flange bolts	4,1 to 5,1	30 to 38
FRONT/REAR AXLE AND FINAL DRIVE		
Differential input flange nut	9,6 to 16,5	70 to 120
Differential crownwheel bolts	5,5 to 6,2	40 to 45
Differential bearing cap bolts	6,9 to 9,0	50 to 65
Bearing sleeve to axle case	6,2	45
Axle flange to hub	4,1 to 5,1	30 to 38
Pinion housing to axle case	3,5 to 4,6	26 to 34
Road wheel to hub – steel wheels	10,0 to 11,7	75 to 85
Road wheel to hub – alloy wheels*	12,5 to 13,35	90 to 95
Swivel bearing housing to axle case	7,6 to 8,6	55 to 62
Oil seal retainer to swivel pin housing	1,0 to 1,2	7 to 9

* Suitable only for latest universal hubs, see 74–4.

+ Coat first three threads with Loctite 572

STEERING

	kgf.m	lbf.ft
Universal joint pinch bolt	3,5	25
Ball joint nuts	4,0	30
Steering box end cover bolts, manual steering	21 to 25	15 to 18
Drop arm nut	17,9	125
Track rod clamp bolts	1,4	10
Drag link clamp bolts	1,4	10
Steering wheel nut	3,8	28
Sector shaft cover bolts and nuts, power steering box	2,2 to 2,9	16 to 20
Flow control valve cap, power steering pump	4,0 to 4,9	30 to 35
Pulley bolt, power steering pump	1,4 to 1,6	10 to 12
Union bolt, inlet adaptor, power steering box	3,8 to 4,0	28 to 30

SUSPENSION

Front

	kgf.m	lbf.ft
Hub cap and driving shaft bolts	4,1 to 5,1	30 to 38
Stub axle bolts	3,0 to 4,2	22 to 30
Swivel pin bolts	7,0 to 8,9	50 to 65
Oil seal retainer to swivel housing bolt	1,0 to 1,2	7 to 9
Swivel housing to axle case bolts	5,2 to 6,6	38 to 48

Rear

	kgf.m	lbf.ft
	4,1 to 5,1	30 to 38
Hub cap and driving shaft bolts	5,2 to 6,6	38 to 48
Hub bearing sleeve bolts	6,0	44
Ball pins to self-levelling unit	3,5	25
Ball joint nuts, self-levelling unit	7,0	50
Pivot bracket ball joint nut, self-levelling unit	17,9	130
Bottom link stem end	12,4	90

BRAKES

	kgf.m	lbf.ft
Caliper mounting bolts	8,3	60
Caliper half joint bolts	8,3	60
Disc to hub bolts	5,0	38
Brake failure switch end plug	2,2	16
Brake failure switch to five-way connector	17,28 cmkg	15 lb in

ELECTRICAL

	kgf.m	lbf.ft
Starter motor to cylinder block bolts	4,0 to 4,9	30 to 35
Alternator shaft nut	3,5 to 4,2	25 to 30
Wiper motor yoke through bolt	1,6 to 2,2	12 to 16
Wiper motor gearbox spacing ring	11,50 kgf.cm	10 lbf.in.

GENERAL FITTING INSTRUCTIONS

Precautions against damage

1. Always fit covers to protect wings before commencing work in engine compartment.
2. Cover seats and carpets, wear clean overalls and wash hands or wear gloves before working inside car.
3. Avoid spilling hydraulic fluid or battery acid on paint work. Wash off with water immediately if this occurs. Use Polythene sheets in boot to protect carpets.
4. Always use a recommended Service Tool, or a satisfactory equivalent, where specified.
5. Protect temporarily exposed screw threads by replacing nuts or fitting plastic caps.

Safety Precautions

1. Whenever possible use a ramp or pit when working beneath car, in preference to jacking. Chock wheels as well as applying hand brake.
2. Never rely on a jack alone to support car. Use axle stands or blocks carefully placed at jacking points to provide rigid location.
3. Ensure that a suitable form of fire extinguisher is conveniently located.
4. Check that any lifting equipment used has adequate capacity and is fully serviceable.
5. Inspect power leads of any mains electrical equipment for damage and check that it is properly earthed.
6. Disconnect earth (grounded) terminal of car battery.
7. Do not disconnect any pipes in air conditioning refrigeration system, if fitted, unless trained and instructed to do so. A refrigerant is used which can cause blindness if allowed to contact eyes.
8. Ensure that adequate ventilation is provided when volatile de-greasing agents are being used.

CAUTION: Fume extraction equipment must be in operation when trachloride, methylene chloride, chloroform, or perchlorethylene are used for cleaning purposes.

9. Do not apply heat in an attempt to free stiff nuts or fittings; as well as causing damage to protective coatings, there is a risk of damage to electronic equipment and brake lines from stray heat.
10. Do not leave tools, equipment, spilt oil etc., around or on work area.
11. Wear protective overalls and use barrier creams when necessary.

Preparation

1. Before removing a component, clean it and its surrounding areas as thoroughly as possible.
2. Blank off any openings exposed by component removal, using greaseproof paper and masking tape.
3. Immediately seal fuel, oil or hydraulic lines when separated, using plastic caps or plugs, to prevent loss of fluid and entry of dirt.
4. Close open ends of oilways, exposed by component removal, with tapered hardwood plugs or readily visible plastic plugs.
5. Immediately a component is removed, place it in a suitable container; use a separate container for each component and its associated parts.

6. Before dismantling a component, clean it thoroughly with a recommended cleaning agent; check that agent is suitable for all materials of component.
7. Clean bench and provide marking materials, labels, containers and locking wire before dismantling a component.

Dismantling

1. Observe scrupulous cleanliness when dismantling components, particularly when brake, fuel or hydraulic system parts are being worked on. A particle of dirt or a cloth fragment could cause a dangerous malfunction if trapped in these systems.
2. Blow out all tapped holes, crevices, oilways and fluid passages with an air line. Ensure that any O-rings used for sealing are correctly replaced or renewed, if disturbed.
3. Mark mating parts to ensure that they are replaced as dismantled. Whenever possible use marking ink, which avoids possibilities of distortion or initiation of cracks, liable if centre punch or scriber are used.
4. Wire together mating parts where necessary to prevent accidental interchange (e.g. roller bearing components).
5. Wire labels on to all parts which are to be renewed, and to parts requiring further inspection before being passed for reassembly; place these parts in separate containers from those containing parts for rebuild.
6. Do not discard a part due for renewal until after comparing it with a new part, to ensure that its correct replacement has been obtained.

Inspection – General

1. Never inspect a component for wear or dimensional check unless it is absolutely clean; a slight smear of grease can conceal an incipient failure.

2. When a component is to be checked dimensionally against figures quoted for it, use correct equipment (surface plates, micrometers, dial gauges, etc.) in serviceable condition. Makeshift checking equipment can be dangerous.

3. Reject a component if its dimensions are outside limits quoted, or if damage is apparent. A part may, however, be refitted if its critical dimension is exactly limit size, and is otherwise satisfactory.

4. Use 'Plastigauge' 12 Type PG-1 for checking bearing surface clearances; directions for its use, and a scale giving bearing clearances in 0.0001 in. (0,0025 mm) steps are provided with it.

Ball and Roller Bearings

NEVER REPLACE A BALL OR ROLLER BEARING WITHOUT FIRST ENSURING THAT IT IS IN AS-NEW CONDITION.

1. Remove all traces of lubricant from bearing under inspection by washing in petrol or a suitable degreaser; maintain absolute cleanliness throughout operations.

2. Inspect visually for markings of any form on rolling elements, raceways, outer surface of outer rings or inner surface of inner rings. Reject any bearings found to be marked, since any marking in these areas indicates onset of wear.

3. Holding inner race between finger and thumb of one hand, spin outer race and check that it revolves absolutely smoothly. Repeat, holding outer race and spinning inner race.

4. Rotate outer ring gently with a reciprocating motion, while holding inner ring; feel for any check or obstruction to rotation, and reject bearing if action is not perfectly smooth.

5. Lubricate bearing generously with lubricant appropriate to installation.

6. Inspect shaft and bearing housing for discolouration or other marking suggesting that movement has taken place between bearings and seatings. (This is particularly to be expected if related markings were found in operation 2). If markings are found, use 'Loctite' in installation of replacement bearing.

7. Ensure that shaft and housing are clean and free from burrs before fitting bearing.

8. If one bearing of a pair shows an imperfection it is generally advisable to renew both bearings: an exception could be made if the faulty bearing had covered a low mileage, and it could be established that damage was confined to it only.

9. When fitting bearing to shaft, apply force only to inner ring of bearing, and only to outer ring when fitting into housing.

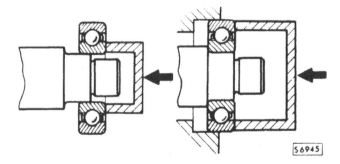

10. In the case of grease-lubricated bearings (e.g. hub bearings) fill space between bearing and outer seal with recommended grade of grease before fitting seal.

11. Always mark components of separable bearings (e.g. taper roller bearings) in dismantling, to ensure correct reassembly. Never fit new rollers in a used cup.

Oil Seals

1. Always fit new oil seals when rebuilding an assembly. It is not physically possible to replace a seal exactly when it has bedded down.

2. Carefully examine seal before fitting to ensure that it is clean and undamaged.

3. Smear sealing lips with clean grease; pack dust excluder seals with grease, and heavily grease duplex seals in cavity between sealing lips.

4. Ensure that seal spring, if provided, is correctly fitted.

5. Place lip of seal towards fluid to be sealed and slide into position on shaft, using fitting sleeve when possible to protect sealing lip from damage by sharp corners, threads or splines. If fitting sleeve is not available, use plastic tube or adhesive tape to prevent damage to sealing lip.

6. Grease outside diameter of seal, place square to housing recess and press into position, using great care and if possible a 'bell piece' to ensure that seal is not tilted. (In some cases it may be preferable to fit seal to housing before fitting to shaft.) Never let weight of unsupported shaft rest in seal.

7. If correct service tool is not available, use a suitable drift approximately 0.015 in. (0,4 mm) smaller than outside diameter of seal. Use a hammer VERY GENTLY on drift if a press is not suitable.

8. Press or drift seal in to depth of housing if housing is shouldered, or flush with face of housing where no shoulder is provided.

   NOTE: Most cases of failure or leakage of oil seals are due to careless fitting, and resulting damage to both seals and sealing surfaces. Care in fitting is essential if good results are to be obtained.

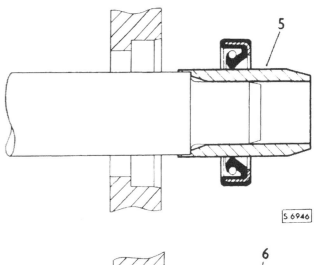

S 6946

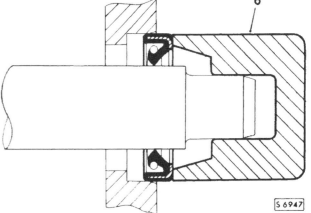

S 6947

Joints and Joint Faces

1. Always use correct gaskets where they are specified.

2. Use jointing compound only when recommended. Otherwise fit joints dry.

3. When jointing compound is used, apply in a thin uniform film to metal surfaces; take great care to prevent it from entering oilways, pipes or blind tapped holes.

4. Remove all traces of old jointing materials prior to reassembly. Do not use a tool which could damage joint faces.

5. Inspect joint faces for scratches or burrs and remove with a fine file or oil stone; do not allow swarf or dirt to enter tapped holes or enclosed parts.

6. Blow out any pipes, channels or crevices with compressed air, renewing any O-rings or seals displaced by air blast.

Flexible Hydraulic Pipes, Hoses

1. Before removing any brake or power steering hose, clean end fittings and area surrounding them as thoroughly as possible.

2. Obtain appropriate blanking caps before detaching hose end fittings, so that ports can be immediately covered to exclude dirt.

continued

3. Clean hose externally and blow through with airline. Examine carefully for cracks, separation of plies, security of end fittings and external damage. Reject any hose found faulty.

4. When refitting hose, ensure that no unnecessary bends are introduced, and that hose is not twisted before or during tightening of union nuts.

5. Containers for hydraulic fluid must be kept absolutely clean.

6. Do not store hydraulic fluid in an unsealed container. It will absorb water, and fluid in this condition would be dangerous to use due to a lowering of its boiling point.

7. Do not allow hydraulic fluid to be contaminated with mineral oil, or use a container which has previously contained mineral oil.

8. Do not re-use fluid bled from system.

9. Always use clean brake fluid to clean hydraulic components.

10. Fit a blanking cap to a hydraulic union and a plug to its socket after removal to prevent ingress of dirt.

11. Absolute cleanliness must be observed with hydraulic components at all times.

12. After any work on hydraulic systems, inspect carefully for leaks underneath the car while a second operator applies maximum pressure to the brakes (engine running) and operates the steering.

Metric Bolt Identification

1. An ISO metric bolt or screw, made of steel and larger than 6 mm in diameter can be identified by either of the symbols ISO M or M embossed or indented on top of the head.

2. In addition to marks to identify the manufacture, the head is also marked with symbols to indicate the strength grade, e.g. 8.8, 10.9, 12.9 or 14.9, where the first figure gives the minimum tensile strength of the bolt material in tens of kg/sq mm.

3. Zinc plated ISO metric bolts and nuts are chromate passivated, a greenish-khaki to gold-bronze colour.

Metric Nut Identification

1. A nut with an ISO metric thread is marked on one face or on one of the flats of the hexagon with the strength grade symbol 8, 12 or 14. Some nuts with a strength 4, 5 or 6 are also marked and some have the metric symbol M on the flat opposite the strength grade marking.

2. A clock face system is used as an alternative method of indicating the strength grade. The external chamfers or a face of the nut is marked in a position relative to the appropriate hour mark on a clock face to indicate the strength grade.

3. A dot is used to locate the 12 o'clock position and a dash to indicate the strength grade. If the grade is above 12, two dots identify the 12 o'clock position.

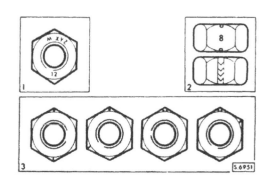

Hydraulic Fittings – Metrication
WARNING: Metric and Unified threaded hydraulic parts. Although pipe connections to brake system units incorporate threads of metric form, those for power assisted steering are of UNF type. It is vitally important that these two thread forms are not confused, and careful study should be made of the following notes.

Metric threads and metric sizes are being introduced into motor vehicle manufacture and some duplication of parts must be expected. Although standardisation must in the long run be good, it would be wrong not to give warning of the dangers that exist while UNF and metric threaded hydraulic parts continue together in service. Fitting UNF pipe nuts into metric ports and vice-versa should not happen, but experience of the change from BSF to UNF indicated that there is no certainty in relying upon the difference in thread size when safety is involved.

To provide permanent identification of metric parts is not easy but recognition has been assisted by the following means. (Illustrations 'A' metric, 'B' unified).

1. All metric pipe nuts, hose ends, unions and bleed screws are coloured black.
2. The hexagon area of pipe nuts is indented with the letter 'M'.
3. Metric and UNF pipe nuts are slightly different in shape.

The metric female nut is **always** used with a trumpet flared pipe and the metric male nut is **always** used with a convex flared pipe.

4. All metric ports in cylinders and calipers have no counterbores, but unfortunately a few cylinders with UNF threads also have no counterbore. The situation is, all ports with counterbores are UNF, but ports not counterbored are most likely to be metric.
5. The colour of the protective plugs in hydraulic ports indicates the size and the type of the threads, but the function of the plugs is protective and not designed as positive identification. In production it is difficult to use the wrong plug but human error must be taken into account.

The Plug colours and thread sizes are:

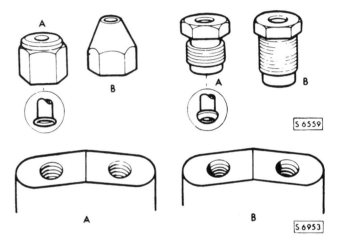

	UNF
RED	3/8 in. x 24 UNF
GREEN	7/16 in. x 20 UNF
YELLOW	½ in. x 20 UNF
PINK	7/8 in. x 18 UNF
	METRIC
BLACK	10 x 1 mm
GREY	12 x 1 mm
BROWN	14 x 1.5 mm

6. Hose ends differ slightly between metric and UNF.

Gaskets are not used with metric hoses. The UNF hose is sealed on the cylinder or caliper face by a copper gasket by the metric hose seals against the bottom of the port and there is a gap between faces of the hose and cylinder.

Pipe sizes for UNF are 3/16 in., ¼ in., and 5/16 in. outside diameter.

continued

25

07–5

Metric pipe sizes are 4.75 mm, 6 mm x 8 mm.

4.75 mm pipe is exactly the same as 3/16 in. pipe. 6 mm pipe is .014 in. smaller than ¼ in. pipe. 8 mm pipe is .002 in. larger than 5/16 in. pipe.

Convex pipe flares are shaped differently for metric sizes and when making pipes for metric equipment, metric pipe flaring tools must be used.

The greatest danger lies with the confusion of 10 mm and 3/8 in. UNF pipe nuts used for 3/16 in. (or 4.75 mm) pipe. The 3/8 in. UNF pipe nut or hose can be screwed into a 10 mm port but is very slack and easily stripped. The thread engagement is very weak and cannot provide an adequate seal.

The opposite condition, a 10 mm nut in a 3/8 in. port, is difficult and unlikely to cause trouble. The 10 mm nut will screw in 1½ or two turns and seize. It has a crossed thread 'feel' and it is impossible to force the nut far enough to seal the pipe. With female pipe nuts the position is of course reversed.

The other combinations are so different that there is no danger of confusion.

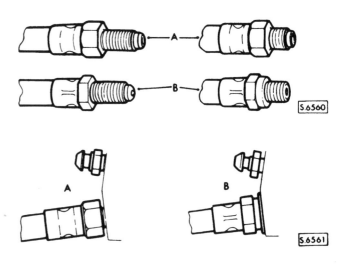

Keys and Keyways
1. Remove burrs from edges of keyways with a fine file and clean thoroughly before attempting to refit key.
2. Clean and inspect key closely; keys are suitable for refitting only if indistinguishable from new, as any indentation may indicate the onset of wear.

Tab Washers
1. Fit new washers in all places where they are used. Always renew a used tab washer.
2. Ensure that the new tab washer is of the same design as that replaced.

Split Pins
1. Fit new split pins throughout when replacing any unit.
2. Always fit split pins where split pins were originally used. Do not substitute spring washers: there is always a good reason for the use of a split pin.
3. All split pins should be fitted as shown unless otherwise stated.

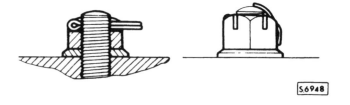

Nuts
1. When tightening a slotted or castellated nut never slacken it back to insert split pin or locking wire except in those recommended cases where this forms part of an adjustment. If difficulty is experienced, alternative washers or nuts should be selected, or washer thickness reduced.
2. Where self-locking nuts have been removed it is advisable to replace them with new ones of the same type.

NOTE: Where bearing pre-load is involved nuts should be tightened in accordance with special instructions.

Locking Wire
1. Fit new locking wire of the correct type for all assemblies incorporating it.
2. Arrange wire so that its tension tends to tighten the bolt heads, or nuts, to which it is fitted.

Screw Threads
1. Both UNF and Metric threads to ISO standards are used. See below for thread identification.
2. Damaged threads must always be discarded. Cleaning up threads with a die or tap impairs the strength and closeness of fit of the threads and is not recommended.
3. Always ensure that replacement bolts are at least equal in strength to those replaced.
4. Do not allow oil, grease or jointing compound to enter blind threaded holes. The hydraulic action on screwing in the bolt or stud could split the housing.
5. Always tighten a nut or bolt to the recommended torque figure. Damaged or corroded threads can affect the torque reading.
6. To check or re-tighten a bolt or screw to a specified torque figure, first slacken a quarter of a turn, then re-tighten to the correct figure.
7. Always oil thread lightly before tightening to ensure a free running thread, except in the case of self-locking nuts.

Five Thread Forms Replaced by – ISO Metric

B.A.	B.S.W.	B.S.F.	U.N.C.	U.N.F.	Metric Size
2	3/16	3/16	10	10	M5
1			12	12	M6
0	1/4	1/4	1/4	1/4	
	5/16	5/16	5/16	5/16	M8
	3/8	3/8	3/8	3/8	M10
	7/16	7/16	7/16	7/16	M12
	1/2	1/2	1/2	1/2	

Unified Thread Identification
1. **Bolts**
   A circular recess is stamped in the upper surface of the bolt head.
2. **Nuts**
   A continuous line of circles is indented on one of the flats of the hexagon, parallel to the axis of the nut.
3. **Studs, Brake Rods, etc.**
   The component is reduced to the core diameter for a short length at its extremity.

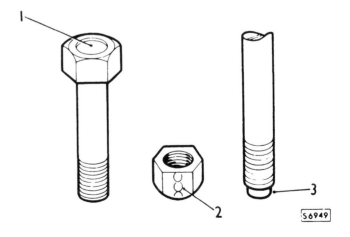

S6949

This page intentionally left blank

RECOMMENDED LUBRICANTS AND FLUIDS UK

These recommendations apply to temperate climates where operational temperatures may vary between -10°C (14°F) and 32°C (90°F)

COMPONENT	BP	CASTROL	DUCKHAMS	ESSO	MOBIL	PETROFINA	SHELL	TEXACO
Engine and carburettor dashpots Oils must meet BL Ltd. specification BLS-OL-02 or the requirements of the CCMC	BP Super Visco-Static 20-50	Castrol GTX 15W/50	Duckhams Q Motor Oil 20W/50	Esso Superlube 10W/40	Mobil Super 15W/40 or 10W/50	Fina Supergrade Motor Oil 15W/40 or 20W/50	Shell Super Motor Oil 15W/40	Havoline Motor Oil 15W/40
Main gearbox, overdrive, transfer gearbox	BP Super Visco-Static 20-50	Castrol GTX 15W/50	Duckhams Q Motor Oil 20W/50	Esso Superlube 10W/40	Mobil Super 15W/40	Fina Supergrade Motor Oil 15W/40	Shell Super Motor Oil	Havoline Motor Oil 15W/40
Final drive units Swivel pin housings Steering box	BP Gear Oil SAE 90EP	Castrol Hypoy SAE90EP	Duckhams Hypoid 90	Esso Gear Oil GX 85W/90	Mobil Mobilube HD 90	Fina Pontonic MP SAE 90	Shell Spirax 90EP	Texaco Multigear Lubricant EP 90
Power steering fluid reservoir, as applicable	BP Autran G	Castrol TQF	Duckhams Q-Matic	Essoglide	Mobil ATF210	Pursimatic 33G	Shell Donax TF	Texamatic Type G or F
Lubrication nipples (ball joints, hubs, propshafts)	BP Energrease L2	Castrol LM Grease	Duckhams LB 10	Esso Multi-Purpose Grease H	Mobilgrease MP	Fina Marson HTL 2	Shell Retinax A	Marfak All-purpose Grease
Windscreen Washers	All Season Screen Washer Fluid.							
Brakes and Clutch reservoirs	Brake fluids having a minimum boiling point of 260 C (500 F) and complying with FMVSS 1,16 DOT3 specification.							
Engine cooling system	Use an ethylene glycol based anti-freeze (containing no methanol) with non-phosphate corrosion inhibitors suitable for use in aluminium engines to ensure the protection of the cooling system against frost and corrosion.							
Inhibitor solution for engine cooling system	Marston Lubricants SQ26 - Coolant inhibitor concebtrate. For summer use only when frost precautions are not necessary.							
Air Conditioning System Refrigerant Compressor Oil	METHYLCHLORIDE REFRIGERANTS MUST NOT BE USED. Use only with refrigerant 12. This includes 'Freon 12' and 'Arcton 12'. Shell Clavus 68 BP Energol LPT 68 Sunisco 4GS Texaco Capella E Wax Free 68							

RECOMMENDED LUBRICANTS AND ANTI-FREEZE SOLUTIONS OTHER THAN UK

COMPONENT	SERVICE CLASSIFICATION		AMBIENT TEMPERATURE °C
	Performance Level	SAE Viscosity	-30 -20 -10 0 10 20 30
Engine and Carburetter Dashpots	Oils conforming to BL Ltd. SPECIFICATION BLS-22-OL-02 or the requirements of CCMC or API-SE	5W/20	
		5W/30	
		5W/40	
		10W/30	
		10W/40	
		10W/50	
		15W/40	
		15W/50	
		20W/40	
		20W/50	
Main Gearbox, and Transfer Gearbox Overdrive where fitted	BLS-22-OL-02 or the requirements of CCMC or API-SE	10W/30	
		10W/40	
		10W/50	
		10W/40	
		10W/50	
		15W/40	
		15W/50	
		20W/40	
		20W/50	
Final Drive Units Swivel Pin Housings Steering Box	API GL4 or MIL-L-2105	90 EP or 80 W, EP	
Power Steering	ATF M2C, 33G or 33F		
Lubrication Nipples (Hubs, Ball Joints, etc.)	NLGI–2 Multipurpose Lithium grease		
Brake & Clutch Reservoirs	Brake Fluids having a minimum boiling point of 260°C (500°F) and complying with FMVSS 1,16 DOT3.		
Engine Cooling System	Use an ethylene glycol based anti-freeze (containing no methanol) with non-phosphate corrosion inhibitors suitable for use in aluminium engines to ensure the protection of the cooling system against frost and corrosion. Where frost precautions are not necessary use Marston Lubricant SQ36 to prevent corrosion of the engine alloy.		
Windscreen Washers	Screen Washer Fluid		
AIR CONDITIONING SYSTEM Refrigerant Compressor Oil	METHYLCHLORIDE REFRIGERANTS MUST NOT BE USED. Use only with refrigerant 12. This includes 'Freon 12' and 'Arcton 12'. Shell Clavus 68 BP Energol LPT 68 Sunisco 4GS Texaco Capella E Wax Free 68		

09–1

ANTI–FREEZE

Anti-Freeze Concentration			50%
Specific Gravity of Coolant at 15.5°C (60°F)			1.076
Anti-Freeze Quantity		Litres	5.7
		Pints Imp.	10.0
		Pints U.S.A.	12.0
Complete protection Vehicle may be driven away immediately from cold			−36°C −33°F
Safe Limit protection Coolant in mushy state. Engine may be started and driven away after short warm-up period			−41°C −42°F
Lower Protection Prevents frost damage to cylinder head, block and radiator. Thaw out before starting engine			−47°C −53°F

RECOMMENDED FUEL

With the exception of the 9.35:1 high compression (emission) engine which is designed to operate on 97 octane fuel (British 4 star rating) all other Range Rover engines are designed for fuel having a minimum octane rating of 91 to 93 (the British 2 star rating).

Where these fuels are not available and it is necessary to use fuels of lower or unknown rating, the ignition timing must be retarded from the specified setting, just sufficiently to prevent audible detonation (pinking) under all operating conditions, otherwise damage to the engine may occur. Use exhaust gas analysis equipment to check the final engine exhaust emissions after resetting. (See 'Engine Tuning', page 05−1 for details).

The use of lower octane fuels will result in the loss of engine power and efficiency.

CAUTION: Do not use oxygenated fuels such as blends of methanol/gasolene or ethanol/gasolene (e.g. 'GASOHOL").

In the interests of public health, and to assist in keeping undesirable exhaust emissions as low as possible, fuels of an octane rating higher than that recommended should not be used.

CAPACITIES (Approx)	Litres	Imperial Unit	US Unit
Engine sump oil	5,1 litres	9 pints	10.5 pints
Extra when refilling after			
fitting new filter	0,56 litres	1 pint	1.25 pints
Main gearbox oil	2,6 litres	4.5 pints	5.5 pints
Transfer gearbox oil	3,1 litres	5.5 pints	6.5 pints
Rear differential oil	1,7 litres	3.0 pints	3.5 pints
Front differential oil	1,7 litres	3.0 pints	3.5 pints
Swivel housing oil (each)	0,26 litres	0.5 pints	0.5 pints
Steering box oil (manual)	0,40 litres	0.75 pints	0.75 pints
Cooling system	11,3 litres	20 pints	24 pints
Fuel tank	81,5 litres	18 gallons	21.5 gallons

MAINTENANCE OPERATIONS

	Maintenance Operation No.
Routine Maintenance Operations – Excluding Australia	
1,000 miles (1,600 km) Free Service	10.10.03
6,000 miles (10,000 km) Service	10.10.12
12,000 miles (20,000 km) Service	10.10.24

Details for these operations are listed in a summary chart on pages 10–2 and 10–3.

ROUTINE MAINTENANCE OPERATIONS – AUSTRALIA ONLY – ADR 27A

MAINTENANCE INTERVALS

Service	Km x 1000	
A	1.6	
B	5.15, 25, 35, 45, 55, 65, 75	10.10.03
C	10, 30, 50, 70	10.10.14
D	20, 60	10.10.26
E	40, 70	10.10.50

Details for these operations are listed in a summary chart on pages 10–4 and 10–5.

NOTE: The service schedules are based on an annual total of approximately 20,000 km. Should the vehicle complete substantially less kilometres than this per annum, it is recommended that a 'C' service is completed at six month intervals and a 'D' service at twelve month intervals.

MAINTENANCE SUMMARY

The following should be checked weekly or before a long journey.

Engine oil level	Windscreen/Tailgate/Headlamp Washer. Reservoir level(s)	Operation of Horn
Brake fluid level	Battery Electrolyte level(s)	Operation of Washers and Wipers
Radiator coolant level	All Tyres for Pressure and Condition	Operation of all lights

SUMMARY CHART – EXCLUDING AUSTRALIA

Maintenance Operation Number *Intervals in Miles x 1000 or time intervals in months. Intervals in Kilometres x 1000	10.10.03 1 1.6	10.10.12 6 10	10.10.24 12 20
Operation Description			
ENGINE			
Check for oil/fuel/fluid leaks	X	X	X
Check/top up engine oil level			
Renew engine oil	X	X	X
Renew engine oil filter		X	X
Renew engine breather filter			X
Renew carburetter air intake cleaner elements			X
Check fuel system for leaks, pipes and unions for chafing and corrosion	X	X	X
Check cooling/heater systems for leaks and hoses for security and condition	X	X	X
Check/top up cooling system	X	X	X
Check/adjust operation of all washers and top up reservoirs	X	X	X
Check driving belts adjust or renew as necessary	X	X	X
Lubricate accelerator control linkage and pedal pivot – check operation	X	X	X
Check security of engine mountings	X		
Check/top up carburetter piston dampers	X	X	X
Check/adjust carburetter idle settings	X	X	X
Check/adjust throttle control vacuum switch			X
Renew fuel filter element/cartridge		X	X
Check air injection system hoses/pipes for condition and security			X
Clean/renew engine flame traps			X
Check air intake temperature control system			X
Check crankcase breathing and evaporative loss systems. Check hoses/pipes and restrictors for blockage, security and condition	X		X
Clean electric fuel pump element	48,000 miles (80,000 km)		
Check exhaust system for leakage and security	X	X	X
IGNITION			
Clean/adjust spark plugs		X	
Renew spark plugs			X
Clean/adjust distributor contact breaker points (sliding contact type)			X
Clean/adjust distributor contact breaker points (non-sliding contact type)	X	X	
Renew distributor contact breaker points (sliding contact type)	24,000 miles (40,000 km)		
Renew distributor contact breaker points (non-sliding contact type)			X
Lubricate distributor	X	X	
Check ignition wiring and high tension leads for fraying, chafing and deterioration	X		X
Clean distributor cap. Check for cracks and tracking			X
Check security of distributor vacuum unit line and operation of vacuum unit	X		X
Check/adjust dwell angle and ignition timing using electronic equipment	X		X
Check coil performance on oscilloscope			X
TRANSMISSION			
Check for oil leaks	X	X	X
Check/top up clutch fluid reservoir	X	X	X
Check clutch pipes for chafing, leaks or corrosion	X	X	X
Check/top up gearbox and transfer box oil levels		X	X
Renew gearbox and transfer box oil	X	24,000 miles (40,000 km)	
Check front and rear axle case breathers			X
Check/top up front and rear axle oil levels		X	X
Renew front and rear axle oil	X	24,000 miles (40,000 km)	
Drain flywheel housing if drain plug is fitted for wading	X	X	X
Check tightness of propeller shaft coupling bolts	X	X	
Lubricate propeller shaft	X	X	X
Lubricate propeller shaft sealed sliding joint	24,000 miles (40,000 km)		

Maintenance Operation Number *Intervals in Miles x 1000 or time intervals in months. Intervals in Kilometres x 1000	10.10.03 1 1.6	10.10.12 6 10	10.10.24 12 20
Operation Description			
STEERING AND SUSPENSION			
Check condition and security of steering unit, joints, relays and gaiters	X	X	X
Check steering rack/gear for oil/fluid leaks	X	X	X
Check shock absorbers for fluid leaks	X	X	X
Check power steering system for leaks, hydraulic pipes and unions for chafing, cracks and corrosion	X	X	X
Check/top up fluid in power steering reservoir or manual steering box oil level	X	X	X
Check/adjust front wheel alignment	X	X	X
Check security of suspension fixings	X		X
Check/adjust steering box	X	X	X
Check/top up swivel pin housing oil levels		X	X
Renew swivel pin housing oil	X	24,000 miles (40,000 km)	
Check suspension self levelling unit for fluid leaks	X	X	X
BRAKES			
Check visually, hydraulic pipes and unions for chafing, leaks and corrosion	X	X	X
Check/top up brake fluid reservoir(s)	X	X	X
Check footbrake operation; (Self adjusting)	X	X	X
Check handbrake for security and operation; adjust if necessary	X	X	X
Inspect brake pads for wear, discs for condition		X	X
Lubricate handbrake mechanical linkage and cable guides (lever pivot)		X	X
Check brake servo hose(s) for security and condition	X	X	X
Renew hydraulic brake fluid	Every 18 months 18,000 miles (30,000 km)		
Renew rubber seals in braking system, flexible hoses and servo air filter	Every 36 months 36,000 miles (60,000 km)		
WHEELS AND TYRES			
Check/adjust tyre pressures including spare wheel	X	X	X
Check tyres for tread depth and visually for external cuts in fabric, exposure of ply or cord structure, lumps or bulges	X	X	X
Check that tyres comply with manufacturers specification		X	X
Check tightness of road wheel fastenings	X	X	X
ELECTRICAL			
Check function of electrical equipment	X	X	X
Check/top up battery electrolyte	X	X	X
Clean and grease battery connections		X	X
Check headlamp alignment, adjust if necessary	X	X	X
Check, if necessary renew, wiper blades		X	X
BODY			
Lubricate all locks and hinges (Not steering lock)	X	X	X
Check operation of window controls	X		X
Check condition and security of seats and seat belts	X	X	X
Check operation of seat belt inertia reel mechanism (where fitted)	X	X	X
Check operation of all door, bonnet and tailgate locks	X	X	X
Check rear view mirror(s) for cracks and crazing		X	X
Ensure cleanliness of controls, door handles, steering wheel	X	X	X
GENERAL			
Road/roller test and check function of all instrumentation	X	X	X
Report additional work required		X	X

*The service schedules above are based on an annual total of approximately 12,000 miles (20,000 km). Should the vehicle complete substantially less miles/kilometres in any year it is recommended that service intervals are based on a proportional time basis.

SUMMARY CHART – AUSTRALIAN – ADR 27A

Maintenance Operation Number		10.10.03	10.10.14	10.10.26	10.10.50
Service	A	B	C	D	E

Operation Description

ENGINE

	A	B	C	D	E
Check for oil/fuel/fluid leaks	X	X	X	X	X
Check/top up engine oil level		X			
Renew engine oil	X		X	X	X
Renew engine oil filter			X	X	X
Renew engine breather filter					X
Renew carburetter air intake cleaner elements				X	X
Check fuel system for leaks, pipes and unions for chafing and corrosion	X	X	X	X	X
Check cooling/heater systems for leaks and hoses for security and condition	X	X	X	X	X
Check/top up cooling system	X	X	X	X	X
Check/adjust operation of all washers and top up reservoirs	X	X	X	X	X
Check driving belts adjust or renew as necessary	X			X	X
Lubricate accelerator control linkage and pedal pivot – check operation	X		X	X	X
Check security of engine mountings	X				
Check/top up carburetter piston dampers	X			X	X
Check/adjust carburetter idle settings	X			X	X
Check/adjust throttle control vacuum switch				X	X
Check security of E.G.R. valve operating lines	X				
Renew fuel filter element/cartridge				X	X
Check air injection system hoses/pipes for condition and security				X	X
Clean/renew engine flame traps				X	X
Check air intake temperature control system				X	X
Check crankcase breathing and evaporative loss systems. Check hoses/pipes and restrictors for blockage, security and condition	X			X	X
Renew adsorption canister					
Check E.G.R. system			80,000 km		
Clean electric fuel pump element					
Check exhaust system for leakage and security	X	X.	X	X	X

IGNITION

	A	B	C	D	E
Renew spark plugs				X	X
Renew distributor contact breaker points (sliding contact type)					X
Renew distributor contact breaker points (non-sliding contact type)					X
Lubricate distributor			X		
Check ignition wiring and high tension leads for fraying, chafing and deterioration	X			X	X
Clean distributor cap. Check for cracks and tracking				X	X
Check security of distributor vacuum unit line and operation of vacuum unit	X			X	X
Check/adjust dwell angle and ignition timing using electronic equipment	X			X	X
Check coil performance on oscilloscope				X	X

TRANSMISSION

	A	B	C	D	E
Check for oil leaks	X	X	X	X	X
Check/top up clutch fluid reservoir	X	X	X	X	X
Check clutch pipes for chafing, leaks or corrosion	X	X	X	X	X
Check/top up gearbox and transfer box oil levels			X	X	
Renew gearbox and transfer box oil	X				X
Check front and rear axle case breathers					X
Check/top up front and rear axle oil levels			X	X	
Renew front and rear axle oil	X				X
Drain flywheel housing if drain plug is fitted for wading		X	X	X	X
Check tightness of propeller shaft coupling bolts	X			X	
Lubricate propeller shaft	X		X	X	X
Lubricate propeller shaft sealed sliding joint					X

Maintenance Operation Number		10.10.03	10.10.14	10.10.26	10.10.50
Service	A	B	C	D	E
Operation Description					
STEERING AND SUSPENSION					
Check condition and security of steering unit, joints, relays and gaiters	X	X	X	X	X
Check steering rack/gear for oil/fluid leaks	X	X	X	X	X
Check shock absorbers for fluid leaks	X	X	X	X	X
Check power steering system for leaks, hydraulic pipes and unions for chafing, cracks and corrosion	X	X	X	X	X
Check/top up fluid in power steering reservoir or manual steering box oil level	X	X	X	X	X
Check/adjust front wheel alignment			X	X	X
Check security of suspension fixings				X	X
Check/adjust steering box	X	X	X	X	X
Check/top up swivel pin housing oil levels			X	X	
Renew swivel pin housing oil	X				X
Check suspension self levelling unit for fluid leaks	X	X	X	X	X
BRAKES					
Check visually, hydraulic pipes and unions for chafing, leaks and corrosion	X	X	X	X	X
Check/top up brake fluid reservoir(s)	X	X	X	X	X
Check footbrake operation (Self adjusting)	X	X	X	X	X
Check handbrake for security and operation; adjust if necessary	X	X	X	X	X
Inspect brake pads for wear, discs for condition		X	X	X	X
Lubricate handbrake mechanical linkage and cable guides (lever pivot)			X	X	X
Check brake servo hose(s) for security and condition	X	X	X	X	X
Renew hydraulic brake fluid		Every 18 months, 30,000 km			
Renew rubber seals in braking system, flexible hoses and servo air filter		Every 36 months, 60,000 km			
WHEELS AND TYRES					
Check/adjust tyre pressures including spare wheel	X	X	X	X	X
Check tyres for tread depth and visually for external cuts in fabric, exposure of ply or cord structure, lumps or bulges	X	X	X	X	X
Check that tyres comply with manufacturers specification		X	X	X	X
Check tightness of road wheel fastenings	X	X	X	X	X
ELECTRICAL					
Check function of electrical equipment	X	X	X	X	X
Check/top up battery electrolyte	X	X	X	X	X
Clean and grease battery connections			X	X	X
Check headlamp alignment, adjust if necessary	X	X	X	X	X
Check, if necessary renew, wiper blades		X	X	X	X
Check output of alternator charging system	X		X	X	X
BODY					
Lubricate all locks and hinges (not steering lock)	X		X	X	X
Check operation of window controls	X			X	X
Check condition and security of seats and seat belts	X	X	X	X	X
Check operation of seat belt inertia reel mechanism (where fitted)	X	X	X	X	X
Check operation of all door, bonnet and tailgate locks	X		X	X	X
Check rear view mirror(s) for cracks and crazing		X	X	X	X
Ensure cleanliness of controls, door handles, steering wheel	X	X	X	X	X
GENERAL					
Road/roller test and check function of all instrumentation	X		X	X	X
Report additional work required		X	X	X	X

The summary charts on pages 10–2 to 10–5 give details of the intervals at which the following operations should be completed. The operations are listed in the sequence shown in the summary chart.

ENGINE

Check for Oil/Fuel/Fluid Leaks.

Check/Top up engine oil level.

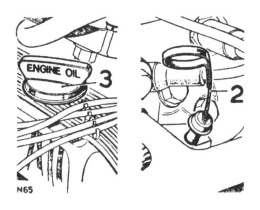

1. Stand the car on level ground and allow the oil to drain back into the sump.
2. Withdraw the dipstick at left-hand side of engine; wipe it clean, re-insert to its full depth and remove a second time to take the reading.
3. Add oil as necessary through the screw-on filler cap marked 'engine oil' on the right-hand front rocker cover. Never fill above the 'High' mark.

Renew Engine Oil.

1. Run the engine to warm up the oil; switch off the ignition.
2. Place an oil tray under the drain plug.
3. Remove the drain plug in the bottom of the sump at left-hand side. Allow oil to drain away completely and replace the plug.
4. Refill the engine sump with the correct quantity and grade of oil – refer to Division 09.
5. Run the engine to check for oil leaks at the drain plug.

Renew Engine Oil Filter.

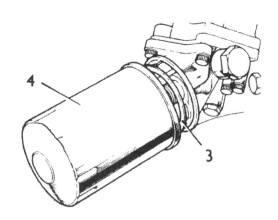

1. Place an oil tray under the engine.
2. Unscrew the filter anti-clockwise by the hexagon on end or casing (early models) or using a strap spanner (later models).
3. Smear clean engine oil on the rubber washer of the new filter.
4. Fill the filter with new oil as far as possible, noting the angle at which the filter is to be fitted.
5. Screw on the filter until the sealing ring touches the oil pump cover face, then tighten it a further half turn by hand only. **Do not overtighten.**
6. Check/Top up the engine oil level.
7. Run the engine to check for oil leaks at the filter.

2RA299A

Renew Engine Breather Filter.

1. Remove the air cleaner 19.10.01.
2. Withdraw rear hose from the filter.
3. Slacken the filter clip.
4. Withdraw the filter from the clip and front hose.
5. Fit the new filter with end marked 'IN' facing forward. Alternatively, if the filter is marked with arrows, they must point rearward. Refit hoses and tighten clip.
6. Refit the air cleaner 19.10.01.

Renew Carburetter Air Intake Cleaner Elements.

Attention to the air cleaner is extremely important. Replace elements every 10,000 km (6,000 miles) under severe dusty conditions, as performance will be seriously affected if the car is run with an excessive amount of dust or industrial deposits in the element.

Two types of air cleaner can be fitted, follow the instructions as appropriate.

European specification

1. Release the hose clips each side of the air cleaner.
2. Withdraw the air cleaner elbows.
3. Detach the choke cable from the clip on the air cleaner.
4. Withdraw the air cleaner from the retaining posts, at the same time disconnecting the hose from the engine breather filter.
5. Release the end plate clips.
6. Withdraw the end plates complete with elements.
7. Remove the wing nut, washer and retaining plate for each element.
8. Then withdraw the elements.
9. Discard old elements and replace with new units.
10. Ensure that sealing washers on end plates and retaining plates are in good condition; if not replace.
11. Reassemble elements to air cleaner, and air cleaner to engine by reversing removal procedure.

Australian specification

1. Disconnect the balance pipe running between the air cleaners.
2. Disconnect the rocker cover breather pipe from the R.H. air cleaner.
3. Slacken the jubilee clips and disconnect the air intake temperature control systems from both air cleaners.
4. Disconnect the H.T. leads from the retaining clips located on the top of both air cleaners.
5. Disconnect the two pipes running into the temperature sensor on the R.H. air cleaner.
6. On both air cleaners, remove the four bolts securing the air cleaners to the air intake adaptor.
7. Remove the air cleaners.
8. Release the 4 clips on each air cleaner and extract the elements.
9. Fit new elements.

    NOTE: when fitting new elements ensure that the plastic bevelled edge faces the air intake aperture.
10. For reassembly reverse the above procedure 1-6.

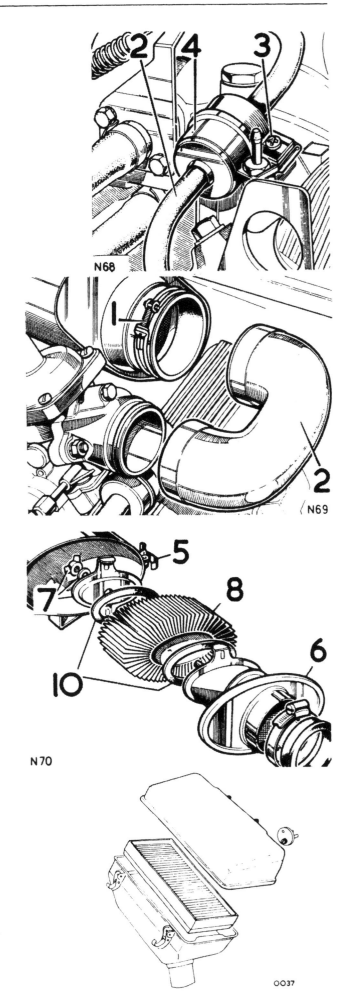

N68

N69

N70

0037

Check fuel system for leaks, pipes and unions, for chafing and corrosion.

Check cooling/heater systems for leaks and hoses for security and condition.

Cooling system hoses should be changed at the first signs of deterioration.

Check/top up cooling system

1. To prevent corrosion of the aluminium alloy engine parts it is imperative that the cooling system is filled with a solution of water and anti-freeze, winter or summer, or water and inhibitor during the summer only. Never fill or top up with plain water.
2. The expansion tank filler cap is under the bonnet.
3. With a cold engine, the correct coolant level should be up to the 'Water Level' plate situated inside the expansion tank below the filler neck.

WARNING. Do not remove the filler cap when engine is hot because the cooling system is pressurised and personal scalding could result.

4. When removing the filler cap, first turn it anti-clockwise a quarter of a turn and allow all pressure to escape, before turning further in the same direction to lift it off.
5. When replacing the filler cap it is important that it is tightened down fully, not just to the first stop. Failure to tighten the filler cap properly may result in water loss, with possible damage to the engine through over-heating. Use soft water whenever possible, if local water supply is hard, rain-water should be used.

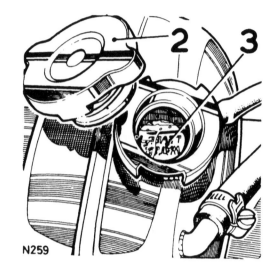

N259

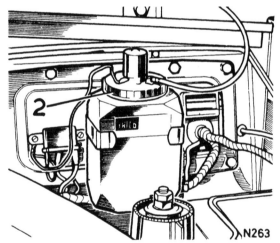

N263

Check/Adjust operation of all washers and top up reservoirs.

1. Check operation of windscreen, tailgate and headlamp washer. Adjust as necessary referring to section 84.
2. Remove reservoir caps by turning anti-clockwise.
3. Top up reservoir to within approximately 25 mm (1 in.) below bottom of filler neck.
   Use 'Clearalex' windscreen washer powder in the bottle, this will remove mud, flies and road film.
   In cold weather to prevent freezing of the water, add 'Isopropyl Alcohol'.

Do NOT use methylated spirits, which has a detrimental effect on the screenwasher impeller.

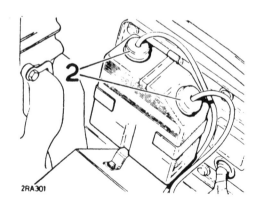

2RA301

Check Driving Belts, Adjust or renew as necessary.

1. Examine the following belts for wear and condition and renew if necessary.
    (i) Crankshaft – water pump – alternator.
    (ii) Water pump – jockey pulley – air pump.
    (iii) Crankshaft – power steering.

2. Each belt should be sufficiently tight to drive the appropriate auxiliary without undue load on the bearings.
    Correct tension: This is measured by allowing 0.4 mm movement on the slack side of the belt per 25.4 mm between pulley centres.
    Ex   Distance between pulley centres = 254 mm
    Tension = $\dfrac{254 \text{ mm} \times 0.4 \text{ mm}}{25.4}$

      = 4.00 mm.

3. Slacken the bolts securing the unit to its mounting bracket.

4. Slacken appropriate pivot bolt and the fixing at the adjustment link.

5. Pivot the unit inwards or outwards as necessary and adjust until the correct belt tension is obtained.

6. Tighten unit adjusting bolts.

Check adjustment again, when a new belt is fitted, after approximately 1500 km (1,000 miles) running.

Lubricate Accelerator Control Linkage and Pedal Pivot – Check Operation.

The throttle linkage will not require adjustment during normal operation.
To ensure complete throttle closure a degree of 'lost motion' or slackness is incorporated into the linkage; no attempt must be made to eliminate this.

Check security of Engine Mountings.

Check/Top up carburetter piston dampers.

1. Unscrew and withdraw the plug and damper assembly from the top of each carburetter.
2. Top-up the damper chambers with the seasonal grade of engine oil.
3. The oil level is correct, when utilising the damper as a dipstick its threaded plug is 6mm above the dash pots and resistance is felt.
4. Screw down the damper plugs.

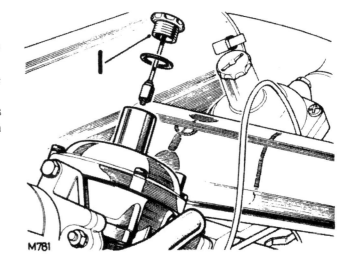

M781

39

Check/Adjust Carburetter Idle Settings

Refer to Operation 19.15.02

Check/Adjust Throttle Control Vacuum Switch (Australian)

When fitted the vacuum switch enables normal advance characteristics to be obtained for quick acceleration, high speed driving and vehicle laden conditions.
The switch is activated by a cam fitted to the left-hand carburetter spindle and is set to interrupt the vacuum line between the manifold and the retard side of the distributor vacuum unit.
The following check and adjustment must only be carried out after the engine idle speed has been set correctly.

1. Ensure that the throttle linkage is fully in the idle position.
2. Push the plunger fully into the switch and hold in this position.
3. Measure the clearance between the plunger and the cam on the throttle linkage. This must be 0,76 mm (0.030 in.).
4. Adjust as necessary by slackening the fixings mounting the switch to the bracket and moving the switch in the required direction.
5. Recheck the clearance after adjustment to ensure that it has not been disturbed when tightening the fixings.

NOTE: After final adjustment ensure that the centre of the cam contacts the centre of the switch button.

Check Security of E.G.R. Valve Operating Lines (Australian)

1. Check the following for security:—
   (i) Line from carburetter to E.G.R. valve.
   (ii) Pipe to inlet manifold from E.G.R. valve.
   (iii) E.G.R. valve locknut.
2. Renew any pipes which show signs of deterioration.

Renew Fuel Filter Element/Cartridge

The element provides a filter between the pump and carburetter and is located on the front LH wing.
Replace as follows:—
1. Unscrew the centre bolt.
2. Withdraw the filter bowl.
3. Remove the small sealing ring and remove element.
4. Withdraw the large sealing ring from the underside of the filter body.
5. Discard the old element and replace with a new unit.
6. Ensure that the centre and top sealing rings are in good condition and replace as necessary.
7. Fit new element, small hole downwards.
8. Refit sealing rings.
9. Replace filter bowl and tighten the centre bolt.

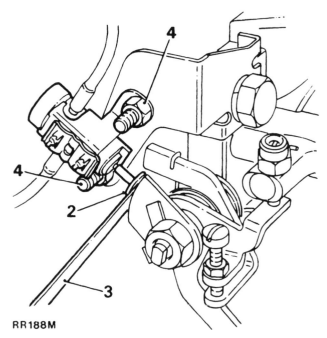

RR188M

6. Start the engine and warm it to normal running temperature.
7. When the engine is idling steadily, disconnect the retard pipe at the distributor. A noticeable rise in engine speed should be apparent if the system is functioning.
8. When satisfied that the facility is operating correctly, reconnect the vacuum pipe to the distributor, ensuring a secure connection.

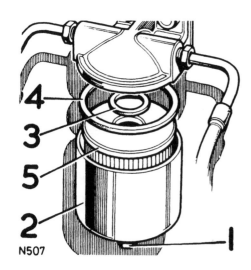

N507

The cartridge (fitted to earlier models only) provides an additional filter between pump and carburetter, located at the front of the LH rocker cover.
Replace as follows:—

1. Disconnect the fuel pipes from each end of the filter.
2. Slacken the securing clip.
3. Withdraw the filter.
4. Fit the new filter with end marked 'IN' downwards. Alternatively if the filter is marked with arrows, they must point upwards. Tighten securing clip and refit fuel pipes.

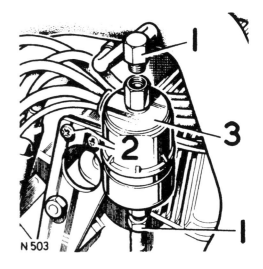
N 503

Check air injection system hoses/pipes for condition and security

1. Check the hoses between the air pump, check valves and relief valve.
2. Check for security, the two bolts securing the air pump outlet adaptor to the air pump.
3. Check the security of the nuts securing the air injection rails to the cylinder heads.

Check/renew engine flame traps

1. Pull the flame trap clip clear of the retainers.
2. Pull the hoses from the flame traps.
3. Withdraw the flame traps.
4. Wash in clean petrol and allow to dry.
5. Refit the flame traps, which are located in position by the hoses and clips.

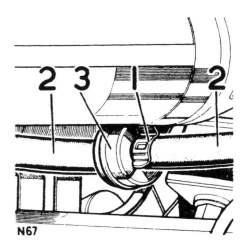
N67

Check Air Intake Temperature Control System

The following check also applies to the Australian specification except that there are two hot boxes and two flap valves with vacuum capsules.

1. Check operation of the mixing flap valve in the air cleaner by starting the engine from cold and observing the flap valve as the engine temperature rises.
2. The valve should start to open slowly within a few minutes of starting and continue to open until a stabilised position is achieved. This position and the speed of operation will be entirely dependent on prevailing ambient conditions.
3. Failure to operate indicates failure of flap valve vacuum capsule or thermostatically controlled vacuum switch or both.
4. Check by connecting a pipe directly from the banjo on No. 8 point inlet manifold to the flap valves, thus by-passing the temperature sensor.
5. If movement of the flap valve is evident the temperature sensor is faulty. If no movement is detected, the vacuum capsule is faulty.
6. Fit new parts where necessary.

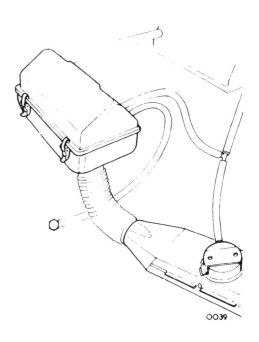
0039

Check Crankcase Breathing and (Australian) Evaporative
Loss Systems

**Check Hoses/Pipes and Restrictors for blockage, security
and condition**

1. Check visually, the security of the following hoses:—
    (i) Fresh air inlet from air cleaner to air filter and
        air filter to rocker cover.
    (ii) Purge lines from canister to carburetter intake
        adaptors.
    (iii) Both carburetter float chamber vent pipes.
    (iv) Fuel tank vent pipe.
2. Disconnect the fresh air inlet pipe from the air
    cleaner.
3. Disconnect the purge lines from the carburetters.
4. Connect a low pressure air pipe to the fresh air inlet
    and blow through the system, checking for any
    restrictions.
5. Clear any blockage.
6. Refit the pipes ensuring that all connections are
    secure. Renew any doubtful hoses.

Renew Adsorption Canister (Australian)

Removing

1. Disconnect from the canister:—
    (i) Canister line to fuel tank.
    (ii) Canister purge lines.
    (iii) Carburetter vent pipe.
2. Slacken the clamp nut screw.
3. Remove the canister.

Refitting

4. Secure the canister in the clamp.
5. Reverse instructions 1 and 2 above.

WARNING: The use of compressed air to clean an adsorption canister or clear a blockage in the evaporative system is
very dangerous. An explosive gas present in a fully saturated canister may be ignited by the heat generated when
compressed air passes through the canister.

Check E.G.R. System (Australian)

1. Disconnect the vacuum control from the top of the E.G.R. valve.
2. Slacken the unions securing the asbestos lagged pipe to the inlet manifold and the E.G.R. valve.
3. Remove the pipe.
4. Slacken the locknut at the base of the valve.
5. Unscrew the valve from the exhaust manifold.
6. Clean the assembly area of the valve with a wire brush. Use a standard spark plug machine to clean the valve and seat. Insert the valve opening into the machine and lift the diaphragm evenly. Blast the valve for approximately 30 seconds; remove and inspect. If necessary repeat until all the carbon deposits are removed. Use compressed air to remove all traces of carbon grit from the valve. Use a flexiwire brush to clean the steel pipe; blow clear of carbon grit.
7. Refit the E.G.R. valve and asbestos lagged pipe by reversing 2-5 above.
8. Reconnect the vacuum control line to the top of the E.G.R. valve.
9. Check the E.G.R. valve operating line and asbestos lagged pipe for security.
10. Renew any pipes which show signs of deterioration.
11. Check the function of the E.G.R. valve as follows:— Warm the engine to normal running temperature and ensure that the autochoke control is fully 'off'. Open and close the throttle several times and observe or feel the E.G.R. valve, which should open and close with the changes in engine speed. This valve should close instantly when the throttle is closed.

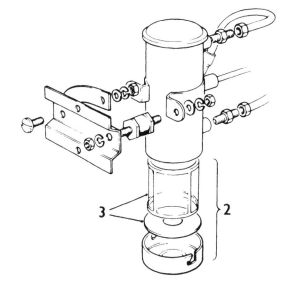

Clean Electric Fuel Pump Element

Alternative fuel pumps can be fitted and may be located on the left chassis member to the front of the rear wheel or on the heelboard beneath the rear seat.

1. From beneath the vehicle disconnect the fuel inlet pipe from the pump and blank the end of the pipe by suitable means to prevent fuel draining from the tank.
2. Release the end cover from the bayonet fixing.
3. Withdraw the filter and clean by using a compressed air jet from the inside of the filter.
4. Remove the magnet (where fitted) from the end cover and clean. Replace the magnet in the centre of the end cover.
5. Reassemble the fuel pump and refit the fuel inlet pipe. Use a new gasket for the end cover if necessary.

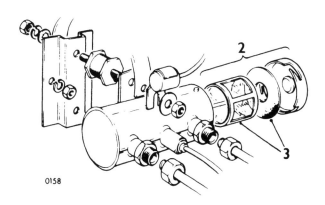

0158

Check Exhaust System for Leaks and Security

IGNITION

Clean/Adjust Spark Plugs

1. Use the special spark plug spanner and tommy bar supplied in the tool kit when removing or refitting spark plugs.
2. Take great care when fitting spark plugs not to cross-thread the plug, otherwise costly damage to the cylinder head will result.
3. Check or replace the spark plugs as applicable. If the plugs are in good condition, clean and reset the electrode gaps refer to Page 05-1. At the same time file the end of the central electrode until bright metal can be seen.
4. It is important that only the correct type of spark plugs are used for replacements.
5. Incorrect grades of plugs may lead to piston over-heating and engine failure.

To remove spark plugs proceed as follows:-

6. Remove the leads from the spark plugs.
7. Remove the plugs and washers.
8. To clean the spark plugs:-
   (a) Fit the plug into a 14 mm adaptor of an approved spark plug cleaning machine.
   (b) Wobble the plug in the adaptor with a circular motion for three or four seconds only with the abrasive blast in operation. Important: Excessive abrasive blasting will lead to severe erosion of the insulator nose. Continue to wobble the plug in its adaptor with air only, blasting the plug for a minimum of 30 seconds: this will remove abrasive grit from the plug cavity.
   (c) Wire-brush the plug threads; open the gap slightly, and vigorously file the electrode sparking surfaces using a point file. This operation is important to ensure correct plug operation by squaring the electrode sparking surfaces.
   (d) Wash new plugs in petrol to remove protective coating.
9. Set the electrode gap to the recommended clearance.
10. Shows dirty plug.
11. Filing plug electrodes.
12. Clean plug set to correct gap.
13. Test the plugs in accordance with the plug cleaning machine manufacturers' recommendations.
14. If satisfactory the plugs can be refitted.
15. When pushing the leads on to the plugs, ensure that the shrouds are firmly seated on the plugs.

Fitting H.T. leads

16. Ensure that replacement HT leads are refitted in their spacing cleats in accordance with the correct layout illustrated.
    Failure to observe this instruction may result in cross-firing between two closely fitted leads which are consecutive in the firing order.

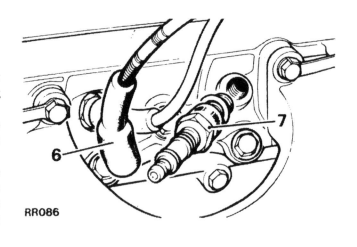

RRO86

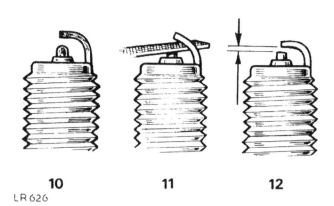

10   **11**   **12**

LR 626

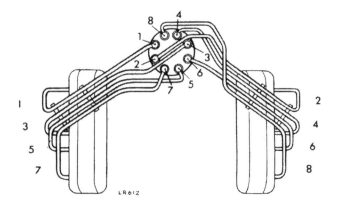

LR612

Renew distributor contact breaker points

Fixed contact type (fitted to distributor with black cap)

1. Remove distributor cap.
2. Remove the nut on the terminal block.
3. Lift off the spring and moving contact.
4. Remove adjustable contact, secured with a screw.
5. Clean the new points with petrol to remove the protective coating.
6. Add a smear of grease to contact pivot before fitting.

Sliding contact type (fitted to distributor with blue cap)

1. Release the clips and remove the distributor cap.
2. Remove the rotor arm from the cam spindle.
3. Remove the retaining screw and washers and lift the complete contact breaker assembly from the moveable plate.
4. Remove the nut and plastic bushes from the terminal post to release the leads and spring.
5. Discard the old contact breaker assembly.
6. Clean the new points with petrol to remove the protective coating.
7. Connect the leads to the terminal post in the following sequence:—
   (a) lower plastic bush
   (b) red lead tab
   (c) contact breaker spring eye
   (d) black lead tab
   (e) upper plastic bush
   (f) retaining nut.
8. Fit the contact breaker assembly to the moveable plate ensuring that the pegs underneath locate in the holes in the moveable plate.
9. The sliding contact actuating fork must also be located over the fixed peg in the adjustable base plate.
10. Fit the retaining screw plain and spring washer to secure the contact breaker assembly to the moveable plate.

Check/adjust distributor contact breaker points

To obtain satisfactory engine performance it is most important that the contact points are adjusted to the dwell angle which is 26° to 28°, using suitable workshop equipment (See operation 86.35.20 for details). However, contact points may be adjusted provisionally or in circumstances where specialised checking equipment is not available, by one of the following methods:—

Feeler gauge method

1. Remove the distributor cap.
2. Turn the engine in direction of rotation until the contacts are fully open.
3. The clearance should be 0,35 to 0,40 mm (0.014 to 0.016 in) with the feeler gauge a sliding fit between the contacts.
4. Adjust by turning the adjusting nut clockwise to increase gap or anti-clockwise to reduce gap.
5. Replace the distributor cap.

Timing lamp method

1. Remove the distributor cap.

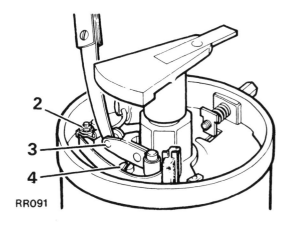

RR091

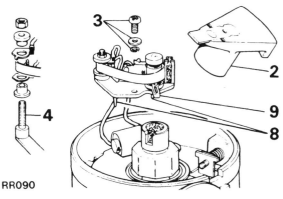

RR090

2. Turn the engine in the direction of rotation until the contact breaker heel is on the peak of number one cylinder cam. Points should be fully open.
3. Connect a 12 volt timing lamp, or suitable voltmeter, across the contact breaker lead terminal and a suitable earth point.
4. Switch on the ignition.
5. Turn the distributor adjusting nut anti-clockwise until the timing lamp goes out or there is no reading on the voltmeter.
6. Continue a further two turns of the adjuster in an anti-clockwise direction.
   During this operation the adjusting nut should be pressed inwards with the thumb to assist the helical return spring.
7. Slowly turn the adjusting nut clockwise until the timing lamp just comes on, or there is a voltage shown on the voltmeter.
8. Noting the position of the flats on the adjusting nut, continue in a clockwise rotation for a further five flats.
9. Remove timing lamp or voltmeter and switch off ignition.
10. Replace the distributor cap.

IMPORTANT: At the first available opportunity after the contacts have been adjusted as detailed above, they must be finally set to the dwell angle using specialised equipment.

NOTE: When new contact points have been fitted the dwell angle must be checked after a further 1500 km (1,000 miles) running.

Lubricate Distributor (black cap)

Fixed contact type

1. Remove distributor cap.
2. Remove rotor arm.
3. Lightly smear the cam with clean engine oil.
4. Add a few drops of thin machine oil to lubricate the cam bearing and distributor shaft.
5. Wipe the inside and outside of the distributor cap with a soft dry cloth.
6. Ensure that the carbon brush works freely in its holder.
7. Replace rotor arm and distributor cap.

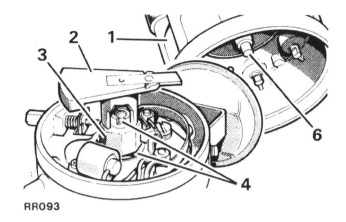

RRO93

Lubricate Distributor (blue cap)

Sliding contact type

1. Remove distributor cap.
2. Remove rotor arm.
3. Lightly smear the cam with clean engine oil.
4. Add a few drops of thin machine oil to lubricate the cam bearing and distributor shaft.
5. Wipe the inside and outside of the distributor cap with a soft dry cloth.
6. Ensure that the carbon brush works freely in its holder.
7. Lubricate the actuator ramps and contact breaker heel ribs with Shell 'Retinax' or equivalent grease.
8. Grease the underside of the heel actuator.
9. Apply grease to the fixed pin and actuator fork.
10. Refit rotor arm and distributor cap.

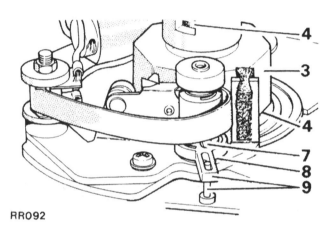

RRO92

Check

Check Ignition Wiring and High Tension Leads for Fraying, Chafing and Deterioration.
Clean Distributor Cap and check for Cracks and Tracking.

Check Security of Distributor Vacuum Unit Line and Operation of Vacuum Unit.

1. Check security of distributor vacuum unit line connections at the carburetter, throttle-controlled vacuum switch and distributor. Ensure that carburetter idle settings are correct.
2. Check/adjust the throttle-controlled vacuum switch as follows:—
   (i) Ensure that the throttle linkage is fully returned to the idle position.
   (ii) Push the plunger fully into the switch and hold it in this position.
   (iii) Measure the clearance between the plunger and the cam on the throttle linkage. This must be 0,6 to 0,7 mm (0.025 to 0.030 in).
   (iv) Adjust as necessary by slackening the fixings, mounting bracket to inlet manifold, and moving the switch and bracket complete in the required direction.

3. Start the engine and warm it to normal running temperature.
4. When the engine is idling steadily, disconnect the retard pipe at the distributor. A noticeable rise in engine speed should be apparent if the system is functioning.
5. When satisfied that the facility is operating correctly, reconnect the vacuum pipe to the distributor ensuring a secure connection.

Check

Check/Adjust Dwell Angle and Ignition Timing using Electronic Equipment (See 86.35.20 for details).

Coil Performance on an Oscilloscope
See 'Engine Tuning' page 05−1 for details.

TRANSMISSION.

Check for Oil leaks.

Check/Top up Clutch Fluid Reservoir.

1. Check the fluid level in the reservoir, mounted on the bulkhead adjacent to the brake servo.
2. Remove the cap, top up if necessary to bottom of filler neck. Refer to Division 09 for recommended fluids.

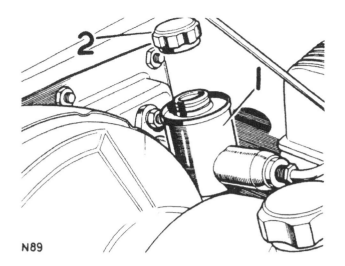

N89

Check clutch pipes for chafing, leaks and corrosion.

Check/top up gearbox and transfer box oil levels.

Check oil levels daily or weekly when operating under severe wading conditions.

Main gearbox oil levels

Check oil level daily or weekly when operating under severe wading conditions.

Earlier Models
1. Remove gearbox cover trim.
2. Remove oil level dipstick, located under the main gear lever sealing rubber, and check that oil level is up to the 'H' level mark on the dipstick. If oil is required, proceed as follows:
3. Remove the oval rubber blanking plug on the gearbox cover.
4. Remove the oil filler cap from the gearbox and top up as necessary. If significant topping up is required check for oil leaks at drain plug and filler cap, all joint faces and through drain hole in bell housing

Later Models
1. From beneath the vehicle remove the filler/level plug at the side of the gearbox.
2. Add oil to the bottom of filler plug orifice.
3. Replace the filler plug.

Continued

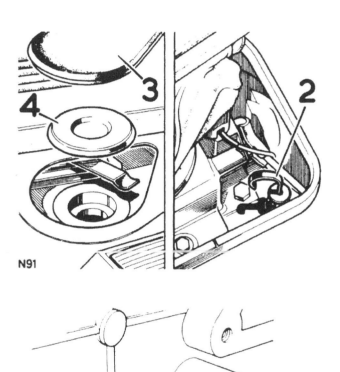

N91

1391

Transfer box oil level

1. To check oil level, remove the oil level plug, located on the rear of the transfer box casing, oil should be level with the bottom of the hole.
2. To top up, remove the round rubber blanking plug from the gearbox cover.
3. Remove the oil filler plug from the transfer box, and top up as necessary.

If significant topping up is required, check for oil leaks at drain and filler plugs.

NOTE: Except where the transfer box is removed from the vehicle it is more convenient to top-up through the level plug hole. Do not overfill otherwise leakage will occur.

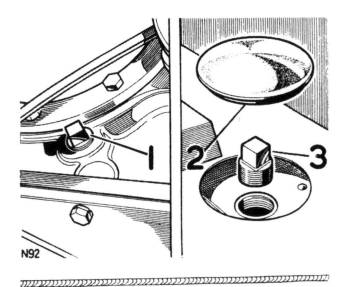

N92

Renew Gearbox and Transfer Box Oil

Drain and refill monthly when operating under severe wading conditions.

Main gearbox oil changes

1. Immediately after a run when the oil is warm, drain off the oil by removing the drain plug and washer from the bottom of the gearbox casing.
2. Remove the oil filter.
3. Wash the filter in clean fuel; allow to dry and replace.
4. Refit drain plug and refill gearbox through the oil level/filler hole with the correct grade of oil to the bottom of the level/filler hole. On early gearboxes fill to the 'H' level mark on the dipstick.

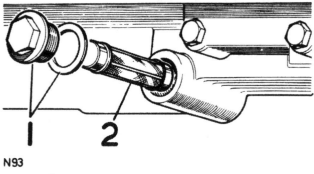

N93

Transfer gearbox oil changes

1. Immediately after a run when the oil is warm, drain off the oil by removing the drain plug in the bottom of the transfer box.
2. Replace the drain plug and refill the transfer box through the oil filler hole, with the correct grade of oil, to the bottom of the oil level plug hole.

Refer to Division 09 for gearbox capacity.

NOTE: Except where the transfer box is removed from the vehicle it is more convenient to refill through the level plug hole. Do not overfill otherwise leakage will occur.

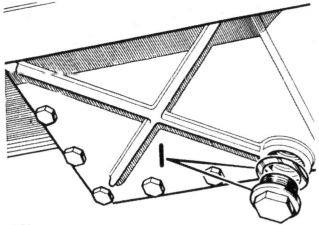

N94

Axle case breathers

Clean the axle case breathers, one in each axle case.

1. Clean off the axle breathers and the surrounding surfaces of the axle cases taking care to remove any gritty foreign matter.
2. Unscrew the axle breathers from their tapered threads in the axle tubes and soak in petrol or a suitable cleaning solvent for several minutes and clean with a soft brush.
3. Shake each breather to ensure the ball valve is free. If it is not the breather valve must be renewed.
4. Lubricate the balls lightly with engine oil before replacing the breathers.

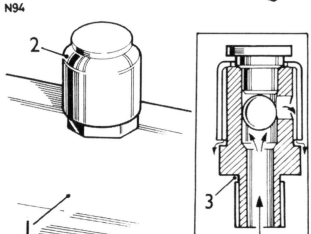

10–18

Check/top up front and rear axle oil levels

Renew front and rear axle oil

1. Remove filler/level plug, check oil level and top up if necessary to the bottom of the filler plug hole. This is located on the side of the pinion housing on both axles.
2. The projecting square plug on the front axle 'banjo' casing should be disregarded.

NOTE: On later axles a recessed square threaded filler/level plug is fitted to front and rear axle 'banjo' casings.

3. If significant topping up is required, check for oil leaks at plugs, joint faces, and oil seals adjacent to axle shaft flanges and propeller shaft driving flange.

To change the differential oil, proceed as follows:—

4. Immediately after a run, when the oil is warm, drain off the oil by removing the respective drain plug.
5. Replace drain plug and refill with oil of the correct grade.

Refer to Division 09.

Important. Do not overfill otherwise damage to seals may occur. Drain plugs have slotted heads, which can be removed with the end of a single-ended spanner.

NOTE: On later axles a recessed square headed drain plug is fitted to front and rear axle banjo casings.

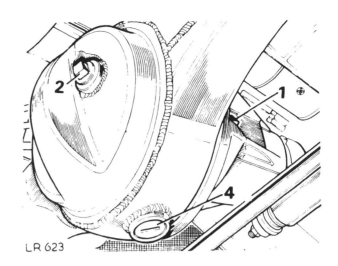

LR 623

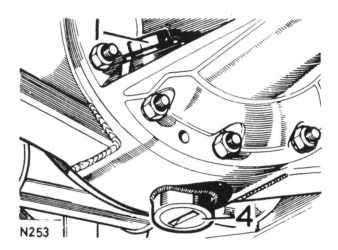

N253

Drain flywheel housing if drain plug is fitted for wading

Flywheel housing drain plug

When in use for wading

1. The flywheel housing can be completely sealed to exclude mud and water under severe wading conditions, by means of a plug fitted in the bottom of the housing.
2. The plug is screwed into the housing adjacent to the drain hole, and should only be fitted when the vehicle is expected to do wading or very muddy work. When the plug is in use it must be removed periodically and all oil allowed to drain off before the plug is replaced.

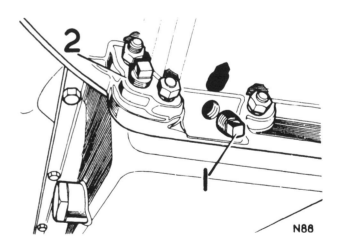

N88

Check Tightness of Propeller Shaft Coupling Bolts.

Lubricate Propeller Shaft.

1. Apply one of the recommended greases at the lubrication nipple on the sliding portion of the rear propeller shaft.
2. To the lubrication nipples fitted to the universal joints of both front and rear shafts.

Lubricate Propeller Shaft Sealed Sliding Joint.

Lubricate the sliding spline of the front propeller shaft, with one of the recommended greases, as follows:

3. Disconnect one end of the propeller shaft.
4. Remove plug in sliding spline and fit a suitable grease nipple.
5. Important. Compress propeller shaft at sliding joint to avoid overfilling then apply grease.
6. Replace grease nipple with plug and reconnect propeller shaft.

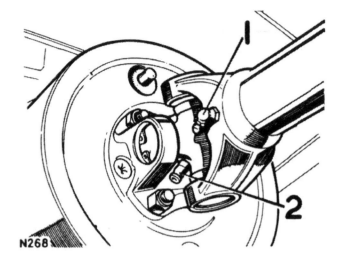

N268

STEERING AND SUSPENSION

Check Condition and Security of Steering Unit, Joints, Relays and Gaiters.

Check Steering Rack/Gear for Oil/Fluid Leaks.

Check Shock Absorbers for Fluid Leaks.

Check Power Steering System for Leaks, Hydraulic Pipes and Unions for Chafing and Corrosion.

continued

Verify that the vehicle is being operated within the specified maximum loading capabilities. Drive the vehicle on to level ground and remove all loads. Should the vehicle lean to one side it indicates a fault with the springs or shock absorbers, **not** the self-levelling unit. If the levelling unit is believed to be at fault, the procedure below should be followed:—

1. Check the levelling unit for excessive oil leakage and if present the unit must be changed. Slight oil seepage is permissible.

2. Remove any excessive mud deposits and loose items from the rear seat and load area.

3. Measure the clearance between the rear axle bump pad and the bump stop rubber at the front outer corner on both sides of the vehicle. The average clearance should be in excess of 67 mm (2.8 in.). If it is less than this figure remove the rear springs and check their free length against the 'Road Spring Data' (see page 04–2). Replace any spring whose free length is more than 20 mm (0.787 inches) shorter than the figure given. If after replacing a spring the average bump clearance is still less than 67 mm (2.8 in.), replace the levelling unit.

4. With the rear seat upright, load 450 kg (992 lb) into the rear of the vehicle, distributing the load evenly over the floor area. Check the bump stop clearance, with the driving seat occupied.

5. Drive the vehicle for approximately 5 km (3 miles) over undulating roads or graded tracks. Bring the vehicle to rest by light brake application so as not to disturb the vehicle loading. With the driving seat occupied, check the bump stop clearance again.

6. If the change in clearance is less than 20 mm (0.787 in.) the levelling unit must be replaced.

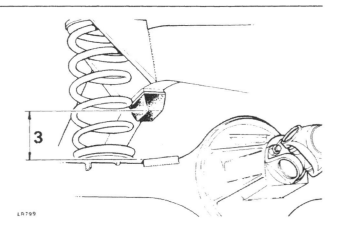

LR799

Check/Top up Fluid in Power Steering Reservoir or Manual Steering Box Oil Level

Power Steering Fluid Reservoir

1. Unscrew the fluid reservoir cap.
2. Check that the fluid is up to the mark on the dipstick.
3. If necessary, top up using one of the recommended grades of fluid.

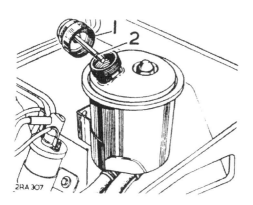

2RA 307

Steering box oil level

1. Check oil level and top-up if necessary to the bottom of the filler plug hole on the top of the cover plate.
2. If significant topping-up is required check for oil leaks at joint faces and rocker shaft oil seal. Access to the filler plug is gained by lifting the bonnet panel.

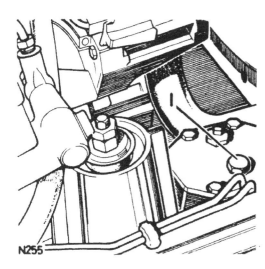

N255

Check/adjust front wheel alignment

Refer to page 04–3 for Data.

To adjust.

1. Set the vehicle on level ground, with the road wheels in the straightahead position, and push it forward a short distance.
2. Slacken the clamps securing the adjusting shaft to the track rod.
3. Turn the adjusting shaft to decrease or increase the effective length of the track rod, as necessary, until the toe-out is correct.
4. Re-tighten the clamps.
5. Push the vehicle rearwards, turning the steering wheel from side to side to settle the ball joints. Then with the road wheels in the straight ahead position, push the vehicle forward a short distance.
6. Rechck the toe-out. If necessary carry out further adjustment.

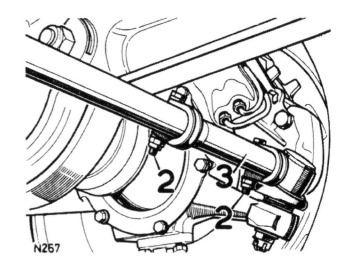

Check security of suspension fixings

Check/adjust steering box

Refer to operation 57.10.13 – Power Steering
Refer to operation 57.35.01 – Manual Steering

Check/top up swivel pin housing oil levels

Renew swivel pin housing oil

Check

1. The front wheel drive universal joints and swivel pins, receive their lubrication from the swivel pin housing.
2. Check oil level by removing the small grub screw at the rear of the swivel pin housing, oil should be level with bottom of hole.
3. Top up if necessary through the filler plug hole at the front of the housing.
   If significant topping up is required, check for oil leaks at plugs, joint faces and oil seals.

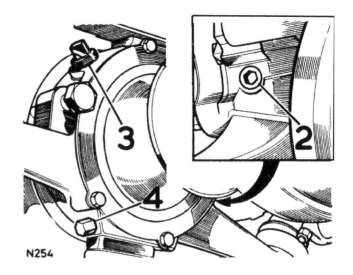

Renew

4. Immediately after a run, when the oil is warm, remove the drain plug from the bottom of each housing.
5. Allow the oil to drain away completely and replace drain plugs.
6. Refit with oil of correct grade.
   The capacity of each housing is approximately 0.26 litres, 5 US pint (.5 Imperial pint).

Check suspension self levelling unit for fluid leaks

10–22

BRAKES

Check visually, hydraulic pipes and unions for chafing, leaks and corrosion.

Check/top up brake fluid reservoir.

The tandem brake reservoir is integral with the servo unit and master cylinder.

1. Remove cap to check fluid level; top up if necessary until the fluid reaches the bottom of the filter neck. Refer to Section 09 for recommended fluids.
2. If significant topping up is required check master cylinder, brake disc cylinders and brake pipes and connections for leakage; any leakage must be rectified immediately.

Caution. When topping up the reservoir, care should be taken to ensure that brake fluid does not come into contact with any paintwork on the vehicle.

N270

Check footbrake operation

If footbrake travel is excessive, check brake pad and caliper condition.
If footbrake is "spongy" bleed brake system. 70.25.02.

Check handbrake for security and operation; adjust if necessary.

The handbrake lever acts on a transmission brake at the rear of the gearbox unit, adjust as follows:-

1. Set the vehicle on level ground.
2. Release the hand brake fully.
3. From beneath the vehicle, remove the rubber blanking plugs from the brake drum.
4. Move the vehicle either forwards or backwards until the adjuster can be seen through one of the apertures.
5. Turn the adjuster until the brake shoes expand to prevent drum rotation.
6. Turn the adjuster back two 'clicks' and replace blanking plugs.
7. Check that the handbrake operates correctly and holds the vehicle.

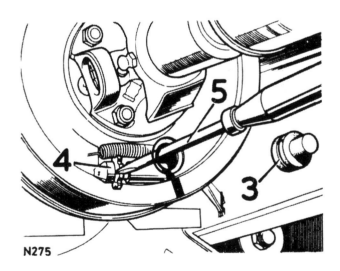

N275

Inspect Brake Pads for wear and Discs for condition.

Hydraulic disc brakes are fitted at the front and rear, and the correct brake adjustment is automatically maintained; no provision is therefore made for adjustment.

1. Check the thickness of the front brake pads and renew if the minimum thickness is less than 3.0 mm (.125 in.).
2. Check the thickness of the rear brake pads and renew if the minimum thickness is less than 1,5 mm (0.062 in.). Refer to Division 70.
3. Also check for oil contamination on brake pads and discs.
4. If replacement or rectification is necessary, this should be carried out by your local Rover Distributor or Dealer.
5. Check brake hoses and pipes for security, fracture, leakage and change as applicable.

When it becomes necessary to renew the brake pads and shoes, it is essential that only genuine components with the correct grade of lining are used. Always fit new pads or shoes as complete axle sets, never individually or as a single wheel set. Serious consequences could result from out of balance braking due to mixing of linings.

Lubricate Handbrake Mechanical Linkage and Cable Guides (lever pivot).

Check Brake Servo Hose(s) for Security and Condition.

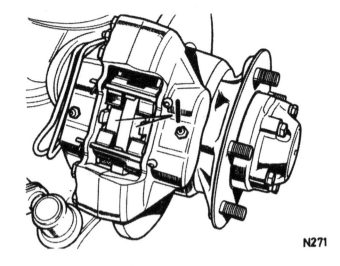
N271

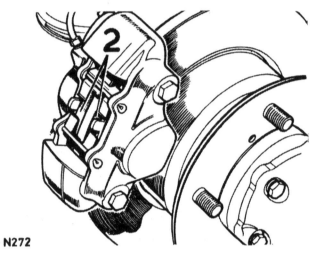
N272

Renew Hydraulic Brake Fluid.

Brake fluid absorbs water and in **time** the boiling point of the fluid will be lowered sufficiently to cause the fluid to be vapourised by the heat generated when the vehicle brakes are applied. This will result in loss of braking efficiency or in extreme cases brake failure.

Therefore all fluid in the brake system should be changed every eighteen months or 18,000 miles (30,000 km) whichever is the sooner. It should also be changed before touring in mountainous areas if not done in the previous nine months. Bleed all fluid from the system and refill – 70.25.02.

Refer to Division 09 for recommended fluids.

Care must be taken always to observe the following points:
(a) At all times use the recommended brake fluid.
(b) Never leave fluid in unsealed containers. It absorbs moisture quickly and can be dangerous if used in the braking system in this condition.
(c) Fluid drained from the system or used for bleeding is best discarded.
(d) The necessity for absolute cleanliness throughout cannot be over-emphasised.

Renew rubber seals in braking system, flexible hoses and servo air filter.

Brake system pressure are very high, up to 2000 lb/in², when the brakes are applied. Brake master cylinder and caliper seals deteriorate with time as do the flexible hoses. Therefore the seals, flexible hoses and the brake servo air filter should be changed at the recommended intervals.

Renew brake master cylinder seals   Refer to 70.30.09

Renew brake caliper seals   Refer to 70.55.13/14

Renewing brake servo filter

1. Remove the servo from the car. 70.50.01.
2. Slide the rubber boot and end cap along the pushrod.
3. Remove the old filter from the neck of the diaphragm housing.
4. Sever the new filter obliquely from the periphery to the centre hole.
5. Fit the filter into the neck of the diaphragm housing.
6. Fit the end cap and rubber boot.

Alternatively, on some models, it is possible to gain access to the filter cover (2) from inside the car. When fitting new filter, slice it diagonally to fit around the brake pedal operating rod before locating the filter in the servo and replacing the cover.

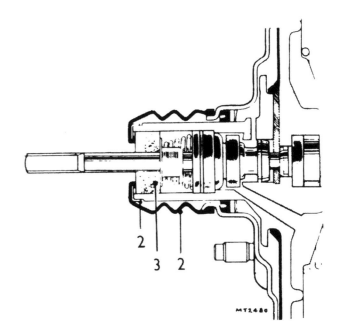

WHEELS AND TYRES

Check/adjust tyre pressures including spare wheel.

1. Maximum tyre life and performance will be obtained only if the tyres are maintained at the correct pressures.

Check/adjust tyre pressures including spare

2. These should be checked at least every month for normal road use and at least weekly, preferably daily, if the vehicle is used off the road or high-speed touring.
3. Whenever possible check with the tyres cold, as the pressure is about 0,2 kg/cm² (3 lb/sq.in.) 0,21 bars higher at running temperature.
4. Replace the valve caps, as they form a positive seal. Check that pressures on all tyres, including the spare, are correct. Any unusual pressure loss in excess of 0,05 to 0,20 kg/cm² (1 to 3 lb/sq.in.) 0,07 to 0,21 bars per week should be investigated and corrected.

Normal on- and off-road use
All speeds and loads

	Front	Rear
kg/cm²	1,8	2,5
lb.in²	25	35
bars	1,72	1,24

NOTE: For extra comfort rear tyre pressures can be reduced to 1,5 kg/cm², 25 lb/in², 1.72 bars, when the rear axle weight does not exceed 1250 kg (2755 lb).

Off-road emergency soft use
Maximum speed of 64 kph (40 mph)

	Front	Rear
kg/cm²	1,1	1,5
lb.in²	15	25
bars	1,03	1,72

As soon as reasonable conditions are reached, pressures should be restored as for normal 'on and off-road use. When high speed touring the tyre pressures should be checked much more frequently, even to the extent of a daily check.

Check tyres for tread depth and visually for external cuts in the fabric, exposure of ply or cord structure

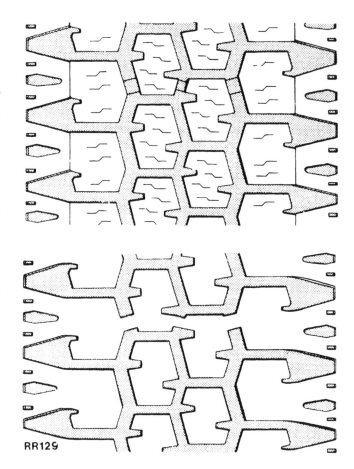

RR129

1. Most tyres fitted to Range Rovers as original equipment include wear indicators in their tread pattern. When the tread has worn to a remaining depth of 1,6 mm (1/16in.) the indicators appear at the surface as bars which connect the tread pattern across the full width of the tyre, as in the Goodyear tyre section illustrated. When the indicator appear in two or more adjacent grooves, at three locations around the tyre, a new tyre should be fitted. If the tyres do not have wear indicators, the tread should be measured at every maintenance inspection and when the tread has worn to a remaining depth of 1,6 mm (1/16in.), new tyres should be fitted. Do not continue to use tyres that have worn to the recommended limit or the safety of the vehicle could be affected and legal regulations governing tread depth may be broken.
2. Check that there are no lumps or bulges in the tyres or exposure of the ply or cord structure.
3. Clean off any oil or grease, using white spirit sparingly. At the same time remove embedded flints, etc. from the treads with the aid of a penknife or similar tool, and check that the tyres have no 'breaks' in the fabric or cuts to sidewalls, etc.

ELECTRICAL

Check Function of all Electrical Equipment

Check/Top up Battery Electrolyte

Clean and grease battery connections.

Battery acid level

The specific gravity of the electrolyte should be checked at every maintenance attention.

Readings should be:—
Temperate climates below 80°F (26.5°C) as commissioned for service, fully charged 1.270 to 1.290 specific gravity.
As expected during normal service three-quarter charged 1.230 to 1.250 specific gravity.
If the specific gravity should read between 1.190 to 1.210, half-charged, the battery must be bench charged and the electrical equipment on the vehicle should be checked.
Tropical climate, above 80°F (26.5°C) as commissioned for service, fully charged 1.210 to 1.230 specific gravity.
As expected during normal service three-quarter charged 1.170 to 1.190 specific gravity.
If the specific gravity should read between 1.130 to 1.150, half-charged, the battery must be bench charged and the electrical equipment on the vehicle should be checked.
The battery is located under the bonnet at the right-hand front side.

Check acid level as follows:–

1. Remove the battery cell cover(s).
2. If necessary add sufficient distilled water to raise the level to the top of the separators; do NOT overfill.
3. Avoid the use of naked lights when examining the cells.
4. In very cold weather it is essential that the vehicle is used immediately after topping up to ensure that the distilled water is thoroughly mixed with the electrolyte. Neglect of this precaution may result in the distilled water freezing and causing damage to the battery.

Battery terminal

6. Remove battery terminals, clean, grease and refit.
7. Replace terminal screw; do not overtighten. Do not use the screw for pulling down the terminal.
8. Do NOT disconnect the battery cables while the engine is running or damage to alternator semi-conductor devices may occur. It is also inadvisable to break or make any connection in the alternator charging and control circuits while the engine is running.
9. It is essential to observe the polarity of connections to the battery, alternator and regulator, as any incorrect connections made when reconnecting cables may cause irreparable damage to the semiconductor devices.

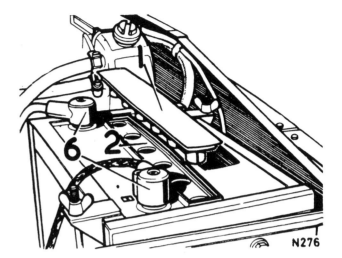

Check Headlamp Alignment, adjust if necessary

This operation requires special equipment.
In an emergency each headlamp unit can be adjusted by means of:
1. The headlamp lateral adjusting screw.
2. The headlamp vertical adjusting screw.

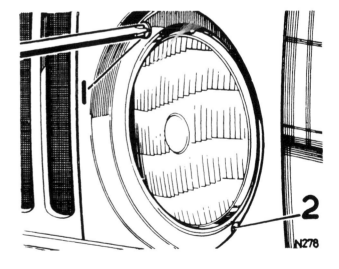

10–27

Check, if necessary renew, wiper blades

Operate the washer and wipers. Renew the wiper blades if they are damaged or the glass is smeared.

To replace windscreen or tailgate wiper blades

1. Pull wiper arm away from the glass.
2. Lift spring clip (arrowed) and withdraw blade from wiper arm.
3. To fit new blade reverse removal procedure.

To replace headlamp wiper blades, refer to 84.25.06.

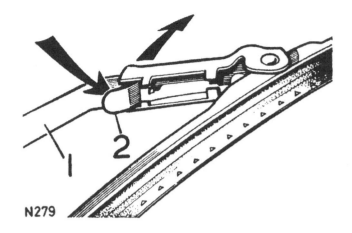

N279

Check output of alternator charging system

1. The alternator is a sealed unit, and requires no lubrication or maintenance.
2. Check and ensure that any dirt or oil which may have collected around the apertures in the slip-ring end bracket and moulded cover is wiped clear.
3. Using proprietary equipment, check the alternator output. Refer to page 04–4.

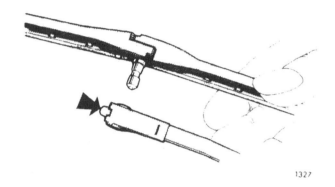

1327

BODY

Lubricate all Locks and Hinges (NOT Steering Lock)

Check Operation of Window Controls

Check Condition and Security of Seats and Seat Belts

Check Operation of Seat Belt Inertia Reel Mechanism

Seats: Check security of seats in runners and runners to floor.

Seat Belts: Check seat belt mounting bolts for security. The wearing of seat belts is advisable for all journeys in the vehicle, no matter how short the journey, and is compulsory in some territories. The belts themselves must be in good condition and the inertia mechanism in sound working order to afford the full designed protection to the seat belt wearer.

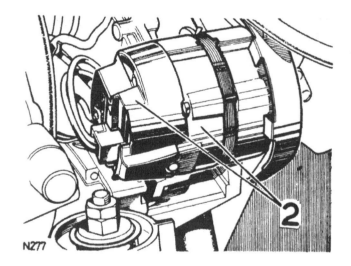

N277

Renew the unit if the belt is damaged, the inertia mechanism is defective or the belt has not been in use when the vehicle was involved in an impact accident.

Check Operation of all Door, Bonnet and Tailgate Locks

Check rear view mirror(s) for cracks and crazing

ROAD TEST

During the road test the function of the engine, transmission, brakes, steering and seat belts should be checked together with the function of all instrumentation.

10–28

ENGINE OPERATIONS

Description	Operation No.
Camshaft—remove and refit	12.13.01
Connecting rods and pistons	
—remove and refit	12.17.01
—overhaul	12.17.10
Crankshaft	
—remove and refit	12.21.33
—overhaul	12.21.46
—spigot bearing—remove and refit	12.53.20
Cylinder	
—heads—remove and refit	12.29.10
—heads—overhaul	12.29.18
—pressures—check	12.25.01
Engine assembly—remove and refit	12.41.01
Flywheel	
—remove and refit	12.53.07
—overhaul	12.53.10
—starter ring gear—remove and refit	12.53.19
Oil	
—filter assembly, external—remove and refit	12.60.01
—pump—remove and refit	12.60.26
—pump—overhaul	12.60.32
—sump—remove and refit	12.60.44
Timing	
—chain and gears—remove and refit	12.65.12
—gear cover	12.65.01
Valve gear—remove and refit	12.29.34

Where emission control equipment is fitted to an engine, refer to section 17 for remove and refit operations.

Engine

FAULT DIAGNOSIS

SYMPTOM	POSSIBLE CAUSE	CURE
ENGINE FAILS TO START	1. Incorrect starting procedure 2. Starter motor speed too slow 3. Faulty ignition system 4. Water or dirt in fuel system 5. Carburetter(s) flooding 6. Defective fuel pump system 7. Defective starter motor 8. Starter pinion not engaging	1. *See Instruction Manual* 2. *Check battery and connections* 3. *See Group 86* 4. *See Group 19* 5. *See Group 19* 6. *See Group 19* 7. *See Group 86* 8. *Remove starter motor and investigate*
ENGINE STALLS	1. Low idling speed 2. Faulty sparking plugs 3. Faulty coil or condenser 4. Faulty distributor points 5. Incorrect mixture 6. Foreign matter in fuel system	1. *Adjust carburetter. See Group 19* 2. *Clean and test, renew if necessary* 3. *Renew* 4. *Rectify or renew. See Group 86* 5. *Adjust carburetter. See Group 19* 6. *See Group 19*
LACK OF POWER	1. Poor compression 2. Badly seating valves 3. Faulty exhaust silencer 4. Incorrect ignition timing 5. Leaks or restrictions in fuel system 6. Faulty sparking plugs 7. Excessive carbon deposit 8. Brakes binding 9 Faulty coil, condenser or battery	1. *If the compression is appreciably less than the correct figure, the piston rings or valves are faulty. Low pressure in adjoining cylinders indicates a faulty cylinder head gasket* 2. *Rectify or renew* 3. *Renew* 4. *Rectify* 5. *See Group 19* 6. *Rectify* 7. *Decarbonise* 8. *See Group 70* 9. *See Group 86*
ENGINE RUNS ERRATICALLY	1. Faulty electrical connections 2. Defective sparking plugs 3. Low battery charge 4. Defective distributor 5. Foreign matter in fuel system 6. Faulty fuel pump 7. Sticking valves 8. Defective valve springs 9. Incorrect ignition timing 10. Worn valve guides or valves 11. Faulty cylinder head gaskets 12. Damaged exhaust system 13. Vacuum pipes disconnected at inlet manifold, distributor or gearbox	1. *Rectify* 2. *Renew or rectify* 3. *Recharge battery* 4. *Rectify* 5. *See Group 19* 6. *See Group 19* 7. *Rectify or renew* 8. *Renew* 9. *Rectify* 10. *Renew* 11. *Renew* 12. *Rectify or renew* 13. *Refit pipes*
ENGINE STARTS, BUT STOPS IMMEDIATELY	1. Faulty electrical connections 2. Foreign matter in fuel system 3. Faulty fuel pump 4. Low fuel level in tank	1. *Check HT leads for cracked insulation; check low tension circuit* 2. *See Group 19* 3. *See Group 19* 4. *Replenish*
ENGINE FAILS TO IDLE	1. Incorrect carburetter setting 2. Faulty fuel pump 3. Sticking valves 4. Faulty cylinder head gasket(s)	1. *See Group 19* 2. *See Group 19* 3. *Rectify or renew* 4. *Renew*
ENGINE MISFIRES ON ACCELERATION	1. Distributor points incorrectly set 2. Faulty coil or condenser 3. Faulty sparking plugs 4. Faulty carburetter 5. Vacuum pipes disconnected at inlet manifold	1. *Rectify. See Group 86* 2. *Renew* 3. *Rectify* 4. *See Group 19* 5. *Check all vacuum connections*
ENGINE KNOCKS	1. Ignition timing advanced 2. Excessive carbon deposit 3. Incorrect carburetter setting 4. Unsuitable fuel 5. Worn pistons or bearings 6. Distributor advance mechanism faulty 7. Defective sparking plugs	1. *Adjust* 2. *Decarbonise* 3. *See Group 19* 4. *Adjust ignition timing. See Group 86* 5. *Renew* 6. *Rectify. See Group 86* 7. *Rectify or renew*

12-2

SYMPTOM	POSSIBLE CAUSE	CURE
ENGINE BACKFIRES	1. Ignition defect 2. Carburetter defect 3. Sticking valve 4. Weak valve springs 5. Badly seating valves 6. Excessively worn valve stems and guides 7. Excessive carbon deposit 8. Incorrect sparking plug gap 9. Air leak in induction or exhaust systems	1. *See Group 86* 2. *See Group 19* 3. *Rectify* 4. *Renew* 5. *Rectify or renew* 6. *Renew* 7. *Decarbonise* 8. *Reset* 9. *Renew faulty gaskets or components*
BURNED VALVES	1. Sticking valves 2. Weak valve springs 3. Excessive deposit on valve seats 4. Distorted valves 5. Excessive mileage between overhauls	1. *Rectify* 2. *Renew* 3. *Re-cut* 4. *Renew* 5. *Decarbonise*
NOISY VALVE MECHANISM	1. Excessive oil in sump, causing air bubbles in hydraulic tappets 2. Worn or scored parts in valve operating mechanism 3. Valves and seats cut down excessively, raising end of valve stem 1,27 mm (.050 in) above normal position 4. Sticking valves 5. Weak valve springs 6. Worn timing chain or chainwheels	1. *Drain and refill to correct level* 2. *Replace faulty parts* 3. *Grind off end of valve stem or replace parts* 4. *Rectify* 5. *Renew* 6. *Renew worn parts*
NOISE FROM HYDRAULIC TAPPETS 1. Rapping noise only when engine is started 2. Intermittent rapping noise 3. Noise on idle and low speed 4. General noise at all speeds 5. Loud noise at normal operating temperature only	1. Oil too heavy for prevailing temperature Excessive varnish in tappet 2. Leakage at check ball 3. Excessive leakdown 4. High oil level in sump Leakage at check ball Worn tappet body Worn camshaft 5. Excessive leakdown rate or scored-lifter plunger	1. *Drain and refill with correct grade* *Replace tappet* 2. *Replace tappet* 3. *Replace tappet* 4. *Drain and refill to correct level* *Replace tappet* *Replace tappet* *Replace camshaft* 5. *Replace tappet*
MAIN BEARING RATTLE	1. Low oil level 2. Low oil pressure 3. Excessive bearing clearance 4. Burnt-out bearings 5. Loose bearing caps	1. *Replenish as necessary* 2. *See next symptom* 3. *Renew bearings; grind crankshaft* 4. *Renew* 5. *Tighten*
LOW OIL PRESSURE WARNING LIGHT REMAINS ON, ENGINE RUNNING	1. Thin or diluted oil 2. Low oil level 3. Choked pump strainer 4. Faulty release valve 5. Excessive bearing clearance 6. Oil pressure switch unserviceable 7. Electrical fault 8. Relief valve plunger sticking 9. Weak relief valve spring 10. Pump rotors excessively worn 11. Excessively worn bearings; main, connecting rod, big end, camshaft, etc.	1. *Drain and refill with correct oil* 2. *Replenish* 3. *Clean* 4. *Rectify* 5. *Rectify* 6. *Renew* 7. *Check circuit* 8. *Remove and ascertain cause* 9. *Renew* 10. *Renew* 11. *Ascertain which bearings and rectify*
RATTLE IN LUBRICATION SYSTEM	1. Oil pressure relief valve plunger sticking.	1. *Remove and clean*
ENGINE OVERHEATING	1. Low coolant level 2. Faulty cooling system 3. Faulty thermostat 4. Incorrect timing 5. Defective lubrication system	1. *Check for leaks* 2. *See Group 26* 3. *Renew* 4. *Rectify* 5. *See Group 12*

12–3

CAMSHAFT

– Remove and refit   12.13.01

Removing

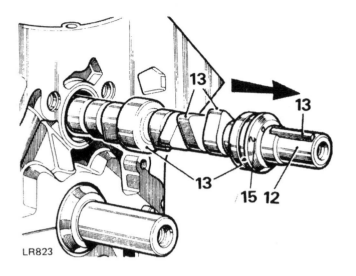

LR823

1. Drain the cooling system. 26.10.01.
2. Remove the fan blades and pulley. 26.25.01.
3. Remove the fan cowl. 26.25.11.
4. Remove the radiator block. 26.40.04.
5. Remove the radiator grille. 76.55.03.
6. Remove the alternator. 86.10.02.
7. Remove the air cleaner. 19.10.01.
8. Remove the induction manifold. 30.15.02.
9. Remove the valve gear. 12.29.34.
10. Remove the timing chain cover. 12.65.01.
11. Remove the timing chain. 12.65.12.

CAUTION: **Do not damage the bearings when withdrawing the camshaft as the camshaft bearings are not serviceable.**

12. Withdraw the camshaft.

Inspecting

13. Check all bearing surfaces for excessive wear and score marks. Also check cam lobes for excessive wear. Check key and keyway.

Refitting

14. Reverse 1 to 12.
15. To identify the low-lift camshaft, used with the high compression engine (9.35:1 compression ratio) a '2' is stamped on the thrust face bearing chamfer at the front of the camshaft.

CAUTION: During reassembly, it is essential that the camshaft key, spacer and distributor drive gear are all refitted as described in operation 12.65.12.

Failure to observe these requirements may result in restriction or total blockage of oil passage to the timing gear.

CONNECTING RODS AND PISTONS

– Remove and refit   12.17.01

Service tool 605351 – Guide bolts for connecting rods

NOTE: There are two designs of pistons in use. These differ slightly in weight and this is compensated for by two standards of crankshaft balance.

The early plain skirt type pistons (A) are fitted with a crankshaft which has a plain face at the starter dog end. Later 'W' slot skirt type pistons (B) are used with a crankshaft which has an identification groove (C) in the face at the starter dog end. The two versions are interchangeable provided that the eight pistons and applicable crankshaft are fitted initially as a set.

To identify the latest 9.35:1 piston used with the high compression engine the compression ratio is stamped in the concave crown.

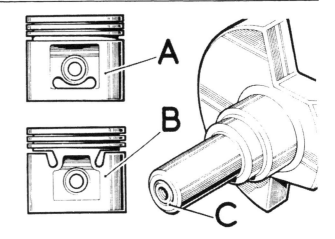

Removing

1.   Drain the cooling system. 26.10.01.
2.   Remove the air cleaner. 19.10.01.
3.   Remove the alternator. 86.10.02.
4.   Remove the induction manifold. 30.15.02.
5.   Remove the valve gear. 12.29.34.
6.   Remove the cylinder heads. 12.29.10.
7.   Remove the oil sump. 12.60.44.
8.   Remove the sump oil strainer.
9.   Remove the connecting rod caps and retain them in sequence for reassembly.
10. Screw the guide bolts 605351 on to the connecting rods.
11. Push the connecting rod and piston assembly up the cylinder bore and withdraw it from the top. Retain the connecting rod and piston assemblies in sequence with their respective caps.
12. Remove the guide bolts 605351 from the connecting rod.

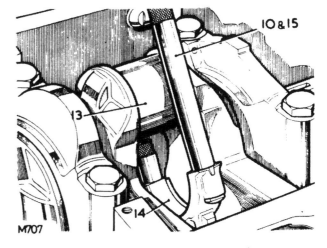

Refitting

13. Locate the applicable crankshaft journal at BDC.
14. Place the bearing upper shell in the connecting rod.
15. Retain the upper shell by screwing the guide bolts 605351 on to the connecting rods.
16. Insert the connecting rod and piston assembly into its respective bore, noting that the domed shape boss on the connecting rod must face towards the front of the engine on the right-hand bank of cylinders and towards the rear on the left-hand bank. When both connecting rods are fitted, the bosses will face inwards towards each other.
17. Position the oil control piston rings so that the ring gaps are all at one side, between the gudgeon pin and piston thrust face. Space the gaps in the ring rails approximately 25 mm (1 in.) each side of the expander ring joint.

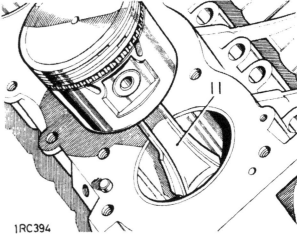

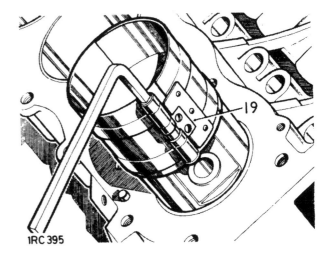

Continued

63

12–5

18. Position the compression rings so that their gaps are on opposite sides of the piston between the gudgeon pin and piston thrust face.

19. Using a piston ring compressor, locate the piston into the cylinder bore.

20. Place the bearing lower shell in the connecting rod cap.

21. Locate the cap and shell on to the connecting rod, noting that the rib on the edge of the cap must be towards the front of the engine on the right-hand bank of cylinders and towards the rear on the left-hand bank.

22. Secure the connecting rod cap. Torque 4,0 to 4,9 kgf.m (30 to 35 lbf.ft).

23. Reverse 1 to 8.

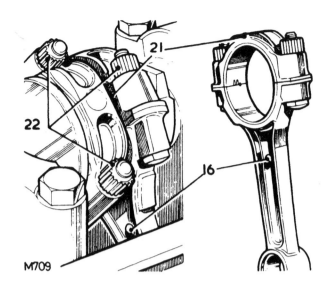

M709

CONNECTING RODS AND PISTONS

– Overhaul   12.17.10

Service tools: 18G 537, 18G 587, 18G 1150, 18G 1150 E or 605350
605238 – Plastigauge

Dismantling

1. Remove the connecting rods and pistons, see 12.17.01.
   NOTE: The connecting rods, caps and bearing shells must be retained in sets, and in the correct sequence.

2. Remove the piston rings over the crown of the piston.

3. If the same piston is to be refitted, mark it relative to its connecting rod to ensure that the original assembly is maintained.

4. Withdraw the gudgeon pin, using tool 18G 1150 as follows:—
   (a) Clamp the hexagon body of 18G 1150 in a vice.
   (b) Position the large nut flush with the end of the centre screw.

(c) Push the screw forward until the nut contacts the thrust race.

(d) Locate the piston adaptor 18G 1150 E with its long spigot inside the bore of the hexagon body.

(e) Fit the remover/replacer bush of 18G 1150 on the centre screw with the flanged end away from the gudgeon pin.

(f) Screw the stop-nut about half-way on to the smaller threaded end of the centre screw, leaving a gap 'A' of 3 mm (1/8 in.) between this nut and the remover/replacer bush.

(g) Lock the stop-nut securely with the lock screw.

(h) Check that the remover/replacer bush is correctly positioned in the bore of the piston.

(i) Push the connecting rod to the right to expose the end of the gudgeon pin, which must be located in the end of the adaptor 'd'.

(j) Screw the large nut up to the thrust race.

(k) Hold the lock screw and turn the large nut until the gudgeon pin has been withdrawn from the piston. Dismantle the tool.

Continued

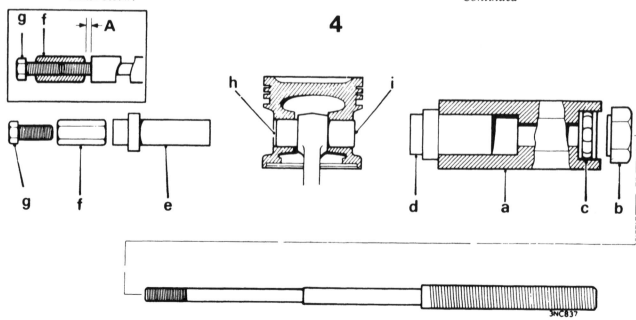

5. As an alternative to tool 18G 1150, press the gudgeon pin from the piston using an hydraulic press and the components which comprise tool 605350 as follows:

   (a) Place the base of tool 605350 on the bed of an hydraulic press which has a capacity of 8 tons (8 tonnes).

   (b) Fit the guide tube into the bore of the base with its countersunk face uppermost.

   (c) Push the piston to one side so as to expose one end of the gudgeon pin and locate this end in the guide tube.

   (d) Fit the spigot end of the small diameter mandrel into the gudgeon pin.

   (e) Press out the gudgeon pin, using the hydraulic press.

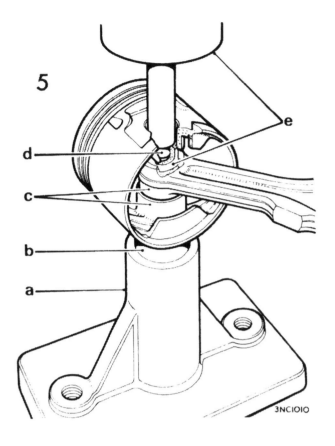

Overhauling pistons

Original pistons

6. Remove carbon and deposits, particularly from the ring grooves.

7. Examine the pistons for damage or excess wear — see under 'New pistons' for clearance dimensions — fit new replacements as necessary.

New pistons

Pistons are available in service standard size and in oversizes of 0,25 mm (0.010 in.) and 0,50 mm (0.020 in.). Service standard size pistons are supplied 0,0254 mm (0.001 in.) oversize. When fitting new service standard size pistons to a cylinder block, check for correct piston to bore clearance, honing the bore if necessary. Bottom of piston skirt/bore clearance should be 0,018 to 0,033 mm (0.0007 to 0.0013 in.).

NOTE: The temperature of the piston and cylinder block must be the same to ensure accurate measurement. When reboring the cylinder block, the crankshaft main bearing caps must be fitted and tightened to the correct torque.

Continued

8. Check the cylinder bore dimension at right angles to the gudgeon pin, 40 to 50 mm (1½ to 2 in.) from the top.

9. Check the piston dimension at right angles to the gudgeon pin, at the bottom of the skirt.

10. The piston dimension must be 0,018 to 0,033 mm (0.0007 to 0.0013 in.) smaller than the cylinder.

11. If new piston rings are to be fitted without reboring, deglaze the cylinder walls with a hone, without increasing the bore diameter.

IMPORTANT: A deglazed bore must have a cross-hatch finish.

12. Check the compression ring gaps in the applicable cylinder, held square to the bore with the piston. Gap limits: 0,44 to 0,56 mm (0.017 to 0.022 in.). Use a fine-cut flat file to increase the gap if required. Select a new piston ring if the gap exceeds the limit.

NOTE: Gapping does not apply to oil control rings.

Continued

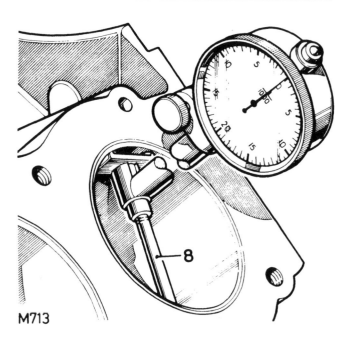

M713

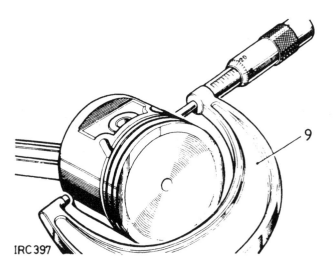

IRC 397

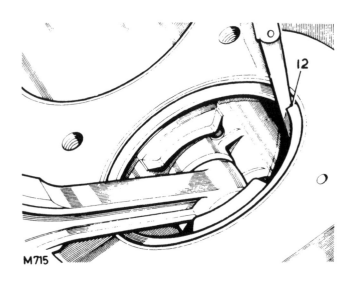

M715

13. Temporarily fit the compression rings to the piston
    with the marking 'T' or 'TOP' uppermost. The
    chrome compression ring in the top groove is un-
    marked (reversible) on 8.13:1 and 9.35:1 type
    pistons.
14. Check the compression ring clearance in the piston
    groove. Clearance limits: 0,05 to 0,10 mm (0.002 to
    0.004 in.).

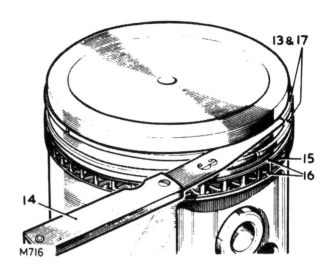

Fitting piston rings

15. Fit the expander ring into the bottom groove making
    sure that the ends abut and do not overlap.
16. Fit two ring rails to the bottom groove, one above
    and one below the expander ring.
17. Fit the compression rings as indicated in paragraph
    13 above.

Connecting rods

18. Check the alignment of the connecting rod.
19. Check the connecting rod small end, the gudgeon pin
    must be an interference fit.

Big-end bearings

20. Locate the bearing upper shell into the connecting
    rod.
21. Locate the connecting rod and bearing on to the
    applicable crankshaft journal, noting that the domed
    shape boss on the connecting rod must face towards
    the front of the engine on the right-hand bank of
    cylinders and towards the rear on the left-hand bank.
    When both connecting rods are fitted, the bosses will
    face inwards towards each other.
22. Place a piece of Plastigauge 605238 across the centre
    of the lower half of the crankshaft journal.
23. Locate the bearing lower shell into the connecting
    rod cap.
24. Locate the cap and shell on to the connecting rod.
    Note that the rib on the edge of the cap must be the
    same side as the domed shape boss on the connecting
    rod.
25. Secure the connecting rod cap. Torque 4,0 to 4,9
    kgf.m (30 to 35 lbf.ft.).

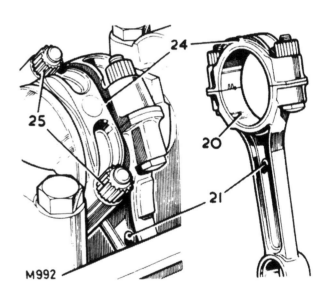

IMPORTANT: Do not rotate the crankshaft while the Plastigauge is fitted.

26.   Remove the connecting rod cap and shell.
27.   Using the scale printed on the Plastigauge packet, measure the flattened Plastigauge at its widest point. The graduation that most closely corresponds to the width of the Plastigauge indicates the bearing clearance.
28.   The correct bearing clearance with new or overhauled components is 0,013 to 0,06 mm (0.0006 to 0.0022 in.).
29.   If a bearing has been in service, it is advisable to fit a new bearing if the clearance exceeds 0,08 mm (0.003 in.).
30.   If a new bearing is being fitted, use selective assembly to obtain the correct clearance.
31.   Wipe off the Plastigauge with an oily rag. DO NOT scrape it off.

IMPORTANT: The connecting rods, caps and bearing shells must be retained in sets, and in the correct sequence.

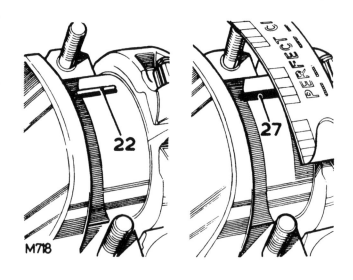

Reassembling

32.   If an hydraulic press and tool 605350 were used for dismantling, refit each piston to its connecting rod as follows:—
  (a)   Check that the base of tool 605350 and the guide tube are fitted as in 5a and 5b.
  (b)   Fit the long mandrel inside the guide tube.
  (c)   Fit the connecting rod into the piston with the markings together if the original pair are being used, then place the piston and connecting rod assembly over the long mandrel until the gudgeon pin boss rests on the guide tube.
  (d)   Fit the gudgeon pin into the piston up to the connecting rod, and the spigot end of the small diameter mandrel into the gudgeon pin.
  (e)   Press in the gudgeon pin until it abuts the shoulder of the long mandrel.

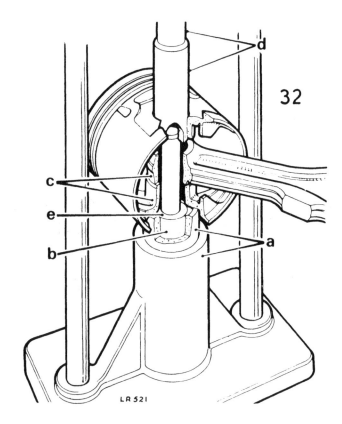

Continued

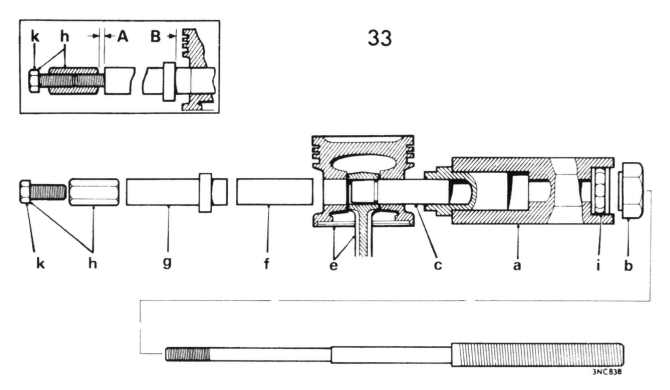

33

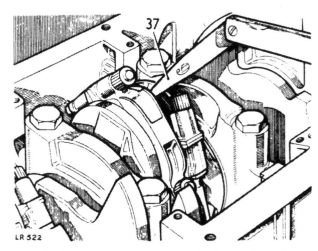

33. If tool 18G 1150 was used for dismantling, refit each piston to its connecting rod as follows:—

(a) Clamp the hexagon body of 18G 1150 in a vice, with the adaptor 18G 1150 E positioned as in 4d.

(b) Remove the large nut of 18G 1150 and push the centre screw approximately 2 in (50 mm) into the body until the shoulder is exposed.

(c) Slide the parallel guide sleeve, grooved end last, on to the centre screw and up to the shoulder.

(d) Lubricate the gudgeon pin and bores of the connecting rod and piston with graphited oil (Acheson's Colloids 'Oildag'). Also lubricate the ball race and centre screw of 18G 1150.

(e) Fit the connecting rod and the piston together on to the tool with the markings together if the original pair are being used and with the connecting rod around the sleeve up to the groove.

(f) Fit the gudgeon pin into the piston bore up to the connecting rod.

(g) Fit the remover/replacer bush 18G 1150/3 with its flanged end towards the gudgeon pin.

(h) Screw the stop-nut on to the centre screw and adjust this nut to obtain a 1 mm (1/32 in.) end-float 'A' on the whole assembly, and lock the nut securely with the screw.

(i) Slide the assembly back into the hexagon body and screw on the large nut up to the thrust race.

(j) Set the torque wrench 18G 537 to 12 lbf.ft. This represents the minimum load for an acceptable interference fit of the gudgeon pin in the connecting rod.

(k) Using the torque wrench and socket 18G 587 on the large nut, and holding the lock screw, pull the gudgeon pin in until the flange of the remover/replacer bush is 0.160 in. (4 mm) 'B' from the face of the piston. Under no circumstances must this flange be allowed to contact the piston.

CAUTION: If the torque wrench has not broken throughout the pull, the fit of the gudgeon pin to the connecting rod is not acceptable and necessitates the renewal of components. The large nut and centre screw of the tool must be kept well-oiled.

(l) Remove the tool.

34. Check that the piston moves freely on the gudgeon pin and that no damage has occurred during pressing.

35. Fit the connecting rods and pistons, see 12.17.01, carrying out the following checks during fitting.

36. Check that the connecting rods move freely sideways on the crankshaft. Tightness indicates insufficient bearing clearance or a misaligned connecting rod.

37. Check the end-float between the connecting rods on each crankshaft journal. Clearance limits: 0,15 to 0,37 mm (0.006 to 0.014 in.).

12-11

DATA

Standard size cylinder bore diameter	88,861 to 88,900 mm (3.4985 to 3.5000 in.)

Connecting rod:

Length between centres	143,71 to 143,81 mm (5.658 to 5.662 in.)
Bearings:	
Clearance on crankshaft	0,015 to 0,055 mm (0.0006 to 0.0022 in.)
End float on crankshaft	0,15 to 0,37 mm (0.006 to 0.014 in.)

Pistons:

Type (up to engine No. 355 009 84 A)	8.25:1 compression ratio, plain skirt
(from engine No. 355 009 85 A to engine No. suffix E)	8.25:1 compression ratio, 'W' slot skirt
(from engine No. suffix F)	8.13:1 compression ratio, 'W' slot skirt
(from engine No. prefix 11D)	9.35:1 compression ratio, 'W' slot skirt
Material	Aluminium alloy, concave topped
Depth of concave – (8.25:1 piston)	4,47 to 4,57 mm (0.176 to 0.18 in.)
(8.13:1 and 9.35:1 piston)	5,21 to 5,31 mm (0.205 to 0.209 in.)
Clearance at skirt bottom	0,018 to 0,033 mm (0.0007 to 0.0013 in.)

Piston rings:

No. 1 compression ring (8.25:1 piston)	Chrome faced and marked 'TOP'
(8.13:1 and 9.35:1 piston)	Chrome faced unmarked, reversible
No. 2 compression ring	Stepped to 'L' shape and marked 'TOP'
Compression ring height (8.25:1 piston)	1,98 to 2,01 mm (0.078 to 0.079 in.)
(8.13:1 and 9.35:1 piston)	1,71 to 1,73 mm (0.0615 to 0.0625 in.)
Compression ring clearance in piston groove	
(8.25:1 piston)	0,05 to 0,10 mm (0.002 to 0.004 in.)
(8.13:1 and 9.35:1 piston)	0,04 to 0,09 mm (0.0015 to 0.0035 in.)
Compression ring gap	0,44 to 0,56 mm (0.017 to 0.002 in.)
Oil control ring	Two oil ring rails with separate spacer

Gudgeon pins:

Length	72,67 to 72,79 mm (2.861 to 2.866 in.)
Diameter	22,112 ro 22,219 mm (0.8745 to 0.8748 in.)
Fit in con rod	Press fit
Clearance in piston	0,005 to 0,007 mm (0.0002 to 0.0003 in.)

CRANKSHAFT

–Remove and refit   12.21.33

Service tools 605351–Guide bolts for connecting rods

Removing

1.   Remove the engine assembly. 12.41.01.
2.   Remove the timing gear cover. 12.65.01.
3.   Remove the timing chain and gears. 12.65.12.
4.   Remove the clutch. 33.10.01.
5.   Remove the flywheel. 12.53.07.
6.   Remove the oil sump. 12.60.44.
7.   Remove the sump oil strainer.
8.   Remove the connecting rod caps and lower bearing shells and retain in sequence.
9.   Remove the main bearing caps and lower bearing shells and retain in sequence. (On later engines the bearing caps may be visually different to those shown).

IMPORTANT: If the same bearing shells are to be refitted, retain them in pairs and mark them with the number of the respective journal.

10.   Withdraw the crankshaft.

Engine

Refitting

NOTE: There are two types of rear oil seal for the crankshaft, early models have a woven rope two piece seal, later models have a circular plastic lipped seal. The seals are not interchangeable, replacements must be the same type as originally fitted. There are also two designs of side sealing for the rear main bearing cap in use, early engines have a plain strip type seal and matching cap, later engines use a cross shaped design. Bearing caps must not be interchanged, but if necessary, the early seal can be used with the latest cap. The following instructions and illustrations include procedures for all engines.

11. Locate the upper main bearing shells into the cylinder block; these must be the shells with the oil drilling and oil grooves.
12. Locate the flanged upper main bearing shell in the centre position.
13. If a rope type of rear oil seal is being fitted, proceed as follows. Carry out instructions (a) and (b) then continue at 14 for all engines.
    (a) Remove the rear oil seal from the cylinder block.
    (b) Place a new oil seal in the groove with both ends projecting above the parting face of the cylinder block.
    (c) Force the seal into the groove by rubbing down with a hammer handle until the seal projects above the groove not more than 1,5 mm (0.031 in.).
    (d) Cut off the ends of the seal flush with the surface of the cylinder block, using a sharp knife.
    (e) Fit the rear oil seal to the rear main bearing cap following the same procedure as (a) to (d) above.
    (f) Lubricate the rear oil seal with heavy engine oil.
14. Place suitable blocks, approximately 12,5 mm (0.500 in.) thick, on to each end of the cylinder block so that they cover the front and rear upper main bearing shells.
15. Lift the crankshaft into position with the ends supported on the blocks.
16. Lubricate the crankshaft journals and bearing shells with engine oil.
17. Holding the connecting rods in position, remove one of the blocks and lower the crankshaft on to the connecting rod bearings. Repeat for the opposite end.
18. Where necessary, use the guide bolt 605351 to draw the connecting rods up to the crankshaft journal.
19. Locate the bearing caps and lower shells on to the connecting rods, noting that the rib on the edge of the cap must be towards the front of the engine on the right-hand bank of cylinders, and towards the rear on the left-hand bank.

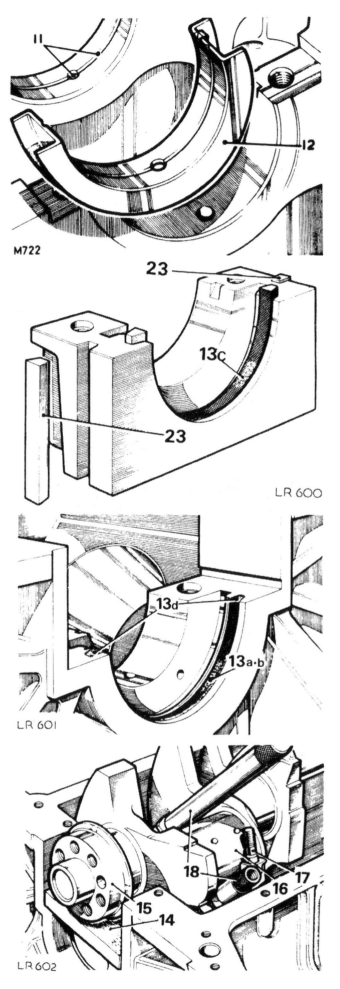

20. Secure the connecting rod caps. Torque: 4,0 to 4,9 kgf.m (30 to 35 lbf.ft.).
21. Lubricate the lower main bearing shells with engine oil.
22. Fit numbers one to four main bearing caps and shells, leaving the fixing bolts finger tight at this stage.

CAUTION: Do not handle the seal lip, visually check that it is not damaged and ensure that the outside diameter remains clean and dry.

23. Fit the side seals to the grooves each side of the rear main bearing cap. Do not cut the side seals to length, they must protrude 1,5 mm (0.062 in.) approximately above the bearing cap parting face.
24. Apply 'Unipart Universal' jointing compound to the rearmost half of the rear main bearing cap parting face or, if preferred, to the equivalent area on the cylinder block as illustrated.
25. Lubricate the bearing half and bearing cap side seals with clean engine oil.
26. Fit the bearing cap assembly to the engine. Do not tighten the fixings at this stage but ensure that the cap is fully home and squarely seated on the cylinder block.

    (a) Tension the rear main bearing cap bolts equally by one-quarter turn approximately, then slacken one complete turn on each fixing bolt.

    (b) Position the seal guide RO.1014 on the crankshaft flange.

    (c) Ensure that the oil seal guide and the crankshaft journal are scrupulously clean, then coat the seal guide and oil seal journal with clean engine oil.

NOTE: The lubricant coating must cover the seal guide outer surface completely to ensure that the oil seal lip is not turned back during assembly.

    (d) Position the oil seal, lipped side towards the engine, on to the seal guide. The seal outside diameter must be clean and dry.

    (e) Push home the oil seal fully and squarely by hand into the recess formed in the cap and block until it abuts against the machined step in the recess.

    (f) Withdraw the seal guide.

27. If the side seals are the plain strip type, use a blunt drift to push the side oil seals into the cap.
28. If a plastic lip type rear oil seal is being fitted, proceed as follows, instructions (a) to (f), then continue at 29 for all engines.

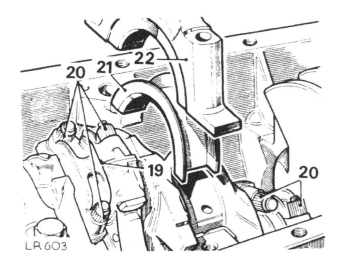

LR 603

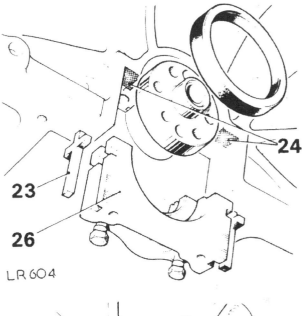

LR 604

LR 605

continued

29. Align the thrust faces of the centre main bearing by tapping the crankshaft with a mallet, rearward and then forward to the limits of its travel.

30. Tighten numbers one to four main bearing cap bolts. Torque: 7,0 to 7,6 kgf.m (50 to 55 lbf.ft.).

31. Tighten the rear main bearing cap bolts. Torque: 9,0 to 9,6 kgf.m (65 to 70 lbf.ft.).

32. Check the crankshaft end-float. Limits: 0,10 to 0,20 mm (0.004 to 0.008 in.). If not correct investigate the cause.

33. Reverse 1 to 7.

CAUTION: Do not exceed 1,000 engine rpm when first starting the engine, otherwise the crankshaft rear oil seal will be damaged.

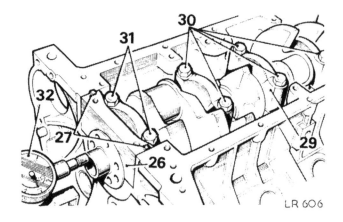

LR 606

CRANKSHAFT

– Overhaul   12.21.46

Service tools 605238 – Plastigauge

1. Remove the crankshaft. 12.21.33.

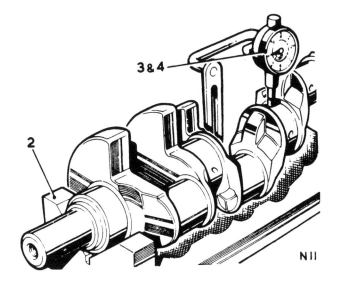

Inspecting

2. Rest the crankshaft on vee-blocks at numbers one and five main bearing journals.
3. Using a dial test indicator, check the run-out at numbers two, three and four main bearing journals. The total indicator readings at each journal should not exceed 0,08 mm (0.003 in.).
4. While checking the run-out at each journal, note the relation of maximum eccentricity on each journal to the others. The maximum on all journals should come at very near the same angular location.
5. If the crankshaft fails to meet the foregoing checks it is bent and is unsatisfactory for service.
6. Check each crankshaft journal for ovality. If ovality exceeds 0,040 mm (0.0015 in.), a reground or new crankshaft should be fitted.
7. Bearings for the crankshaft main journals and the connecting rod journals are available in the following undersizes:
    0,25 mm (0.010 in.)
    0,50 mm (0.020 in.)
8. The centre main bearing shell, which controls crankshaft thrust, has the thrust faces increased in thickness when more than 0,25 mm (0.010 in.) undersize, as shown on the following chart.
9. When a crankshaft is to be reground, the thrust faces on either side of the centre main journal must be machined in accordance with the dimensions on the following charts.

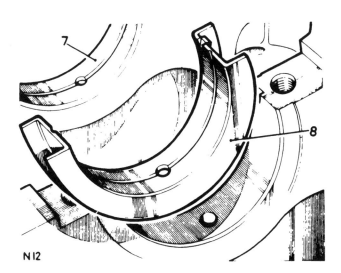

Main bearing journal size	Thrust face width
Standard	Standard
0,25 mm (0.010 in.) undersize	Standard
0,50 mm (0.020 in.) undersize	0,25 mm (0.010 in.) oversize

For example: If a 0,50 mm (0.020 in.) undersize bearing is to be fitted, then 0,12 mm (0.005 in.) must be machined off each thrust face of the centre journal, maintaining the correct radius.

Crankshaft dimensions 10 to 14

10. The radius for all journals except the rear main bearing is 1,90 to 2,28 mm (0.075 to 0.090 in.).
11. The radius for the rear main bearing journal is 3,04 mm (0.120 in.).
12. Main bearing journal diameter, see the following charts.
13. Thrust face width, see the following charts.
14. Connecting rod journal diameter, see the following charts.

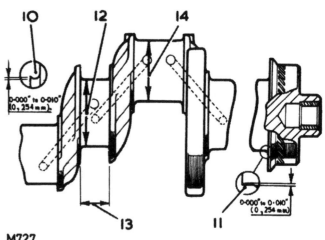

M727

Crankshaft dimensions – metric

Crankshaft Grade	Diameter '12'	Width '13'	Diameter '14'
Standard	58,400-58,413 mm	26,975-27,026 mm	50,800-50,812 mm
0,254 mm U/S	58,146-58,158 mm	26,975-27,026 mm	50,546-50,559 mm
0,508 mm U/S	57,892-57,904 mm	27,229-27,280 mm	50,292-50,305 mm

Crankshaft dimensions – inches

Crankshaft Grade	Diameter '12'	Width '13'	Diameter '14'
Standard	2.2992-2.997 in.	1.062-1.064 in.	2.0000-2.0005 in.
0.010 in. U/S	2.2892-2.2897 in.	1.062-1.064 in.	1.9900-1.9905 in.
0.020 in. U/S	2.2792-2.2797 in.	1.072-1.074 in.	1.9800-1.9805 in.

Checking the main bearing clearance

15. Remove the oil seals from the cylinder block and the rear main bearing cap.
16. Locate the upper main bearing shells into the cylinder block. These must be the shells with the oil drilling and oil grooves.
17. Locate the flanged upper main bearing shell in the centre position.
18. Place the crankshaft in position on the bearings.

19. Place a piece of Plastigauge 605238 across the centre of the crankshaft main bearing journals.
20. Locate the bearing lower shell into the main bearing cap.
21. Fit numbers one to four main bearing caps and shells. Torque: 7,0 to 7,6 kgf.m (50 to 55 lbf.ft.).
22. Fit the rear main bearing cap and shell. Torque: 9,0 to 9,6 kgf.m (65 to 70 lbf.ft.).

IMPORTANT: Do not rotate the crankshaft while the Plastigauge is fitted.

23. Remove the main bearing caps and shells.
24. Using the scale printed on the Plastigauge packet, measure the flattened Plastigauge at its widest point. The graduation that most closely corresponds to the width of the Plastigauge indicates the bearing clearance.
25. The correct bearing clearance with new or overhauled components is 0,023 to 0,065 mm (0.0009 to 0.0025 in.).
26. If the correct clearance is not obtained initially, use selective bearing assembly.
27. Wipe off the Plastigauge with an oily rag. Do NOT scrape it off.

IMPORTANT: The bearing shells must be retained in sets and in the correct sequence.

28. If required, check the connecting rod big-end bearing clearance. 12.17.10.
29. Refit the crankshaft. 12.21.33.

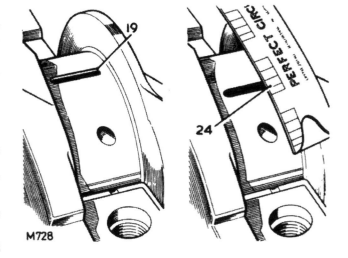

M728

DATA

Crankshaft

Material	Iron, spheroidal graphite
No. of main journals	5
End-thrust	Taken on Number 3
End-play	0,10 to 0,20 mm (0.004 to 0.008 in.)
Crankpin journal diameter (standard)	50,800 to 50,812 mm (2.0000 to 2.0005 in.)
Main bearing:	
Material and type	Vandervell lead indium
Clearance	0,023 to 0,061 mm (0.0009 to 0.0024 in.)
Journal diameter (standard)	58,399 to 58,412 mm (2.2992 to 2.2997 in.)
Bearing overall length	20,24 to 20,49 mm (0.797 to 0.807 in.). Nos. 1, 2, 4 and 5
	26,82 to 26,87 mm (1.056 to 1.058 in.) Number 3
Crankshaft vibration damper type	Torsional

CYLINDER PRESSURES

–Check   **12.25.01**

Checking

1. Run the engine until it attains normal operating temperature.
2. Remove all the sparking plugs.
3. Secure the throttle in the fully open position.
4. Check each cylinder in turn as follows:
5. Insert a suitable pressure gauge into the sparking plug hole.
6. Crank the engine with the starter motor for several revolutions and note the highest pressure reading obtainable.
7. If the compression is appreciably less than the correct figure or varies greater than 10% between cylinders, the piston rings or valves may be faulty.
8. Low pressure in adjoining cylinders may be due to a faulty cylinder head gasket.

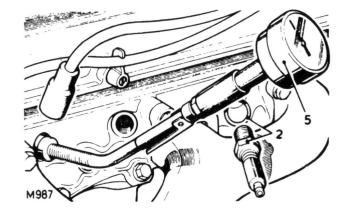

DATA

Starter motor cranking speed at 15 degrees C (60 degrees F) approximate ambient
temperature   150 to 200 engine rpm
Compression ratio   8.25:1 or 8.13:1
Compression pressure (minimum)   9,5 kg/cm^2 (135 lb/in.^{2})

CYLINDER HEADS

–Remove and refit	12.29.10
Left-hand cylinder head	12.29.11
Right-hand cylinder head	12.29.12

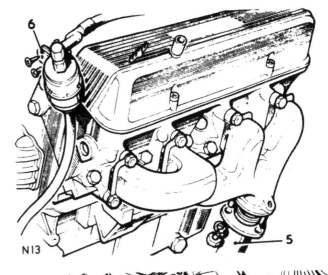

Removing

1. Drain the cooling system. 26.10.01.
2. Remove the air cleaner. 19.10.01.
3. Remove the induction manifold. 30.15.02.
4. Remove the valve gear. 12.29.34.
5. Disconnect the front exhaust pipes from the manifolds.
6. LH cylinder head–Disconnect the fuel line filter bracket from the engine lifting hook (where fitted).
7. RH cylinder head–Remove the alternator. 86.10.02.
8. Slacken the cylinder head bolts evenly.
9. If both cylinder heads are being removed, mark them relative to LH and RH sides of the engine.
10. Remove the cylinder heads and discard the gaskets.
11. If required, remove the exhaust manifolds. 30.15.10, 30.15.11.

Refitting

12. If removed, fit the exhaust manifolds. 30.15.10, 30.15.11.
13. Fit new cylinder head gaskets with the word 'TOP' uppermost. Do NOT use sealant.
14. Locate the cylinder heads on the block dowel pins.
15. Clean the threads of the cylinder head bolts then coat them with Thread Lubricant-Sealant 3M EC776,
16. Locate the cylinder head bolts in position:
    Long bolts – 1, 3 and 5.
    Medium bolts – 2, 4, 6, 7, 8, 9 and 10.
    Short bolts – 11, 12, 13 and 14.
17. Tighten the cylinder head bolts a little at a time in the sequence shown. Final torque:
    Bolts 1 to 10, 9,0 to 9,6 kgf.m (65 to 70 lbf.ft).
    Bolts 11 to 14, 5,6 to 6,2 kgf.m (40 to 45 lbf.ft).
18. When all bolts have been tightened, recheck the torque settings.
19. Reverse 1 to 7.

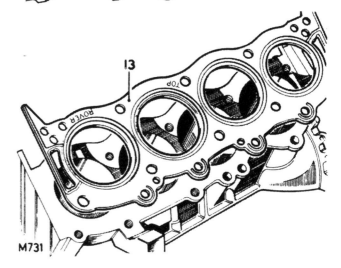

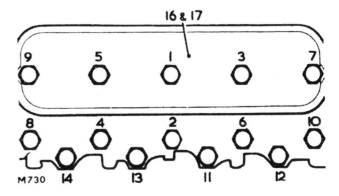

Cylinder head thread insert salvage instructions

19. These three holes may be drilled 0.3906 in. dia. x 0.937 + 0.040 in. deep. Tapped with Helicoil Tap No. 6 CPB or 6CS x 0.875 in. (min) deep (3/8 UNC 1½D insert).

20. These eight holes may be drilled 0.3906 in. dia. x 0.812 + 0.040 in. deep. Tapped with Helicoil Tap No. 6 CBB 0.749 (min) deep (3/8 UNC 1½D insert).

21. These four holes may be drilled 0.3906 in. dia. x 0.937 + 0.040 in. deep. Tapped with Helicoil Tap No. 6 CPB or 6CS x 0.875 (min) deep (3/8 UNC 1½D insert).

22. These four holes may be drilled 0.261 in. dia. x 0.675 + 0.040 in. deep. Tapped with Helicoil Tap No. 4CPB or 4CS x 0.625 (min) deep (¼ UNC 1½D insert).

23. These six holes may be drilled 0.3906 in. dia. x 0.937 + 0.040 in. deep. Tapped with Helicoil Tap No. 6 CPB or 6CS x 0.875 (min) deep (3/8 UNC 1½D insert).

CAUTION: Any attempt to salvage the sparking plug threads in the cylinder head may result in breaking into the water jacket, rendering the head scrap.

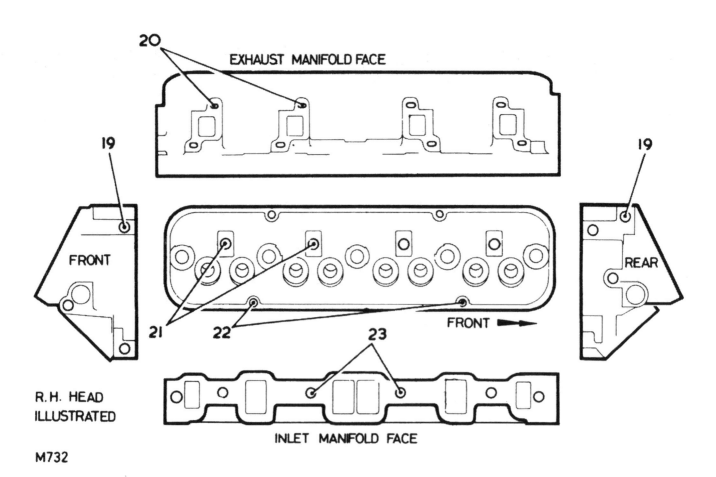

EXHAUST MANIFOLD FACE

20 19 19

FRONT REAR

21 22 23 FRONT ▶

R.H. HEAD ILLUSTRATED

INLET MANIFOLD FACE

M732

DATA

Cylinder heads

Material	Aluminium alloy
Type	Two heads with separate alloy inlet manifold
Inlet and exhaust valve seat material	Piston ring iron
Inlet and exhaust valve seat angle	46 + ¼ degrees

Valves

Valves, inlet

Overall length	116,58 to 117,34 mm (4.590 to 4.620 in.)
Actual overall head diameter – Engines, suffix 'A' to 'E'	37,97 to 38,22 mm (1.495 to 1.505 in.)
– Engines, suffix 'F'	39,75 to 40,00 mm (1.565 to 1.575 in.)
Angle of face	45 degrees
Stem diameter	8,640 to 8,666 mm (0.3402 to 0.3412 in.) at the head and increasing to 8,653 to 8,679 mm (0.3407 to 0.3417 in.)
Stem clearance in guide	Top 0,02 to 0,07 mm (0.001 to 0.003 in.) Bottom 0,013 to 0,063 mm (0.0005 to 0.0025 in.)

Valves, exhaust

Overall length	116,58 to 117,34 mm (4.590 to 4.620 in.)
Actual overall head diameter – Engines, suffix 'A' to 'E'	33,215 to 33,466 mm (1.3075 to 1.3175 in.)
– Engines, suffix 'F'	34,226 to 34,480 mm (1.3475 to 1.3575 in.)
Angle of face	45 degrees
Stem diameter	8,628 to 8,654 mm (0.3397 to 0.3407 in.) at the head and increasing to 8,640 to 8,666 mm (0.3402 to 0.3412 in.)
Stem clearance in guide	Top 0,038 to 0,088 mm (0.0015 to 0.0035 in.) Bottom 0,05 to 0,10 mm (0.002 to 0.004 in.)
Valve lift	9,9 mm (0.39 in.) both valves

Valve spring length:

Engines, suffix 'A' to 'E' – Outer	40,6 mm (1.6 in.) at pressure of 17,69 to 20,41 kg (39 to 45 lb)
Inner	41,2 mm (1.63 in.) at pressure of 9,75 to 12,02 kg (21.5 to 26.5 lb)
Engines, suffix 'F'	40,0 mm (1.577 in.) at pressure of 30,16 to 33,34 kg (66.5 to 73.5 lb)

CYLINDER HEADS

— Overhaul	12.29.18
Left-hand cylinder head	12.29.19
Right-hand cylinder head	12.29.30

Service tools: 276102 – Valve spring compressor
274401 – Valve guide remover
600959 – Valve guide drift
605774 – Distance piece for valve guide drift

1. Remove the cylinder heads. 12.29.10.

Dismantling

2. Using the valve spring compressor 276102 remove the valves and springs and retain in sequence for refitting.

NOTE: Engines up to suffix 'E' are fitted with double valve springs. Engines from suffix 'F' are fitted with single valve springs.

3. Clean the combustion chambers with a soft wire brush.
4. Clean the valves.
5. Clean the valve guide bores.
6. Regrind or fit new valves as necessary.
7. If a valve must be ground to a knife-edge to obtain a true seat, fit a new valve.
8. The correct angle for the valve face is 45 degrees.
9. The correct angle for the seat is 46 + ¼ degrees, and the seat witness should be towards the outer edge.
10. Check the valve guides and fit replacements as necessary. 11 to 15.
11. Using the valve guide remover 274401, drive out the old guides from the combustion chamber side.

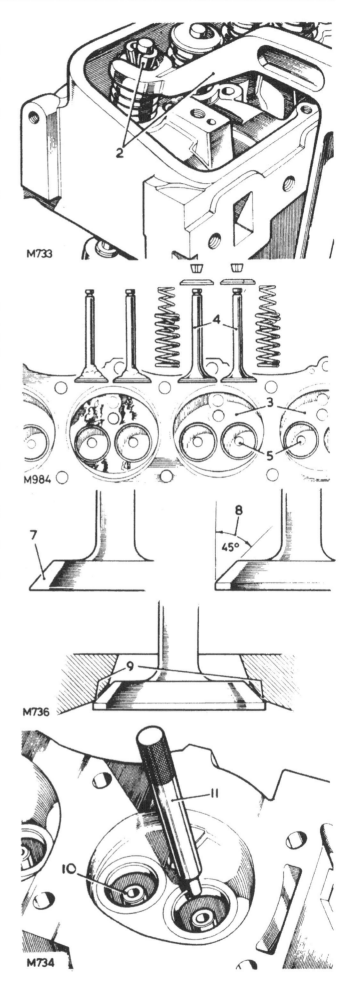

M733

M984

M736

M734

12. Engines up to suffix 'E' — fit the valve guides as follows:—
   (a) Locate the distance piece for the valve guide drift 605774 on the valve spring seat in the top of the cylinder head.
   (b) Lubricate the new valve guide and insert it into the distance piece.
   (c) Using the valve guide drift 600959, drive the valve guide into the cylinder head until the drift bottoms on the distance piece.

13. Engines from suffix 'F' onwards — fit the valve guides as follows:—
   (a) Lubricate the new valve guide and locate it in position on the cylinder head.
   (b) Using the valve guide drift 600959, drive the valve guide into the cylinder head until the dimension in the following instruction is achieved.

14. The fitted guide should stand 19 mm (¾ in.) above the step surrounding the valve guide boss in the cylinder head.

NOTE: Service valve guides are 0,02 mm (0.001 in.) larger on the outside diameter than the original equipment to ensure interference fit.

15. Check the valve seats and fit replacements as necessary. 17 to 19.

16. Remove the old seat inserts by grinding them away until they are thin enough to be cracked and prised out.

17. Heat the cylinder head evenly to approximately 65 degrees C (150 degrees F).

18. Press the new insert into the recess in the cylinder head.

NOTE: Service valve seat inserts are available in two over-sizes: 0,25 and 0,50 mm (0.010 and 0.020 in.) larger on the outside diameter to ensure interference fit.

19. If necessary, cut the valve seats to 46 + ¼ degrees.

20. The nominal seat width is 1,5 mm (0.031 in.). If the seat exceeds 2,0 mm (0.078 in.) it should be reduced to the specified width by the use of 20 and 70 degree stones.

21. The inlet valve seat is 35,25 mm (1.388 in.) diameter and the exhaust seat is 30,48 mm (1.200 in.) diameter for earlier engines, 37,03 mm (1.458 in.) diameter and 31,50 mm (1.240 in.) diameter respectively for later engines.

22. Check the height of the valve stems above the valve spring seat surface of the cylinder head. This MUST NOT exceed 47,63 mm (1.875 in.). If necessary grind the end of the valve stem or fit new parts.

23. Lubricate the valve stems and guides with engine oil and fit each valve as follows:—

24. Insert the valve into its guide.

25. Place the valve springs in position.

CAUTION: On later engines it is essential that the bottom of the valve spring is correctly located in the cylinder head recess.

continued

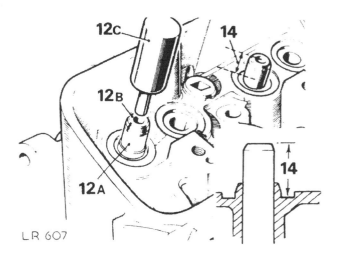

LR 607

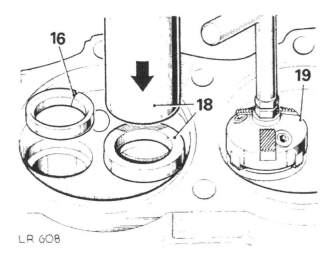

LR 608

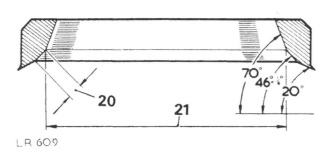

LR 609

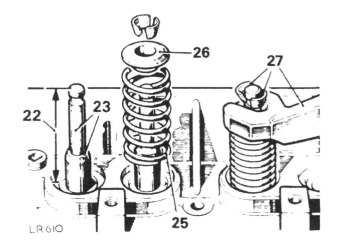

LR 610

26. Locate the cap on the springs.
27. Using the valve spring compressor 276102, fit the valve collets.
28. Refit the cylinder heads. 12.29.10.

For data refer to page 12–23.

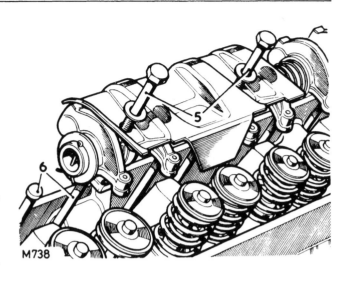

M738

VALVE GEAR

–Remove and refit   12.29.34

ROCKER SHAFTS

–Remove and refit 2, 4, 5 and 30 to 37   12.29.54

Removing

1. Drain the cooling system. 26.10.01.
2. Remove the air cleaner. 19.10.01.
3. Remove the induction manifold. 30.15.02.
4. Remove the rocker covers.
5. Remove the rocker shaft assemblies.
6. Withdraw the pushrods and retain in the sequence removed.
7. Withdraw the tappets and retain with respective pushrods.

NOTE: If a tappet cannot be withdrawn, remove the camshaft and withdraw the tappet from the bottom.

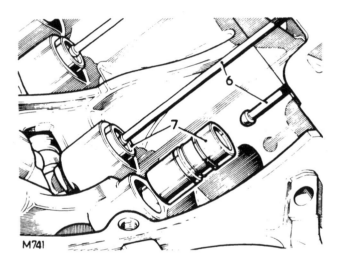

M741

Dismantling rocker shafts

8. Remove the split pin from one end of the rocker shaft.
   Withdraw the following components and retain them in the correct sequence for reassembly:
9. A plain washer.
10. A wave washer.
11. Rocker arms.
12. Brackets.
13. Springs.

Inspection of hydraulic tappets and pushrods

14. Hydraulic tappet; inspect inner and outer surfaces of body for blow holes and scoring. Replace hydraulic tappet if body is roughly scored or grooved, or has a blow hole extending through the wall in a position to permit oil leakage from lower chamber.
15. The prominent wear pattern just above lower end of body should not be considered a defect unless it is definitely grooved or scored; it is caused by side thrust of cam against body while the tappet is moving vertically in its guide.

Continued.

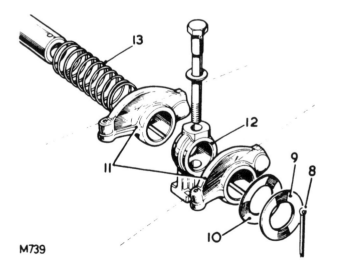

M739

16. Inspect the cam contact surface of the tappets. Fit new tappets if the surface is excessively worn or damaged.

17. A hydraulic tappet body that has been rotating will have a round wear pattern and a non-rotating tappet body will have a square wear pattern with a very slight depression near the centre.

18. Tappets MUST rotate and a circular wear condition is normal, and such bodies may be continued in use if the surface is free of defects.

19. In the case of a non-rotating tappet, fit a new replacement and check camshaft lobes for wear; also ensure new tappet rotates freely in the cylinder block.

20. Fit a new hydraulic tappet if the area where the pushrod contacts is rough or otherwise damaged.

21. Pushrod. Replace with new, any pushrod having a rough or damaged ball end or seat.

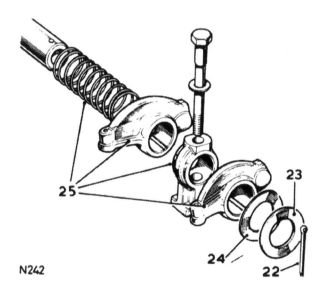

M742

Refitting

Assembling rocker shafts 22 to 28

22. Fit a split pin to one end of the rocker shaft.
23. Slide a plain washer over the long end of the shaft to abut the split pin.
24. Fit a wave washer to abut the plain washer.

NOTE: Two different rocker arms are used and must be fitted so that the valve ends of the arms slope away from the brackets.

25. Assemble the rocker arms, brackets and springs to the rocker shaft.
26. Compress the springs, brackets and rockers, and fit a wave washer, plain washer and split pin to the end of the rocker shaft.
27. Locate the oil baffle plates in place over the rockers furthest from the notched end of the rocker shaft.
28. Fit the bolts through the brackets and shaft so that the notch on the one end of the shaft is uppermost and towards the front of the engine on the right hand side, and towards the rear on the left-hand side.
29. Fit the tappets and pushrods in the original sequence.

N242

Continued.

IMPORTANT: The rocker shafts are handed and must be fitted correctly to align the oilways.

30. Each rocker shaft is notched at one end and on one side only. The notch must be uppermost and towards the front of the engine on the right-hand side, and towards the rear on the left-hand side.

31. Fit the rocker shaft assemblies. Ensure that the pushrods engage the rocker cups and that the baffle plates are fitted to the front on the left-hand side, and to the rear on the right-hand side. Tighten the bolts evenly. Torque: 3,5 to 4,0 kgf.m (25 to 30 lbf. ft.).

If it is necessary to fit a new rocker cover gasket, proceed as follows. 32 to 36.

32. Clean and dry the gasket mounting surface, using Bostik cleaner 6001.

33. Apply Bostik 1775 impact adhesive to the seal face and the gasket, using a brush to ensure an even film.

34. Allow the adhesive to become touch-dry, approximately fifteen minutes.

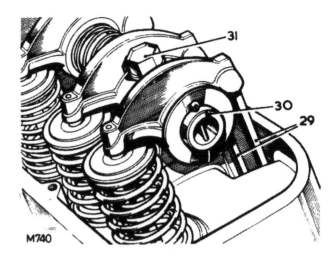

M740

NOTE: The gasket fits one way round only and must be fitted accurately first time; any subsequent movement would destroy the bond.

35. Place one end of the gasket into the cover recess with the edge firmly against the recess wall; at the same time hold the remainder of the gasket clear; then work around the cover, pressing the gasket into place ensuring that the outer edge firmly abuts the recess wall.

36. Allow the cover to stand for thirty minutes before fitting it to the engine.

37. Reverse 1 to 4.

NOTE: Tappet noise

It should be noted that tappet noise can be expected on initial starting up after an overhaul due to oil drainage from the tappet assemblies or indeed if the vehicle has been standing over a very long period. If excessive noise should be apparent after an overhaul, the engine should be run at approximately 2,500 rpm for a few minutes, when the noise should be eliminated.

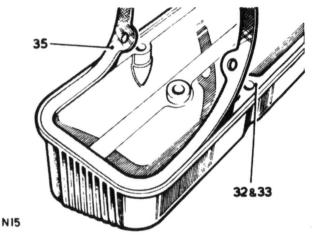

N15

ENGINE ASSEMBLY

—Remove and refit   12.41.01

Removing

1. Remove the bonnet. 76.16.01.
2. Disconnect the battery earth lead.
3. Drain the cooling system. 26.10.01.
4. Remove the fan blades. 26.25.06.
5. Remove the fan cowl, 26.25.11.
6. Remove the radiator block. 26.40.04.
7. Remove the air cleaner. 19.10.01.
8. Disconnect the inlet hose to the heater.
9. Disconnect the return hose from the heater.
10. Disconnect the throttle cable from the LH carburetter and induction manifold.
11. Disconnect the vacuum pipe to the gearbox.
12. Disconnect the choke cable from the LH carburetter.
13. Disconnect the fuel spill return pipe from the RH carburetter.
14. Disconnect the vacuum pipe for the brake servo.
15. Disconnect the leads from the alternator.
16. Disconnect the lead from the choke thermostat switch.
17. Disconnect the lead from the water temperature transmitter.
18. Disconnect the leads from the ignition coil.
19. Unclip the engine harness and draw it clear.
20. Disconnect the lead from the oil pressure switch.
21. Disconnect the leads from the starter motor.
22. Disconnect the earth strap from the engine.
23. Disconnect the fuel supply pipe from the fuel pump.
24. Disconnect the exhaust pipes from the manifolds.
25. Remove all the fixings from the engine front mounting rubbers.
26. Remove the cover plate from the bell housing.

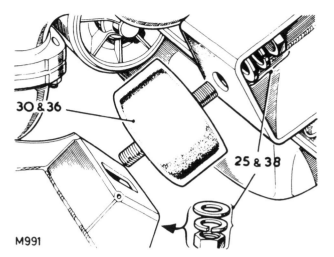

M991

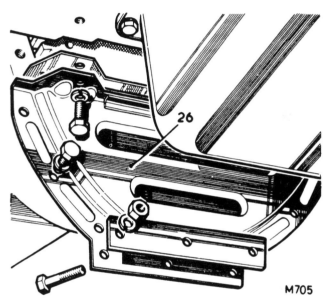

M705

27. Remove the fixings securing the bell housing to the engine. The lower fixings are accessible from under the vehicle, and the upper fixings are accessible from the engine compartment.
28. Attach a suitable lifting chain and hoist to the engine lifting hooks.
29. Tension the hoist sufficient to lift the engine just clear of the front mounting rubbers.
30. Withdraw the engine front mounting rubbers.
31. Draw the engine forward to release it from the dowelled location to the bell housing, and to clear the primary pinion from the clutch.
32. Lift the engine clear.

Refitting

Before refitting the engine

Smear the splines of the primary pinion, the clutch centre and the withdrawal unit abutment faces with molybdenum disulphide grease, Rocol MTS.1000. Smear the engine to gearbox joint faces with Universal jointing compound.

33. Attach a lifting chain and hoist to the engine lifting hooks.
34. Lower the engine into position, locating the primary pinion into the clutch and engage the bell housing dowels.
35. Secure the engine to the bell housing with at least two bolts.
36. Locate the engine front mounting rubbers in position.
37. Lower the engine on to the mountings and remove the lifting chain.
38. Secure the fixings at the engine front mounting rubbers.
39. Fit the remaining engine to bell housing fixings.
40. Apply a coating of 'Universal' jointing compound to the vertical joint face of the bell housing cover plate.
41. Locate the seal on to the bell housing cover plate.
42. Fit the bell housing cover plate and seal ensuring that the fillet is filled with jointing compound.
43. Reverse 1 to 24.

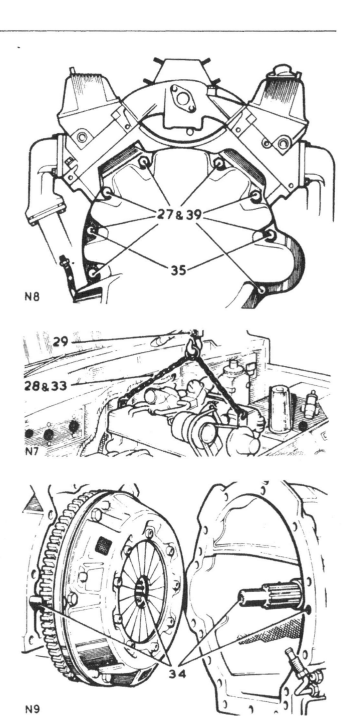

N8

N7

N9

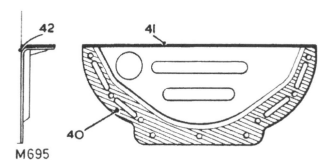

M695

FLYWHEEL

–Remove and refit   12.53.07

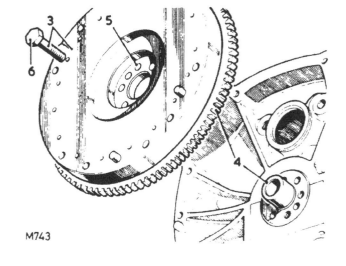

M743

Removing

1. Remove the engine assembly. 12.41.01.
2. Remove the clutch assembly. 33.10.01.
3. Remove the flywheel.

Refitting

4. Locate the flywheel in position on the crankshaft spigot, with the ring gear towards the engine.
5. Align the flywheel fixing bolt holes which are off-set to prevent incorrect assembly.
6. Fit the flywheel fixing bolts and before finally tightening, take up any clearance by rotating the flywheel against the direction of engine rotation. Torque: 7,0 to 8,5 kgf.m (50 to 60 lbf.ft.).
7. Reverse 1 and 2.

FLYWHEEL

–Overhaul   12.53.10.

1. Remove the flywheel. 12.53.07.

Procedure

2. Check the flywheel type. Two types of flywheel are in use and can be identified by the thickness of the ring gear:
    Early flywheel–ring gear thickness 9,52 to 9,65 mm (.375 to .380 in.).
    Later flywheel–ring gear thickness 10,97 to 11,22 mm (.432 to .442 in.).

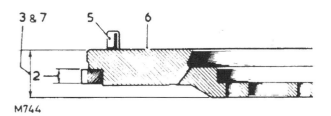

M744

3. Measure the overall thickness of the flywheel. Fit a new flywheel if it is less than the applicable minimum thickness:
    Early flywheel–minimum overall thickness 38,35 mm (1.510 in.).
    Later flywheel–minimum overall thickness 39,93 mm (1.572 in.).
4. If the flywheel is above the minimum thickness, the clutch face can be refaced as follows.
5. Remove the dowels.
6. Reface the flywheel over the complete surface.
7. Check the overall thickness of the flywheel to ensure that it is still above the minimum thickness.
8. Refit the flywheel. 12.53.07.

STARTER RING GEAR

—Remove and refit   12.53.19

Removing

1. Remove the flywheel. 12.53 07.
2. Drill a 10 mm (.375 in.) diameter hole axially between the root of any tooth and the inner diameter of the starter ring sufficiently deep to weaken the ring. Do NOT allow the drill to enter the flywheel.
3. Secure the flywheel in a vice fitted with soft jaws.
4. Place a cloth over the flywheel to protect the operator from flying fragments.

WARNING: Take adequate precautions against flying fragments as the starter ring gear may fly asunder when being split.

5. Place a chisel immediately above the drilled hole and strike it sharply to split the starter ring gear.

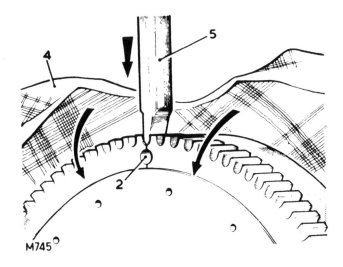

M745

Refitting

6. Heat the starter ring gear uniformly to between 170 degrees and 175 degrees C (338 degrees to 347 degrees F) but do not exceed the higher temperature.
7. Place the flywheel, flanged side down, on a flat surface.
8. Locate the heated starter ring gear in position on the flywheel, with the chamfered inner diameter towards the flywheel flange. If the starter ring gear is chamfered both sides, it can be fitted either way round.
9. Press the starter ring gear firmly against the flange until the ring contracts sufficiently to grip the flywheel.
10. Allow the flywheel to cool gradually. Do NOT hasten cooling in any way and thereby avoid the setting up of internal stresses in the ring gear which may cause fracture or failure in some respect.
11. Fit the flywheel. 12.53.07.

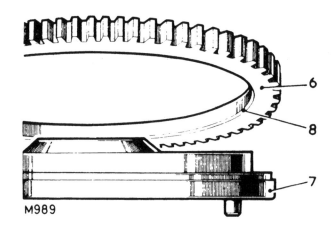

M989

SPIGOT BEARING

– Remove and refit   12.53.20

Removing

1. Remove the engine assembly. 12.41.01.
2. Remove the clutch assembly. 33.10.01.
3. Remove the spigot bearing.

Refitting

4. Fit the spigot bearing flush with, or to a maximum of 1,6 mm (0.063 in.) below the end face of the crank-shaft.
5. Ream the spigot bearing to 19,151 + 0,025 mm (0.7504 + 0.001 in.) inside diameter. Ensure all swarf is removed.
6. Reverse 1 and 2.

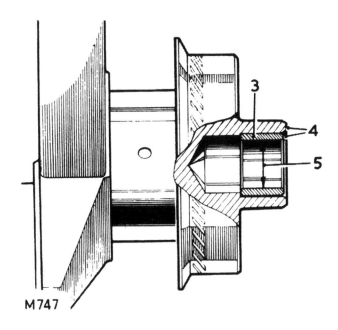

M747

OIL FILTER ASSEMBLY – EXTERNAL

– Remove and refit   12.60.01

Removing

1. Unscrew the filter and discard.

NOTE: Later filters are not fitted with an end nut. If the filter is difficult to remove, use a strap spanner.

2. Withdraw the sealing washer and discard.

CAUTION: Do NOT delay fitting a new filter, otherwise the oil pump may drain and require priming (12.60.26) before running the engine.

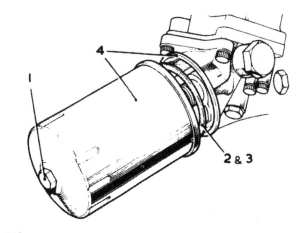

M748

Refitting

3. Place a new sealing washer on to a new filter.
4. Fit the filter BY HAND until the sealing washer touches the oil pump cover face, then give a further half turn – do NOT overtighten.

NOTE: The hexagon on the filter casing is for removal purposes only.

5. Check and if necessary, replenish the engine oil sump.
6. Run the engine and check the filter joint for leaks.
7. Check the oil sump level after the engine has been stopped for a few minutes and replenish if necessary.

CAUTION: Two type of oil filter have been fitted, the later type from engine number suffix 'F', incorporates a relief valve. These oil filters are not interchangeable.

OIL PUMP

–Remove and refit 12.60.26

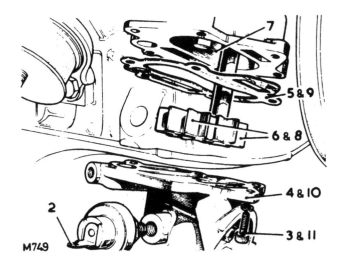

Removing

1. Remove the oil filter assembly. 12.60.01.
2. Disconnect the electrical leads from the switches.
3. Remove the bolts from the oil pump cover.
4. Withdraw the oil pump cover.
5. Lift off the cover gasket.
6. Withdraw the oil pump gears.

Refitting

7. Fully pack the oil pump gear housing with petroleum jelly. Use only petroleum jelly; no other grease is suitable.
8. Fit the oil pump gears so that the petroleum jelly is forced into every cavity between the teeth of the gears.

IMPORTANT: Unless the pump is fully packed with petroleum jelly it may not prime itself when the engine is started.

9. Place a new gasket on the oil pump cover.
10. Locate the oil pump cover in position.
11. Fit the special fixing bolts and tighten alternatively and evenly. Torque: 1,4 to 2,0 kgf.m (10 to 15 lbf. ft.).
12. Reverse 1 and 2.
13. Check the oil level in the engine sump and replenish as necessary.

OIL PUMP

—Overhaul   12.60.32

1. Remove the oil pump. 12.60.26.

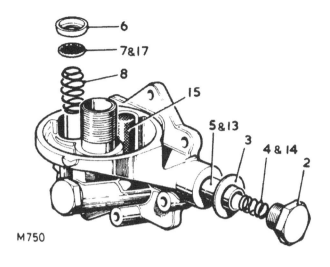

M750

Dismantling

2. Unscrew the plug from the pressure relief valve.
3. Lift off the joint washer for the plug.
4. Withdraw the spring from the relief valve.
5. Withdraw the pressure relief valve.
6. Earlier engines -- Prise out the seat for the oil filter by-pass valve.
7. Withdraw the by-pass valve.
8. Withdraw the spring.

Inspecting

9. Check the oil pump gears for wear or scores.
10. Fit the oil pump gears and shaft into the front cover.
11. Place a straight edge across the gears.
12. Check the clearance between the straight edge and the front cover. If less than 0,05 mm (.0018 in.) check the front cover gear pocket for wear.
13. Check the oil pressure relief valve for wear or scores.
14. Check the relief valve spring for wear at the sides or signs of collapse.
15. Clean the gauze filter for the relief valve.
16. Check the fit of the relief valve in its bore. The valve must be an easy slide fit with no perceptible side movement.
17. Check the oil filter by-pass valve for cracks, nicks or scoring.

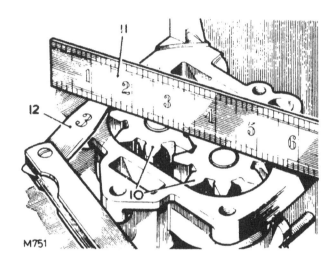

M751

Assembling

18. Lubricate the oil pressure relief valve and fit it into its bore.
19. Insert the relief valve spring.
20. Locate the sealing washer on to the relief valve plug.
21. Fit the relief valve plug. Torque: 4,0 to 4,9 kgf.m (30 to 35 lbf ft.).
22. Earlier engines -- Insert the by-pass spring into its bore.
23. Earlier engines -- Place the by-pass valve on the spring.
24. Earlier engines -- Press in the by-pass valve seat, concave side outward, until the outer rim is between 0,5 and 1,0 mm (.020 and .040 in.) below the surface of the surrounding casting.
25. Refit the oil pump. 12.60.26.

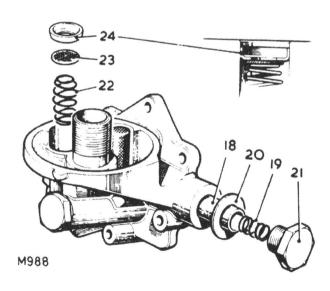

M988

93   12–35

OIL SUMP

— Remove and refit   12.60.44

Removing

1. Remove the sump drain plug.
2. Allow all the oil to drain, then refit the plug and sealing washer.
3. Remove the sump.

Refitting

4. Clean the sump mating surfaces at the join between the timing chain cover and the cylinder block.
5. Apply a coating of Unipart Universal jointing compound across the join.
6. Place the sump gasket in position (Early models).
   Later models
   (i)   Remove any remaining sealant from around mating faces.
   (ii) Ensure that mating faces are free from dirt and grease.
   (iii) Apply Hylosill RTV liquid sealant to the sump and cylinder block joint faces.
   (iv) Fit the sump.
   NOTE: Hylosill RTV liquid sealant is marketed by Marstons Lubricants Limited.
7. Fit the sump.
8. Unscrew the oil filler cap.
9. Using the correct grade oil, see Lubrication – Group 09, replenish the sump.
10. Use the sump dipstick to set the final level. Do NOT fill above the 'HIGH' mark.
11. Start the engine and check that the oil pressure warning light goes out. If the light remains on, the engine must be stopped and the oil pump dismantled and primed 12.60.26.
12. Run the engine and check the sump joint for leaks.
13. Check the sump oil level after the engine has been stopped for a few minutes and replenish if necessary.

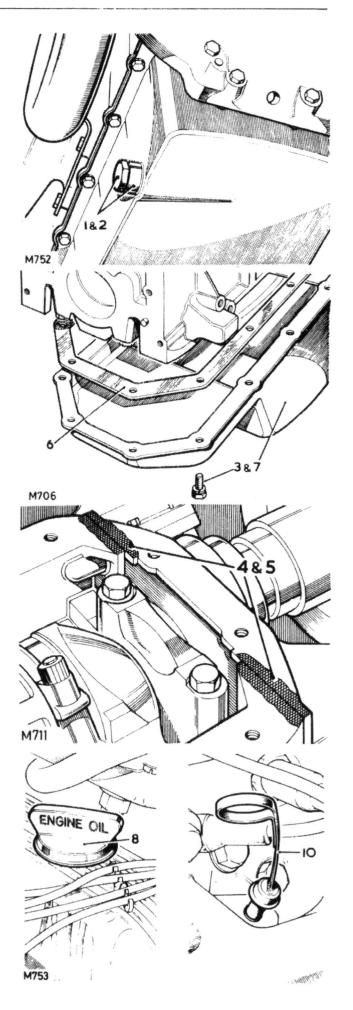

M752

M706

M711

ENGINE OIL

M753

TIMING GEAR COVER

–Remove and refit   12.65.01

Includes

CRANKSHAFT FRONT OIL SEAL

–Remove and refit   12.21.14

Removing

1. Drain the cooling system. 26.10.01.
2. Remove the fan blades and pulley. 26.25.01.
3. Remove the fan cowl. 26.25.11.
4. Disconnect the by-pass hose from the thermostat.
5. Disconnect the heater return hose from the water pump.
6. Disconnect the inlet hose from the water pump.
7. Release the alternator adjusting link from the water pump housing.
8. Disconnect the vacuum pipe from the distributor.
9. Release the distributor cap, unclip the leads and move the cap to one side.
10. Disconnect the low-tension lead from the ignition coil.
11. Disconnect the lead from the oil pressure switch.
12. Earlier engines – Remove the fixings from the fuel pump and move the pump to one side. Do NOT disconnect the fuel pipes.
13. Remove the starter dog.
14. Withdraw the crankshaft pulley and, on later engines, the mud deflector.
15. Mark the distributor body relative to the centre line of the rotor arm.
16. If the distributor is to be removed, make corresponding marks on the distributor and timing cover.
17. Remove the timing cover fixings, including two from the sump.
18. Withdraw the timing cover complete.
19. Remove the joint washer.

continued

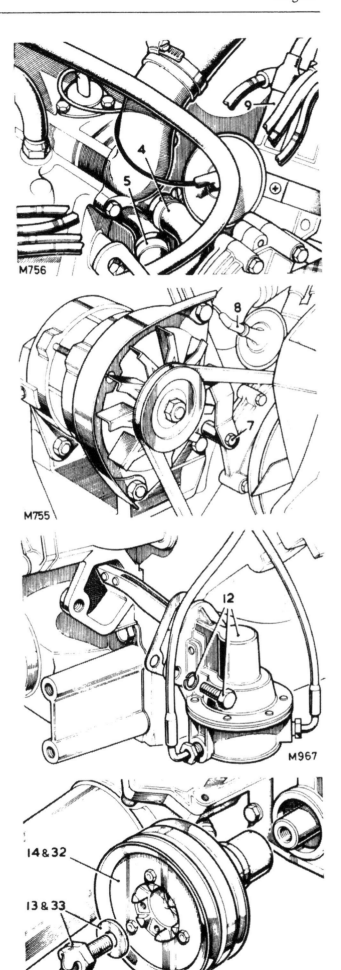

95

12–37

Oil seal, remove and refit 20 to 24

Suffix 'A' Engine

20. Remove the oil seal and oil thrower.
21. Coil a new seal into a new oil thrower.
22. Fit the oil seal and thrower assembly into the front cover, ensuring that the ends of the seal are at the top.
23. Stake the oil thrower in place at equidistant points.
24. Rotate a hammer handle around the seal until the crankshaft pulley can be inserted.

Suffix 'B' Engines onwards

20a. Remove the fixings and withdraw the mudshield.
21a. Remove the oil seal.
22a. Position the gear cover with the front face uppermost and the underside supported across the oil seal housing bore.
23a. Enter the oil seal, open side first, into the housing bore.
24a. Press in the oil seal until the plain face is 1,5 mm (0.062 in.) approximately below the gear cover face. Reverse instruction 20a.

continued

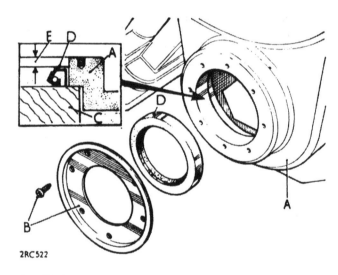

2RC522

A – Timing gear cover
B – Mudshield and fixings
C – Support block
D – Oil seal
E – Dimension 1,5 mm (0.062 in.) approx.

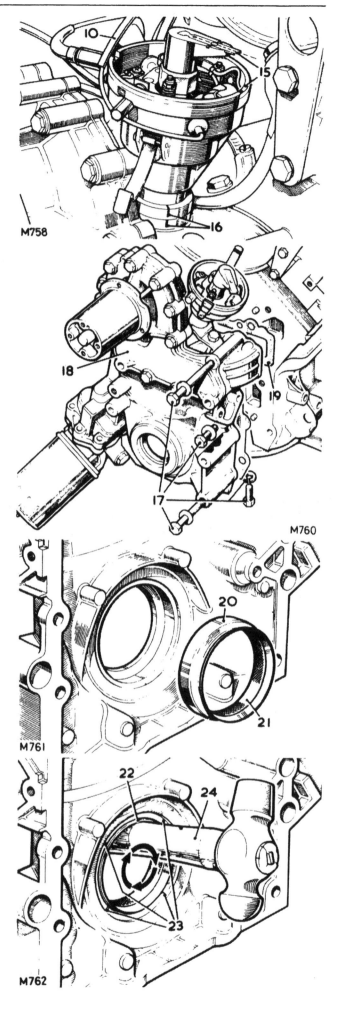

M758

M760

M761

M762

Refitting

25. Place a new timing cover joint washer in position.
26. Prime the oil pump by injecting engine oil through the suction port.
27. Set the distributor rotor arm approximately 30 degrees before the final positioning mark, to compensate for the skew gear engagement.
28. Locate the timing cover in position, applying a coating of 'Unipart Universal' about 13 to 19 mm (½ to ¾ in.) wide in the area shown.
29. Check that the distributor marking alignment is correct.
30. Clean the threads of the timing cover securing bolts, then coat them with Thread Lubricant-Sealant 3M EC776, Rover Part No. 605764.
31. Fit the timing cover securing bolts. Torque: 2,8 to 3,5 kgf.m (20 to 25 lbf.ft.).
32. Fit the crankshaft pulley.
33. Fit the starter dog. Torque: 19,3 to 22,3 kgf.m (140 to 160 lbf.ft.).
34. Reverse 1 to 14.
35. Check, and if necessary adjust the ignition timing. 86.35.20.

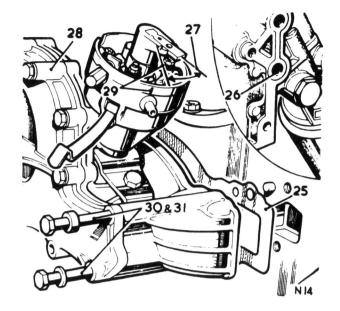

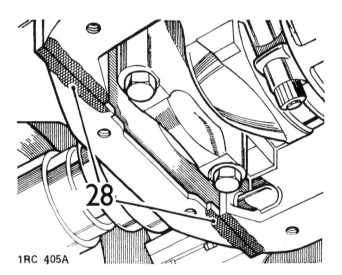

1RC 405A

TIMING CHAIN AND GEARS

– Remove and refit   12.65.12

Removing

1. Set the engine – number one piston at TDC.
2. Remove the timing chain cover. 12.65.01.
3. Check that number one piston is still at TDC.
4. Early engines – withdraw the oil thrower.
5. Remove the distributor drive gear.
6. *Early engines* – withdraw the fuel pump cam.
   Later engines – withdraw the spacer.
7. Withdraw the chain complete with the chainwheels.

CAUTION: Do NOT rotate the engine if the rocker shafts are fitted, otherwise the valve gear and pistons will be damaged.

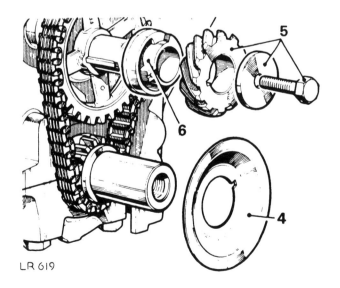

LR 619

Refitting

NOTE: If the crankshaft and/or camshaft have not been rotated, commence at item 12. If they have been rotated, commence at item 8.

8. Remove the rocker shafts. 12.29.54.

9. Set the engine – number one piston at TDC.

10. Temporarily fit the camshaft chainwheel with the marking 'FRONT' outward.

11. Turn the camshaft until the mark on the chainwheel is at the six o'clock position, then remove the chainwheel without disturbing the camshaft.

12. Locate the chainwheels to the chain with the timing marks aligned.

13. Engage the chainwheel assembly on the camshaft and crankshaft key locations and check that the camshaft key is parallel to the shaft axis to ensure adequate lubrication of the distributor drive gear.

CAUTION: The space between the key and keyway acts as an oilway for lubrication of the drive gear as indicated in the illustration.

The key must be a tight fit and be seated to the full depth of the keyway. Ensure that the key is parallel to the shaft and that the overall dimension of the shaft and key does not exceed 30,15 mm (1.187 in.).

14. Check that the timing marks are in line.

15. *Early engines* – Fit the fuel pump cam with the marking 'F' outwards.

    Later engines – Fit the spacer with the flange to the front, as illustrated.

16. Fit the distributor drive gear. On later engines ensure that the annular grooved side of the gear is fitted to the rear, i.e. towards the spacer, as illustrated.

17. Secure the drive gear with the washer and bolt. Torque: 5,5 to 6,2 kgf.m (40 to 45 lbf.ft.).

18. Fit the timing chain cover. 12.65.01.

DATA

Timing chain and wheels

Timing chain type	Inverted tooth
Number of links	54
Width	22,22 mm (0.875 in.)
Pitch	9,52 mm (0.375 in.)
Crankshaft chainwheel	Sintered iron
Camshaft chainwheel	Aluminium alloy, teeth covered with nylon.
Valve timing	See page 04–1

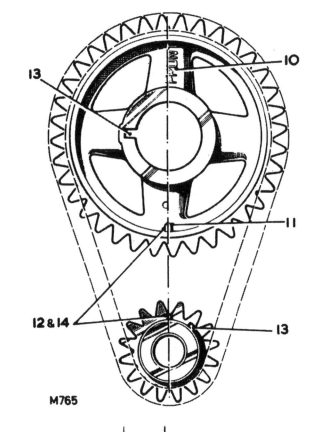

M765

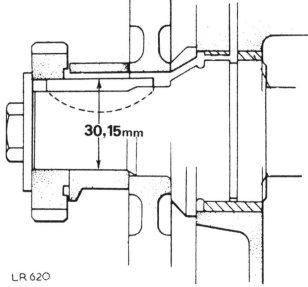

LR 620

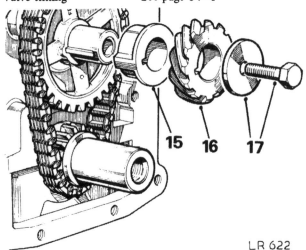

LR 622

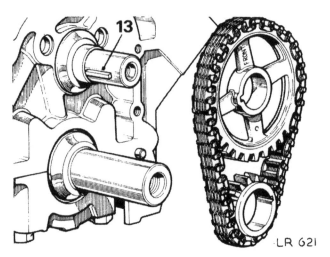

LR 621

EMISSION AND EVAPORATIVE CONTROL OPERATIONS

Range Rover vehicles are fitted during manufacture with various items of emission and evaporative control equipment to meet individual territory requirements. Therefore some operations in this section of the manual may not be applicable to all vehicles.

Unauthorised replacement or modification of the emission or evaporative control equipment may contravene local territory legislation and render the vehicle user and/or repairer liable to legal penalties.

CRANKCASE BREATHING SYSTEM

Description   17.10.00

The 'blow-by' gases from the crankcase are vented into the combustion system to be burned with the fuel/air mixture. The system provides positive emission control under all conditions. During engine running, crankcase fumes which may collect in the crankcase are vented to the carburetter via hoses and flame traps.

NOTE. Filters may have alternative locations according to terriotory build specification.

ENGINE BREATHER FILTER

-Remove and refit   17.10.02

Removing

1. Remove the air cleaner 19.10.01
2. Withdraw the filter top hose.
3. Slacken the filter clip.
4. Withdraw the filter from the bottom hose.

Refitting

5. Fit the filter with the end marked 'IN' facing forward. Alternatively, if the filter is marked with arrows, they must point rearward.
6. Connect the filter bottom hose.
7. Connect the filter top hose.
8. Secure the filter retaining clip.
9. Fit the air cleaner 19.10.02.

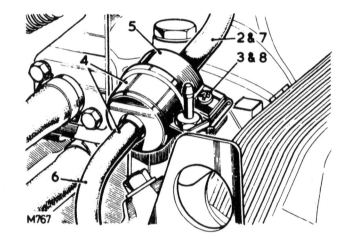

ENGINE BREATHER FILTER (Australian)

Removing

1. Disconnect the fresh air intake pipe, leading from the air cleaner to the filter, at the filter end.
2. Disconnect the pipe at the other side of the filter.
3. Remove the filter.

Refitting

4. Fit new filter and reconnect both pipes.
NOTE. Fit new filter with end marked 'IN' facing forward. Alternatively, if the filter is marked with arrows, they must point rearwards.
5. Refit hoses and tighten clip.

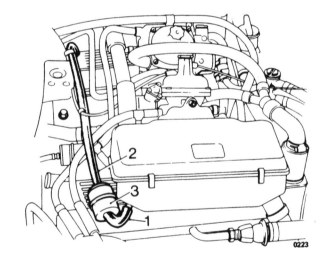

ENGINE FLAME TRAPS

NOTE: Flame traps may have alternative locations according to territory build specification.

– Remove and refit   **17.10.03**

Removing

1. Pull the flame trap clip clear of the carburetter/air cleaner.
2. Remove the hoses from the flame traps.
3. Withdraw the flame traps.

Cleaning

4. Wash the flame traps in clean petrol and allow to dry.

Refitting

5. Reverse 1 to 3.

EVAPORATIVE LOSS CONTROL SYSTEM (Australian)

Description   **17.15.00**

The Evaporative Loss Control System reduces the amount of fuel vapour vented to the atmosphere. An adsorption canister receives vapour from the carburetter float chambers and, on some models, from the fuel tank. The canister is purged of these vapours by inlet manifold or carburetter depression.

ADSORPTION CANISTER

– Remove and refit   **17.15.13**

NOTE: Illustration number 0225 shows early models supplied to Australia. Illustration number RR747M shows later models supplied to Australia and Saudi Arabia.

Removing

1. Disconnect from the canister:–
        (i) Canister line to fuel tank
        (ii) Canister purge line
        (iii) Carburetter vent pipe
2. Slacken the clamp nut screw.
3. Remove the canister.

Refitting

4. Secure the canister in the clamp.
5. Reverse instructions 1 and 2 above.

WARNING: The use of compressed air to clean an adsorption canister or clear a blockage in the evaporative system is very dangerous. An explosive gas present in a fully saturated canister may be ignited by the heat generated when compressed air passes through the canister.

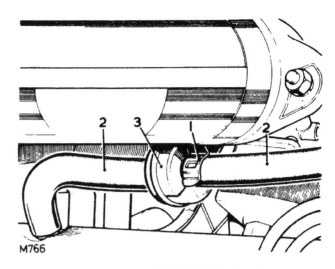

M766

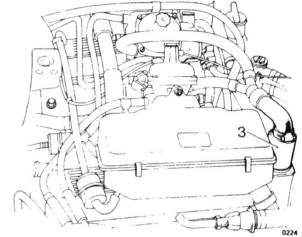

0224

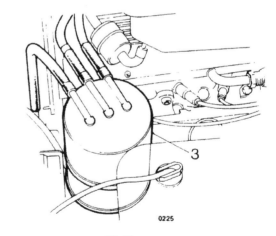

0225

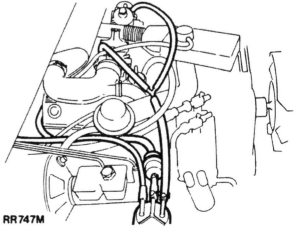

RR747M

PULSAIR AIR INJECTION 17.25.00

"Pulsair" is a system of self-induced air injection. The induced air which is taken from the air cleaner at high manifold depressions passes through one-way valves and a configuration of pipes which inject clean air into the exhause gases via the manifold, thereby reducing carbon monoxide emission to the atmosphere.

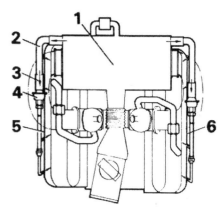

1. Air Cleaner
2. Connecting Pipes
3. Connecting Hoses
4. Pulsair valves
5. Right hand air manifold
6. Left hand air manifold

"Pulsair" - Early system above, later system below.

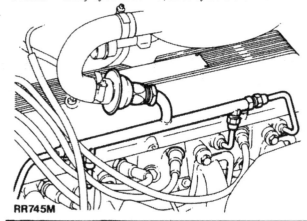

RR745M

AIR INJECTION SYSTEM (Australia)

Description 17.25.00

The air pump, driven by the engine, delivers air through an air manifold and open ended stainless steel tubes into each exhaust port. A one-way valve prevents damage to the pump should backfire or breakage occur.

A relief valve dumps part of the air delivery at high pump speeds to prevent damage.

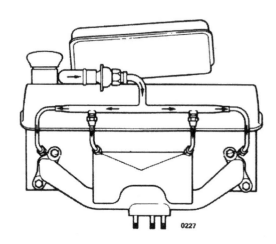

0227

Air Pump

The rotary vane type air pump is fitted at the front of the engine and driven by a belt from a crankshaft pulley.

The pump delivers air under pressure to each of the four exhaust ports via a diverter valve (where fitted) a relief valve, check valve and air inlet manifold.

The pressurized air combines with the exhaust gases to continue and assist in making more complete the oxidization process in the exhaust system.

1. Inlet port
2. Exhaust port
3. Vanes
4. Rotor
5. Carbon sealing shoes
6. Bearings

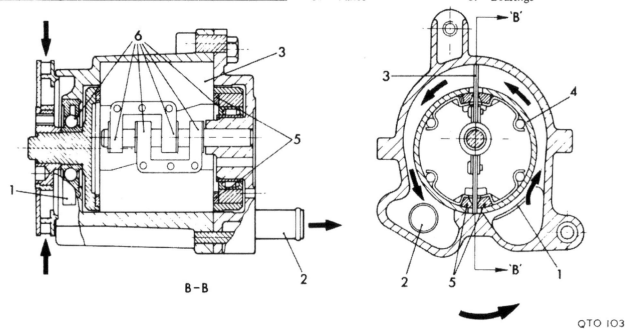

B - B

QTO.103

AIR PUMP (Australian)

– Remove and refit　　　　　　　17.25.07

Removing

1. Release the air hoses from the pump.
2. Slacken the idler pulley mountings and lift the drive belt from the air pump pulley.
3. Loosen but do not remove the air pump mounting bolts.
4. Take the weight of the air pump and remove the mounting bolts.
5. Lift the pump from the engine.

Refitting

6. Reverse instructions 1 to 5.
7. Check and adjust the drive belt tension 17.25.13.

RELIEF VALVE (Australian)

– Check operation　　　　　　　17.25.10

1. The relief valve allows excessive air pressure at high engine speed to discharge to the atmosphere.
2. Run the engine at high speeds and check the valve operation.
3. The valve cannot be adjusted and a defective unit should be replaced.

RELIEF VALVE (Australian)

– Remove and refit　　　　　　　17.25.11

Removing

1. Disconnect the hoses from the valve.
2. Remove the valve.

Refitting

3. Reverse instructions 1 and 2.

AIR PUMP DRIVE BELT (Australian)

– Check and adjust tension　　　　17.25.13

The belt should be sufficiently tight to drive the appropriate auxiliary without undue load on the bearings.
1. This tension is measured by allowing 0,4 mm movement on the slack side of the belt per 25,4 mm between pulley centres.

If adjustment is necessary:–
2. Slacken the idler pulley mountings.
3. Apply gentle leverage to the idler pulley until the belt tension is correct.
4. Tighten the idler pulley mountings.

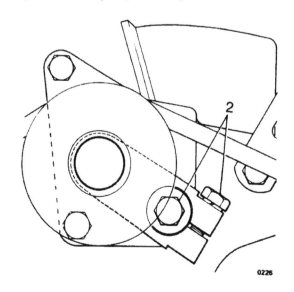

0226

AIR PUMP DRIVE BELT (Australian)

– Remove and refit   **17.25.15**

Removing

1. Remove the alternator/power steering/air conditioning pump drive belts as applicable to gain access to the air pump drive belt.
2. Slacken the idler pulley mountings.
3. Move the idler pulley towards the engine and remove the drive belt.

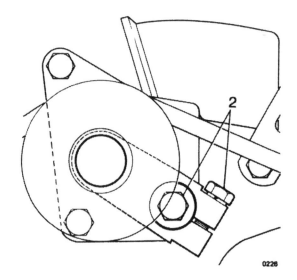

Refitting

4. Reverse instructions 1 to 3, adjusting belt tensions as necessary.
5. Check and adjust the drive belt tension 17.25.13.

AIR MANIFOLD (Australian)

– Remove and refit   **17.25.17**

Removing

1. Remove the carburetter air cleaners. 19.10.01.
2. Release the manifold from the check valve.
3. Release the manifold from the cylinder head.
NOTE: It may be more convenient to release the centre branches at the manifold junction.
4. Lift off the manifold.

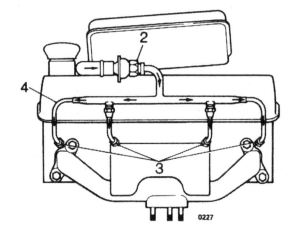

Refitting

5. Reverse instructions 1 to 4.
6. Run the engine and check for air leaks at the manifold.

CHECK VALVE (Australian)

PULSE AIR VALVE (European)

– Remove and refit   17.25.21

Removing

1. Disconnect the air hose from the check valve/pulse air valve.
2. Using two open-ended spanners – one on the air distribution manifold hexagon, to support the manifold, and the other to remove the valve anti-clockwise.

CAUTION: Do not impose any strain on the air manifold.

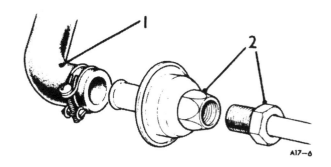

Refitting

3. Reverse instructions 1 and 2.

NOTE: The pulse-air valve is identified by a pink paint spot and the part number 4974-196 on the face.

CHECK VALVE (Australian)

– Test   17.25.22

The check valve is a one-way valve positioned to protect the pump from back-flow of exhaust gases. The valve closes if the pump pressure falls while the engine is running, should, for example, the drive belt break.

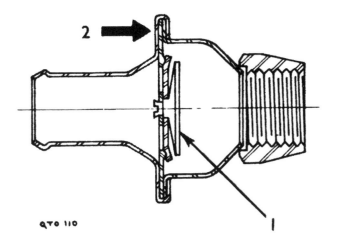

Testing

CAUTION: Do not use a pressure air supply for this test.

1. Remove the check valve. 17.25.21.
2. Blow through the valve orally in both directions in turn. Air should only pass through the valve when blown from the hose connection end. Should air pass through the valve when blown from the air manifold end, renew the valve.
3. Refit the check valve. 17.25.21.

DIVERTER AND RELIF VALVE (Australian)

– Remove and refit   17.25.25

Removing

1. Disconnect the battery.
2. Disconnect the air hose from the valve to the check valve.
3. Disconnect the diverter valve triggering vacuum pipe.
4. Remove the two bolts retaining the valve to the air pump.
5. Remove the valve.

Refitting

6. Fit the valve to the air pump using a new gasket.
7. Connect the check valve to diverter hose.
8. Connect the diverter air valve triggering vacuum pipe to the valve.
9. Reconnect the battery.

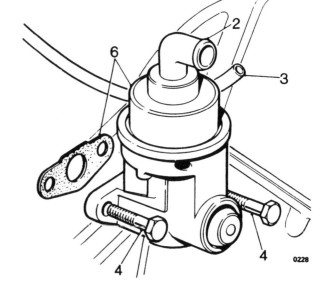

DIVERTER AND RELIEF VALVE (Australian)

– Check operation   17.25.26

This combined valve is incorporated to divert the air from the pump during deceleration to prevent backfire. The relief valve allows excessive air pressure at high engine speeds to discharge to the atmosphere.
Check by means of a functional test.

AIR INTAKE TEMPERATURE CONTROL SYSTEM

Description   17.30.00

To maintain an efficient air intake temperature, a sensor is incorporated in the air cleaner. The sensor allows inlet manifold vacuum to operate a flap in the air cleaner intake. The flap permits cold air, from forward of the radiator and/ or hot air, from a scoop around the exhaust manifold, to mix in varying amounts to provide the required air temperature.

This system also applies to the Australian specification except that there are two air cleaners, hot boxes and flap control valves. The single temperature sensing device is situated in the right hand air cleaner on the clean side of the element.

The European specification comprises:

1. A hot box surrounding the exhaust manifold.
2. A vacuum operated thermostatically controlled flap valve in the air cleaner and silencer intake.
3. The flap valve controls the source of the intake air supply which may be warm air drawn entirely from the hot box or cold air drawn from the under bonnet area or a combination of both.
4. The hot box is connected via a hose to the flap valve in the air intake.
5. The temperature sensing device is situated in the air cleaner on the clean side of the element.
6. A pipe from the manifold is attached to the temperature sensing device via a non-return valve.
7. From the other side of the temperature sensing device is a pipe connecting the vacuum capsule operating the flap valve.

CONTROL VALVE

Check Operation   17.30.01

1. Check operation of the mixing flap valve in the air cleaner by starting the engine from cold and observing the flap valve as the engine temperature rises.
2. The valve should start to open slowly within a few minutes of starting and continue to open until a stabilised position is achieved. This position and the speed of operation will be entirely dependent on prevailing ambient conditions.
3. Failure to operate indicates failure of the flap valve vacuum capsule or the thermostatically controlled vacuum switch or both.
4. Check by connecting a pipe directly to the flap valve, thus by-passing the temperature sensor.
5. If movement of the flap valve is evident the temperature sensor is faulty. If no movement is detected, the vacuum capsule is faulty.
6. Fit new parts where necessary.

AIR INTAKE TEMPERATURE CONTROL – FAULT FINDING

SYMPTON	POSSIBLE CAUSE	ACTION
Poor or erratic idle Hesitation or flat spot (cold engine) Excessive fuel consumption Lack of power Engine overheating	Hot air inlet hose loose, adrift or blocked	Check hot air inlet hose for condition and security. Renew if necessary
Poor or erratic idle Hesitation or flat spot (cold engine) Excessive fuel consumption Lack of power Engine cuts-out or stalls (at idle) Engine 'runs-on' Engine 'knocks or pinks' Rich running (excess CO)	Flap valve jammed	Check operation of flap valve 17.30.01 If fault cannot be rectified renew air cleaner outer cover which includes flap valve. See 19.10.01
Poor or erratic idle Hesitation or flat spot (cold engine) Excessive fuel consumption Lack of engine power Engine cuts-out or stalls (at idle) Engine misfires Lean running (low CO)	Vacuum pipes disconnected or leaking	Check the vacuum pipes for security and deterioration. Renew if necessary
Hesitation or flat spot (cold engine) Excessive fuel consumption Lack of power Engine overheating Engine cuts-out or stalls (at idle) Engine 'runs-on' Engine knocks or pinks Rich running (excess CO)	One-way valve faulty	Blow through valve to check 'one-way' action. If the valve leaks fit a new valve. 17.30.05.
Poor or erratic idle Hesitation or flat spot (cold engine) Excessive fuel consumption Lack of power Engine overheating Engine cuts-out or stalls (at idle) Engine 'runs-on' Engine knocks or pinks Rich running (excess CO)	Temperature sensor faulty, leaking or jammed	Check and renew if necessary. 17.30.10.
Poor or erratic idle Hesitation or flat spot (cold engine) Excessive fuel consumption Lack of power Engine cuts-out or stalls Rich running	Flap valve diaphragm leaking	Check with a distributor vacuum test unit. If leakage is apparent, renew air cleaner outer cover which includes the servo motor.

AIR INTAKE TEMPERATURE CONTROL VALVE

– Remove and refit   **17.30.15**

NOTE: Alternative valves may be fitted according to territory build specification.

Removing

1. Disconnect the pipes and leads from the valve unit, noting their position for refitting.
2. Disconnect the hose to the air cleaner.
3. Disconnect the hot air box clamp (where fitted) and release the valve from the hot air box.
4. Release the valve from its support and lift it from the engine compartment.

Refitting

5. Reverse instructions 1 to 4.

HOT AIR BOX (Australian)

– Remove and refit   **17.30.31**

Removing

1. Remove the air intake temperature control valve 17.30.15 or remove the pipe from the hot air box, as applicable.
2. Remove two bolts/nuts securing the hot air box to the exhaust manifold.
3. Withdraw the hot air box from the manifold.

Refitting

4. Reverse instructions 1 to 3.

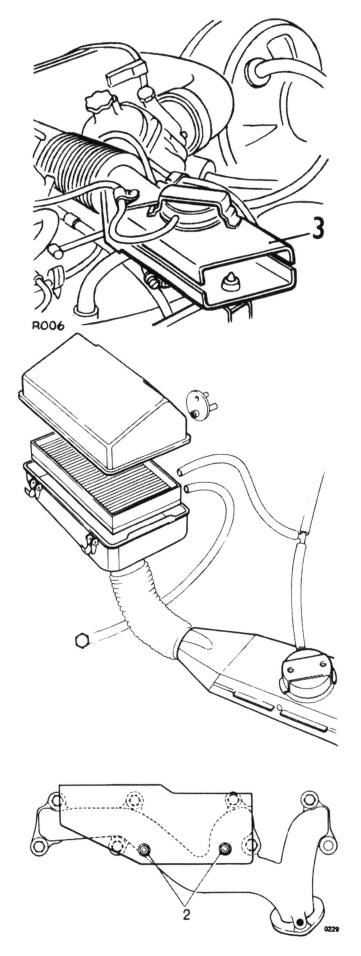

EXHAUST GAS RECIRCULATION (E.G.R.) VALVE
(Australian)

– Description   17.45.00

To reduce the nitric oxide content in the exhaust, the peak combustion temperatures are lowered by recirculating a controlled quantity of the exhaust gases through the combustion process.

The E.G.R. valve is mounted on the exhaust manifold. A control signal, taken from a throttle edge tapping in the carburetter, gives no recirculation at idle or full load, but does allow an amount of recirculation, dependent on the vacuum signal and a metering profile of the valve, under part load conditions.

EXHAUST GAS RECIRCULATION (E.G.R.) VALVE
(Australian)

– Remove and refit   17.45.01

Removing

1. Disconnect the vacuum pipe from the valve.
2. Disconnect the asbestos lagged pipe from the valve.
3. Unscrew the valve from the manifold.

Refitting

4. Reverse instructions 1 to 3, ensuring that the valve is securely sealed to the manifold.

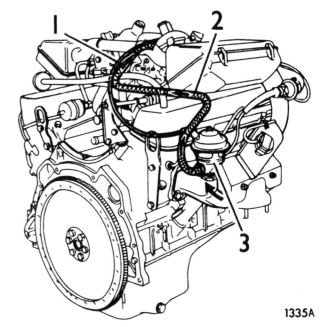

1335A

EXHAUST GAS RECIRCULATION (E.G.R.) VALVE
(Australian)

– Check operation   17.45.20

1. Warm the engine to normal running temperature.
2. Ensure that the autochoke control is fully "off".
3. Open and close the throttle several times and observe or feel the E.G.R. valve, which should:–
    (a)   open and close with changes in engine speed.
    (b)   close instantly when the throttle is closed.

FUEL SYSTEM OPERATIONS

Operation No.

19-1

Air cleaner
        -remove and refit   19.10.01
        -renew elements   19.10.08

Carburetters
        -auto choke — remove and refit   19.15.38
        -deceleration (poppet) valve — remove and refit   19.15.64
        -description   19.15.01
        -diaphragm — remove and refit   19.15.35
        -float chamber levels — check and adjust   19.15.32
        -float chamber needle valve — check and adjust   19.15.24
        -overhaul   19.15.18
        -remove and refit   19.15.11
        -temperature compensator — remove and refit   19.15.59
        -throttle (butterfly) disc — remove and refit   19.20.04
        -tune and adjust   19.15.02

Choke control cable - remove and refit   19.20.13

Fuel line filter - remove and refit   19.25.01

Fuel main filter
        -remove and refit   19.25.02
        -element — remove and refit   19.25.07

Fuel pipes
        -main line, tank end section — remove and refit   19.40.02
        -main line, centre section   19.40.03
        -main line, carburetter end section —remove and refit   19.40.04
        -spill return - remove and refit   19.40.08

Fuel pump
        -remove and refit   19.45.08
        -overhaul   19.45.15

Fuel tank — remove and refit   19.55.01

Thermostat switch — remove and refit   19.15.50

Throttle cable — remove and refit   19.20.06

Throttle linkage — remove and refit   19.20.07

Throttle pedal — remove and refit   19.20.01

19-1

FAULT DIAGNOSIS

SYMPTOM	POSSIBLE CAUSE	CURE
DIFFICULT STARTING WHEN COLD	Insufficient choke action	Check action of cold start unit to ensure that the choke is being applied fully—adjust choke cable Check position of cold start adjuster—move outward
	Fast idle adjustment in-correct	Check and adjust fast idle setting. Check linkage between choke and throttle for distortion
	Float chamber level too low	Check needle valve for sticking—(closed). Check float level setting. Check inlet connection filter for blockage. Check external fuel system in accordance with fuel system fault diagnosis
	Carburetter flooding	Check needle valve for sticking—(open) Float punctured Fuel pump pressure too high Float level too high
	No fuel supply to carburetter	Check filters and pump for blockage Check fuel tank breather and fuel lines for blockage Remove fuel pump and check operation. Overhaul or fit new pump
DIFFICULT STARTING WHEN HOT	Choke sticking 'on'	Check to ensure choke is returning to fully 'off' position, reset as necessary
	Blocked air cleaner	Fit new air cleaner elements
	Float chamber level too high	Check float level setting. Check float arms for distortion. Check needle valve for sticking. Punctured float, fuel pump pressure too high .
LACK OF ENGINE POWER	No oil in damper or oil too thin	Check level of oil in damper, and fill to correct level with oil of a viscosity of SAE 20
	Piston sticking	Check piston assembly moves freely and returns under spring load— centre jet assembly Check diaphragm for cracks or porosity
	Water in fuel	If water is present in float chamber, the complete fuel system should be drained, fuel components should be dismantled, inspected for contamination, paying particular attention to filters

SYMPTOM	POSSIBLE CAUSE	CURE
ERRATIC SLOW-RUNNING OR STALLING ON DECELERATION	Float level too low	Check float chamber level. Check for needle valve sticking
	Incorrect jet setting	Check and reset jet settings in accordance with carburetter overhaul instructions
	Carburetter air leaks	Check throttle spindle and bearings for wear
	Manifold air leaks	Check inlet manifold gasket for leakage. Check inlet manifold for cracks and distortion of mating faces. Check gasket between carburetter and manifold. Check condition of vacuum advance pipe and connections. Check vacuum servo pipes and connections
	Damper oil too thick. No oil in damper	Check and refill to correct level with oil specified
EXCESSIVE IDLE SPEED WHEN ENGINE IS HOT	Malfunction of decelerating (poppet) valve	If fault confirmed renew throttle (butterfly) disc which includes the decelerating valve. (See 19.20.04).
EXCESSIVE FUEL CONSUMPTION	Blocked air cleaner	Fit new air cleaner elements
	Damper oil too thick	Replace with correct grade
	Incorrectly adjusted carburetter	Check and reset slow running in accordance with carburetter tune and adjust instructions
	Float level too high	Check and reset float level
	Worn jets and needle	Check and replace as necessary
	Incorrect needle	Check needle type
	Choke sticking 'on'	Check to ensure choke is returning to fully 'off' position, reset as necessary
	Engine fault	See Section 12

AIR CLEANER (European)

–Remove and refit   19.10.01

Removing

1. Release the hose clips each side of the air cleaner.
2. Withdraw the air cleaner elbows.
3. Detach the choke cable from the clip on the air cleaner.
4. Withdraw the air cleaner from the retaining posts, at the same time disconnecting the hose from the engine breather filter.

Refitting

5. Fit the air cleaner, locating the rubber mountings over the retaining posts.
6. Connect the engine breather hose at the underside of the air cleaner.
7. Smear the 'O' rings at the carburetter intakes with MS4 grease.
8. Fit the air cleaner elbows.
9. Secure the hose clips.
10. Retain the choke cable in the clip at the front of the air cleaner.

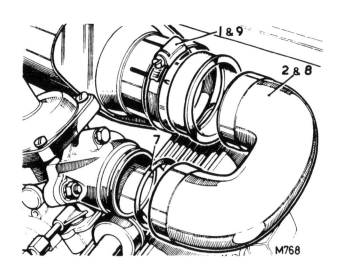

M768

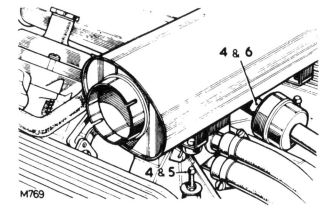

M769

AIR CLEANER (Australian)

– Remove and refit   19.10.01

Removing

1. Disconnect the balance pipe running between the air cleaners.
2. Disconnect the rocker cover breather pipe from the R.H. air cleaner.
3. Slacken the jubilee clips and disconnect the air intake temperature control systems from both air cleaners.
4. Disconnect the H.T. leads from the retaining clips located on the top of both air cleaners.
5. Disconnect the two pipes running into the temperature sensor on the R.H. air cleaner.
6. On both air cleaners, remove the four bolts securing the air cleaners to the air intake adaptor.
7. Remove the air cleaners.

Refitting

8. Reverse instructions 1 to 7.

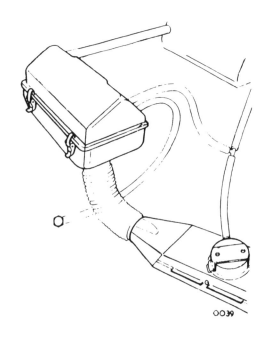

0039

AIR CLEANER (European)

–Renew elements   19.10.08

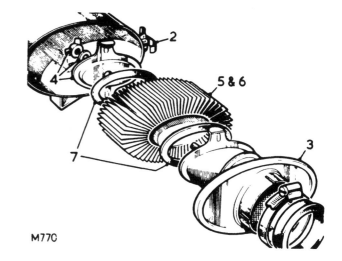

Removing

1.   Remove the air cleaner 19.10.01.
2.   Release the end plate clips.
3.   Withdraw the end plates.
4.   Remove the wing nut, washer and retaining plate.
5.   Withdraw the air cleaner elements.

Fitting

6.   Fit new air cleaner elements.
7.   Fit new sealing washers.
8.   Reverse 1 to 4.

M770

AIR CLEANER (Australian)

– Renew elements   19.10.08

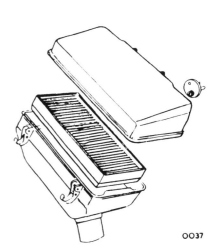

Removing

1.   Release the 4 clips on each air cleaner.
2.   Raise the cover and remove the elements.

Refitting

3.   Fit new elements.
NOTE: when fitting new elements ensure that the plastic bevelled edge faces the air intake aperture.

3.   Replace the cover and clips.

0037

CARBURETTERS

– Description   19.15.00

Variations in carburetters may be fitted to meet local territory legislation.

Tamperproofing

These carburetters may be externally identified by a tamperproof sealing tube fitted around the slow running adjustment screw.

The purpose of these carburetters is to more stringently control the air fuel mixture entering the engine combustion chambers and in consequence the exhaust gas emissions leaving the engine.

For this reason the only readily accessible external adjustment on these carburetters is to the throttle settings for fast idle speed and, on some later carburetters, this may require the use of a special tool to adjust the settings.

On all but early carburetters a decelerating (poppet) valve is incorporated in the throttle (butterfly) disc (A) and consists of a precisely set, spring-loaded plate valve (B). With low manifold depression, the valve remains closed. Under high induction manifold depression conditions, for instance during over-run with the throttle closed, the valve opens thereby slightly reducing the depression, allowing a correct quantity of fuel and air mixture to enter the engine. This improves the combiston of fuel during these conditions and helps to prevent high-value hydrocarbon emissions.

The decelerating valve is not adjustable (See Fault diagnosis page 19–3).

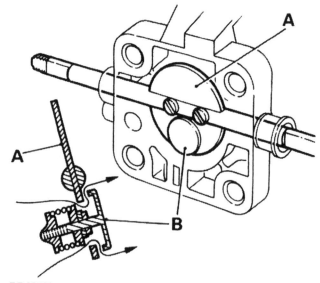

RR181M

Emission Specifications

All carburetters fitted to the Range-Rover conform at the time of manufacture to particular territory requirements in respect of exhaust and evaporative emissions control. However, in some cases changes to the basic carburetters themselves have been necessary to achieve this, for example, the replacing of a manual choke with a temperature actuated auto choke. Therefore some operations included in this section of the manual may not be applicable to all carburetters fitted to the model.

CAUTION- Unauthorised breaking of tamperproofing devices, adjustment of carburetter settings or the fitting of incorrectly related parts may render the vehicle user liable to legal penalties according to local territory legislation.

Whenever adjustments are made to the settings of tamperproofed or emission specification carburetters an approved type CO meter must be used to ensure that the final exhaust gas analysis meets with local territory requirements.

CARBURETTERS WITH MANUAL CHOKE

–Tune and adjust   19.15.02

Service tools: 605330 Carburetter balancer

The service tool 605330 carburetter balancer must be used to adjust the carburetters. Primarily this instrument is for balancing the air flow through the carburetters, but it also

gives a good indication of the mixture setting. Investigation has shown that incorrect mixture setting causes either stalling of the engine or a considerable drop in engine rpm if the balancer is fitted when the mixture is too rich or a considerable increase in rpm when used with the mixture setting too weak. Before balancing the carburetters it is most important therefore that the following procedure be carried out:

1. Check that the throttle control between the pedal and the carburetters is free and has no tendency to stick.

2. Check the throttle cable setting with the throttle pedal in the released position. The throttle linkage must not have commenced movement, but commences with the minimum depression of the pedal.

3. Run the engine until it attains normal operating temperature, that is, thermostat open.

4. Remove the air cleaner 19.10.01.

5. Slacken the screws securing the throttle adjusting levers on both carburetters.

6. Start the engine and check the idle speed. If necessary, adjust the throttle stop screws to give the correct idle speed. Refer to section 05. If a tamper-proof sleeve is fitted over this screw, the slow running speed can only be adjusted using a special tool supplied to authorised service outlets.

7. *Early models only* – Check the mixture on each carburetter separately, by lifting the air valve 0,8 mm (0.031 in.). If the engine speed increases immediately, the mixture is too rich. If the engine speed decreases immediately, the mixture is too weak.

8. *Early models* – Screw the jet orifice adjusting screw into the carburetter to weaken the mixture, unscrew it to enrich.

   Later models – Remove the piston damper plug and using special tool S.353 illustrated adjust the mixture. Locate the outer sleeve of the tool to engage a machined slot to prevent the air valve twisting. Turn the inner tool clockwise to enrich the mixture and anti-clockwise to weaken it.

   1981 models – Use special tool MS.80 to adjust the mixture: This has a slotted end to engage a raised blade on the mixture screw.

9. When the mixture is correctly adjusted, the engine speed will remain constant or may fall slowly a slight amount as the air valve is lifted.

10. Check and, if necessary, zero the gauge on tool 605330.

11. Place tool 605330 on to the carburetter adaptors, ensuring that there are no air leaks. If the engine stalls or decreases considerably in speed, the mixture is too rich. If the engine speed increases, the mixture is too weak.

12. If necessary, remove tool 605330 and readjust the mixture, then refit the tool.

13. Check tool 605330 gauge reading.

14. If the gauge pointer is in the 'ZERO' sector, no adjustment is required.

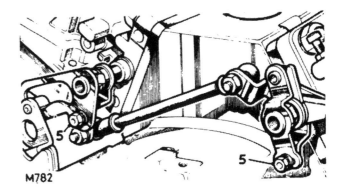

M782

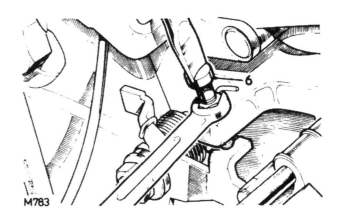

M783

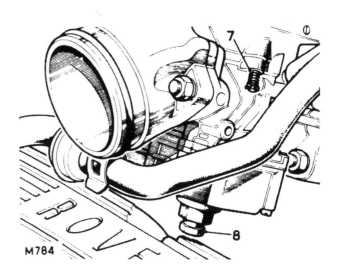

M784

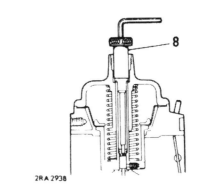

2RA2938

Continued

15. If the gauge pointer moves to the right, descrease the air flow through the left hand carburetter by unscrewing the throttle stop screw or increase the air flow through the right hand carburetter by screwing in the throttle stop screw. Reverse the procedure if the pointer moves to the left.

16. If the engine idle speed rises too high or drops too low during balancing, adjust to the correct idle speed, maintaining the gauge pointer in the 'ZERO' sector.

17. Remove tool 605330. With the mixture setting and carburetter balance correctly adjusted the difference in engine rpm with the tool 605330 on or off will be negligible, approximately plus or minus 25 rpm.

NOTE: Using a recognised type CO meter, the exhaust gas analysis reading should not exceed 4% carbon monoxide or any other levels fixed by local territory legislation.

18. On the left hand carburetter, place a 0,15 mm (.006 in.) feeler between the underside of the roller on the countershaft lever and the throttle lever.

19. Apply pressure to the throttle lever to hold the feeler.

20. Tighten the screw to secure the throttle adjusting lever, then withdraw the feeler.

21. On the right hand carburetter, place a 0,15 mm (.006 in) feeler between the left leg of the fork on the adjusting lever and the pin on the throttle lever.

22. Apply light pressure to the linkage to hold the feeler.

23. Tighten the screw to secure the throttle adjusting lever, then withdraw the feeler.

24. Refit the air cleaner 19.10.01.

Continued

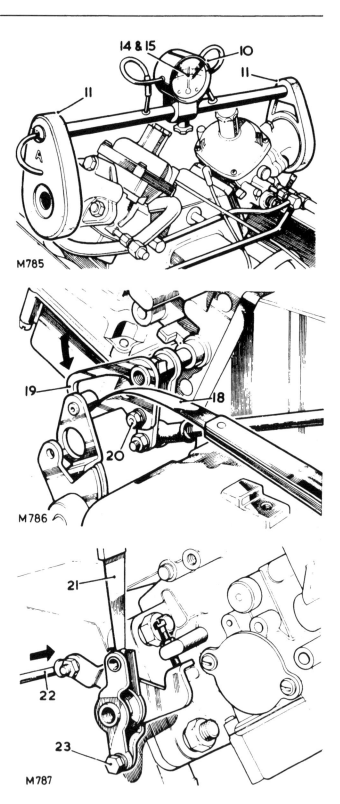

M785

M786

M787

Fast idle adjustment

The fast idle adjustment is pre-set on the left hand carburetter and should not normally require adjustment. If adjustment is required, the correct procedure is to remove the left hand carburetter and carry out items 82 to 87 detailed in 19.15.18. Alternatively, the fast idle can be approximately set as follows, but this alternative method is not recommended for ambient temperatures below 8°C (10°F).

25. Set the fast idle adjustment screw against the cam to give an engine speed of 1000 to 1200 rpm when the choke warning light just goes out.

NOTE: On some later carburetters a special tool, supplied to authorised service outlets, may be necessary to adjust the fast idle speed setting.

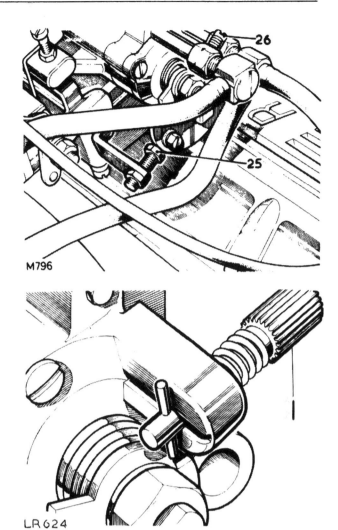

M796

LR624

Cold start unit

26. Set the cold start adjuster as necessary.
27. When operating the vehicle in ambient temperatures above 18°C (0°F), push the adjuster screw inward and turn it so that the peg is at right angles to the slot. Leave in the position illustrated.
28. When operating the vehicle in ambient temperatures below 18°C (0°F) turn the screw until the peg is recessed in the slot.

CARBURETTERS WITH AUTO CHOKE (Australian)

— Tune and adjust   19.15.02

Special tools: CO exhaust gas analyser and mixture adjusting tool S.353.

CAUTION: To ensure compliance with exhaust emission legislative requirements the following items must not be changed or modified in any way.

The fuel jet assembly.

The piston assembly.

The depression cover.

The following items must not be adjusted in service but should be replaced completely by factory-set units:

The temperature compensator.

The piston assembly return spring.

NOTE: During the course of the following instructions do not allow the engine to idle for longer than three minutes without a 'clear-out' burst of one minute duration at 2000 r.p.m.

Continued

1. Run the engine until normal operating temperature is reached.
2. Remove the air cleaner. 19.10.01.
3. Disconnect the throttle linkage so that each carburetter operates independantly.

Adjusting idle speed and air flow for balance.

4. Ensure that the fast idle screw is clear of the fast idle cam.
5. Using an air flow meter check that the air flow through both carburetters is the same. If not, adjust as necessary.

NOTE. On these carburetters the adjacent slow running and fast idle adjustment screws should be carefully studied. It will be seen that the fast idle screw is in the location normally occupied by the slow running adjustment screw on non auto choke carburetters.

6. If necessary turn the throttle adjusting screws on both carburetters an equal amount to maintain the correct idle speed. Refer to Page 05—6.
7. Increase the engine speed to 1600 r.p.m. and check the balance with the air flow meter. If necessary turn the throttle adjusting screws as required to achieve a balance.
8. Re-check the air flow balance at idle speed.

Checking and adjusting CO level at idle

9. Disconnect and plug the outlet hose from the air rail.
10. Maintain the engine at normal operating temperature and check that the idle speed is correct.
11. Check and if necessary adjust the ignition timing. 86.35.15.
12. Re-check the idle speed.
13. Insert the gas analyser probe as far as possible into the exhaust pipe.
14. Check the CO reading.
15. Adjust the mixture if necessary — see mixture adjustment.
16. Check and if necessary adjust the idle speed.
17. Withdraw the analyser probe.
18. Switch off the ignition.
19. Unplug air injection hose and reconnect to air rail.

Continued

Mixture adjustment

CAUTION. The setting MUST ALWAYS be checked by means of a non-dispersive infra-red exhaust gas analyser. For significant deviation outside the specified CO limits the mixture should be adjusted as follows.

20. Remove the piston damper from both carburetters.
21. Carefully insert special tool S353 into the dashpot until the outer tool engages in the air valve and the inner tool engages the hexagon in the needle adjuster plug.

CAUTION. The outer tool must be correctly engaged and held in position otherwise damage to the diaphragm may result.

22. Holding the outer tool turn the inner tool:
    a) Clockwise to enrich the mixture.
    b) Anti-clockwise to weaken.
23. Repeat instruction 21 and 22 on the remaining carburetter ensuring that the adjustment made is by the same amount.
24. Top-up the carburetter dampers, refer to section 10.
25. Re-check the CO reading — adjust until the CO reading is within the specified limits.
26. Set the fast idle. 19.15.38.

Setting the throttle linkage.

27. Ensure that the throttle lever clamp bolt is slackened.
28. Disconnect the ball end link.
29. Screw both throttle stop screws away from the stops.
30. Screw each throttle stop screw until the lever contacts the fast idle screw and then move each screw three further turns.
31. Check carburetter balance with the air flow meter and adjust as necessary.
32. Set the ball end link to a gauge length of 3.4521" (77.683 mm).
33. Fit the ball end link noting that the left hand carburetter throttle will open slightly.
34. Push the left hand throttle lever back to the fully closed position and hold it there while taking up the clearance with the spring lever screw.
35. Tighten the throttle lever clamp bolt.
36. Check and adjust the idle speed by turning the adjustment screw an equal amount on each carburetter.
37. Refit the air cleaner 19.10.01.

CARBURETTERS

-Remove and refit   19.15.11

Removing

1. Remove the air cleaner 19.10.01.
2. Disconnect the choke cable (where fitted)
3. Disconnect the emission control pipes.
4. Disconnect the distributor vacuum pipe.
5. Disconnect the throttle linkage.
6. Disconnect the main fuel supply pipe.
7. Disconnect the choke fuel supply pipe.
8. Remove the carburetters.
9. If required withdraw the joint washers, insulator and liner.

Refitting

10. Locate a joint washer on the inlet manifold.
11. Fit the insulator, aligning the arrows.
12. Fit the liner fully into the insulator, engaging the three tabs into the recesses.
13. Locate a joint washer on the insulator.
14. Reverse 1 to 8.
15. Fit the air cleaner 19.10.01.
16. Tune and adjust the carburetters 19.15.02.

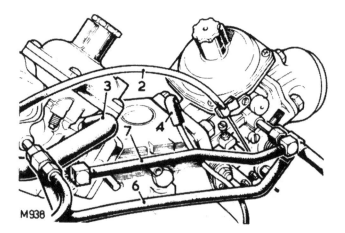

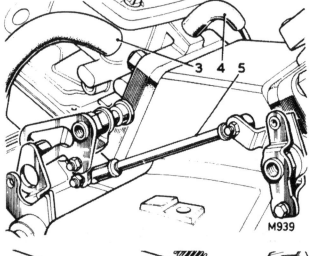

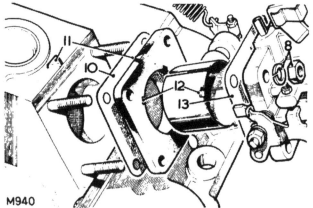

CARBURETTERS

·Overhaul   19.15.18

1.   Remove the carburetters 19.15.11.

Dismantling

Removing the piston assembly

2.   Remove the oil cap and damper.
3.   Remove the top cover and spring.
4.   Withdraw the air valve, shaft and diaphragm assembly.
5.   Remove the metering needle, retained by a locking screw.
6.   Remove the diaphragm from the air valve.

Removing the float chamber

7.   Remove and dismantle the jet assembly – **Early models** only.
     NOTE: On later models the jet is a one piece unit pressed into the carburetter body.
8.   Remove the float chamber and gasket.
9.   Unclip the float and arm complete with the spindle.
10. Remove the needle valve and washer from the carburetter body.

Dismantling the carburetter body

11. Add location marks to the throttle (butterfly) disc and spindle.
12. Remove the throttle butterfly, taking care not to damage the deceleration poppet valve on later models.
13. Left hand carburetter. Remove the throttle levers.
14. Withdraw the throttle spindle.
15. If required, remove the throttle stop and fast idle lever.
16. Remove the cold start assembly.
17. Dismantle the cold start assembly, but DO NOT remove the discs from the spindle.

Cleaning and inspection.

Carburetter cleaning.

18. When cleaning fuel passages do not use metal tools (files, scrapers, drills, etc) which could cause dimensional changes in the drillings or jets. Cleaning should be effected using clean fuel and where necessary a moisture-free air blast.

Joint Faces.

19. Examine the faces for deep scores which would lead to leakage taking place when assembled.

Continued

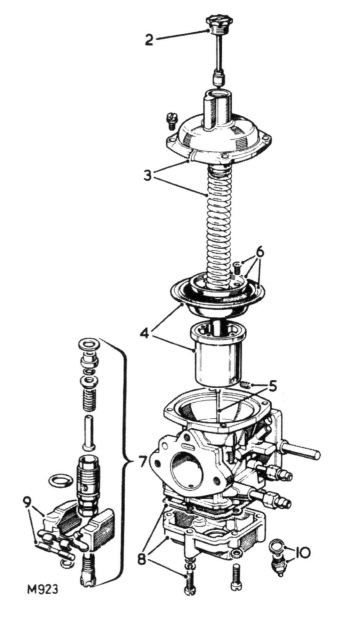

M923

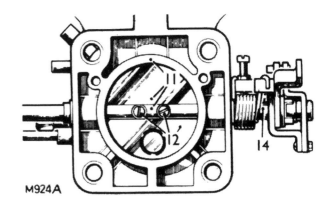

M924A

123

Joint gasket and seals

20. New gaskets and seals should be used throughout carburetter rebuild. A complete set of gaskets is available for replacement purposes.
21. Inspect metering needle, it is machined to very close limits and should be handled with care. Examine for wear, bend and twist, renew if necessary.

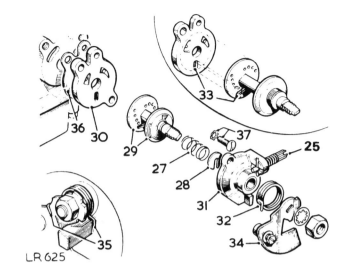

LR 625

Diaphragm

22. In common with other products made from rubber compounds, any contact of the diaphragm with volatile cleaners should be avoided, use only CLEAN RAG. Examine for damage and deterioration.
23. Examine float, for puncture or damage and chamber for corrosion, retaining clips for wear.
24. Examine cold start bushes for wear, renew starter cover as necessary.
25. Examine cold start adjuster for two positions, renew as necessary.
26. Examine lifting pin or air valve for correct operation.

Assembling

Assembling the cold start, L H Carburetter

27. Place the spring on the cold start spindle.
28. Fit the spring retaining clip.
29. Check that the discs slide easily on the spindle.
30. Place the cold start spindle on the starter face.
31. Place the starter cover in position.
32. Fit the return spring over the spindle.
33. Rotate the spindle until the oval port in the end disc is aligned with the oval port in the starter face.
34. Fit the cold start lever.
35. Engage the return spring over the lug on the starter cover and the back of the cold start lever.
36. Place the cold start gasket on to the carburetter body.
37. Fit the cold start assembly to the carburetter body, then check for ease of operation.

Continued

Assemble the throttle spindles

Left hand carburetter 38 and 39

38. Place the return spring over either end of the spindle.
39. Fit the throttle stop and fast idle lever.

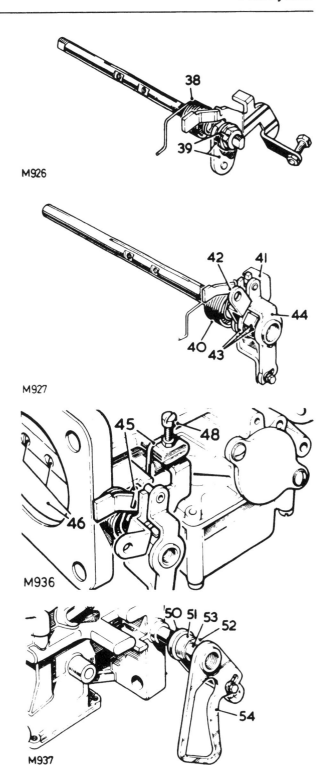

M926

Right hand carburetter 40 to 44

40. Place the return spring over the recessed end of the spindle.
41. Fit the throttle stop and fast idle lever.
42. Fit the throttle lever.
43. Secure the assembly with a bushed washer, tab washer and nut. Engage the tab washer.
44. Fit the throttle adjusting lever.
45. Insert the throttle spindle from the cold start side of the carburetter body (blank plate side on R H carburetter) fitting the throttle return spring on the fast idle adjustment holder, tension the spring half a turn.
46. Fit the throttle butterfly, maintaining the previously marked alignment. Leave the retaining screws loose.
47. Actuate the throttle several times to centralise the butterfly, then tighten the retaining screws and lock by peening ends.
48. Fit the throttle stop adjusting screw until it touches the stop, then turn a further one and a half turns and secure the locknut.

M927

M936

Left hand carburetters

49. Fit the fast idle adjustment screw and adjust to give slight clearance from the cold start lever, then secure the locknut.
50. Fit the throttle lever to the spindle.
51. Place the spacer on the spindle.
52. Place the tab washer on the spindle.
53. Fit the sleeve nut, sleeve end last, and engage the tab washer.
54. Fit the throttle adjusting lever.

Continued

M937

125

19-15

Assembling the float chamber

55. Fit the neddle valve and washer.
56. Locate the spindle into the float arm and engage the assembly in the retaining clips.
57. With the needle valve on its seating and the tab on the float carrier contacing the needle valve, measure the distance between the carburetter flange face and the highest point on the floats.
58. The dimension required for correct float lever is 17 to 18 mm (.67 to.71 in). Adjust by bending the tab on the float carrier or fitting on additional washer under the needle seating.

NOTE: The float carrier tube must be maintained at right angles to the needle in the closed position.

59. Fit the float chamber and gasket but do not fully tighten the screws at this stage.

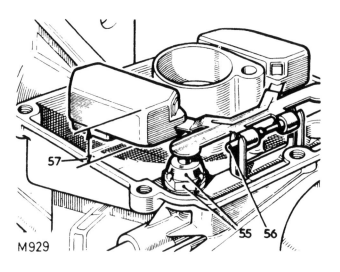

M929

Assembling the jet assembly – *Early models only*

60. Fit the 'O' ring into the guide bush.
61. Fit the 'O' ring over the jet orifice carrier.
62. Fit the 'O' ring over the adjusting screw.
63. Place the spring over the jet orifice.
64. Fit the guide bush (thin flanged) onto the jet orifice.
65. Fit the top bush to the jet orifice.
66. Place a plain washer onto the top bush.
67. Place the jet orifice assembly into the carrier. Insert the assembly through the float chamber and fully tighten, then tighten the float chamber screws.
68. Fit the adjusting screw and adjust the jet orifice until it is in line with the top of the bushing.

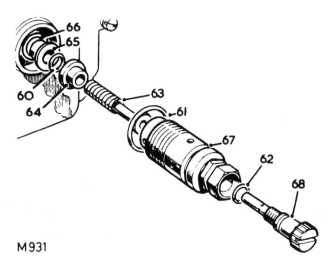

M931

Assembling air valve, shaft and diaphragm

69. Locate the diaphragm onto the air valve with the tab engaged in the recess.
70. Fit the diaphragm retaining ring.
71. Locate the metering needle into the air valve.
72. Align the needle shoulder with the top surface of the air valve shaft –Earlier models only.
73. Secure the needle in position.

Continued

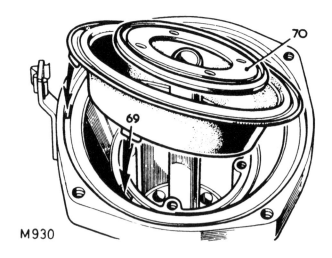

M930

Jet centralisation – Early models only

74. Locate the air valve and needle assembly into the carburetter and allow the air valve to bottom on the jet bridge, DO NOT push the valve down. If the valve does not bottom, unclamp the jet assembly sufficient to allow the valve to bottom. In this position, ensure that the locating tab on the diaphragm locates in the recess on the carburetter body.

75. Fit the air valve return spring and carburetter top cover.

76. Lift the air valve and tighten the jet assembly fully.

77. Slacken off the whole jet assembly approximately half a turn to release the orifice bush.

78. Allow the air valve to fall, if necessary assist by inserting a pencil in the dash pot. The needle will automatically centralise the jet orifice.

79. Slowly tighten the jet assembly, checking frequently that the needle remains free in the orifice. Check by raising the air valve approximately 6 mm (¼ in) and allowing it to fall freely. The air valve should stop firmly on the bridge.

80. Fill the dashpot in the air valve to within 6 mm (¼ in) of the top of the air valve shaft with S A E 20 engine oil.

81. Fit the damper assembly.

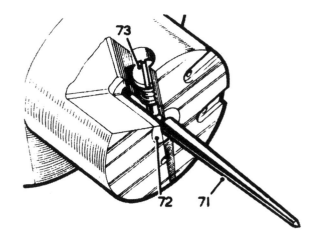

Fast idle adjustment – as applicable

NOTE: On some later carburetters a special tool is required for the adjustment.

82. Set the cold start adjustor fully outward.

83. Slacken the fast idle adjusting screw.

84. Hold the cold start cam lever in the maximum position.

85. Adjust the fast idle adjusting screw against the cam lever until there is 0,61 to 0,66 m m (.024 in to .026 in) gap between the top edge of the throttle butterfly and the carburetter barrel wall.

86. Use feeler gauges or a 0,65 m m diameter (no 72) drill to measure the gap at the top edge of the throttle butterfly.

87. Secure the locknut on the fast idle adjusting screw without disturbing the adjustment.

88. Refit the carburetters 19.15.11.

89. Tune and adjust the carburetters 19.15.02.

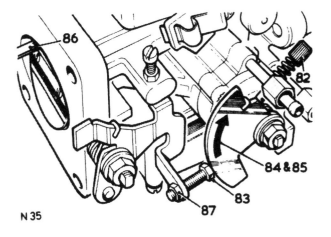

N 35

For data, refer to page 05–1.

FLOAT CHAMBER NEEDLE VALVE

-Remove and refit   19.15.24

Removing

1. Remove the carburetters. 19.15.11.   OR
2. Remove the six screws securing the float chamber to the body. (On later carburetters the drain plug illustrated is deleted).
3. Remove the float chamber.
4. Remove the gasket.
5. Remove the float assembly by gently prising the spindle from the locating clips.
6. Remove the needle valve and washer.

Refitting

7. Fit the needle valve and renew the washer.
8. Fit the float assembly.
9. Check and if necessary, adjust the height of both floats. Instruction 6. 19.15.32.
10. Renew the gasket and refit the float chamber.
11. Refit the carburetters.

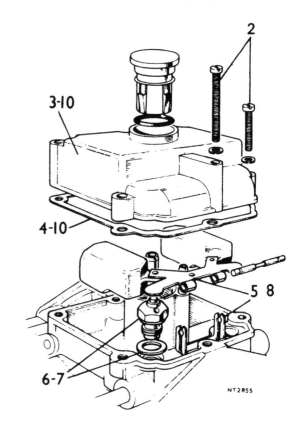

FLOAT CHAMBER LEVELS

-Check and adjust   19.15.32

Check

1. Remove the carburetters. 19.11.15.
2. Remove the six screws securing the float chamber to the body.
3. Remove the float chamber.
4. Remove the gasket.
5. With the carburetter in the inverted position check the distance between the gasket face on the carburetter body to the highest point of each float A.

NOTE. The height of both floats must be the same i.e. 0.625 to 0.627 in (16 to 17 mm).

Adjust

6. Bend the tab that contacts the needle valve but ensure that it sits at right angles to the valve to prevent the possibility of sticking.
7. Fit a new gasket and reverse instructions 1 to 3.

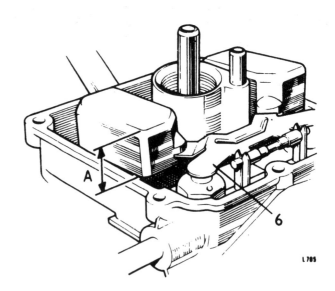

DIAPHRAGM

-Remove and refit   **19.15.35**

Removing

1. Remove the four screws securing the top cover to the carburetter body.
2. Lift off the top cover.
3. Remove the diaphragm spring.
4. Remove the diaphragm retaining plate.
5. Remove the diaphragm.

Refitting

6. Fit the diaphragm, locating the inner tag in the air valve recess.
7. Fit the retaining plate and ensure the correct diaphragm seating and tighten the screws.
8. Locate the diaphragm outer tag in the recess in the carburetter body.
9. Fit the top cover and evenly tighten the screw.
10. Check and if necessary top up damper — see special instructions, Section 10.

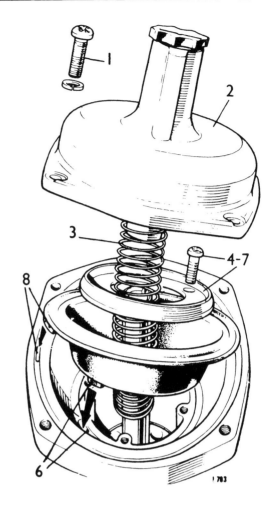

AUTOMATIC CHOKE (Australian)

-Remove and refit   **19.15.38**

CAUTION. The automatic choke (auto choke) must only be renewed as a complete unit.

Removing

1. Remove the air cleaner.
2. Remove the carburetter from the engine.
3. Open the throttle and prevent closure by inserting a suitable stop (plastic, rubber or soft wood) between the throttle bore and butterfly.
4. Remove the three retaining screws securing the auto choke to the carburetter, noting that the lower screw is shorter.
5. Lift off the auto choke and gasket.

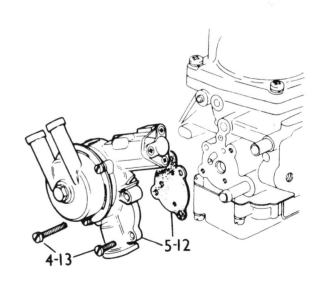

Continued

Refitting

6. Clean the carburetter and auto choke mating faces.
7. Remove the central bolt and washer on the water jacket and remove:
    a) The water jacket.
    b) The rubber sealing ring.
8. Remove the three screws and spring washers retaining the clamp ring.
9. Remove the clamp ring.
10. Carefully remove the finned aluminium heat mass, ensuring that the attached temperature sensitive bi-metal coil does not become strained.
11. Remove the heat insulator.
12. Fit a new gasket to the carburetter.
13. Fit the auto choke body to the carburetter tightening the retaining screws progressively and evenly to 40–45 in lb (46–52 cm kg). Note the lower screw is shorter.
14. Adjust the fast idle screw until the gap between the base circle of the cam and the fast idle pin is within 0.025 in (0,635 mm) dimension A.
15. Position the heat insulator ensuring that the bi-metal lever protrudes through the slot in the insulator.

NOTE. Provided the auto choke is in the fully 'ON' position, this insulator can only be fitted in one position; the back of it locating in the auto-choke body and the three holes aligned with the threaded holes in the starter body.

16. Carefully position the aluminium heat mass with the fins facing outwards and in such a way that the rectangular loop on the outer end of the temperature sensitive bi-metal coil fits over the bi-metal lever.
17. Without lifting the heat mass from its location (this is necessary to prevent the rectangular loop disengaging from the bi-metal lever) rotate it in both directions through a small angle and in both cases it should spring back to its static position. If not repeat instruction 16.
18. Fit the clamp ring locating it with the three screws and spring washers, but leaving the screws slack.
19. Rotate the heat-mass in an anti-clockwise direction to align the scribed line on its outermost edge with the datum mark on the insulator and auto-choke body. Whilst holding it in this position evenly tighten the clamp ring retaining screws to lock the heat mass in this position.

CAUTION: It should only be necessary to turn the heat mass through a small angle. Rotation in excess of 30° must be avoided to prevent permanent damage being caused to the temperature sensitive coil.

Continued

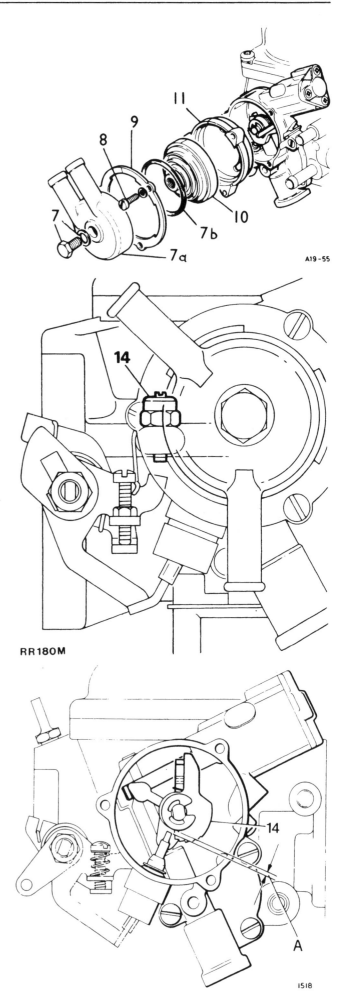

RR180M

20. Fit the sealing ring.
21. Fit the water-jacket leaving the centre retaining bolt slack until the carburetter is fitted to the engine so that the water jacket may be rotated to line-up with the water pipes.
22. Refit the carburetter to the engine.
23. Connect up the water pipes to the water jacket and tighten the centre retaining bolt.
24. While the engine is cold check and if necessary top-up the cooling system.
25. Operate the throttle before attempting to start the engine to enable the auto-choke to reset itself.
26. Fit the air cleaner.
27. Start the engine and run until normal operating temperature is reached.
28. Adjust the engine idling speed setting screw to give an idling speed of 800 r.p.m.
29. Check and if necessary top-up damper — see special instructions page 10—9.

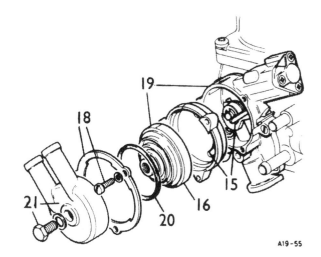

A19-55

THERMOSTAT SWITCH

-Remove and refit   19.15.50

Removing

1. Disconnect the battery earth lead.
2. Disconnect the lead from the switch.
3. Remove the thermostat switch.

Refitting

4. Using a new joint washer, reverse 1 to 3.

M771

TEMPERATURE COMPENSATOR

-Remove and refit   **19.15.59**

CAUTION. This component must only be renewed as a complete new unit.

Removing

1. Remove the air cleaner assembly. 19.10.01.
2. Remove the two screws and shake proof washers securing the temperature compensator to the carburetter.
3. Withdraw the compensator complete.
4. Remove and discard the outer rubber washer.
5. Remove the inner rubber washer from the carburetter body and discard.

Refitting

6. Clean the carburetter and temperature compensator mating faces.
7. Insert a new inner rubber washer into the bore in the carburetter body.
8. Fit a new outer rubber washer.
9. Fit the compensator to the carburetter and secure with the two screws and shakeproof washers.
10. Refit the air cleaner.

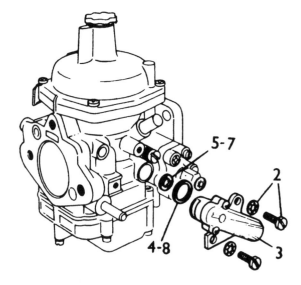

A19-65

DECELERATION (POPPET) VALVE

-Remove and refit   **19.15.64**

Removing

1. This component is an integral part of the throttle (butterfly) disc assembly and is not adjustable.
2. If the deceleration valve does not function correctly renew the complete throttle disc assembly. See 19.20.04.

THROTTLE PEDAL

-Remove and refit   19.20.01

Removing

1. Remove the lower facia panel 76.46.03.
2. Remove the clevis pin from the throttle pedal.
3. Release the tension from the pedal return spring.
4. Remove the circlip from the pedal pivot pin.
5. Withdraw the pivot pin.
6. Withdraw the throttle pedal.

Refitting

7. Reverse 1 to 6.

THROTTLE (BUTTERFLY) DISC

-Remove and refit   19.20.04

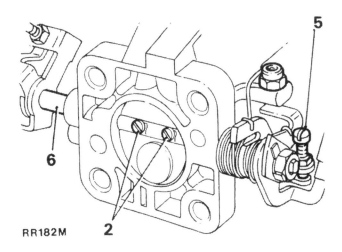

Removing

1. Remove the carburetters 19.15.11.
2. Remove the two screws retaining the throttle (butterfly) disc and remove it from the throttle spindle.

Refitting

3. Fit a new throttle (butterfly) disc.
4. Fit the two retaining screws loosely.
5. Unscrew the throttle stop to the fully closed position.
6. Operate the throttle spindle to centre the disc assembly and tighten the retaining screws.
7. Using the screwdriver spread the split-ended screws. Refit the carburetters 19.15.11.

NOTE: A deceleration (poppet) valve is fitted in the throttle disc on all but early carburetters.

THROTTLE CABLE

Remove and refit   19.20.06

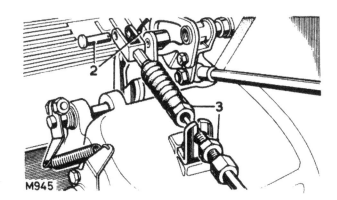

Removing

1. Remove the air cleaner 19.10.01.
2. Disconnect the cable from the carburetter.
3. Release the cable from the adjustment bracket.
4. Remove the lower facia panel 76.46.03.
5. Remove the clevis pin from the throttle pedal.
6. Release the cable from the dash bracket.
7. Withdraw the cable complete.

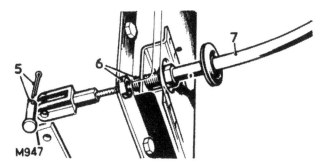

Refittings

8. Reverse 2 to 7.
9. Adjust the cable with the throttle pedal in the released position. The throttle linkage must not have commenced movement, but commences with the minimum depression of the pedal.
10. Fit the air cleaner 19.10.01.

THROTTLE LINKAGE

-Remove and refit   19.20.07

Removing

1.. Disconnect the throttle cable.
2. Remove the throttle adjusting lever.
3. Disconnect the throttle return spring.
4. Remove the return spring lever.
5. Remove the circlip and plain washer from the countershaft.
6. Withdraw the countershaft assembly.
7. Remove the throttle link from between the carburetters.

NOTE: For details of carburetter linkage 19.15.18 refers.

Refitting

8. If the throttle link ball joints have been disturbed, they should be set at 122,55 mm (4.825 in) centres or 77.683 mm (3.4521 in) on later units.
9. Fit the throttle link.
10. Fit the countershaft assembly.
11. Secure the countershaft with a plain washer and circlip.
12. Fit the lever for the throttle return spring to abut the circlip.
13. Position the lever and countershaft assembly so that the holes for the throttle cable and return spring connection are at $172\frac{1}{2}^\circ + \frac{1}{2}^\circ$.
14. Reverse 1 to 3.
15. Adjust the carburetter linkage 19.15.02.

CHOKE CONTROL CABLE

-Remove and refit   19.20.13

Removing

1. Disconnect the choke cable from the carburetter.
2. Release the cable from the clips in the engine compartment.
3. Disconnect the leads from the choke control switch.
4. Release the nut securing the outer cable.
5. Withdraw the choke control knob and cable complete.

Refitting

6. Reverse 1 to 5, ensuring that the nut and spring washer are fed over the cable before passing it through the dash panel.
7. Reconnect the choke cable with approximately 1,5m m (.031 in) free movement.
8. Run the engine and check that the choke control operates correctly.

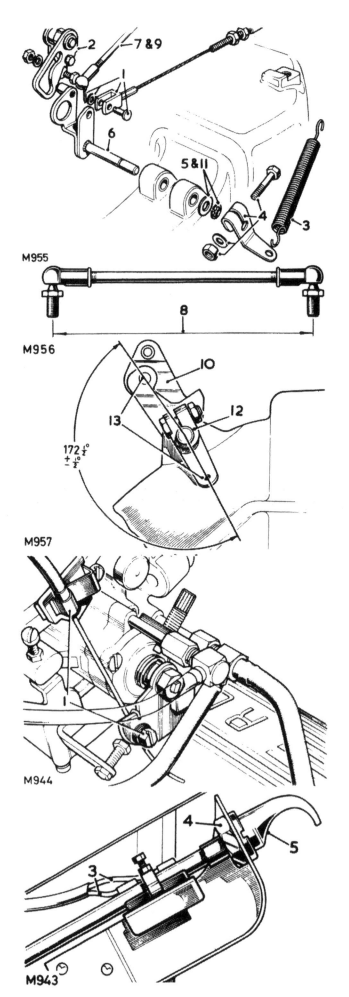

M955

M956

M957

$172\frac{1}{2}^\circ \pm \frac{1}{2}^\circ$

M944

M943

FUEL LINE FILTER (early models)

-Remove and refit   19.25.01

Removing

1. Disconnect the fuel pipes.
2. Slacken the clip securing the filter.
3. Withdraw the filter.

Refitting

4. Fit the filter with the end marked In downwards. Alternatively, if the filter is marked with arrows, they must point upwards.
5. Connect the fuel pipes.
6. Secure the filter clip.

FUEL MAIN FILTER

-Remove and refit   19.25.02

Removing

1. Disconnect the fuel pipes.
2. Take precautions against fuel leaking from the tank.
3. Remove the filter complete.

Refitting

4. Reverse 1 and 3.

FUEL MAIN FILTER ELEMENT

-Remove and refit   19.25.07

Removing

1. Unscrew the centre bolt.
2. Withdraw the filter bowl.
3. Remove the small sealing ring.
4. Remove the element.
5. Withdraw the large sealing ring from the underside of the filter body.

Refitting

6. Fit new centre sealing rings as necessary.
7. Fit new top sealing ring.
8. Fit new element small hole downward.
9. Fit new small sealing ring.
10. Reverse 1 and 2.

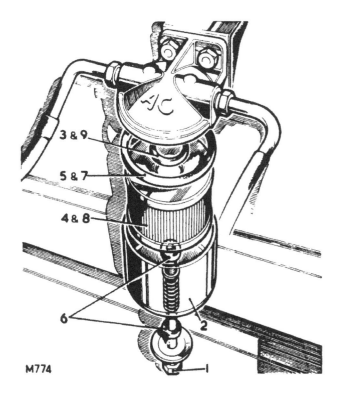

FUEL PIPES

Remove and refit

Main line, tank end section	19.40.02
Main line, centre section	19.40.03
Main line, carburetter end section	19.40.04
Spill return	19.40.08

Removing

1. Take precautions against fuel spillage. If removing the main line tank end section or the spill return pipe-drain the fuel tank.
2. Disconnect the pipe unions.
3. Release the pipe from the retaining clips.
4. Withdraw the fuel pipe.

Refitting

5. Reverse 1 to 4.

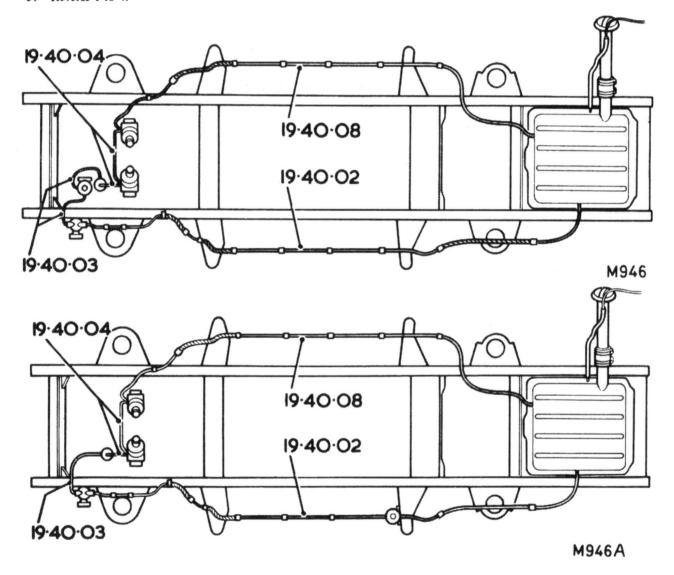

M946

M946A

FUEL PUMP – MECHANICAL

Early models

– Remove and refit   19.45.08

Removing

1. Disconnect the fuel pipes from the pump.
2. Take precautions against fuel leaking from the tank.
3. Remove the fuel pump.

Refitting

4. Using a new joint washer, reverse 1 and 3.
5. Run the engine and check that the pump operates and that there are no leaks.

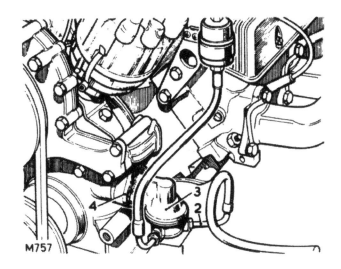

FUEL PUMP – ELECTRICAL

Later models

– Remove and refit   19.45.08

Removing

1. Remove the pump cover (where fitted).
2. Remove the fuel pipes from the pump plugging them to prevent fuel loss.
3. Disconnect the electrical leads.
4. Remove the nuts and washers holding the fuel pump bracket.
5. Remove the pump complete with bracket from the vehicle.
6. Remove the bracket from the pump noting the position of the earth braid for refitting.

Refitting

7. Reverse instructions 1 to 6. Ensure that a shakeproof washer is fitted each side of the earth strap on the vehicle fixing.

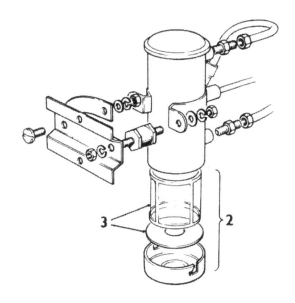

FUEL PUMP – ELECTRICAL

Later models

– Overhaul   19.45.15

1. Remove the pump. 19.45.08.
2. Release the end cap and withdraw the gasket and filter.
3. Renew the gasket and filter as necessary.
4. Reverse instructions 1 and 2.

NOTE: The electrical components of the pump are sealed and are not repairable.

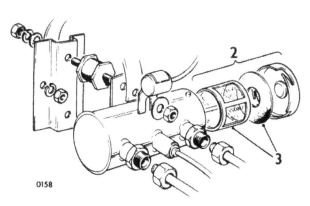

FUEL PUMP – MECHANICAL

Early models

-Overhaul 19.45.15

1. Remove the fuel pump 19.45.08.

Dismantling

2. Add alignment marks between the pump body and the fuel cover.
3. Remove the fuel cover.
4. Remove the pulsator cover.
5. Withdraw the pulsator diaphragm.
6. Scrape out the burs produced by staking the valves and remove both valves and gaskets.
7. Remove the return spring from the rocker arm.
8. Hold the link depressed.
9. Depress the diaphragm and tilt the lower end away from the link to unhook the pull rod.
10. Withdraw the diaphragm.
11. Withdraw the spring.
12. Withdraw the oil seal.
13. Drive out the rocker arm pivot pin.
14. Withdraw the rocker arm and link.

Inspecting

15. Clean all parts to be re-used in solvent and blow out all passages with an air line.
16. Inspect the pump body and the fuel cover for cracks, breakage or distorted flanges. Examine screw holes for stripped or cross threads.
17. Inspect the rocker arm for wear at the pad end and at the point of contact with the link. Check for excessive rocker arm side play due to wear on pivot pin.
18. If a damaged casing or loose rocker pivot pin are found, fit a new fuel pump complete.

Continued

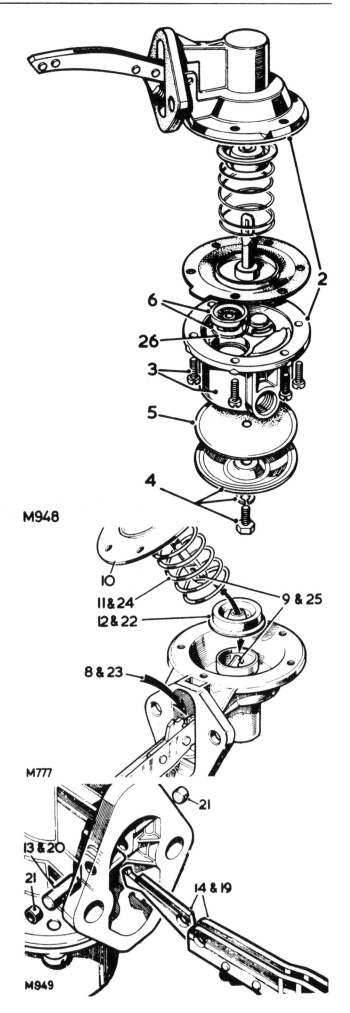

Assembling,

Fitting the rocker arm 19 to 21

19. Locate the link and rocker arm into the pump body.
20. Fit the rocker pivot pin.
21. Smear jointing compound on each end of the pin and fit the retaining caps.
22. Fit a new oil seal to the pump body.
23. Hold the link depressed.
24. Place the diaphragm spring in position.
25. Locate the diaphragm pull rod through the spring and oil seal and engage it over the hook on the link.
26. Place a gasket in each valve seat in the fuel cover.
27. Place a valve in the seat nearest the inlet port (marked with an arrow) with the spring cage facing up.
28. Place a valve in the outlet seat with the spring cage down.
29. Seat valves firmly against gaskets and secure by staking the cover in four places around the edge of each valve.
30. Fit a new pulsator diaphragm.
31. Fit the pulsator cover.
32. Fit the fuel cover, maintaining the original alignment, but do not fully tighten the screws.
33. Hold the pump rocker arm fully depressed while tightening the fuel cover screws alternately and evenly.
34. Fit the rocker arm return spring.
35. Refit the fuel pump 19.45.08.

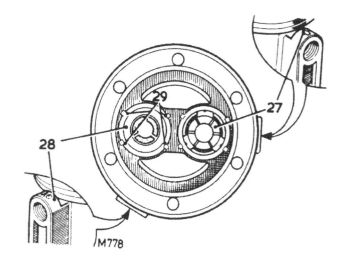

DATA

Fuel pump

TypeA C-mechanical
Delivery pressure 0,25 to 0,35 Kgf/cm2
(3.5 to 5.0 lbf/in^2)

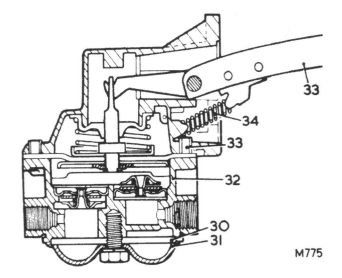

M775

19—29

FUEL TANK

-Remove and refit   19.55.01

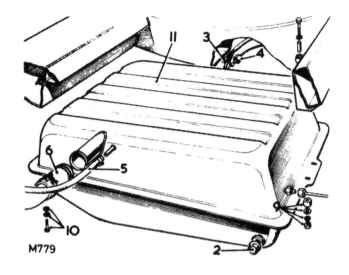

Removing

1. Disconnect the battery earth lead.
2. Drain the fuel tank.
3. Disconnect the electrical leads from the tank gauge unit.
4. Disconnect the outlet pipe.
5. Disconnect the breather pipe.
6. Disconnect the filler neck.
7. Disconnect the spill return pipe.
8. Support the tank.
9. Remove the front fixings.
10. Remove the rear fixings.
11. Withdraw the fuel tank.

Refitting

12. Reverse 1 to 11.
13. Replenish the fuel tank and check for leaks.

M779

COOLING SYSTEM OPERATIONS

SYMPTOM	POSSIBLE CAUSE	CURE
A–EXTERNAL LEAKAGE	1. Loose hose clips 2. Defective rubber hose 3. Damaged radiator seams 4. Excessive wear in the water pump 5. Loose core plugs 6. Damaged gaskets 7. Leaks at the heater connections or plugs 8. Leak at the water temperature gauge plug	1. *Tighten* 2. *Renew* 3. *Rectify* 4. *Renew* 5. *Renew* 6. *Renew* 7. *Rectify* 8. *Tighten*
B–INTERNAL LEAKAGE	1. Defective cylinder head gasket 2. Cracked cylinder wall 3. Loose cylinder head bolts	1. *Renew. Check engine oil for contamination and refill as necessary* 2. *Renew cylinder block* 3. *Tighten. Check engine for oil contamination and refill as necessary*
C–WATER LOSS	1. Boiling 2. Internal or external leakage 3. Restricted radiator or inoperative thermostat	1. *Ascertain the cause of engine overheating and correct as necessary* 2. *See Items A and B* 3. *Flush radiator or renew the thermostat as necessary*
D–POOR CIRCULATION	1. Restriction in system 2. Insufficient coolant 3. Inoperative water pump 4. Loose fan belt 5. Inoperative thermostat	1. *Check hoses for crimps, reverse flush the radiator, and clear the system of rust and sludge* 2. *Replenish* 3. *Renew* 4. *Adjust* 5. *Renew*
E–CORROSION	1. Excessive impurity in the water 2. Infrequent flushing and draining of system 3. Incorrect anti-freeze mixtures	1. *Use only soft, clean water together with correct anti-freeze or inhibitor mixture.* 2. *The cooling system should be drained and flushed thoroughly at least once a year.* 3. *Certain anti-freeze solutions have a corrosive effect on parts of the cooling system. Only recommended solutions should be used.*
F–OVERHEATING	1. Poor circulation 2. Dirty oil and sludge in engine 3. Radiator fins choked with chaff, mud, etc. 4. Incorrect ignition timing 5. Insufficient coolant. 6. Low oil level 7. Tight engine 8. Choked or damaged exhaust pipe or silencer 9. Dragging brakes 10. Overloading vehicle 11. Driving in heavy sand or mud 12. Engine labouring on gradients 13. Low gear work 14. Excessive engine idling 15. Inaccurate temperature gauge 16. Defective thermostat	1. *See Item D* 2. *Refill* 3. *Use air pressure from the engine side of the radiator and clean out passages thoroughly* 4. *See Group 86* 5. *See item D.* 6. *Replenish* 7. *New engines are very tight during the 'running-in' period and moderate speeds should be maintained for the first 1,000 miles (1.500 km)* 8. *Rectify or renew* 9. *See Group 70 – adjust brakes.* 10. *In the hands of the operator* 11. *In the hands of the operator* 12. *In the hands of the operator* 13. *In the hands of the operator* 14. *In the hands of the operator* 15. *Renew* 16. *Renew*
G–OVERCOOLING	1. Defective thermostat 2. Inaccurate temperature gauge	1. *Renew* 2. *Renew*

COOLANT

– Drain and refill   26.10.01

Draining

WARNING: Do not remove the expansion tank filler cap when the engine is hot because the cooling system is pressurised and personal scalding could result.

1. Remove the expansion tank filler cap by first turning it anti-clockwise a quarter of a turn to allow pressure to escape, then turn it further in the same direction and lift off.
2. Remove the plug and drain the radiator, or disconnect the bottom radiator hose. As the system is filled with a solution of anti-freeze or inhibitor, use a clean container if the coolant is to be reused.
3. Refit the drain plug and washer, or reconnect the bottom radiator hose.
4. Remove the drain plugs, one each side of the cylinder block, (located between the exhaust pipes and the cylinder block), and drain the engine.
5. Replace both drain plugs.
   Note: Early engines – fitted with drain taps.

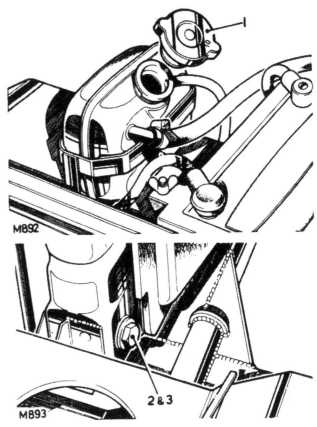

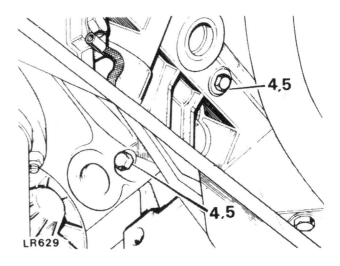

Coolant requirements

Frost precautions and engine protection

The engine cooling system MUST ALWAYS be filled and topped up with a solution of water and anti-freeze, winter and summer, or, where frost precautions are not required, water and inhibitor. NEVER use water alone as this may corrode the aluminium alloy.

Recommended solutions

Anti-freeze:–
Unipart Universal
Inhibitor:–
Marston Lubricants SQ35 Coolant Inhibitor Concentrate.
Use soft water wherever possible. If the local water supply is hard, use rain water.
Anti-freeze will provide adequate protection for two years provided that the specific gravity of the coolant is checked before the onset of the second winter and topped up with new anti-freeze as required.
After the second winter the system should be drained and thoroughly flushed. Before adding new anti-freeze examine all joints and renew defective hoses to make sure that the system is leakproof.

Use the correct anti-freeze mixture according to local climatic conditions, as follows:

Cooling system capacity	Frost precaution		Proportion of anti-freeze
11,0 litres			
20 Imperial pints	−36 degrees C (−33 degrees F)	50%	6 litres 10 Imperial pints 12 US pints
24 US pints			

M895

If frost precautions are not required, use a 10% solution of inhibitor i.e. 1 part inhibitor to 9 parts water.

Refilling

6. Remove the radiator filler plug.
7. Pour 4½ litres (1 gallon) of water into the radiator.
8. Add the recommended quantity of anti-freeze or inhibitor.
9. Top up radiator with water.
10. Fit the radiator filler plug and washer.
11. Add water to the expansion tank, up to the 'WATER LEVEL' plate.
12. Fit the expansion tank filler cap.
13. Run the engine until normal operating temperature is attained, that is, thermostat open.
14. Allow the engine to cool, then check the coolant level and top up if necessary.

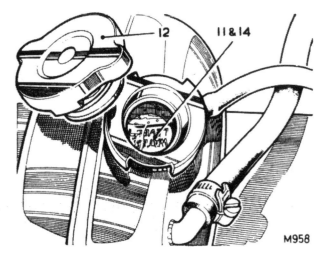

M958

EXPANSION TANK

—Remove and refit   26.15.01

Removing

WARNING. Do not remove the expansion tank filler cap when the engine is hot because the cooling system is pressurised and personal scalding could result.

1. Remove the expansion tank filler cap by first turning it anti-clockwise a quarter of a turn to allow pressure to escape, then turn it further in the same direction and lift off.
2. Disconnect the hose to the radiator
3. Disconnect the overflow pipe.
4. Remove the pinch bolt.
5. Lift out the expansion tank.

Refitting

6. Reverse 2 to 5.
7. Replenish the cooling system. 26.10.01.

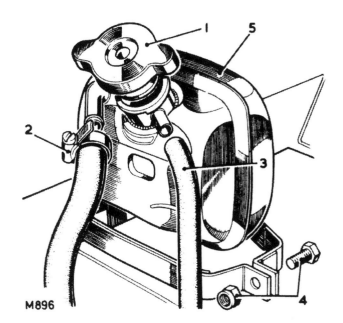

M896

FAN BELT

—Check and adjust tension, 1 and 5 to 6   26.20.01

—Remove and refit, 1 to 6   26.20.07

Removing

1. Slacken the alternator fixings.
2. Pivot the alternator inwards.
3. Lift off the fan belt.

Refitting

4. Locate the fan belt on the pulleys.
5. Using the alternator slotted fixing, adjust the fan belt tension to give 11 to 14 mm (.437 to .562 in.) free movement when checked midway between the alternator and crankshaft pulleys, by hand.
6. Secure the alternator fixings.

   Note: Where air conditioning is fitted the alternator is mounted on the left hand side of the engine.

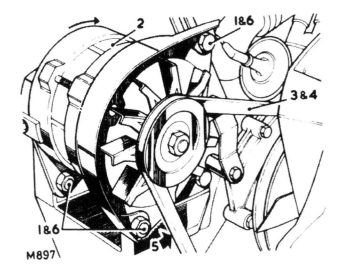

M897

FAN BLADES AND PULLEY

| –Remove and refit, 1 to 6 | 26.25.01 |

FAN BLADES

| –Remove and refit, 2 and 5 | 26.25.06 |

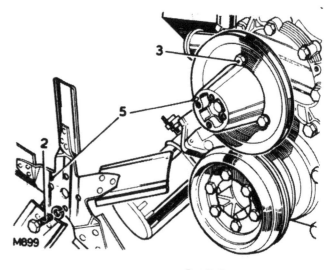

Removing

1. Remove the fan belt. 26.20.07.
2. Remove the fan blades.
3. Standard type pulley. Remove the pulley fixings.
4. Viscous coupling type pulley. Remove the coupling fixings.
5. Slacken the alternator fixings, remove the fan belt and lift off the viscous coupling and/or the fan pulley.

Refitting.

6. Reverse 1 to 5, noting the following.
7. Standard type pulley. An off-set dowel location ensures that the fixing bolt holes only align when the blades are the correct way round.
8. Viscous coupling type pulley. Fit the fan blades with the larger diameter fixing bosses to the front.
9. Adjust the fan belt. 26.20.01.

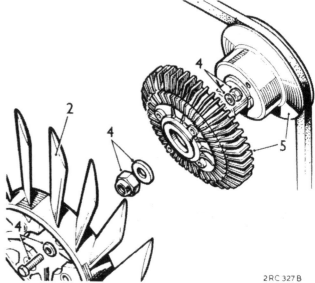

2RC 327 B

FAN COWL

| –Remove and refit | 26.25.11 |

Removing

1. Remove the fan blades. 26.25.06.
2. Remove the fixings from the top of the fan cowl.
3. From inside the engine compartment remove the the lower cowl fixings.
4. Lift out the fan cowl.

Refitting

5. Reverse 1 to 4. Note that the bottom of the fan cowl locates into support brackets, and that the fan blades have an off-set dowel location which ensures that the fixing bolt holes only align when the blades are the correct way round.

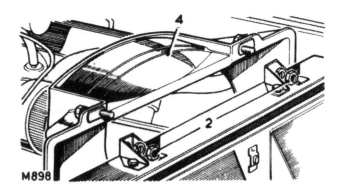

RADIATOR BLOCK

—Remove and refit   26.40.04

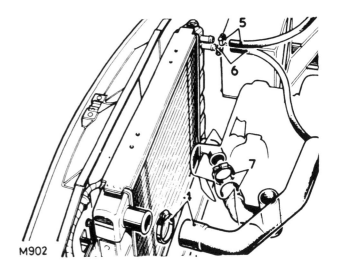

M902

Removing

1. Drain the cooling system. 26.10.01.
2. Remove the fan blades. 26.25.06.
3. Remove the fan cowl. 26.25.11.
4. Disconnect the top hose from the radiator.
5. Disconnect the hose to the expansion tank.
6. Disconnect the hose to the induction manifold.
7. Disconnect the hose from the bottom of the radiator.
8. Remove the fixings from each side of the radiator.
9. Withdraw the radiator from the rubber-mounted spigots.

Refitting

10. Reverse 1 to 9 noting the assembly order of the radiator side fixings and ensuring that the radiator sealing strips (later models) are correctly located and secure.

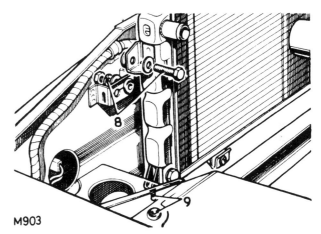

M903

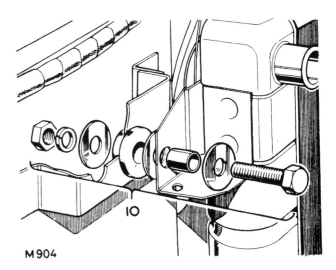

M904

THERMOSTAT

–Remove and refit.   26.45.01.

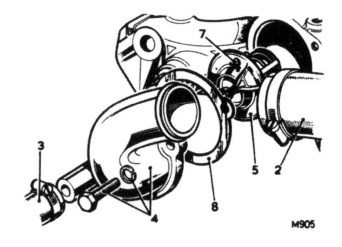

M905

Removing

1. Drain the cooling system. 26.10.01, sufficient to drain the induction manifold.
2. Disconnect the hose to the radiator.
3. Disconnect the thermostat by-pass hose.
4. Remove the outlet elbow.
5. Withdraw the thermostat.

Testing

6. When immersed in hot water, the thermostat should commence opening between 78 to 83 degrees centigrade (173 to 182 degrees farenheit).

Refitting

7. Insert the thermostat with the jiggle pin uppermost (12 o'clock).
8. Using a new joint washer, fit the outlet elbow.
9. Reverse 1 to 3.

26–8

WATER PUMP

– Remove and refit   26.50.01

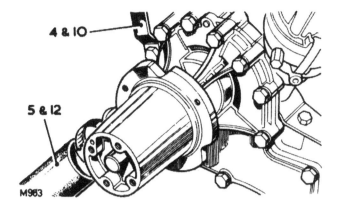

Removing

1. Drain the engine cooling system. 26.10.01.
2. Remove the fan belt. 26.20.07.
3. Remove the fan blades and pulley. 26.25.01.
4. Release the alternator adjusting link from the water pump and where applicable, the power steering pump bracket.
5. Disconnect the inlet hose from the water pump.
6. Remove the water pump.

Refitting

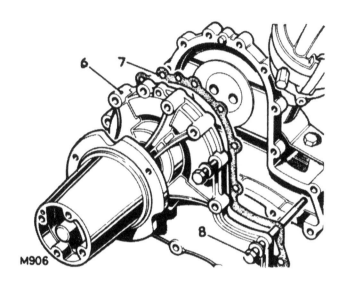

7. Lightly grease a new joint washer and place it in position on the timing cover.
8. Clean the threads of the four long bolts and smear them with 3M-EC776 thread lubricant-sealant, Rover Part No. 605674.
9. Locate the water pump in position.
10. Locate the alternator adjusting link on the water pump.
11. Leave the alternator adjusting link and power steering pump bracket loose and tighten the remaining water pump fixings gradually.
    Torque:– 7/16 in. AF bolts 0,8 to 1,0 kgf.m (6 to 8 lbf.ft).
    1/2 in. AF bolts 2,8 to 3,5 kgf.m (20 to 25 lbf.ft).
12. Connect the inlet hose to the water pump.
13. Fit the fan pulley.
14. Fit and adjust the fan belt 26.20.07 and where applicable, power steering pump belt 57.20.01.
15. Fit the fan blades. 26.25.06.
16. Refill the cooling system. 26.10.01.

WATER PUMP

—Overhaul   26.50.06

1. Remove the water pump. 26.50.01.

Dismantling

2. Using a bearing extractor, remove the pulley hub.
3. Place the water pump assembly in very hot water for a short period.
4. Press out the spindle and impellor assembly and discard.

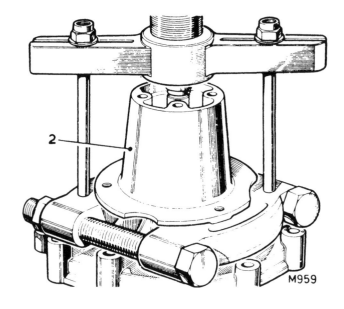

M959

Assembling

5. Place the water pump housing in very hot water for a short period.
6. Press the new spindle assembly, short end first, into the front end of the water pump housing until the spindle protrudes 13,2 mm plus or minus 0,25 mm (.520 in, plus or minus .010 in.) beyond the rear face.

    NOTE. The press load must only be applied to the bearing outer race and the press tool must not contact the housing.

7. Check that the flanged bush is in place on the spindle.
8. Smear the outside diameter of the seal with sealer stag 'A' or Pettmans cement.
9. Smear the seal face with Shell 'Donax C'.
10. Fit the seal ensuring that the flange is seated on the pump housing.
11. Fit the seal mating face.
12. Press the impellor on to the spindle until the rear face of the impellor is 13,2 mm plus or minus 0,25 mm (.520 in. plus or minus .010 in.) from the rear face of the pump housing.
13. Press the pulley hub on to the spindle until the front face of the pulley flange is 66,67 mm plus or minus 0,25 mm (2.625 in. plus or minus .010 in. from the rear face of the pump housing. Ensure that the impellor and spindle are not disturbed.
14. Refit the water pump. 26.50.01.

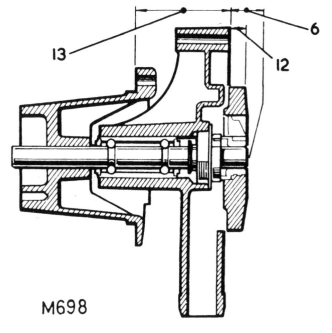

M698

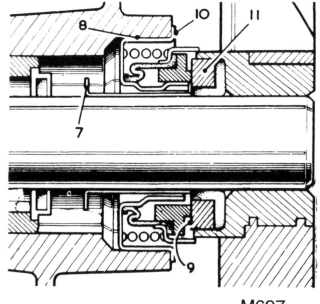

M697

MANIFOLDS AND EXHAUST SYSTEM OPERATIONS

Operation No.

151

30-1

Exhaust manifold
-left hand - remove and refit   30.15.10
-right hand - remove and refit   30.15.11

Exhaust system
-front pipe, left hand - remove and refit   30.10.09
-front pipe, right hand - remove and refit   30.10.10
-silencer - remove and refit   30.10.14
-intermediate pipe - remove and refit   30.10.18
-remove and refit   30.10.01
-tail pipe - remove and refit   30.10.22

Induction manifold - remove and refit   30.15.02

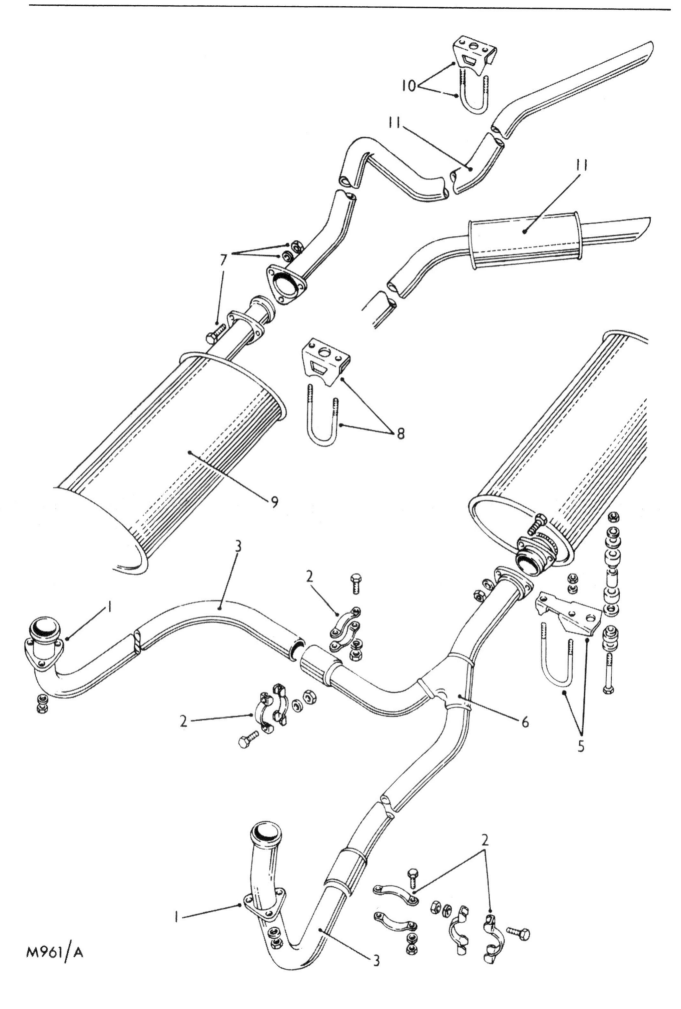

M961/A

EXHAUST SYSTEM COMPLETE

-Remove and refit, 1 to 12	**30.10.01**
Front pipe, left hand 1 to 3 and 13	**30.10.09**
Front pipe, right hand 1 to 3 and 13	**30.10.10**
Silencer 4, 7 to 9 and 14	**30.10.14**
Intermediate pipe 1 to 6 and 15	**30.10.18**
Tail pipe 7, 10, 11 and 16	**30.10.22**

Removing

1. Disconnect the front pipe(s) from the manifold(s).
2. Slacken the clamp between the front and intermediate pipes
3. Withdraw the front pipe
4. Disconnect the intermediate pipe from the silencer
5. Remove the U-bolt from the pipe mounting bracket
6. Withdraw the intermediate pipe
7. Disconnect the silencer from the tail pipe
8. Remove the U-bolt from the pipe mounting bracket
9. Withdraw the silencer
10. Remove the U-bolt from the tail pipe mounting bracket
11. Withdraw the tail pipe

Refitting

12. Complete system, reverse 1 to 11
13. Front pipe, reverse 1 to 3
14. Silencer, reverse 4 and 7 to 9
15. Intermediate pipe, reverse 1 to 6
16. Tail pipe, reverse 7, 10 and 11

NOTE. According to territory specification, some vehicles may be fitted with a twin exhaust tailpipe unit.

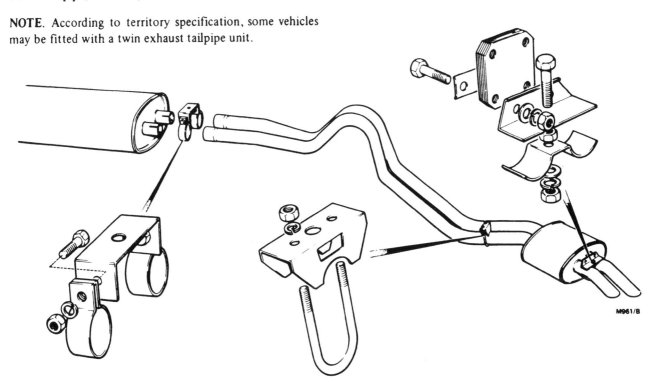

M961/B

INDUCTION MANIFOLD

-Remove and refit   30.15.02

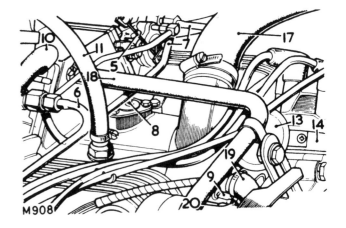

Removing

1. Drain the cooling system 26.10.01
2. Remove the air cleaner 19.10.01
3. Remove the engine breather filter 17.10.02
4. Disconnect the throttle cable from the carburetter and manifold
5. Disconnect the choke cable from the carburetter
6. Disconnect the fuel spill return pipe from the RH carburetter
7. Remove the fuel supply pipe from the carburetters
8. Disconnect the lead from the choke thermostat switch and (where fitted) the E.G.R. valve connections.
9. Disconnect the lead from the water temperature transmitter
10. Disconnect the flame trap hoses from the carburetters
11. Disconnect the vacuum pipe for the brake servo
12. Disconnect the vacuum pipe for the gearbox
13. Disconnect the vacuum pipe from the distributor
14. Release the distributor cap
15. Disconnect the inlet hose to the heater
16. Disconnect the return hose from the heater
17. Disconnect the return hose to the radiator
18. Disconnect the return hose from the top of the induction manifold
19. Disconnect the by-pass hose from the thermostat
20. Disconnect the heater return pipe from the manifold
21. Remove the induction manifold
22. Wipe away any coolant lying on the manifold gasket
23. Remove the gasket clamps
24. Lift off the gasket
25. Withdraw the gasket seals

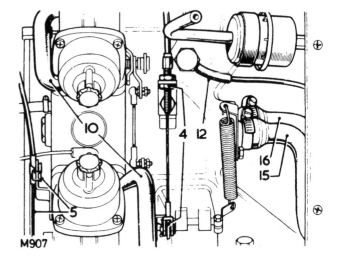

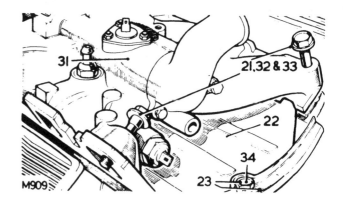

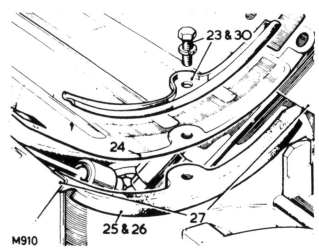

Refitting

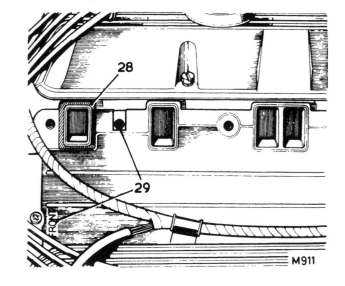

26. Using new seals, smear them on both sides with silicon grease

27. Locate the seals in position with their ends engaged in the notches formed between the cylinder head and block

28. Apply Universal jointing compound on the corners of the cylinder head, manifold gasket and manifold, around the water passage joints.

29. Fit the manifold gasket with the word 'FRONT' to the front and the open bolt hole at the front RH side

30. Fit the gasket clamps but do not fully tighten the bolts at this stage

31. Locate the manifold onto the cylinder head

32. Clean the threads of the manifold securing bolts and then coat them with Thread Lubricant-sealer 3M EC776.

33. Fit all the manifold bolts and tighten them a little at a time, evenly, alternate sides working from the centre to each end. Torque 3,5 to 4,0 kgf.m (25 to 30 lb.ft).

34. Tighten the gasket clamp bolts. Torque 1,4 to 2,0 kgf.m (10 to 15 lb.ft).

35. Reverse 1 to 20

36. Where fitted – reconnect the E.G.R. valve.

37. Run the engine and check for water leaks

EXHAUST MANIFOLD

-Remove and refit

Left hand	30.15.10
Right hand	30.15.11

Removing

1. Disconnect the front exhaust pipe from the manifold and (where fitted) remove the hot air box 17.30.31.
2. Tap back the bolt locking tabs and remove eight bolts with lock taps and washers (later models).
3. Remove and manifold.

Refitting

4. Ensure that the mating surfaces of the cylinder head and exhaust manifold are clean and smooth.
5. Coat the exhaust manifold (cylinder head mating faces) with 'Foliac J 166' or 'Moly Paul' anti-seize compound.
   'Foliac J 166' is manufactured by Rocol Ltd., Rocol House, Swillington, Leeds, England.
   'Moly Paul' is manufactured by K.S. Paul Products Ltd., Nobel Road, London N18.
6. Place the manifold in position on the cylinder head and fit the securing bolts, lockplates and plain washers. The plain washers are fitted between the manifold and lockplates.
7. Evenly tighten the manifold bolts to 2,0 kgf.m (15 lbf.ft).
8. Bend over the lockplate tabs.

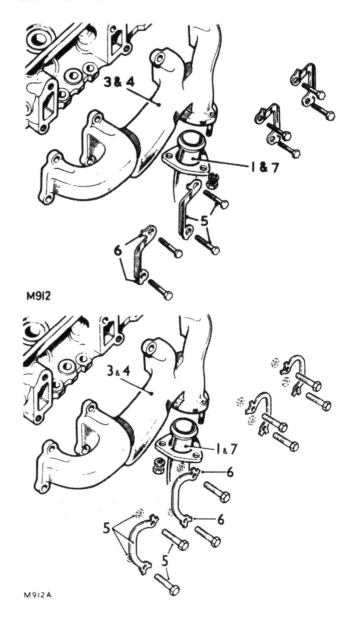

M9I2

M9I2A

CLUTCH OPERATIONS

Description	Operation No.
Clutch assembly	
– remove and refit	33.10.01
– overhaul	33.10.08
Clutch pedal-remove and refit	33.30.02
Hydraulic system-bleed	33.15.01
Master cylinder	
– remove and refit	33.20.01
– overhaul	33.20.07
Release bearing assembly	
– remove and refit	33.25.12
Slave cylinder	
– remove and refit	33.35.01
– overhaul	33.35.07

CLUTCH ASSEMBLY

– Remove and refit   33.10.01

Service tool: 18G79 Clutch centralising tool

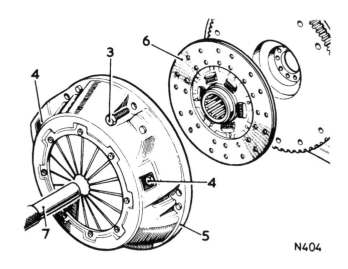

N404

Removing.

1. Remove the engine. 12.41.01.
2. Mark the clutch cover fitted position relative to the flywheel.
3. Where clutch cover fixing bolts heads vary obviously in thickness, note their fitted positions for reference during refitting.
4. Do not disturb the three bolts located in the apertures in the clutch cover.
5. Remove the clutch assembly.
6. Withdraw the clutch driven plate.

Refitting

7. Reverse 5 and 6, aligning the assembly marks. Centralising tool 18G79.
8. Secure the cover fixings evenly, using diagonal selection. Torque load 4,9 to 5,0 kgf.m (35 to 38 lbf.ft).
9. Fit the engine. 12.41.01.
    NOTE. As a precaution against the clutch plate sticking, lubricate the splines using Rocol MV 3 or Rocol MTS 1000 grease.

CLUTCH ASSEMBLY

– Overhaul   33.10.08

Clutch assembly.

The clutch assembly is of the diaphragm spring type and no overhaul procedures are applicable. Repair is by replacement only.

Clutch driven plate

Examine clutch driven plate for wear and signs of oil contamination. Examine all rivets for pulling and distortion, rivets must be below the friction surface. If oil contamination is present on the friction linings or if they are appreciably worn, renew the clutch driven plate assembly complete or alternatively, renew the friction linings following standard workshop practices.

DATA.

Clutch driven plate diameter   266,7 mm.(10.5 in)

HYDRAULIC SYSTEM

– Bleed 33.15.01.

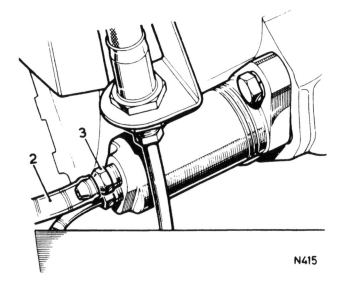

N415

Procedure

Note. During the following procedure, keep the fluid reservoir topped up to avoid introducing air to the system. Use only the recommended hydraulic fluid. Group 09 refers.

1. Attach a length of suitable tubing to the slave cylinder bleed screw.
2. Place the free end of the tube in a glass jar containing clutch fluid.
3. Slacken the bleed screw.
4. Pump the clutch pedal, pausing at the end of each stroke, until the fluid issuing from the tubing is free of air with the tube free end below the surface of the fluid in the container.
5. Hold the tube free end immersed and tighten the bleed screw when commencing a pedal down stroke

MASTER CYLINDER

– Remove and refit 33.20.01

Removing.

1. Remove the cleat securing the accelerator cable to the fluid pipe.
2. Disconnect the fluid pipe at the master cylinder.
3. Remove the lower facia panel.76.46.03.
4. Remove the master cylinder fixings at the dash panel.
5. Remove the pivot bolt and sleeves to free the push rod from the clutch pedal.
6. Withdraw the master cylinder.

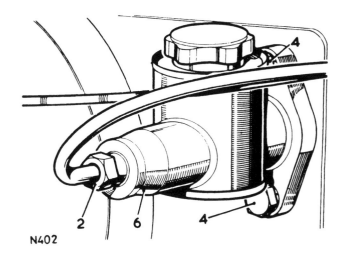

N402

Refitting.

7. Fit the master cylinder and dash fixings.
8. Fit the push rod to the pedal. Do not tighten the pivot bolt nut at this stage.
9. Check the brake pedal setting.70.35.01.
10. Back off the upper stop bolt.
11. Back off the lower stop bolt.
12. Align the clutch pedal to the same angle as the brake pedal by turning the pivot bolt and integral cam.
13. Tighten the pivot bolt securing nut.
14. Adjust the upper stop bolt to touch the pedal then continue a further half turn.
15. Fully depress the pedal.
16. Adjust the lower stop bolt to touch the pedal then continue a further turn.
17. Fit the fluid pipe to the master cylinder.
18. Bleed and replenish the hydraulic system.33.15.01.
19. Fit the lower facia panel.76.46.03.
20. Secure the accelerator cable to the fluid pipe.

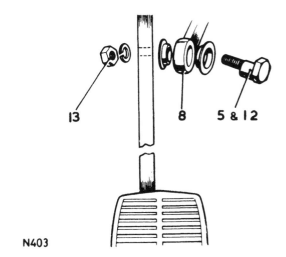

N403

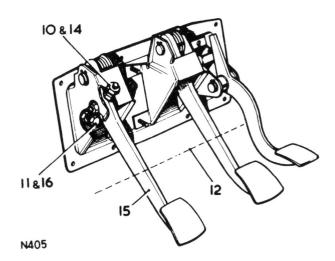

N405

MASTER CYLINDER

– Overhaul   33.20.07

1. Remove the master cylinder. 33.20.01.

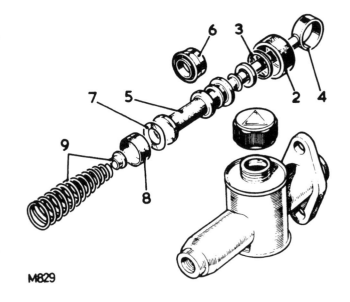

M829

Dismantling

2. Lift the dust cover.
3. Remove the circlip.
4. Withdraw the push rod assembly.
5. Withdraw the piston.
6. Remove the piston seal.
7. Lift out the piston washer.
8. Expel the main seal, using a low pressure air blast.
9. Tip out the spring and seat.

Inspection

10. Clean all components in Girling cleaning fluid and allow to dry.
11. Examine the cylinder bore and piston, ensure that they are smooth to the touch with no corrosion, score marks or ridges. If there is any doubt, fit new replacements.
12. The seals and dust cover should be replaced with new components. These items are all included in the master cylinder overhaul kit.
13. Ensure the reservoir cap vent is clear.

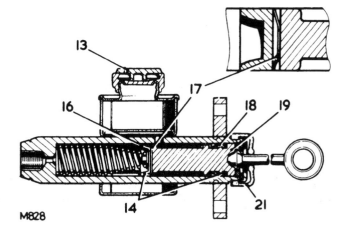

M828

NOTE: An improved master cylinder incorporating a new primary seal has been introduced on later models. Although this master cylinder is interchangeable as a complete assembly it differs internally, necessitating a different overhaul kit.

The later version is easily identified by an annular groove adjacent to the fixing flange.

Assembling

14. Smear the seals with Castrol Girling Rubber Grease.
15. Lubricate the remaining items with clean clutch fluid of correct type. Page 09–1 refers.
16. Reverse 8 and 9. Ensure the seal lip does not fold back.
17. Fit the piston washer, convex face toward the piston.
18. Fit the piston seal.
19. Fit the piston and seal.
20. Reverse 3 and 4.
21. Fill the dust cover with Castrol Girling Rubber Grease and refit.
22. Fit the master cylinder. 33.20.01.

RELEASE BEARING ASSEMBLY

– Remove and refit   33.25.12

Removing

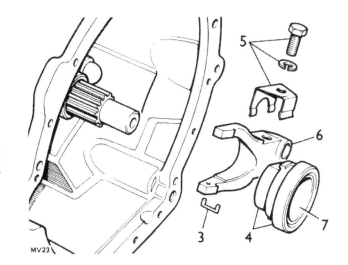

1. Remove the engine. 12.41.01.
2. Remove the clutch slave cylinder, see 33.35.01.
3. Withdraw the retainer staple.
4. Withdraw the bearing and sleeve. If required, press the bearing off the sleeve. Fit the replacement bearing with the domed face outwards from sleeve.
5. Remove the spring clip and fixings.
6. Withdraw the release lever assembly.

Refitting

7. Reverse 1 to 6. Lubricate the bearing sleeve inner diameter with a thin film of molybdenum disulphide base grease.

CLUTCH PEDAL

– Remove and refit   33.30.02.

Removing.

1. Remove the lower facia panel 76.46.03.
2. Remove the pedal bracket fixings at the cab dash panel.
3. Disconnect the brake fluid pipes at the master cylinder.
4. Disconnect the fluid pipe at the clutch master cylinder.
5. Withdraw the pedal bracket assembly into the engine compartment.
6. Disconnect the accelerator control cable at the pedal.
7. Withdraw the pedal bracket assembly from the vehicle.
8. Remove the pivot bolt nut.
9. Withdraw the pivot bolt and bearing sleeves which retain the master cylinder push rod.
10. Remove the pedal spindle circlip.
11. Withdraw the spindle.
12. Lift out the return spring.
13. Withdraw the pedal.
14. If required, press out the spindle bushes. Press in replacements and lubricate.

Refitting.

15. Reverse 9 to 13.
16. Loosely fit the pivot bolt nut
17. Where fitted – Screw in the upper stop bolt clear of the clutch pedal.
18. Align the clutch pedal to the same angle as the brake pedal by turning the pivot bolt and integral cam.
19. Tighten the pivot bolt securing nut.
20. Where fitted – Adjust the upper stop bolt to touch the pedal then continue a further half turn.
21. Fit the accelerator cable.
22. Offer the pedal bracket assembly and joint washer to the dash panel. Avoid damaging the brake light switch.
23. Reverse 1 to 4.
24. Bleed the brake system. 70.25.02.
25. Bleed the clutch system. 33.15.01.

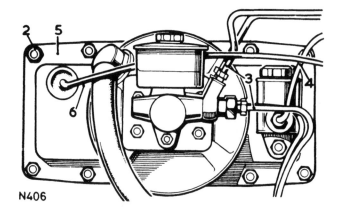

N406

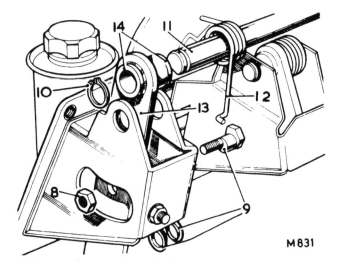

M831

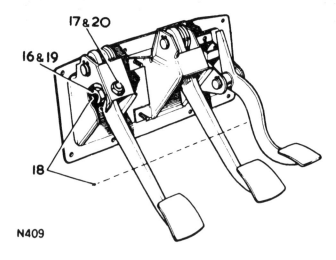

N409

SLAVE CYLINDER

– Remove and refit   33.35.01

Removing

1. Evacuate the clutch system fluid at the slave cylinder bleed valve.
2. Disconnect the fluid pipe.
3. *Early models* – Remove the slave cylinder, and (where fitted) remove the slave cylinder sealing gasket and plate.
    Later models – remove the plate.
4. If the dust cover is not withdrawn with the slave cylinder, withdraw it from the bell housing.

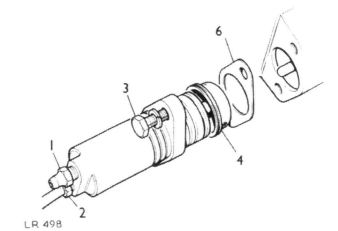

LR 498

Refitting

5. Withdraw the dust cover and backing plate from the slave cylinder.
6. Coat both sides of the backing plate with Unipart Universal jointing compound.
7. Locate the backing plate and dust cover in position on the slave cylinder.
8. Fit the slave cylinder, engaging the push-rod through the centre of the dust cover and with the bleed screw uppermost. Tightening torque for securing bolts:– 2.75 kgf.m (20 lbf.ft).
9. Reconnect the fluid pipe.
10. Replenish and bleed the clutch hydraulic system, see 33.15.01.
11. Check for fluid leaks with the pedal depressed and also with the system at rest.

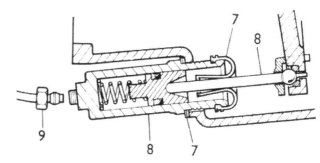

LR 501

SLAVE CYLINDER

– Overhaul   33.35.07.

1. Remove the slave cylinder 33.35.01.

Dismantling.

2. Withdraw the dust cover.
3. Withdraw the push rod.
4. Remove the circlip.
5. Expel the piston and sleeve, applying low pressure air to the fluid inlet.
6. Withdraw the spring seat and spring.
7. Remove the bleed valve.

N414

Inspecting

8. Clean all components in Girling cleaning fluid and allow to dry.
9. Examine the cylinder bore and piston which must be free from corrosion, scores and ridges.
10. Replace the seals and dust cover.

Assembling

11. Fit the bleed valve. Do not overtighten.
12. Lubricate the seals, using Castrol Girling Rubber grease.
13. Lubricate the remaining components, using Castrol Girling Brake and Clutch fluid.
14. Fit the seal, flat side first, to the piston.
15. Locate the spring smaller end in the spring seat and assemble to the piston.
16. Enter the piston assembly, spring first, into the cylinder bore. Ensure that the seal lip does not fold back.
17. Secure with the circlip.
18. Fill the dust cover with Castrol Girling Rubber Grease.
19. Reverse 1 to 3.

GEARBOX OPERATIONS

Description	Operation No.
MAIN GEARBOX	
Bell housing—remove and refit	37.12.07
Bearing plate remove and refit	37.12.22
Front cover and oil pump assembly	
—remove and refit	37.12.34
—overhaul	37.12.37
Gearbox—remove and refit	37.20.01
Gearchange selectors—remove and refit	37.16.31
Layshaft	
—remove and refit	37.20.19
—bearings—remove and refit	37.20.22
Mainshaft assembly—remove and refit	37.20.25
Mainshaft assembly—overhaul	37.20.31
Primary pinion—remove and refit	37.20.16
Reverse idler gear and shaft—remove and refit	37.20.13
Reverse light switch—remove and refit	37.27.01
TRANSFER GEARBOX	
Differential lock actuator assembly	
—remove and refit	37.29.19
—overhaul	37.29.22
Differential lock actuator switch—remove and refit	37.27.05
Differential unit assembly	
—remove and refit	37.29.13
—overhaul	37.29.16
Front output shaft and housing	
—remove and refit	37.10.05
—overhaul	37.10.06
Front output shaft oil seal—remove and refit	37.23.06
Gear lever and cross-shaft—remove and refit	37.29.01
Gear selectors and shaft—remove and refit	37.29.04
Intermediate gears assembly—remove and refit	37.29.10
Mainshaft transfer gear—remove and refit	37.20.28
Rear output shaft oil seal—remove and refit	37.23.01
Speedometer drive housing	
—remove and refit	37.25.09
—overhaul	37.25.13

Gearbox

FAULT DIAGNOSIS

SYMPTOM	POSSIBLE CAUSE	CURE
Gearbox noisy in neutral.	Insufficient oil in gearbox.	Top up as necessary.
	Incorrect grade of oil.	Drain and replenish.
	Primary pinion bearing worn.	Renew bearing.
	Constant mesh gears worn.	Renew primary pinion and layshaft.
	Layshaft bearings worn.	Renew bearings.
Gearbox noisy in all gears except top.	Layshaft, mainshaft or primary pinion bearings worn.	Renew bearings.
	Constant mesh gears worn.	Renew primary pinion and layshaft.
Gearbox noisy in one gear only.	Worn or damaged gears or bearings.	Renew gears and/or bearings.
Gearbox noisy in all gears.	Worn bearings on primary pinion, mainshaft or layshaft.	Renew bearings.
Oil leaks from gearbox.	Gearbox over-filled with lubricating oil.	Rectify oil level with vehicle standing on level floor.
	Loose or damaged drain or level plugs.	Tighten plugs. If damaged, fit new plugs and joint washer as required.
	Obstructed breather.	Clean breather.
	Joint washers damaged, incorrectly fitted or missing.	Fit new joint washer with general purpose grease smeared on both sides.
	Oil seals damaged or incorrectly fitted.	Fit new oil seal with Unipart Universal jointing compound smeared on the outside diameter.
	Cracked or broken gearbox casings.	Fit new casings.
Difficulty in engaging forward gears.	Weak springs or worn parts in synchormesh units.	Renew faulty parts.
	Worn selector forks and/or interlock pins.	Renew components as necessary.
	Faulty clutch operation; clutch fluid leakage.	Check clutch master and slave cylinders. Renew clutch components as necessary.
Difficulty in engaging reverse gear.	Reverse gear bearings worn or damaged.	Renew bearings and shaft as necessary.
	Faulty clutch operation, clutch fluid leakage.	Check clutch master and slave cylinders. Renew clutch components as necessary.
Difficulty in disengaging forward gears.	Synchromesh cones worn; damaged gear dogs.	Renew faulty parts.
	Distorted or damaged splines.	Renew components as necessary.
Difficulty in disengaging reverse gear.	Reverse gear seized on shaft.	Renew parts as necessary.
Gear lever going into reverse too easily and not into first.	Weak reverse stop hinge plate spring.	Renew the spring.

FAULT DIAGNOSIS

SYMPTOM	POSSIBLE CAUSE	CURE
Transfer of oil between main gearbox and transfer gearbox.	Faulty 'O' ring seal on reverse idler shaft.	Renew seal.
	Faulty mainshaft oil seal.	Renew seal.
	Obstructed main gearbox breather.	Clean breather.
Oil leakage from gearbox to bell housing.	Faulty joint washer/s on gearbox front cover and oil pump.	Renew joint washer/s.
	Faulty oil seal, primary pinion to front cover.	Renew oil seal.
	Damaged or porous gearbox front cover.	Renew front cover.
Transfer gearbox noisy.	Insufficient oil in transfer box.	Replenish.
	Incorrect grade of lubricating oil.	Drain and replenish with the correct grade oil. Refer to recommended lubricants Division 09.
	Excessive end float on intermediate gears assembly	Adjust to lower tolerance 0,152 mm (0.006 in.).
	Worn components in gearbox differential unit.	Renew components
	Worn bearings in intermediate gears assembly.	Renew bearings.
Differential lock warning switch bulb fails to light up.	Switch bulb failure.	Renew bulb.
	Air leakage in vacuum circuit.	Renew leaking components.
	Actuator housing not seated square on front output shaft housing.	Slacken fixings, reseat housing, hold in position and tighten fixings.

FRONT OUTPUT SHAFT AND HOUSING

–Remove and refit   37.10.05

Remove the front floor 76.10.12.

Removing

1. Drain off the transfer gearbox oil. See Maintenance, Group 10.
2. Disconnect the front propeller shaft at the gearbox.
3. Remove the differential lock actuator assembly. 37.29.19.
4. Remove the six fixings.
5. Withdraw the output shaft and housing complete.
6. Lift out the lock-up dog clutch.

Refitting

7. Reverse items 1 to 6; note that the housing is dowel located.

Refit the front floor 76.10.12.

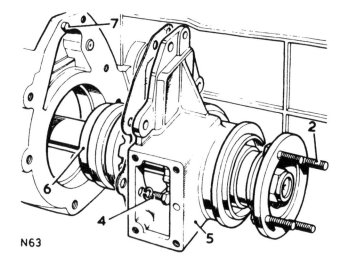

N63

FRONT OUTPUT SHAFT AND HOUSING

–Overhaul   37.10.06

Dismantling

1. Remove the front output shaft and housing. 37.10.05.
2. Remove the locking nut and washer.
3. Withdraw the coupling flange complete with mudshield.
4. If required, press off the mudshield.
5. Press out the shaft toward the rear.
6. Withdraw the oil seal.
7. Remove the circlip.
8. Withdraw the output shaft bearing.

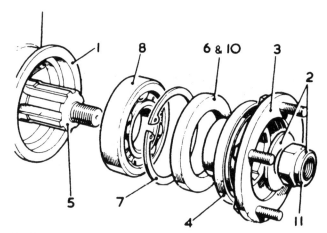

N111

Assembling

9. Reverse items 7 and 8.
10. Fit the oil seal, with the open side toward the bearing.
11. Reverse items 2 to 5. Torque loading for locking nut is 11.75 kgf.m (85 lbf. ft.).
12. Refit the front output shaft and housing. 37.10.05.

BELL HOUSING

—Remove and refit   37.12.07

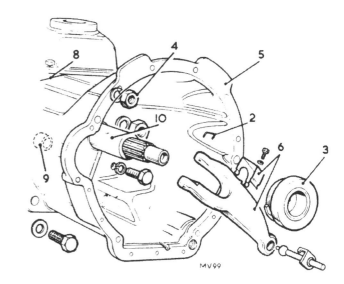

MV99

Removing

1. Remove the gearbox 37.20.01.
2. Withdraw the locating staple from the clutch release sleeve and release lever.
3. Lift out the release sleeve and bearing assembly.
4. Remove the bell housing fixings.
5. Withdraw the bell housing complete with clutch release lever.
6. If required, remove the push rod clip and the spring clip and withdraw the clutch release lever.

Refitting

7. If removed, reverse instruction 6.
8. Apply a thin film of Universal or other suitable jointing compound around the three selector shaft holes in the bell housing rear face.
9. Fit the bell housing, locating on the dowels.
10. Apply a thin film of molybdenum disulphide grease on to the front cover extension sleeve.
11. Reverse 1 to 3.

BEARING PLATE ASSEMBLY

–Remove and refit   37.12.22.

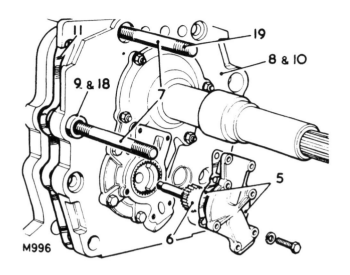

Removing

1. Drain the gearbox oil.
2. Remove the gearbox. 37.20.01.
3. Remove the bell housing. 37.12.07.
4. Position the gearbox with the front end uppermost.
5. Remove the oil pump gears cover and joint washer.
6. Withdraw the oil pump drive gear.
7. Temporarily remove the four fixing studs from the gearbox front face.
8. Ease the bearing plate away from the gearbox.
9. Withdraw the two dowel sleeves which locate the bearing plate.
10. Withdraw the bearing plate assembly complete with primary pinion and layshaft.
11. Withdraw the joint washer.
12. Withdraw the layshaft.

Refitting

NOTE: To replace a bearing plate, a bearing plate and gearbox casing mated assembly must be fitted.

13. Locate the cone into the third/fourth-speed synchromesh unit.
14. Lubricate the oil tube, using clean gearbox oil.
15. Position the joint washer.
16. Engage the layshaft with the primary pinion and front bearing outer member.
17. Fit the bearing plate and layshaft.
18. Align the bearing plate with the gearbox casing and slide home the dowel sleeves.
19. Refit the studs. Smear Loctite 'Studlock' grade CVX, Rover part no. 601168 on the two upper studs securing threads before fitting.
20. Reverse 1 to 6.

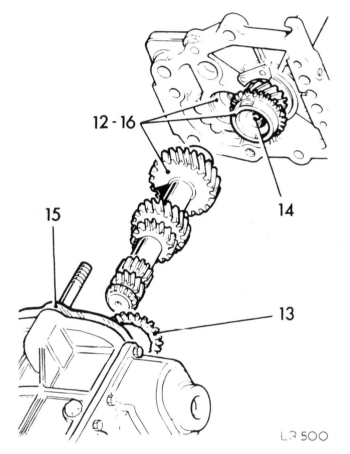

FRONT COVER AND OIL PUMP ASSEMBLY

–Remove and refit   37.12.34.

Service tools RO.1005–Centralising tool for primary pinion.

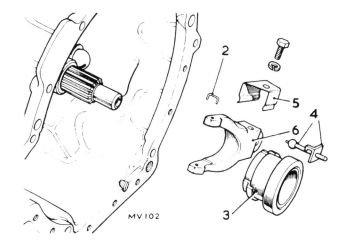

MV102

Removing

1.   Remove the gearbox 37.20.01.
2.   Lift out the retainer staple from the clutch release bearing assembly and the release lever.
3.   Withdraw the release bearing assembly.
4.   Remove the slave cylinder pushrod.
5.   Remove the spring clip.
6.   Withdraw the clutch release lever.
7.   Remove the front cover assembly, complete with oil pump and joint washer.
8.   Remove the shim washer located between the front cover and the layshaft front bearing.

Refitting

9.   Remove the oil pump cover.
10.   Withdraw the oil pump drive gear.
11.   Position the layshaft bearing shim washer.
12.   Position the front cover assembly and loosely fit the fixings.
13.   Fit the oil pump drive gear to engage the drive square in the layshaft.
14.   Fit the oil pump cover and joint washer.
15.   Fit the gauge RO.1005 to align the primary pinion with the bell housing.
16.   Visually check that the front cover is concentric about the primary pinion. Adjust the front cover position about its fixings to suit.
17.   When satisfactory, tighten the front cover fixings.
18.   Reverse 1 to 6.

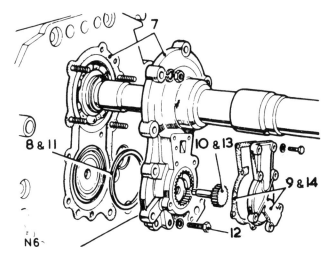

N6

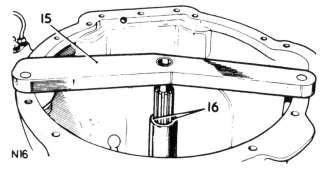

N16

FRONT COVER AND OIL PUMP ASSEMBLY

– Overhaul 37.12.37

Service tools: 18G 134 guide and 18G 134 DG adaptor – assembly tool for fitting oil seal and oil feed ring

Dismantling

1. Remove the front cover and oil pump, see 37.12.34.
2. Remove the pump cover and gasket.
3. Withdraw the pump gears.
4. Remove the oil feed ring.
5. Withdraw the oil seal.
6. Remove the plug and withdraw the ball and spring from the relief valve housing.
7. If required, drift off the extension sleeve. Fit a replacement using Loctite 'AVV' grade.

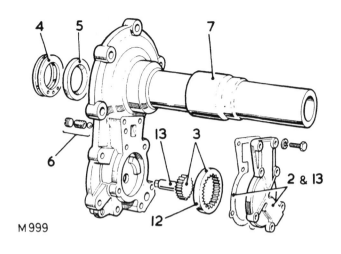

M 999

Reassembling

8. Press in the oil seal, plain face first, using 18G 134 guide and 18G 134 DG adaptor.
9. Align the centre hole of three in the oil feed ring with the oil delivery hole in the front cover.
10. Press in the oil feed ring, using 18G 134 guide and 18G 134 DG adaptor.
11. Fit the ball, spring and plug. When fitted, the plug must be flush with, or not more than, 0,25 mm (0.010 in.) below the front cover rear face.
12. Fit the oil pump ring gear.
13. Fit the front cover and oil pump, see 37.12.34. During this operation the pump drive gear, cover and joint washer are fitted.

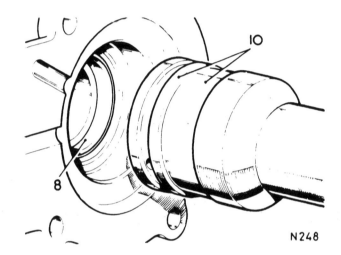

N 248

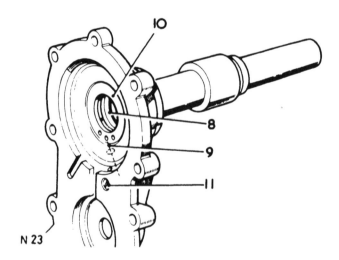

N 23

GEAR-CHANGE SELECTORS, MAIN GEARBOX

– Remove and refit   37.16.31

Removing

1. Remove the gearbox, see 37.20.01.
2. Remove the bell housing, see 37.12.07.
3. Select neutral, remove the reverse light switch and remove the gearbox top cover and joint washer.
4. Lift out the detent springs.
5. Withdraw the detent balls, using a small magnet or an air blast.
6. On early gearboxes, drive out the spring pins securing the first and second speed jaw, first and second selector, third and fourth selector, reverse jaw, and on gearboxes up to No. 35505925A, the reverse finger
   On later gearboxes the reverse selector finger is held by a pinch bolt which should be slackened.
7. Drive out the retaining pins until the shafts are free in the selectors.
8. Tap out the selector shafts, commencing with an outer shaft.
9. Withdraw the selector jaws and forks. Prior to gearbox No. 35527419B selector forks were fixed. Later gearboxes have swivel type selector pads.
10. Withdraw the two interlock plungers from the cross-drilling.
11. Remove the lock-wired pivot bolt.
12. Lift out the reverse cross-over lever.

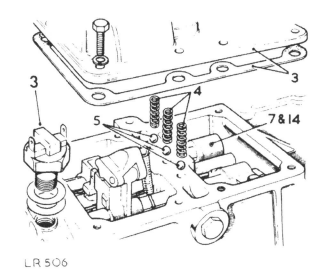

LR 506

Refitting

NOTE Ensure that the retaining pins are an interference fit. Renew if necessary.
Pins fitted to latest gearboxes vary in length and diameter and should be fitted in the following sequence:–

1. Secure Reverse Selector Jaw – retain with extra long pin of increased diameter.
2. Secure 3rd/4th Jaw – retain with two identical pins (as earlier type).
3. Secure both ends of 1st/2nd Jaw – retain with extra long pin.

The Reverse Selector Shaft and Jaw has been modified to take the larger diameter pin.

13. Withdraw the retaining pins from the selector jaws.
14. Position the reverse cross-over lever in the gearbox and locate the lever foot in the groove in the reverse idler gear.
15. Fit the pivot bolt and engage the cross-over lever tapping. Apply Loctite Studlock grade CVX to the bolt threads before screwing fully in. The Loctite must not enter the casing or run on the exposed bolt threads.
16. Locate the first/second gear selector fork in the groove in the outer member, with the boss on the fork to the rear. Position the boss to the R.H. side of the box.
17. Locate the third/fourth gear selector fork in the

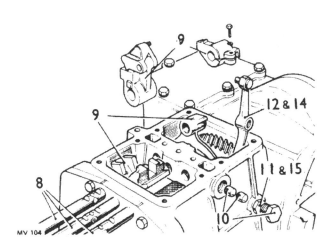

MV 104

groove in the outer member. Position the fork with the retaining pin entry hole at the top R.H. side.

18. Fit the third/fourth gear selector shaft and interlock pin assembly and secure to the selector fork with a retaining pin.

19. Fit the two interlock pins to engage in the grooves each side of the third/fourth gear selector shaft.

20. Position the reverse stop hinge plate and selector jaw in the gearbox, adjacent to the third/fourth gear selector jaw.

21. Fit the reverse gear selector shaft and engage the selector jaw and hinge spring.

22. Push the shaft home and engage the reverse cross-over lever selector finger.

23. Secure the reverse gear selector jaw to the shaft with a retaining pin.

    Early models pin the reverse selector finger.

    Later models loosely fit the reverse selector finger pinch bolt.

24. Position the first/second gear selector jaw in the gearbox.

25. Fit the first/second gear selector shaft; engage the selector jaw and selector fork as the shaft is pushed home.

26. Fit the retaining pins, fitting the jaw pin first.

    Reverse shaft adjustment – later models with reverse selector finger pinch bolt.

27. Move the reverse shaft forward until the selector jaw abuts the casing.

28. Holding the reverse shaft as described in the previous item, move the reverse selector finger forward on the shaft until it abuts the casing, then move it rearward until it is just clear of the casing.

29. Place a 0,25 mm (0.010 in.) feeler gauge between the upper edges of the reverse and third/fourth selector jaws.

    NOTE: The edges of the selector jaws taper slightly, therefore, it is important that the feeler gauge is positioned between the upper edges.

30. Hold the reverse and third/fourth selector jaws together to retain the feeler gauge, then rotate the reverse selector finger until it abuts the third/fourth selector shaft and tighten the pinch bolt.

31. Check the operation of the reverse gear selectors assembly. Ensure there is clearance between the cross-over lever and selector finger sufficient to prevent fouling during operation. If necessary, the 0,25 mm (0.010 in.) clearance obtained in instruction 30 can be increased up to 0,5 mm (0.020 in.) to produce a smooth gear-change.

32. Wire lock the cross-over lever pivot bolt.

33. When fitting the hinge spring to the reverse stop hinge, first engage the large hook around the selector shaft, as illustrated, before fitting the small hook to the reverse stop hinge pin.

    NOTE: A stronger spring, identified by a yellow paint mark, has been introduced to prevent the reverse selector detent ball sticking. The stronger spring can be fitted in place of the original on earlier gearboxes.

34. Reverse instructions 1 to 5.

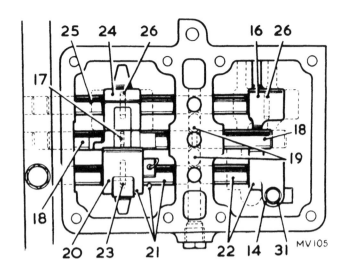

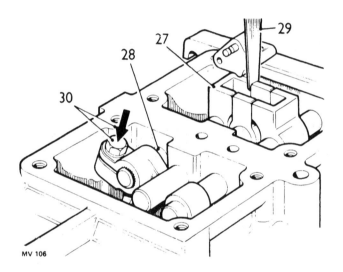

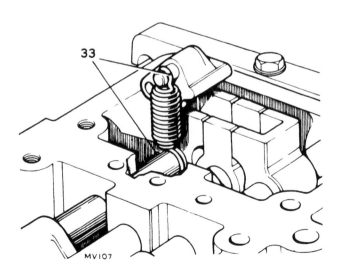

GEARBOX ASSEMBLY

-- Remove and refit   37.20.01

Service tools: RO 1001 -- Lifting bracket for gearbox. An hydraulic or mechanical spreader is also required.

The following instructions detail the procedure for removing the gearbox from underneath the vehicle using suitable lifting equipment. The latest type gearbox is illustrated.

WARNING: It is essential, because of the considerable weight and offset position of the centre of gravity of the gearbox, that a hydraulic transmission hoist of adequate strength and stability is used. Failure to observe this precaution could result in the hoist tipping over and causing serious personal injury or damage to the gearbox.

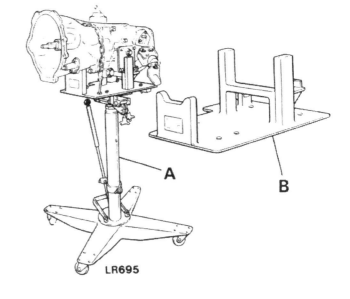

Suitable hydraulic lifting equipment (A) is made by Blackhawk International Limited and Harlem Wilcomatic Limited. A special adaptor plate (B) is also produced by Harlem Wilcomatic for holding the Range Rover and Land Rover V8 gearboxes. This adaptor plate can be used with various transmission hoists provided they meet the safety requirements indicated above.

Removing

1. Drive the vehicle on to a ramp.
2. Prop the bonnet open.

From inside the engine compartment

3. Disconnect the battery.
4. Remove the air cleaner assembly.
5. Release the two plastic breather tubes from the clamp on the engine rear lifting bracket (nut & bolt).

From inside the passenger compartment

6. Lift out the ashtray and remove its base bezel from the gearbox tunnel (four screws).
7. Unscrew the maingear lever knob and remove the gearbox tunnel carpet.
8. Unclip the rubber gaiter covering the gearbox tunnel.
9. Remove the main gear lever and retaining plate (three bolts). **For security, cover the exposed aperture in the top of the gearbox.**
10. Release the differential lock actuator switch from the gearbox tunnel (two nuts and bolts).
11. Unscrew the transfer gear lever knob and remove its rubber gaiter (four self-tapping screws).
12. Remove the handbrake lever rubber gaiters (eight self tapping screws).
13. Raise the ramp as required.

Continued

From underneath the vehicle

14. Drain oil from the main and transfer gearboxes.
15. Remove the **nuts only** from the eight bolts (four each side) which retain the front box cross member to the chassis.
16. Remove the **four upper bolts only** from the front box cross member. The four lower bolts **must** remain to prevent the cross member from falling.
17. Use a spreader on the chassis side members to free the front cross member. Finally, release the four loose bolts and remove the front cross member.
18. Remove the heat shield from the right side down pipe; this is attached to the right side engine mounting bracket (two nuts and bolts).
19. Release both down pipe flanges from the exhaust manifolds (six nuts with spring washers).
20. Release the exhaust pipe flange from the front of the silencer (three nuts and bolts).
21. Release the rubber mounting holding the down pipe to the gearbox (nut and bolt, two steel and two fibre washers).
22. Remove front twin down pipe assembly.
23. Release rubber mounting holding the twin exhaust pipes at the rear of the silencer (nut and bolt, two steel and two fibre washers).
24. Release rubber mounting holding the twin exhaust pipes to the chassis nut and bolt, two steel and two fibre washers.
25. The silencer and rear tail pipes can then be lowered sufficiently and secured to allow the gearbox to be subsequently moved rearwards.

continued

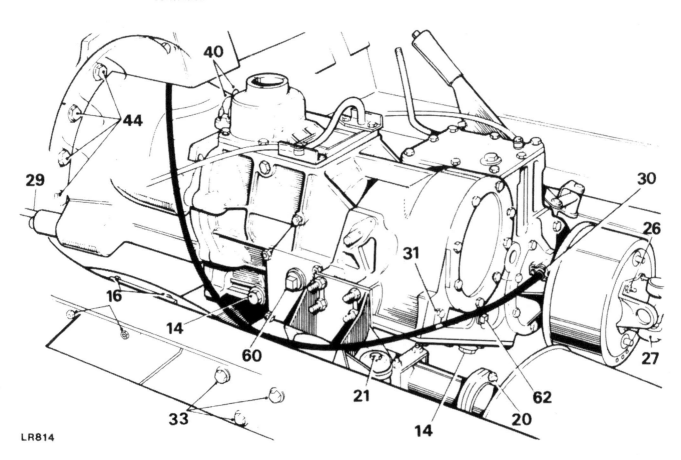

LR814

26. Disconnect the front and rear propeller shafts at the gearbox universal joint flanges (four nuts on each flange).

27. Tie up the propeller shafts to the chassis side members.

28. Remove the bell housing cover plate (four bolts and five nuts and bolts with spring washers) On early gearboxes two additional bolts were fitted upwards into the cylinder block.

29. Remove the clutch slave cylinder from the bell housing (two bolts with spring washers).
   Additionally, if a replacement gearbox with a new clutch slave cylinder is to be fitted, disconnect the hydraulic pipe union from the existing slave cylinder, plugging the pipe to retain the fluid. Also slacken the large nut retaining the pipe to the bracket on the bell housing.

30. Remove the speedometer cable fixing bracket from the rear output shaft housing (one nut only) and disengage the cable from the drive.

31. Remove the speedometer cable retaining clip from the transfer box lower cover (one nut only). On early gearboxes the clip was attached to the rear of the transfer box.

32. Position the transmission lift under the gearbox and support it.

33. Release the two rear gearbox mounting brackets from the chassis side members (three nuts and bolts each side). One of the bolts also holds the remaining speedometer cable clip.

continued

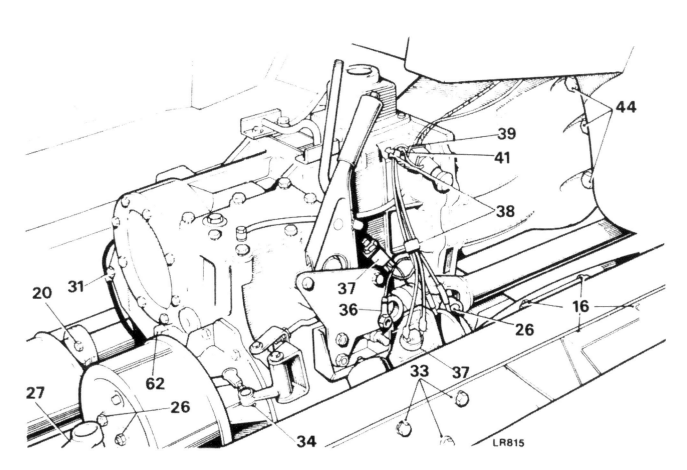

179

34. Disconnect the handbrake link (split pin and clevis pin).
35. Lower the gearbox approximately 150 mm (about 6 inches) on the transmission lift, ensure that the handbrake lever is raised to clear its aperture in the gearbox tunnel and check that the engine fan blades are not fouling the fan cowl before proceeding.
36. Remove the lower front nut securing the handbrake assembly to the gearbox and disconnect the earth tag from the stud.
37. Disconnect the two leads from the handbrake warning light switch and the two leads from the differential warning light switch.
38. Release the two rubber cleats holding the plastic vacuum pipes to the harness.

From inside the passenger compartment

39. Release the harness retaining clip from the gearbox top cover (one bolt).
40. Disconnect the two leads from the reverse light switch.
41.. Disconnect the inlet manifold vacuum pipe union from the differential switch.

From underneath the vehicle

42. Support the engine under the sump.
43. Pass the safety chain around the gearbox and make fast.
44. Remove bell housing bolts (eight bolts).
   NOTE: Two of these bolts are also used to retain the slave cylinder pipe bracket and the harness clip respectively.
45. Withdraw the gearbox from the engine.
46. Fit the lifting bracket RO 1001 to facilitate any subsequent removal from the transmission lift.

Refitting

47. This is virtually a reversal of the removal procedure. However, to facilitate the re-alignment of the gearbox primary shaft with the clutch plate splines the following procedure should be observed.
48. Pull off the plastic vacuum pipe from the **front** of the vacuum unit and apply a compressed air line to the vacuum unit. A short low pressure burst will be sufficient to engage the differential lock which will then emit an audible click.
49. Replace the plastic vacuum pipe.
50. Engage third or fourth gear by using a large screwdriver in the open top of the gearbox.
51. Smear the splines of the primary pinion, the clutch centre and the withdrawal unit abutment faces with molybdenum disulphide grease, such as Rocol MTS 1000.
52. Raise the gearbox into position.

continued

53. Rotate the brake drum until the primary shaft splines are correctly aligned with the clutch plate and move the gearbox forwards to engage the bell housing dowels.

54. Secure the bell housing to the engine. Torque 3,5 kgf.m (25 lb.ft).

55. Pull off the plastic vacuum pipe from the **rear** of the vacuum unit and apply the compressed air line to disengage the differential lock.

56. Replace the plastic vacuum pipe.

57. Reverse instructions 28 to 43.

58. If necessary, fit a new seal to the bell housing cover plate, using a cement such as Holdite 88.

59. Apply 'Universal' jointing compound to the cover plate and seal, for the joints between the bell housing, cylinder block and rear main bearing cap.

60. Secure the cover plate to the engine and bell housing. Torque 1,0 kgf.m (8 lbf.ft).
Reverse instructions 1 to 27.

61. Refit the drain plug, washer and filter to the main gearbox and fill with oil to the bottom of the **level/filler** plug hole. Capacity 2,6 litres (4.5 Imperial Pints).

62. Refit the drain plug and washer to the transfer gearbox and fill with oil through the **filler** plug hole to the bottom of the **level** plug hole. Capacity 3,1 litres (5.5 Imperial Pints).

Should you require further information regarding the hydraulic transmission hoist and the special **Range Rover** gearbox adaptor plate, please contact the equipment manufacturers direct. Their addresses are as follows:—

Transmission Hoist — type 67556 (T-3)

Blackhawk Automotive Limited
Brookfield Industrial Estate
Leacon Road
Ashford
Kent
Telephone: Ashford (0233) 32151
Telex: 96562i BLUK G

**Transmission Hoist — type D S6 HD complete with
Gearbox Adaptor Plate — type DS RT**

Harlem Wilcomatic Limited
Mortimer Road
Narborough
Leicester
England
Telephone: 0533 86651
Telex: 341 981 HARWIL G

GEARBOX ASSEMBLY

– Remove and refit   37.20.01

Service tools: RO 1001 – Lifting bracket for gearbox

Removing

The following instructions detail the procedure for removing the gearbox from the interior of the vehicle when suitable lifting equipment is not available. The earlier type gearbox is illustrated.

1. Disconnect the battery and remove the front floor. 76.10.12.
2. Drain the gearbox oils.
3. Disconnect the rear propeller shaft at the gearbox.
4. Disconnect the front propeller shaft at the gearbox.
5. Remove the speedometer cable retainer clip fixing, tilt and move aside the retainer.
6. Disconnect the speedometer cable securing clips at the gearbox rear and at the chassis LH side member.
7. Withdraw the speedometer cable from the speedometer drive housing.

8. Remove the fixings, gearbox bracket to exhaust pipe bracket.
9. Disconnect the clutch fluid pipe at the slave cylinder.
10. Remove the fixings, cross-member to chassis side-member.
11. Lower the cross-member away from the gearbox.
12. Note the colours and disconnect the reverse light switch leads.
13. Support the engine weight.
14. Fit the lifting bracket RO1001 and lifting tackle to the gearbox.
15. Remove the gearbox mounting brackets fixings at the chassis side members.
16. Remove the fixings, bell housing to engine.
17. Withdraw the gearbox until clear of the flywheel housing dowels and remove from the vehicle.

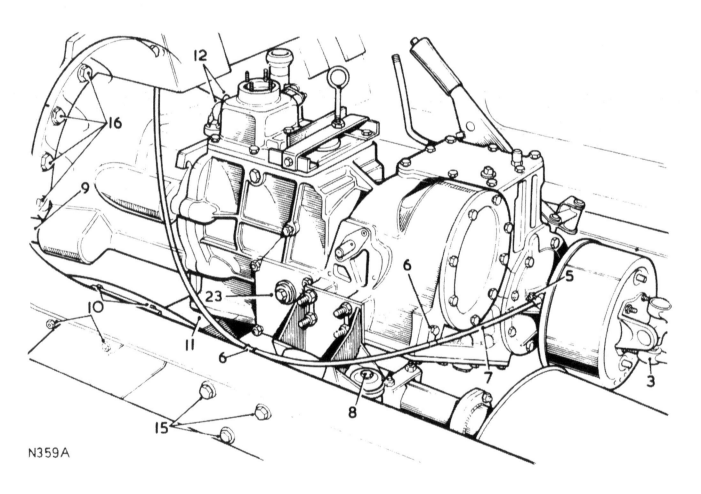

N359A

Refitting

18. Offer the gearbox to the engine. If necessary, select a gear and rotate the transmission brake to engage the primary pinion and clutch splines.
19. Reverse 15 to 17.
20. Apply a coating of 'Universal' jointing compound to the vertical joint face of the bell housing cover plate.
21. Fit the cover plate and seal, ensuring that the fillet is filled with sealing compound.

22. Reverse 2 to 15.
23. Replenish the gearbox oils. See Maintenance – Division 10. Using the dipstick filler (earlier models) or the side filler/level plug (later models).
24. Bleed and replenish the clutch hydraulic system. 33.15.01.
25. Refit the front floor. 76.10.12.

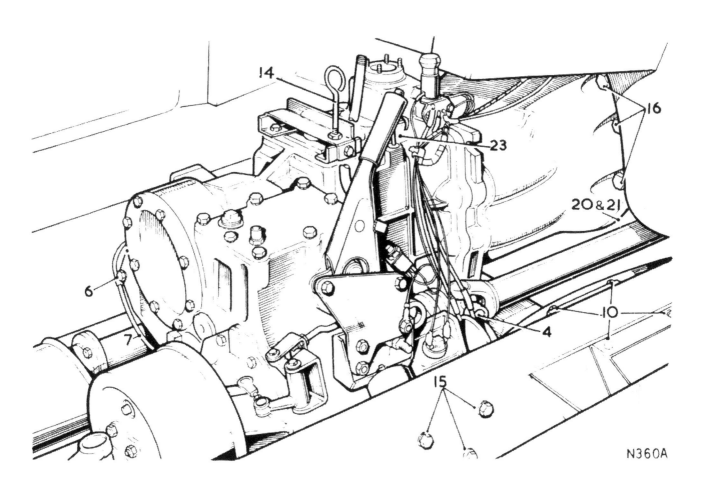

N360A

REVERSE IDLER GEAR AND SHAFT

– Remove and refit   37.20.13

Service tool: 18G 1335 extractor for reverse shaft

Removing

1. Drain the main and transfer gearbox oils.
2. Remove the gearbox side cover.
3. Remove the gearbox bottom cover.
4. Remove the bolt securing the idler gear shaft in the gearbox casing. (A pin is used on early gearboxes).
5. Withdraw the idler gear shaft, utilizing extractor 18G 1335.
6. Remove the 'O' ring seal.
7. Lift out the reverse idler gear assembly.
8. Remove the circlip and plain washer.
9. Lift out the needle roller bearings and further plain washer.
10. Withdraw the remaining circlip.
11. If required, withdraw the shaft support bush.

Refitting

12. If removed, fit the shaft support bush, using Locquic primer grade 'T' and 'AVV' grade.
13. Reverse instructions 6 to 11.
14. Offer the idler shaft to the gearbox and align the retaining bolt/pin holes.
15. Smear clean gearbox oil on to the 'O' ring seal.
16. Position the reverse idler assembly in the casing.
17. Engage the selector foot in the idler gear groove.
18. Drive in the idler gear shaft until the retaining bolt holes are aligned.
19. Early gearboxes – Fit a new pin to secure the reverse shaft.

     Later gearboxes – Before fitting the retaining bolt, treat the threads with Locquic grade 'T' and allow to dry. Then fit the bolt using Loctite Studlock grade.
20. Reverse instructions 1 to 3.

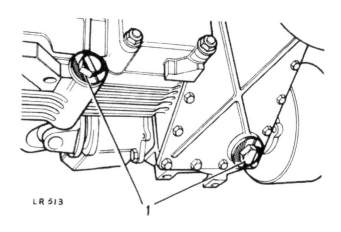

LR 513

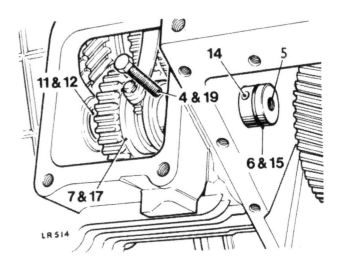

LR 514

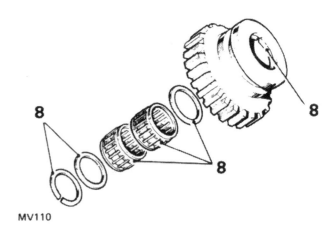

MV110

PRIMARY PINION

–Removing and refit   **37.20.16.**

Removing

1. Remove the front floor. 76.10.12.
2. Remove the gearbox. 37.20.01.
3. Remove the bell housing. 37.12.07.
4. Remove the front cover and oil pump assembly. 37.12.34.
5. Remove the bearing plate assembly. 37.12.22.
6. Remove the circlip.
7. Lift off the shim washer.
8. Press out the primary pinion.
9. Withdraw the bearing retaining plates and serrated bolts.
10. Press out the primary pinion bearing.

Refitting

11. Check that the orifice drilled in the oil tube is clear. During refitting, take care to avoid damage to the oil tube. Rotate the shaft in the bearing to ensure that the oil tube is straight.
12. Support the bearing plate using suitable wooden blocks. Position the blocks across the bearing housing aperture to act as assembly stops.
13. Press in the bearing until flush with the bearing plate.
14. Press in the primary pinion. Check that the bearing remains flush with the bearing plate.
15. Fit the retaining plates and serrated bolts.
16. Fit the shim washer and circlip.
17. Measure the clearance between the circlip and the shim washer. There must be a clearance of 0,05 mm (0.002 in.) maximum.
18. If required, adjust the clearance by fitting a replacement shim washer. Shim range is 2,0 to 2,15 mm (0.079 to 0.085 in.) in 0.05 mm (0.002 in.) stages.
19. Reverse 1 to 5.

DATA

End float, primary pinion
to bearing   0,05 mm (0.002 in.) maximum

LAYSHAFT

–Remove and refit   **37.20.19.**

Removing

1. Remove the bearing plate assembly. 37.12.22.
2. Withdraw the layshaft.

Refitting

3. Reverse instructions 1 and 2.

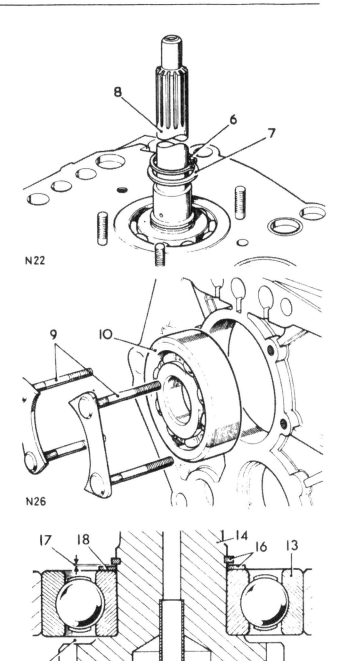

185

LAYSHAFT BEARINGS

—Remove and refit   37.20.22.

Service tools RO.1004
> Extractor for mainshaft spacer
> 18G284 extractor and 18G284AR adaptor
> Extractor layshaft rear bearing outer member
> 18G47 press and 18G47BA collars
> Extractor for layshaft bearing inner members.

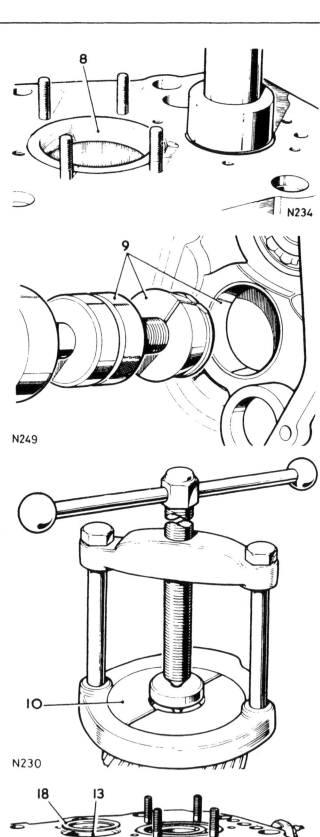

N234

N249

Removing

1. Remove the front floor. 76.10.12.
2. Remove the gearbox. 37.20.01.
3. Remove the bell housing. 37.12.07.
4. Remove the front cover and oil pump assembly. 37.12.34.
5. Remove the front bearing plate assembly. 37.12.22.
6. Remove the main gears selectors. 37.16.31.
7. Remove the mainshaft assembly. 37.20.25.
8. Press out the layshaft front bearing outer member from the front bearing plate.
9. Extract the layshaft rear bearing outer member from the gearbox casing, extractor 18G284 and adaptor 18G284AR.
10. Withdraw the bearing inner members from the layshaft. Extractor 18G47 press and 18G47BA collars.

Checking the bearing pre-load

Replacement bearings inner and outer members are supplied as matched pairs and not as separate items.

The replacement bearings must not be degreased. Before fitting, lubricate with correct grade gearbox oil. Division 09 refers.

11. Press the bearing inner members on to the layshaft.
12. Press the rear bearing outer member into the gearbox casing.
13. Enter the front bearing outer member into the front bearing plate. Do not fit fully in at this stage.
14. Remove the primary pinion from the bearing plate. 37.20.16.
15. Position the layshaft in the gearbox casing.
16. Temporarily fit the front bearing plate and joint washer.
17. Press in the front bearing outer member until there is no end-float on the layshaft and no end-load on the bearings.
18. On the bearing outer member position a shim washer of a thickness suitable to stand 0,25 mm (0.020 in.) approximately proud of the front bearing plate. This shim thickness may be subsequently adjusted depending on the amount of bearing pre-load it affords.

N230

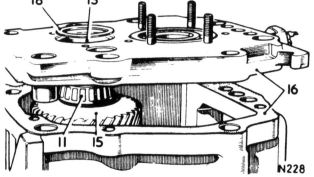

N228

continued

19. Temporarily remove the oil pump top cover and withdraw the pump drive gear.
20. Temporarily fit the front cover and new joint washer to the bearing plate. Ensure that the shim washer remains in position.
21. Measure the rolling resistance of the layshaft, using a spring balance and a cord coiled around the layshaft larger diameter.
22. The rolling resistance must be 2,75 to 3,8 kg (6 to 8.5 lb).
23. To adjust the pre-load, fit a replacement shim of suitable thickness to the front bearing outer member. Shim range is from 1,55 mm (0.059 in.) to 2,50 mm (0.098 in.) in 0,025 mm (0.001 in.) increments.
24. When the pre-load is satisfactory, remove the front cover assembly.
25. Remove the front bearing plate.
26. Fit the primary pinion. 37.20.16.
27. Fit the oil pump cover and drive gear.

N 229

Refitting

28. Reverse 1 to 6.

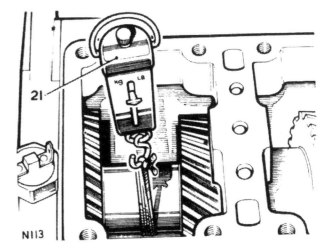

N 113

DATA

Layshaft rolling resistance   2,75 to 3,8 kg (6 to 8.5 lb)

MAIN SHAFT ASSEMBLY

–Remove and refit   **37.20.25.**

Service tools RO.1004   Extractor for mainshaft spacer.

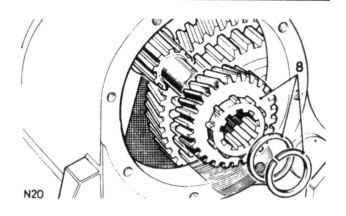

N20

Removing

1. Remove the front floor. 76.10.12.
2. Remove the gearbox. 37.20.01.
3. Remove the bell housing. 37.12.07.
4. Remove the front bearing plate. 37.12.22.
5. Remove the main gearchange selectors. 37.16.31.
6. Remove the mainshaft rear bearing housing and roller bearing.
7. Remove the bottom cover from the transfer gearbox.
8. Remove the snap ring, shim washer and mainshaft transfer gear.
9. Fit extractor RO.1004 to transfer gear spacer.
10. Withdraw the spacer along the mainshaft until the larger diameter on the spacer reaches the transfer gear lever cross-shaft.
11. Alternately tap the mainshaft forward and withdraw the spacer.
12. When the spacer is free on the mainshaft remove the extractor.
13. Withdraw the mainshaft assembly, allowing the first-speed gear to remain behind to avoid fouling on the casing.
14. Lift out the first-speed gear.
15. Refit the first-speed gear, scalloped thrust washer, thrust needle bearing and stepped thrust washer, stepped face outwards.
16. Withdraw the mainshaft spacer.

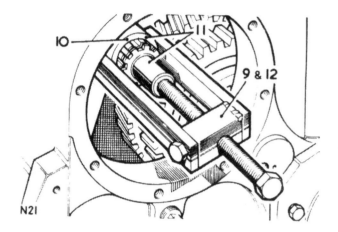

N21

Refitting

17. Position the gearbox with the RH side downwards to aid refitting.
18. Remove the gearbox side cover.
19. Temporarily move the first-speed gear toward the rear of the shaft.
20. Offer the assembled mainshaft to the gearbox and manoeuvre the first-speed gear past the reverse idler gear.
21. Engage the shaft into the main bearing.
22. Engage the first/second gear synchromesh outer member and the reverse idler gear.
23. Push the mainshaft home sufficient to allow the mainshaft spacer to be located on the rear end, with the spacer larger diameter forward of the transfer gear lever cross-shaft.
24. Re-position the first-speed gear, thrust washers and thrust needle bearing correctly on the mainshaft.
25. Push the mainshaft fully home, ensuring that the thrust washers and needle bearing remain correctly located against the first-speed gear.
26. Move the mainshaft spacer along the shaft, and into the oil seal, to abut the main bearing.

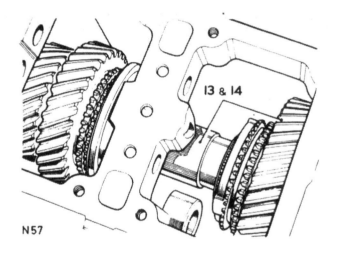

N57

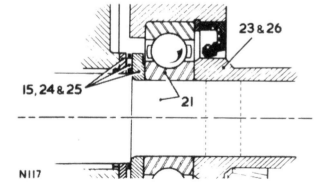

N117

continued

27. Temporarily fit the mainshaft transfer gear.
28. Position the snap ring in the groove in the mainshaft.
29. Hold the mainshaft fully to the rear and measure the clearance between the snap ring and the transfer gear.
30. Select a shim washer to allow 0,050 mm (0.002 in.) max. clearance between the snap ring and transfer gear when fitted. Shim range 1,8 to 2,0 mm (0.071 to 0,079 in.) in 0,05 mm (0.002 in.) increments.
31. Temporarily remove the snap ring and the mainshaft transfer gear.
32. Slide back the mainshaft spacer as far as the transfer gear lever cross-shaft will allow.
33. Apply a thin coating of Loctite AVV grade to the exposed area of the mainshaft.
34. Push home the mainshaft spacer.
35. Fit the mainshaft transfer gear.
36. Fit the previously selected shim washer and secure with the snap ring.
37. Reverse 1 to 7.

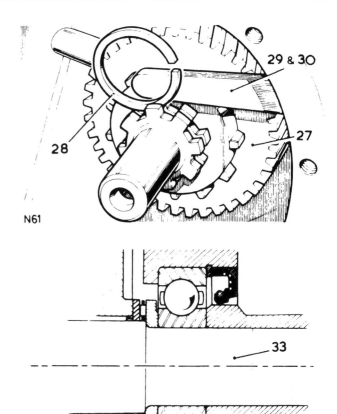

N61

DATA
Transfer gear end float 0,05 mm (0,002 in.) max.

N 318

MAINSHAFT TRANSFER GEAR

–Remove and refit   37.20.28.

Removing

1. Remove the mainshaft rear bearing housing.
2. Lift out the roller bearing.
3. Remove the snap ring.
4. Withdraw the shim washer.
5. Lift out the transfer gear.

Refitting

6. Fit the transfer gear to the mainshaft.
7. Fit the shim washer and snap ring.
8. Check the end float between the shim washer and snap ring. End float to be 0,050 mm (0,002 in.) max. Shim washer range 1,8 to 2,0 mm (0,071 to 0,079 in.) in 0,05 mm (0,002 in.) stages.
9. Fit the roller bearing and rear bearing housing.

DATA
Transfer gear end float 0,05 mm (0,002 in.) max.

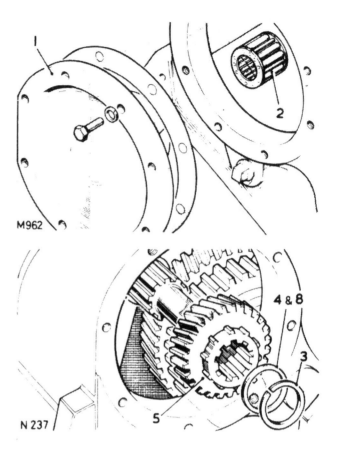

MAINSHAFT ASSEMBLY

– Overhaul   37.20.31

1. Remove the mainshaft assembly. 37.20.25.

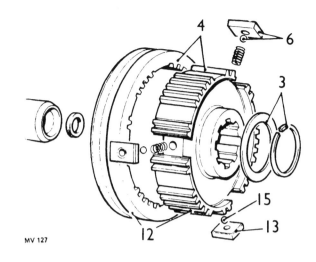

MV 127

Dismantling

2. Withdraw the first speed gear, thrust washers and roller bearings from the rear of the shaft.
3. Remove the snap ring and shim washer from the front of the shaft.
4. Lift off the third/fourth gears synchromesh assembly.
5. Withdraw the third and second speed gears and the associated thrust washers and needle roller bearings.
6. Dismantle the third/fourth gears synchromesh assembly, first pushing down the sliding blocks to free the synchromesh balls from the retaining groove in the outer member.
7. Dismantle the first/second gears synchromesh assembly in a similar manner, particularly noting their position for refitting.
8. Withdraw the oil seal from the bore in the mainshaft front end.
   NOTE: The synchromesh tooth form illustrated is interchangeable with early versions.

Assembling

9. Replacement thrust washers and roller bearings must not be degreased.
10. Lubricate all items before assembly, using clean main gearbox oil. Division 09 refers.
11. Fit the oil seal to the mainshaft front end.

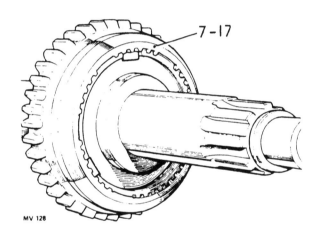

MV 128

Assembling the synchromesh units

12. Fit together the third/fourth gear synchromesh outer and inner members, outer member coned face toward inner member plain face.
13. Fit the sliding blocks, radiused face outward.
14. Locate the springs through the sliding blocks into the housing bores in the inner member.
15. Position the balls on the spring ends; press home in sequence and retain by hand.
16. Lift the outer member to retain the balls. Continue lifting until the balls spring home into the annular groove in the outer member.
17. Assemble the first/second gear synchromesh unit in the manner described for third/fourth gear unit. Fit the outer member coned face toward the front end of the mainshaft.

continued

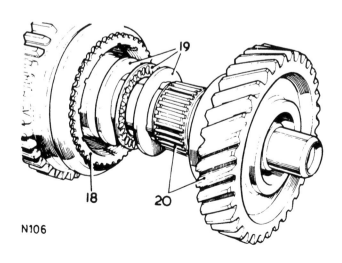

N106

Assembling the mainshaft front end

18. Fit a synchromesh cone to the first/second gear synchromesh outer member.
19. Position a chamfered thrust washer, a thrust needle bearing and a scalloped thrust washer on the mainshaft.
20. Fit a radial needle bearing and the second speed gear.
21. Fit a scalloped thrust washer, a thrust needle bearing and a further scalloped thrust washer.
22. Fit a radial needle bearing and the third speed gear.
23. Fit a scalloped thrust washer, a thrust needle bearing and a further scalloped thrust washer.
24. Position a synchromesh cone on to the third speed gear.
25. Fit the synchromesh unit, coned face to rear.

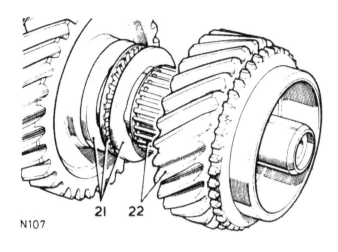

N107

Setting the gears end-float

26. Position the mainshaft assembly vertical, front end uppermost.
27. Apply a light loading on the gears to remove end-float.
28. Position the snap-ring in the mainshaft groove.
29. Measure the distance between the snap-ring lower edge and the synchromesh unit inner member.
30. Select a shim to reduce the measured clearance to 0,025 to 0,150 mm (0.001 to 0.006 in.) when fitted. Shim range is 1,85 to 2,45 mm (0.073 to 0.096 in.) in 0,15 mm (0.006 in.) increments.
31. Fit the selected shim washer and the snap-ring.

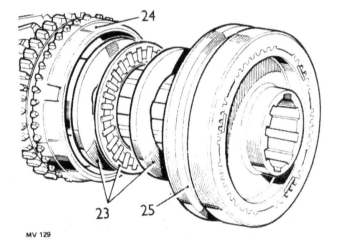

MV 129

Assembling the mainshaft rear end

32. Fit a synchromesh cone to the first/second gear synchromesh outer member.
33. Position a chamfered thrust washer, a thrust needle bearing and a scalloped thrust washer on the mainshaft.
34. Fit the first speed gear and bearing.
35. Fit a scalloped thrust washer, a thrust needle bearing and the stepped thrust washer, stepped face outwards.
36. The mainshaft spacer, transfer gear, shim washer and snap-ring are fitted during mainshaft refitting.
37. Refit the mainshaft, see 37.20.25.

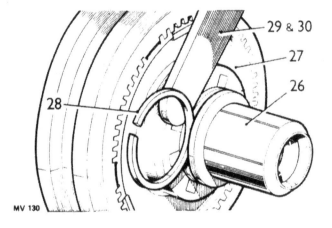

MV 130

DATA

End float on gears . . . . 0,025 to 0,150 mm
(0.001 to 0.006 in.)

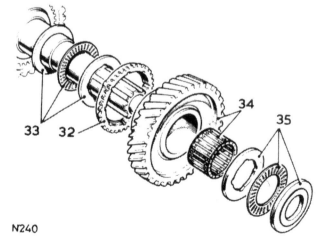

N240

REAR OUTPUT SHAFT OIL SEAL

– Remove and refit   37.23.01

Removing

1. Disconnect the rear propeller shaft at the transmission brake.
2. Remove the locking nut, washer and (later models) the felt rubber oil seal.
3. Withdraw the transmission brake drum complete with rear coupling flange.
4. Remove the oil catcher.
5. Prise off the oil shield.
6. Withdraw the oil seal.

Refitting

7. Press in the oil seal, open face first, until the seal plain face just clears the chamfer on the seal housing bore.
8. Fit the oil shield, which must be a close fit on the speedometer housing.
9. Fit the oil catcher, applying Bostik compound 771 to seal the oil catcher against the brake back plate.
10. Reverse instructions 1 to 3. Torque loading for locking nut: 16,5 kgf.m (120 lbf.ft).

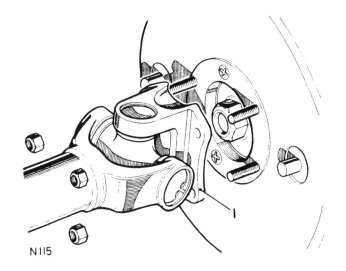

N 115

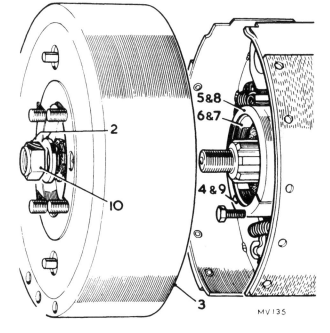

MV 135

FRONT OUTPUT SHAFT OIL SEAL

– Remove and refit   37.23.06

Removing

1. Disconnect the front propeller shaft.
2. Remove the coupling flange locknut and washer.
3. Withdraw the coupling flange complete with mudshield.
4. Withdraw the oil seal.

Refitting

5. Fit the oil seal, open side first.
6. Reverse items 1 to 3. Torque load for locknut: 16,5 kgf.m (120 lbf.ft).

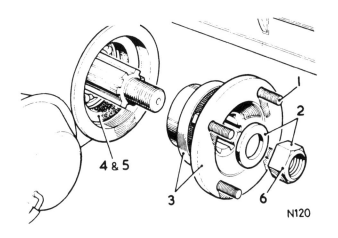

N120

37–27

SPEEDOMETER DRIVE HOUSING

– Remove and refit   37.25.09

Removing

CAUTION: Chock the vehicle wheels before commencing this operation as it is necessary to release the parking brake during the following procedure.

1. Disconnect the rear propeller shaft at the transmission brake.
2. Remove the clevis pin to disconnect the handbrake linkage.
3. Disconnect the speedometer drive cable.
4. Remove the fixings, speedometer drive housing to gearbox casing.
5. Withdraw the speedometer drive housing complete with transmission.

Refitting

NOTE: If a replacement speedometer drive housing is being fitted, carry out the 'Differential bearings pre-load check' in 'Speedometer drive housing – overhaul', 37.25.13.

6. Position the joint washer.
7. Offer the drive housing to the gearbox and engage the rear output shaft splines in the differential unit.
8. Position the flat on the drive housing adjacent to the flat on the intermediate shaft.
9. Reverse instructions 1 to 4. Torque load for the propeller shaft fixings is 4,8 kgf.m (35 lbf.ft).
   Torque load for speedometer drive housing fixings is 3,1 kgf.m (22 lbf.ft).

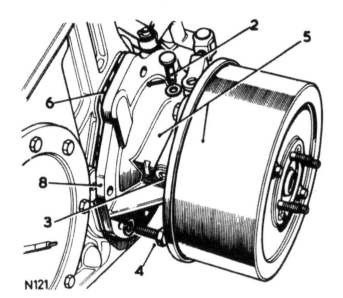

SPEEDOMETER DRIVE HOUSING

– Overhaul   37.25.13

Dismantling

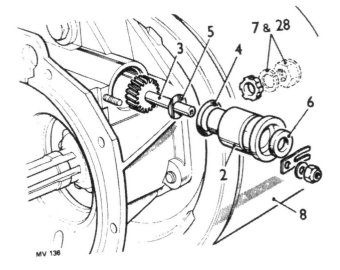

MV 136

1. Remove the speedometer drive housing, see 37.25.09.
2. Remove the speedometer spindle housing.
3. Lift out the driven gear and spindle.
4. Take off the 'O' ring seal.
5. Remove the thrust washer.
6. Withdraw the oil seal.
7. Remove the locking nut, washer and felt seal, output coupling flange to output shaft.
8. Withdraw the brake drum and coupling flange complete.
9. Drive out the rear output shaft, using a hide mallet on the threaded end.
10. Slide off the spacer and speedometer worm.
11. Remove the oil catcher.
12. Withdraw the oil shield.
13. Withdraw the oil seal.
14. Remove the circlip.
15. Tap out the ball bearing.

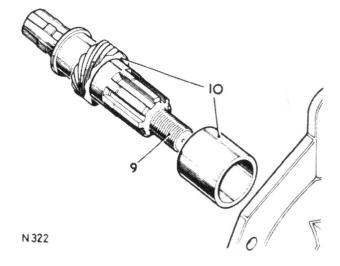

N 322

Differential bearing pre-load check, instructions 16 to 22 inclusive

This check must be carried out if a replacement speedometer drive housing is to be fitted. The check is also required if a replacement gearbox, differential unit or differential unit bearing is being fitted.

16. Measure and record the thickness of the new joint washer for the speedometer drive housing.
17. Offer the speedometer housing, less joint washer, to the gearbox.
18. Engage the differential unit bearing inner member with the outer member in the drive housing.
19. Measure the clearance between the drive housing and gearbox joint faces. This must be 0,05 mm (0.002 in.) more than the recorded thickness of the new joint washer.
20. To adjust the joint face clearance, adjust the thickness of shimming fitted behind the rear bearing outer face as follows: instructions 21 and 22.
21. Drive out the bearing outer race.
22. Withdraw the shim washer and select a replacement of the required thickness. Shim thickness range is 1,65 to 2,80 mm (0.065 to 0.110 in.) in 0,05 mm (0.002 in.) stages.

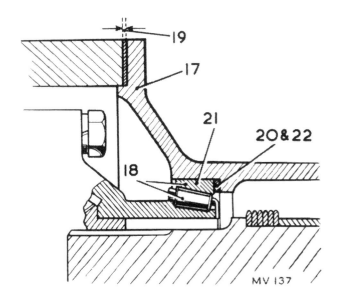

MV 137

continued

Reassembling

23. Reverse instructions 14 and 15.
24. Press in the output coupling flange oil seal, open face first, until the seal plain face just clears the chamfer on the seal housing bore.
25. Fit the oil shield, which must be a close fit on the speedometer housing.
26. Fit the oil catcher, applying Bostik compound 771 to seal the oil catcher against the brake backplate.
27. Reverse instructions 8, 9 and 10.
28. Fit the felt seal, plain washer and locking nut to secure the output flange. Torque:— 16,5 kgf.m (120 lbf.ft).
29. Reverse instructions 1 to 6.

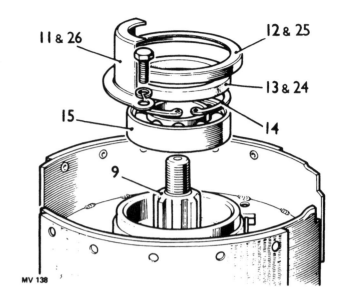

MV 138

DATA

Bearing pre load . . . . 0,050 to 0,10 mm
(0.002 to 0.004 in.)

Alternative rolling resistance method in situ.

With front propellor shaft disconnected and rear drive flange removed a rolling resistance of 14 to 16 lb (6 to 7 kg) pull should be obtained using a spring balance (cord wound around differential).

REVERSE LIGHT SWITCH

–Remove and refit   37.27.01.

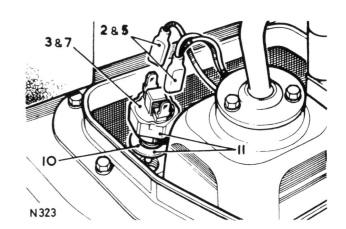

N 323

Removing

1. Lift aside the main gear lever grommet.
2. Disconnect the electrical leads.
3. Unscrew the reverse light switch.

Refitting

4. Engage reverse gear.
5. Connect the electrical leads to the switch.
6. Switch the ignition 'ON'.
7. Screw in the switch, less shim washers, until the switch contacts are made.
8. Screw in a further half turn.
9. Measure the clearance between the switch lower face and the gearbox.
10. Select shim washers to suit the clearance. Shim thicknesses are 0,5 mm (.020 in.) and 0,127 mm (.005 in.)
11. Fit the selected shim washer/s and switch. Tighten to a torque of 1.4 to 2.0 kgf.m (15 to 20 lbf.ft.).
12. Fit the main-gear lever grommet.

DIFFERENTIAL LOCK ACTUATOR SWITCH

–Remove and refit   37.27.05.

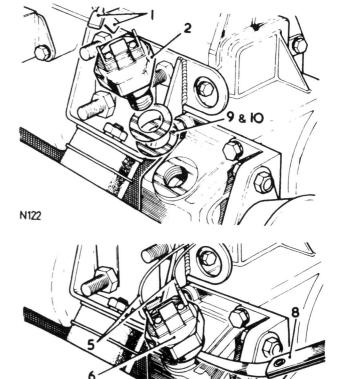

N122

N 123

Removing

1. Disconnect the electrical leads at the switch.
2. Unscrew the switch.

Refitting

3. Start the engine.
4. Move the differential lock vacuum control valve to the 'up' position.
5. Connect the electrical leads to the actuator switch.
6. Screw in the switch, less shim washers, until the switch contacts are made.
7. Screw in a further half turn.
8. Measure the clearance between the switch lower face and the housing.
9. Select shim washers to suit the clearance. Shim thicknesses are 0,5 mm (.020 in.) and 0,127 mm (.005 in.).
10. Fit the selected shim washer/s and the switch.
11. Reverse 3 and 4.

GEAR LEVER AND CROSS-SHAFT, TRANSFER GEARBOX

–Remove and refit   37.29.01.

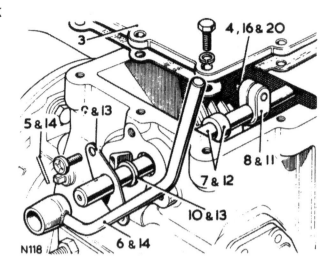

Removing

1. Remove the front floor. 76.10.12.
2. Remove transmission brake lever and linkage. 70.45.01.
3. Remove the top cover.
4. Slacken the selector finger pinch bolt.
5. Drive out the retaining pin, gear lever to cross-shaft.
6. Withdraw the gear lever.
7. Withdraw the cross-shaft and distance collar.
8. Lift out the selector finger.
9. Remove the retaining plates.
10. Withdraw the sealing rings.

Refitting

11. Position the selector finger in the gearbox.
12. Fit the cross-shaft and spacing collar and engage the selector finger.
13. Fit the sealing ring and retaining plate at the RH side of the gearbox.
14. Fit the gear lever and retaining pin.
15. Fit the remaining sealing ring and retaining plate.
16. Tighten the selector finger pinch bolt.
17. Select 'High' transfer range, that is, the larger intermediate gear engaged.
18. Slacken the selector finger pinch bolt.
19. Rotate the xross-shaft until the gear lever is inclined ten degrees forward of the vertical position.
20. Tighten the selector finger pinch bolt.
21. Reverse 1 to 3.

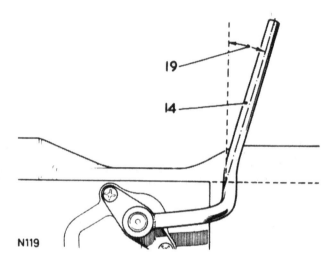

Removing damper (early models)

22. Unscrew and remove the knob and rubber sleeve.
23. Unscrew the upper gear lever.
24. Remove the metal and rubber sleeve assembly.
25. Withdraw the rubber spacer.
26. If necessary, replace the rubber sleeves.

Refitting damper

27. Reverse 22 to 25.

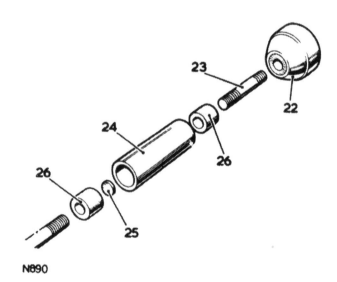

GEAR SELECTORS & SHAFT, TRANSFER GEARBOX

—Remove and refit   **37.29.04**

Removing

1. Remove the front floor. 76.10.12.
2. Remove the speedometer drive housing. 37.25.09.
3. Remove the transfer gearbox top cover.
4. Select 'Low' range transfer gear.
5. Drive out the retaining pin from the front selector fork sufficient to free the fork.
6. Ease the differential unit to the rear.
7. Push the forward selector fork forward on the shaft.
8. Pull to the rear on the rear selector fork to move the selector shaft out of engagement with the detent balls in the casing rear face.
9. Remove the pinch bolt on the rear fork.
10. Partially withdraw the selector shaft and lift out the selector forks.
11. Remove the retaining pin from the front fork.
12. Withdraw the selector shaft, closing the shaft housing by hand to prevent the detent balls from dropping into the casing.
13. Withdraw the two detent balls.
14. Lift out the spacing rod and spring.
15. Remove the closing plug.
16. Withdraw the detent spring from the cross drilling.

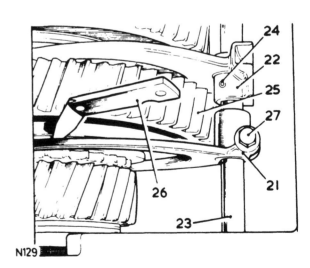

Refitting

17. Position the detent spring in the inner bore in the cross drilling.
18. Locate the detent ball on the spring.
19. Enter the selector shaft, push the ball against the spring and push in the shaft.
20. Fit the detent ball, spring and spacing rod to the vertical drilling.
21. Position the rear selector fork, plain face to rear, in the gearbox.
22. Position the front selector fork, extended boss to the rear, in the gearbox.
23. Align the retaining pin holes and engage the selector shaft in the selector forks.
24. Fit the retaining pin, front fork to shaft.
25. Set transfer gears in 'Neutral' position.
26. Adjust the rear fork position until there is 0,25 mm (.010 in.) clearance between the front face of the rear fork and the rear face of the input gear inner member.
27. Tighten the rear fork pinch bolt.
28. Fit the closing plug to the cross drilling.
29. Reverse 1 to 3.

DATA

Clearance for selectors   0,25 mm (.010 in.)

INTERMEDIATE GEARS ASSEMBLY

– Remove and refit   **37.29.10**

Service tool: RO 1003 slave intermediate shaft

Removing

1. Drain the transfer gearbox oil.
2. Remove the speedometer drive housing, see 37.25.09.
3. If necessary, remove the intermediate exhaust pipe.
4. Remove the gearbox bottom cover.
5. Screw a suitable extractor into the 8 mm threaded hole provided in the intermediate gear shaft.
6. Hold the intermediate gear cluster in position and withdraw the shaft.
7. Insert the slave shaft RO 1003 to retain together the gears assembly.
8. Withdraw the intermediate gears assembly.
9. Slide the thrust washers, bearings and gears from the slave shaft.
10. The input gear and outer member is a riveted assembly and no dismantling is permitted.
11. Remove the 'O' ring seal from the intermediate gear shaft.

continued

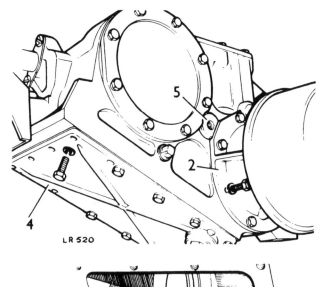

LR 520

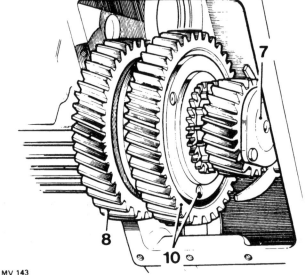

MV 143

Inspecting

12. Examine all parts for wear, damage and general condition. Renew as necessary.

Refitting

13. Place the slave shaft on the bench, extractor thread end uppermost.
14. Fit a pear-shaped thrust washer, inner ring and a thrust bearing washer to the shaft (ring grooved face downwards).
15. Fit a needle roller bearing and the 'high' gear (plain face first) to the shaft.
16. Position a thrust bearing washer on the 'high' gear.
17. Fit a spacer, needle roller bearing and a further spacer to the input gear inner member.
18. Position the assembled input gear on the shaft and engage the lower spacer in the previously positioned thrust bearing washer.
19. Locate a thrust bearing washer over the upper spacer.
20. Fit a needle roller bearing and the 'low' gear (plain side last) to the shaft.
21. Fit the remaining thrust needle bearing, inner ring and thrust washer (ring grooved face upwards).

continued

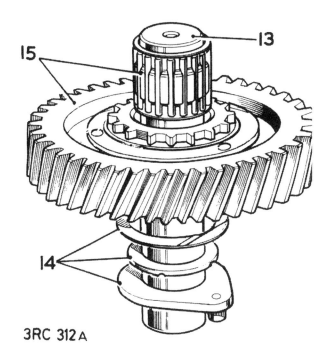

3RC 312 A

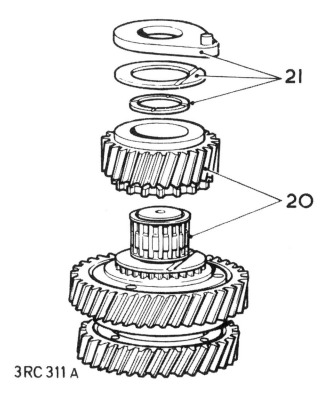

3RC 311 A

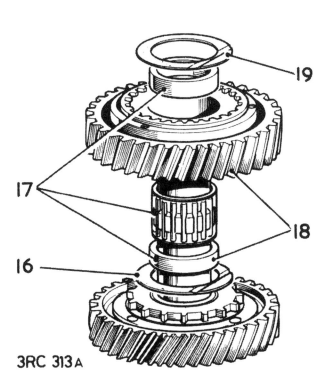

3RC 313 A

37–35

The following procedure, instructions 22 to 27, must be completed before refitting the intermediate gears into the transfer gearbox, to ensure that there is sufficient running clearance for the bearings.

22.　With the intermediate transfer gear assembly located on the slave shaft RO 1003, place the complete assembly on a surface plate with the low gear uppermost.

23.　Remove the two pear-shaped thrust washers, one situated at each end of the gear assembly.

24.　Place a suitable straight-edge across the thrust bearing washer.

25.　Check that a clearance (end-float) exists between the straight-edge and the inner ring, to ensure a running clearance when the assembly is installed.

　　　CAUTION: DO NOT refit the assembly with the needle roller bearings in a pre-load condition.

26.　If there is no clearance between the straight-edge and the inner ring, use selective assembly of alternative components to obtain the required condition.

27.　In event of selective assembly not giving clearance, it is permissible to face down each spacer on a surface plate to a maximum of 0,13 mm (0.005 in.).

28.　When the foregoing bearing clearance check has been completed, slide the gears and slave shaft assembly into the transfer gearbox and engage the selector forks.

29.　Withdraw the slave shaft and lubricate the bearings through the shaft aperture.

30.　Fit the intermediate shaft and 'O' ring seal with the flat on the shaft toward the differential unit.

31.　Measure the clearance between the rear thrust washer and the gear casing. This must be 0,22 to 0,36 mm (0.009 to 0.014 in.).

32.　Adjustment is carried out by substituting one or both of the thrust washers. The washers are available in 3,55 mm (0.139 in.), 3,63 mm (0.143 in.) and 3,74 mm (0.147 in.) thicknesses.

33.　Refit the gearbox bottom cover.

34.　Refit the speedometer drive housing, see 37.25.09.

35.　Refit the intermediate exhaust pipe.

36.　Refill the transfer gearbox to the correct level.

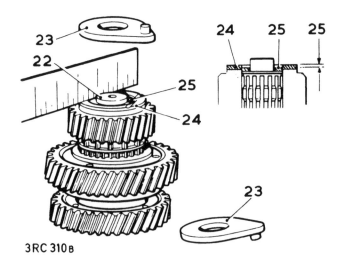

3RC 310B

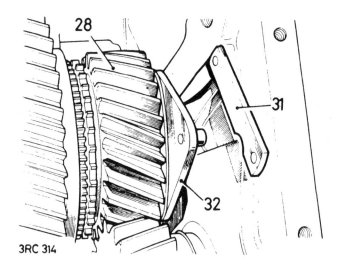

3RC 314

DATA

Gears end-float . . . . 0,15 to 0,23 mm
(0.006 to 0.09 in.)

DIFFERENTIAL UNIT ASSEMBLY

– Remove and refit   37.29.13

Remove the front floor 76.10.12.

Removing

1. Remove the differential lock actuator assembly 37.29.19.
2. Remove the front output shaft and housing 37.10.05.
3. Remove the speedometer drive housing 37.25.09.
4. Withdraw the differential unit.
    NOTE: The early type of differential, shown at the top of the page, is not serviceable and, if worn, must be replaced by the later type, shown below, complete with front and rear output shafts.

Refitting

5. Reverse 1 to 4.

NOTE: If a replacement differential unit is being fitted, carry out the 'Differential bearing pre-load check', described in 'Speedometer drive housing – overhaul'. 37.25.13.

6. Refit the front floor. 76.10.12.

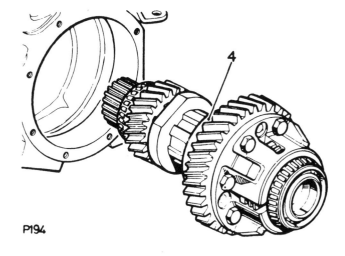

P194

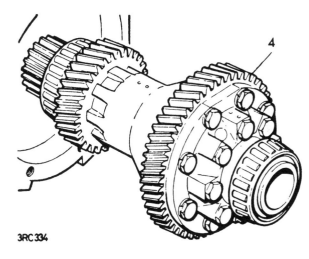

3RC334

DIFFERENTIAL UNIT

– Overhaul   37.29.16

Service tool: 18G 47 BB bearing extractor.

Dismantling

NOTE: During dismantling it is essential that all components are marked in their original position and relative to other components, so that if original components are refitted, their initial setting is maintained.

1. Remove the differential unit, see 37.29.13.
2. Press off the roller bearings using tool 18G 47 BB.
3. Withdraw the high speed gear.
4. Remove the fixings, low speed gear to casing.
5. Withdraw the gear.
6. Remove the casing securing bolts.
7. Lift off the rear case assembly.
8. Withdraw the side gear.
9. Slide out the cross-shafts and remove the bevel pinions and thrust washers from the front case assembly.
10. Withdraw the side gear to dismantle the front case assembly.

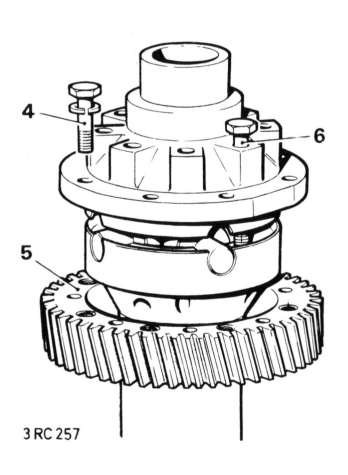

3 RC 257

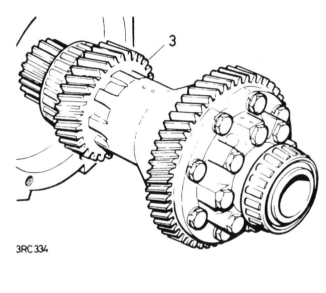

3RC 334

Inspecting

NOTE: If replacements are required, replace the following items 11 and 12 as sets.

11. Bevel pinions and side gears (set of six).
12. Cross-shafts (set of two).
13. Check the gear teeth for damage.
14. Check all parts for satisfactory general condition.
    NOTE: The differential case halves are a matched pair, and halves must not be changed individually.

Reassembling

15. Fit a side gear into the rear casing.
16. Fit a side gear into the front casing.
17. Fit the bevel pinions, thrust washers and cross-shafts into the front casing.
18. Fit the rear casing to the front casing. Tighten the bolts evenly in sequence. Torque load 5,3 to 6,3 kgf.m (40 to 45 lbf.ft).
19. Offer the low gear to the differential casing.
20. Align the fixing holes and fit the bolts evenly in sequence. Torque load 5,8 to 6,5 kgf.m (44 to 47 lbf.ft).
21. Fit the roller bearings and refit the differential unit.
    NOTE: If the differential case or bearings have been replaced, carry out the 'Differential bearing pre-load check', 37.25.13.

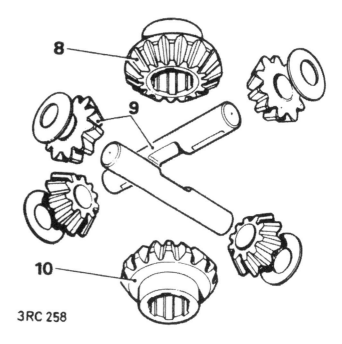

3RC 258

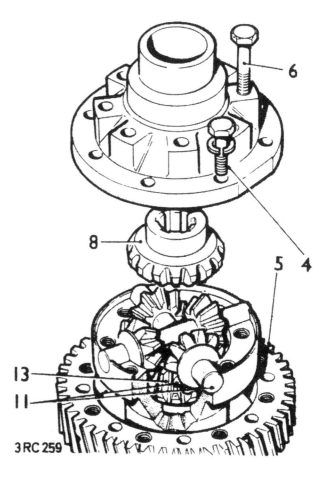

3RC 259

37-39

DIFFERENTIAL LOCK ACTUATOR ASSEMBLY

Remove and refit	37.29.19.
Overhaul. Instructions 7 to 12	37.29.22.

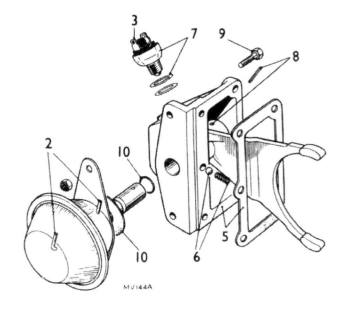

MV144A

Removing.

1. Raise the vehicle on a hoist.
2. Note the vacuum hose positions then disconnect the hoses.
3. Disconnect the leads from the differential lock warning switch.
4. Using a 'Pozidriv' screwdriver bit with a suitable adaptor remove the screws securing the actuator assembly to the gearbox.
5. Withdraw the differential lock actuator assembly and gasket from the gearbox.
6. Withdraw the detent spring and ball.

Dismantling.

7. Remove the differential lock warning switch and shim(s).
8. Remove the pin securing the selector fork to the actuator shaft.
9. Remove the vacuum actuator fixings.
10. Withdraw the vacuum unit, shaft and gasket.
11. Remove the 'O' ring seal.
12. Examine all components and renew as necessary.

Assembling.

13. Reverse instructions 7 to 11. Coat the gasket fitted between the vacuum unit and the housing with Universal jointing compound.

Refitting.

14. Reverse instructions 1 to 6. Coat the gasket fitted between the differential lock actuator assembly and the gearbox with Universal jointing compound.

PROPELLER SHAFT OPERATIONS

Operation No.

Front propeller shaft
    -remove and refit   47.15.02
    -overhaul   47.15.11

Rear propeller shaft
    -remove and refit   47.15.03
    -overhaul   47.15.12

FAULT DIAGNOSIS

SYMPTOM	POSSIBLE CAUSE	CURE
Vibrating propellor shaft	Fixings loose	Tighten the fixings evenly and securely
	Incorrectly assembled propellor shaft	Reassemble propellor shaft correctly aligned
	Worn needle roller bearings	Fit new bearings
	Worn splines	Fit new propellor shaft complete
	Shaft out of balance	Fit new propellor shaft complete
Noisy universal joints	Lack of lubrication	Lubricate propellor shaft
	Fixing loose	Tighten the fixings evenly and securely
	Worn needle roller bearings	Fit new bearings
	Worn splines	Fit new propellor shaft complete

PROPELLER SHAFT

Remove and refit

Front propeller shaft	**47.15.02**
Rear propeller shaft	**47.15.03**

Removing

1. Disconnect the propeller shaft from the axle and gearbox.
2. Withdraw the propeller shaft.

Refitting

3. Locate the propeller shaft in position ensuring that the sliding member is fitted towards the front of the vehicle.
4. Fit the securing nuts and bolts to the axle and gearbox flanges. Torque 4,1 to 5,2 kgf.m (30 to 38 lbf.ft).

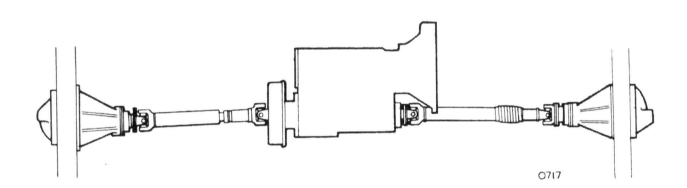

O717

PROPELLER SHAFT

Overhaul

Front propeller shaft	**47.15.11**
Rear propeller shaft	**47.15.12**

Dismantling

1. Remove the propeller shaft from the vehicle.
2. If a gaiter encloses the sliding member release the two securing clips. Slide the gaiter along the shaft to expose the sliding member.
3. Note the alignment markings on the sliding member and the propeller shaft.
4. Unscrew the dust cap and withdraw the sliding member.
5. Clean and examine the splines for wear. Worn splines or excessive backlash will necessitate propeller shaft renewal.
6. Remove paint, rust, etc. from the vicinity of the universal joint bearing cups and circlips.
7. Remove the circlips, and grease nipple.

continued

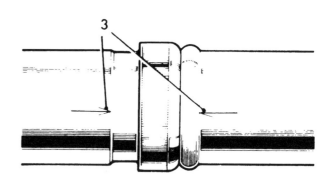

2RC632A

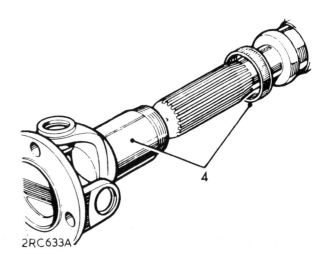

2RC633A

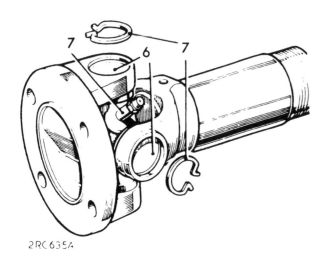

2RC635A

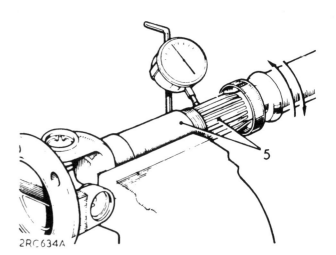

2RC634A

8. Tap the yokes to eject the bearing cups.
9. Withdraw the bearing cups and spider and discard.
10. Repeat instructions 5 to 8 at opposite end of propeller shaft.
11. Thoroughly clean the yokes and bearing cup locations

Assembling

12. Remove the bearing cups from the new spider.
13. Check that all needle rollers are present and are properly positioned in the bearing cups.
14. Ensure bearing cups are one third full of fresh lubricant.
15. Enter the new spider complete with seals into the yokes of the sliding member flange. Ensure that the grease nipple hole faces away from the flange.
16. Partially insert one bearing cup into a flange yoke and enter the spider trunnion into the bearing cup taking care not to dislodge the needle rollers.
17. Insert the opposite bearing cup into the flange yoke. Using a vice, carefully press both cups into place taking care to engage the spider trunnion without dislodging the needle rollers.
18. Remove the flange and spider from the vice.
19. Using a flat faced adaptor of slightly smaller diameter than the bearing cups press each cup into its respective yoke until they reach the lower land of the circlip grooves. Do not press the bearing cups below this point or damage may be caused to the cups and seals.
20. Fit the circlips.
21. Engage the spider in the yokes of the sliding member. Fit the bearing cups and circlips as described in instructions 15 to 19.
22. Lubricate the sliding member splines and fit the sliding member to the propeller shaft ensuring that the markings on both the sliding member and propeller shaft align.

continued

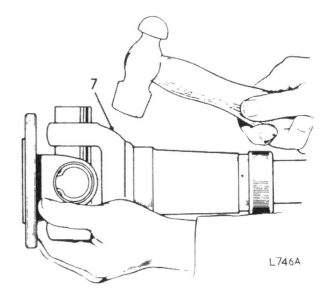

L746A

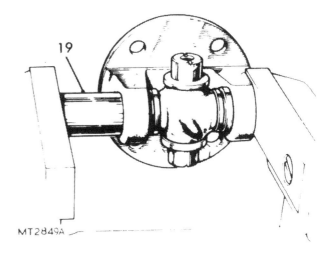

MT2849A

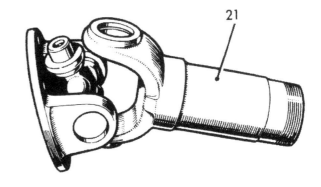

2RC638A

23. Fit and tighten the dust cap.
24. Fit the grease nipples to the spider and the sliding member and lubricate.
25. Slide the gaiter (if fitted) over the sliding member.
26. Locate the sliding member in the mid-position of its travel and secure the gaiter clips. Note that the gaiter clips must be positioned at 180 degrees to each other so that they will not influence propeller shaft balance.
27. Apply instructions 15 to 19 to the opposite end of the propeller shaft.
28. Fit the grease nipple and lubricate.
29. Fit the propeller shaft to the vehicle.

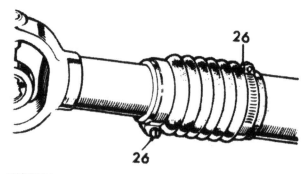

4RC 873A

DATA

Type	Hardy Spicer needle bearing
Tubular shaft	50,8 mm (2 in) diameter
Overall length (face to face in midway position):	
Front propeller shaft	643,4 mm (25.33 in.)
Rear propeller shaft	887,4 mm (34.937 in.)

VIBRATION

In vehicles where drive line imbalance has been identified on road or roller test and when wheels and tyres have been checked the propeller shaft(s) should be turned through 180 degrees relative to the axle flange and the vehicle retested.

REAR AXLE AND FINAL DRIVE OPERATIONS

	Operation No.	
Axle shaft – remove and refit	51.10.01.	51–1
Differential assembly		
–remove and refit	51.15.01.	
–overhaul	51.15.07.	
Pinion oil seal – remove and refit	51.20.01.	
Rear axle assembly – remove and refit	51.25.01.	

DIFFERENTIAL ASSEMBLY

—Remove and refit   1 to 15   **51.15.01**

AXLE SHAFT

—Remove and refit   1 to 5 and 13 to 15   **51.10.01**

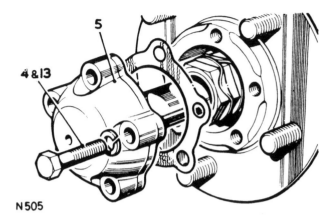

N505

Removing

1. Jack up rear and support chassis.
2. Drain the oil from the differential.
3. Remove the road wheel.
4. Remove the hub cap fixings.
5. Tap on the hub cap to break adhesion and withdraw the hub cap, joint washer and axle shaft.
6. Disconnect the propeller shaft.
7. Remove the fixings at the differential flange.
8. Move aside the brake pipes and mounting bracket.
9. Withdraw the differential assembly and joint washer.

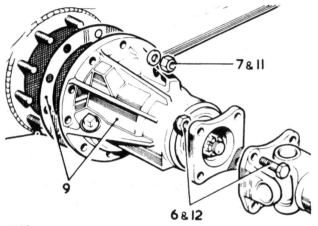

N136

Refitting

10. Reverse 8 and 9.
11. Fit the differential fixings. Torque 3,5 to 4,6 kgf.m (26 to 34 lbf.ft).
12. Fit the propeller shaft. Fixing torque 4,1 to 5,1 kgf.m (30 to 38 lbf.ft).
13. Fit the hub cap and axle shaft. Torque 4,1 to 5,1 kgf.m (30 to 38 lbf.ft).
14. Reverse 1 to 3.
15. Replenish the differential oil, see Maintenance, page 10–19.

DIFFERENTIAL ASSEMBLY

—Overhaul 51.15.07

Special tools

Part No.	262757	Bearing Extractor
Part No.	530105	Differential spanner
Part No.	262761	Height gauge for
Part No.	600299	differential pinion.
Part No.	601998	1 off, as required
Part No.	605004	(See text)
Part No.	530106	Bracket for Dial Test Indicator

1. Remove the differential assembly 51.15.01 or 54.10.01.

N134

Dismantling

NOTE. During dismantling it is essential that all components are marked in their original position and relative to other components, so that if original components are refitted, their initial setting is maintained. Two types of differential cross shaft have been fitted. On earlier vehicles the cross shaft is circular throughout its length and is secured in position with pins. On later vehicles the cross shaft has flats towards either end and is secured in position with circlips.
Bearings may differ between earlier and later units.

2. Remove the bearing caps.
3. Remove the serrated nuts.
4. Withdraw the crownwheel and differential assembly.
5. Withdraw the differential bearings outer tracks.
6. Remove the driving flange.
7. Withdraw the pinion.
8. Withdraw the shim washers.
9. Prise out the oil seal.
10. Withdraw the spacer.
11. Withdraw the drive flange roller bearing.
12. Press off the pinion head bearing.
13. Locate the tool 262757 in the pinion housing. Ensure that the projections on the extractor bar fit the cast slots at the rear of the bearing outer race. If necessary, grind the projections until a sliding fit is obtained, otherwise the pinion housing may be damaged.
14. Extract pinion head bearing outer race together with its shim.
15. Press out flange end outer race.
16. Remove the crownwheel from the differential case.

NOTE. On later models lockplates are replaced by a plain washer.

continued

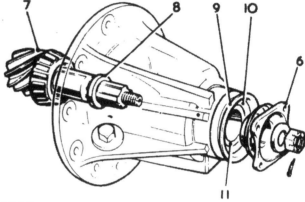

N135

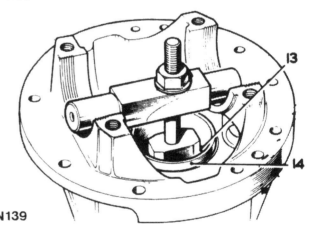

N139

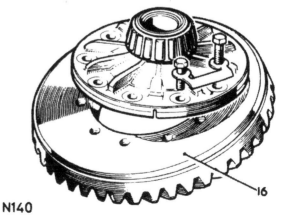

N140

51–3

17. Remove the split pin (earlier models) or circlip (later models) from the differential spindle.
18. Withdraw the spindle.
19. Withdraw the pinions by rotating the wheels.
20. Withdraw the wheels.
21. Withdraw thrust washers.
22. Extract the roller bearings.

Inspecting

23. Examine all components for obvious wear or damage.
24. All bearings must be a press fit, except the flange end pinion bearing, which must be a slide fit on the shaft.
25. Crownwheel and pinion is only supplied as a matched. set and MUST NOT be interchanged separately.
26. Bevel pinion housing and bearing caps are matched sets, and MUST NOT be interchanged separately.
27. Check the differential pinion seatings in the case, as follows—
28. The spherical seats must be finished flush.
29. The seat must not be stepped or recessed.
30. If a step is present, it must be ground away to prevent the pinion teeth rubbing the casing.

continued

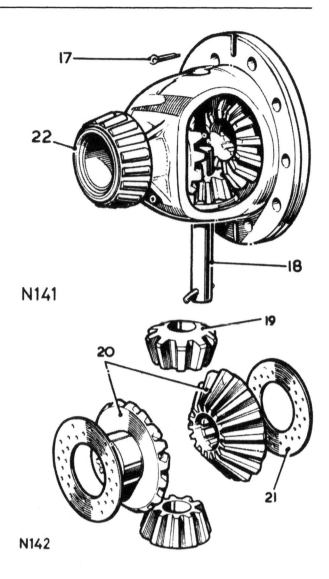

N141

N142

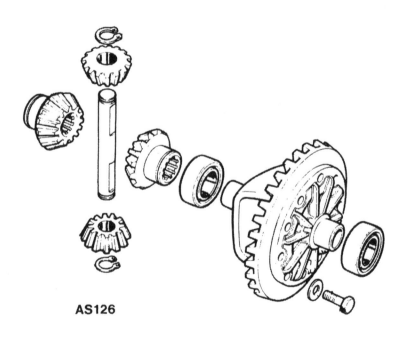

AS126

Assembling

31. Fit shim of same thickness removed during dismantling, in pinion head bearing seat.

NOTE. If original shim has been mislaid, use new shim of at least 1,27 mm (.050 in.) thickness.

32. Press in the pinion head bearing outer race.
33. Press in the flange end bearing outer race.
34. **Press the pinion head roller bearing onto the pinion.**
35. Locate the pinion shaft into the case together with the bearing pre-load adjustment shim removed during dismantling.

NOTE. If original shim has been mislaid, use new shim of at least 4,06 mm (.160 in.) thickness.

36. Fit the flange end roller bearing.
37. Fit the distance washer.

NOTE. Do not fit the oil seal at this stage.

38. Fit the driving flange.
39. Fit the washer and nut. Torque 11,7 kgf.m (85 lbf. ft.). While tightening the nut, check that the pinion can rotate. If the pinion becomes excessively stiff, use a thicker pre-load adjustment shim.

Check and adjust pinion bearing pre-load

40. Tie a length of cord to the driving flange, then coil it around the flange hub.
41. Attach a spring balance to the loose end of the cord.
42. Apply a steady pull on the spring balance and note the force required to rotate the pinion shaft, after having overcome inertia.

    Bearing pre-load is correct when a figure of 2,7 to 11,3 kg (6 to 25 lb.) is recorded on the spring balance

43. Adjustment can be made by changing the shims located on the pinion shaft between the bearings, shims are available in a range of thicknesses. Thicker shimming will reduce bearing pre-load, thinner shimming will increase pre-load.

continued

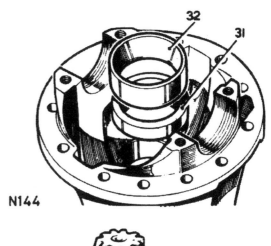

N144

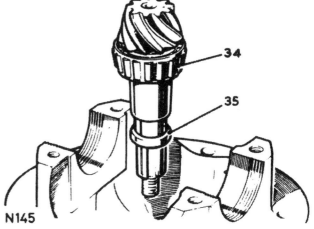

N145

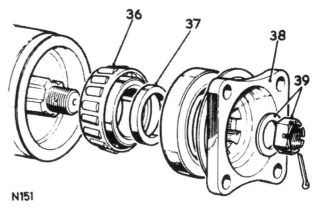

N151

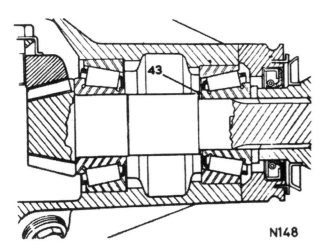

N148

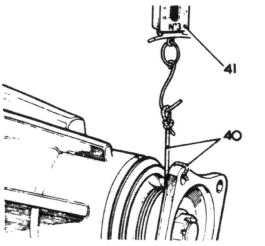

N147

Check and adjust pinion height setting

44. Locate the pinion height gauge into the pinion housing and secure with the bearing caps.

NOTE. There are four variations of height gauge in use, and either one may be used.

45. Place the slip gauge on to the pinion face and hold firmly in place.

46. Using feeler gauges, measure the clearance between the height gauge and the slip gauge. Depending on the height gauge used, the following clearance must be obtained.

    0,07 mm to 0,10 mm (.003 in. to .004 in.) with height gauges Part Nos. 601998, 262761 and 600299.
    0,28 mm to 0,30 mm (.011 in. to .012 in.) with height gauge Part No. 605004.

47. If necessary, adjust the thickness of shims between the pinion head bearing outer race and the pinion case to obtain the correct clearance. Use tool 262757 to remove outer race.

IMPORTANT. Any adjustment of the pinion height will affect the pinion bearing pre-load. When the pinion height is correct, repeat items 40 to 43.

48. When the pinion height and bearing pre-load is correct, remove the height gauge.
49. Remove the pinion driving flange.
50. Smear the outside diameter of the pinion oil seal with jointing compound.
51. Fit the seal, lipped side inward.
52. Fit the driving flange.
53. Fit the flange securing nut and washer. Torque 9,6 to 16,5 kgf.m (70 to 120 lbf.ft.).
54. Secure nut with split pin.

Differential wheel and pinion backlash

55. Place a thrust washer in position on the rear face of each differential wheel.
56. Locate the two differential wheels and thrust washers into the differential case.
57. Insert the differential pinions at exactly opposite points, then rotate the wheel and pinion assembly to align the holes in the pinions and case for the pinion spindle.

NOTE. If original components are being refitted, ensure that the wheel and pinion assembly is in its original position before fitting the spindle.

continued

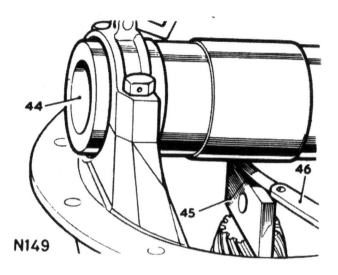

N149

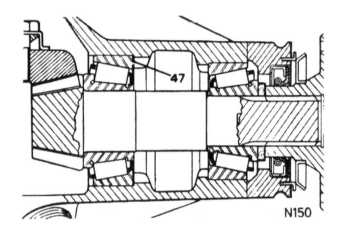

N150

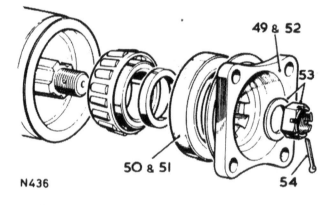

N436

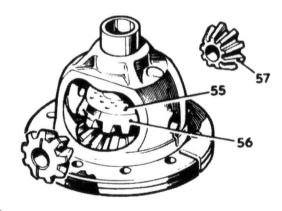

N152

58. Ensure that the plain pin is secure in the pinion spindle. (Suffix 'A' models only).

59. Fit the spindle and later models — fit two new circlips to secure the spindle in the case.

60. Check for backlash between the differential wheels and pinions, a manual check is sufficient, no actual measuring is necessary. There must be backlash, but this must be the minimum obtainable consistent with smooth running wheels and pinions. Adjustment can be made by changing the thrust washers for the differential wheels, which are available in a range of thicknesses. Note that Tufnell washers are fitted only to Suffix 'A' models.

61. Secure the spindle with a split pin. (Suffix 'A' models) or with circlips (From Suffix 'B').

62. Align the crownwheel with the differential case.

63. Suffix 'A' only. Locate lockplates in place. Plain washers are fitted on models from Suffix 'B'.

64. Fit the crownwheel securing bolts and on Suffix 'A' models, the lock plates. Note, on Suffix 'A' models the two fitted bolts which must be fitted diametrically opposite.
   Use Loctite Studloc on the crownwheel bolts on models from Suffix 'B'.
   Torque — Suffix 'A' 4,8 kgf.m (35 lbf.ft).
   — from Suffix 'B' 5,5 to 6,2 kgf.m (40 to 45 lbf.ft.)

65. Press on the differential roller bearings.

66. Fit the bearing outer races and locate the differential into the pinion housing.

67. Fit the serrated nuts.

68. Fit the bearing caps. Tighten the securing bolts firmly but not fully.

69. Using tool 530105, tighten both serrated nuts to remove all bearing end float without introducing pre-load.

70. Using a dial test indicator, measure the run-out on the rear face of the crownwheel, this must not exceed 0,10 mm (0.004 in.). If excessive run-out is recorded, the crownwheel and differential must be removed from the bevel pinion housing and the crownwheel repositioned on the differential case. Re-assemble and recheck. If necessary, this procedure must be repeated until the run-out is correct.

71. When the crownwheel run-out is correct, ensure that the lockplates are fully engaged over the crownwheel securing bolts.

72. Using a dial test indicator, check the crownwheel to bevel pinion backlash. This must be 0,20 mm to 0,25 mm (0.008 in. to 0.010 in.). Where necessary, adjust the crownwheel backlash by alternately slackening and tightening the serrated nuts until the backlash is correct.

continued

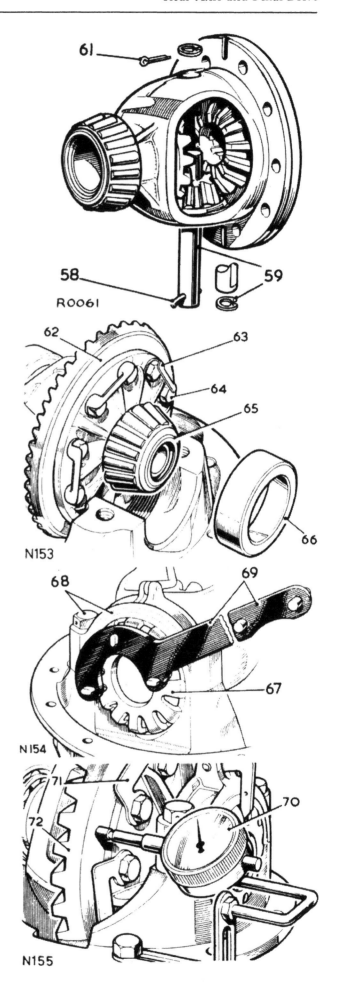

R0061

N153

N154

N155

219

51–7

73. Move serrated nuts as indicated to reduce backlash.
74. Move serrated nuts as indicated to increase backlash.
75. With the backlash correct and no bearing end-float or pre-load, tighten both serrated nuts by half a serration only, to pre-load the taper roller bearings.
76. Engage the lockers into the serrated nuts. If either locker is not opposite a serration, bend it to fit
77. Fit the spring pins to retain the lockers.
78. Tighten the bearing cap bolts. Torque 6,9 to 9,0 kgf.m (50 to 65 lbf.ft).
79. Refit the differential assembly 51.15.01 or 54.10.01.

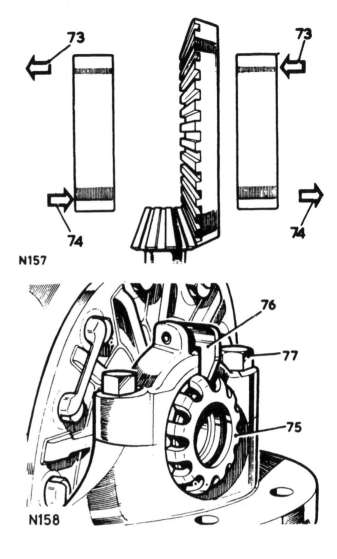

N157

N158

DATA

Pinion bearing pre-load	2,7 to 11,3 kgf.m (6 to 25 lbf.ft.) torque resistance
Pinion height setting, clearance between height gauge and slip gauge	0,07 to 0,10 mm. (.003 to .004 in.) using gauges 601998, 262761 or 600299. 0,28 to 0,30 mm. (.011 to .012 in.) using gauge 605004.
Crownwheel run-out	0,10 mm. (.004 in.)
Crown wheel to bevel pinion backlash	0,20 to 0,25 mm. (.008 to .010 in.)

51−8

PINION OIL SEAL

Remove and refit 51.20.01

Removing

1. Drain the lubricating oil from axle case.
2. Disconnect the propeller shaft at the differential.
3. Remove the pinion driving flange.
4. Prise out the oil seal.

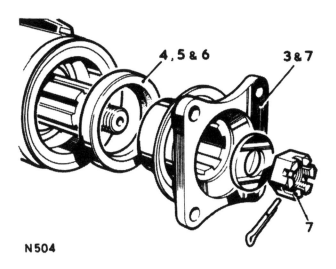

N 504

Refitting

5. Smear outside diameter of the oil seal with jointing compound.
6. Fit the seal, lipped side inward.

NOTE. Before fitting the driving flange, examine outside diameter for roughness or damage which may have caused failure of original seal, and rectify or renew as necessary. Replacement leather oil seals should be soaked in engine oil before fitting.

7. Fit the pinion driving flange. Tighten the securing nut. Torque 9,6 to 16,5 kgf.m (70 to 120 lbf.ft).
8. Fit the propeller shaft. Torque load 4,1 to 5,1 kgf.m (30 to 38 lbf.ft).
9. Replenish the differential lubricating oil. See Maintenance, 09.
10. Ensure the 'axle case breather is clear. A blocked breather could cause failure of oil seals fitted in axle assembly.

REAR AXLE ASSEMBLY

—Remove and refit   51.25.01

Removing

1. Jack up and support the chassis.
2. Jack under the axle.
3. Remove the road wheels.
4. Disconnect the shock absorbers.
5. Disconnect the brake hose at the connection under the floor.
6. Disconnect the lower links at the axle.
7. Disconnect the propeller shaft.
8. Disconnect the pivot bracket ball joint at the axle bracket.
9. Remove the road springs retaining plates.
10. Lower the axle and withdraw the road springs.
11. Withdraw the axle assembly.

Refitting

12. Position the axle and fit the lower links.
13. Position the road springs and fit the retaining plates.
14. Raise the axle and fit the pivot bracket ball joint. Torque load 17,9 kgf.m (130 lbf.ft).
15. Connect the shock absorbers.
16. **Fit the propeller shaft. Torque load 4,1 to 5,1 kgf.m (30 to 38 lbf.ft).**
17. Connect the brake hose.
18. Bleed the braking system 70.25.02.
19. Reverse 1 to 3.

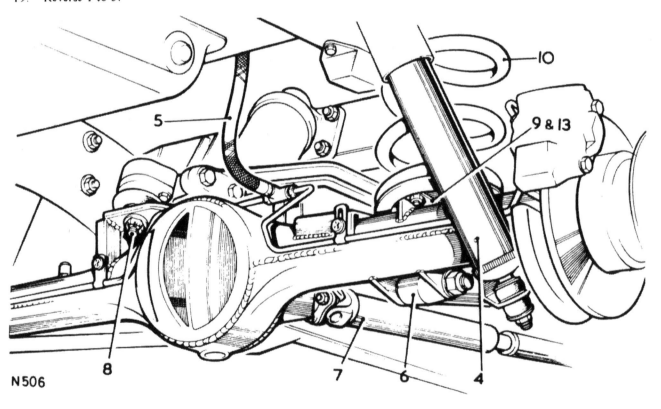

N506

FRONT AXLE AND FINAL DRIVE OPERATIONS

Description	Operation No.
Axle case oil seal	54–1
– remove and refit	54.15.04.
Differential assembly	
– remove and refit	54.10.01.
– overhaul	*Refer to* 51.15.07.
Front axle	
– remove and refit	54.15.01.
Half shaft	
– remove and refit	54.20.07.
Pinion oil seal	
– remove and refit	*Refer to* 51.20.01.

DIFFERENTIAL ASSEMBLY

– Remove and refit   **54.10.01**

Special tools 18G1063:   **Extractor for ball joint**

Removing

1. Jack up and support chassis.
2. Remove the road wheels.
3. Drain the oil from the differential.
4. Disconnect the brake hoses at the wing valance brackets.
5. Disconnect the track road ball joints. Extractor 18G1063.
6. Disconnect the drag link ball joint at the swivel arm. Extractor 18G1063.
7. Disconnect the steering damper at the differential casing bracket.
8. Remove the fixings, swivel bearing housing to axle case.
9. Withdraw the wheel hub assemblies until the axle shafts are disengaged from the differential assembly.
10. Disconnect the propeller shaft at the differential.
11. Remove the differential flange fixings.
12. Withdraw the differential assembly complete with joint washer.

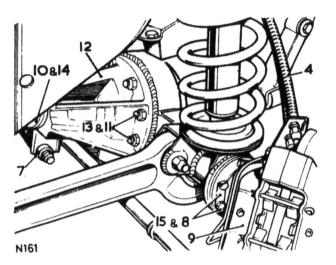

N161

Refitting

13. Reverse 11 and 12. Torque 3,5 to 4,6 kgf.m (26 to 34 lbf.ft).
14. Reverse 10. Torque 4,1 to 5,1 kgf.m (30 to 38 lbf. ft).
15. Reverse 8 and 9. Torque load 7,6 to 8,6 kgf.m (55 to 62 lbf.ft).
16. Fit the steering damper.
17. Reverse 6. Torque load 4,0 kgf.m (30 lbf. ft.).
18. Reverse 5. Torque load 4,0 kgf.m (30 lbf. ft.).
19. Connect the brake hoses.
20. Bleed and replenish the brake system. 70.25.02.
21. Replenish the differential oil. See Maintenance, 09.
22. Reverse 1 and 2.

FRONT AXLE ASSEMBLY

– Remove and refit   54.15.01

Service tools 18G1063:   Extractor for ball joints

Removing

1. Jack up and support the chassis.
2. Remove the road wheels.
3. Support the axle weight.
4. Remove the radius arms 60.10.16.
5. Disconnect the shock absorbers at the lower fixings.
6. Disconnect the brake pipe bracket from the swivel housing.
7. Withdraw the bracket complete with brake pipes.
8. Refit the bracket fixings to prevent oil leakage from the swivel housing.
9. Remove and tie aside the front brake caliper assemblies.
10. Disconnect the drag link at the hub arm. Extractor – 18G1063.
11. Disconnect the Panhard rod at the axle bracket.
12. Disconnect the propeller shaft at the final drive unit.
13. Lower the axle and withdraw the road springs.
14. Withdraw the axle assembly.

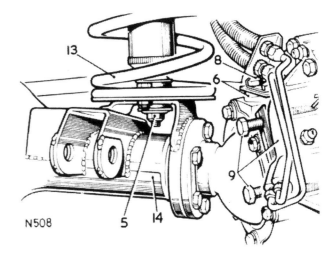

N508

Refitting

15. Position the axle under the vehicle.
16. Fit the Panhard rod to the axle bracket.
17. Raise the axle and support the LH side.
18. Fit the radius arms. 60.10.16.
19. Locate the road springs.
20. **Reverse 12. Torque load 4,1 to 5,1 kgf.m (30 to 38 lbf.ft).**
21. Reverse 10. Torque load 4,0 kgf.m (30 lbf. ft.).
22. Reverse 6 to 9. Torque load for swivel pin and bracket fixings is 8,5 kgf.m (60 lbf. ft.).
23. Connect the shock absorbers.
24. Reverse 1 to 3.

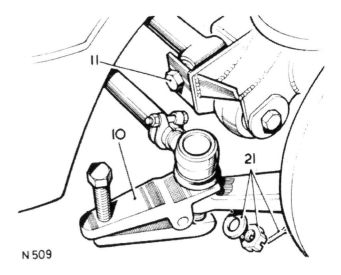

N 509

AXLE CASE OIL SEAL

– Remove and refit   54.15.04

Service tools 18G1063:   Extractor for ball joint

Removing

1. Remove the front hub assembly 60.25.01.
2. Remove the hub stub axle. 60.25.22.
3. Disconnect the track road ball joint at the swivel housing. Extractor 18G1063.
4. LH side only. Disconnect the drag link ball joint at the swivel housing. Extractor 18G1063.
5. Withdraw the constant velocity joint.
6. Withdraw the axle shaft.
7. Remove the swivel bearing housing fixings.
8. Withdraw the bearing housing and joint washer.
9. Withdraw the oil seal.

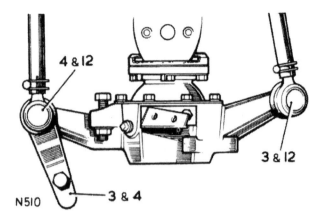

Refitting

10. Fit the oil seal, open side last, to the bearing housing.
NOTE. Replacement leather oil seals should be soaked in engine oil for one hour before fitting.
11. Fit the bearing housing and joint washer. Fixings torque is 7,6 to 8,6 kgf.m (55 to 62 lbf.ft).
12. Reverse 3 to 6. Torque load for ball joints is 4,0 kgf.m (30 lbf. ft.).
13. Reverse 1 and 2.
14. Ensure that the axle case breather is clear.

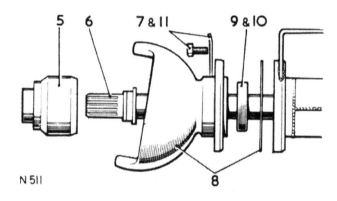

AXLE HALF SHAFT

– Remove and refit   54.20.07

Removing

1. Remove the front hub assembly. 60.25.01.
2. Remove the hub stub axle 60.25.22.
3. Lift out the constant velocity joint.
4. Withdraw the axle shaft.
5. If required, renew the sleeve on the shaft.

Refitting

6. Reverse 1 to 4.

STEERING OPERATIONS

Description	Operation No.
Drag link	
– remove and refit	57.55.17
– ends, remove and refit	57.55.16
Drop arm	
– remove and refit	57.50.14
Front wheel alignment	
– check and adjust	57.65.01
Lock stops	
– check and adjust	57.65.03
Manual steering box	
– adjust	57.35.01
– remove and refit	57.30.01
– overhaul	57.30.07
Power steering box	
– adjust	57.10.13
– remove and refit	57.10.01
– overhaul	57.10.07
Power steering fluid reservoir – remove and refit	57.15.08
Power steering pump	
– remove and refit	57.20.14
– overhaul	57.20.20
– drive belt – adjust	57.20.01
– drive belt – remove and refit	57.20.02
Power steering system	
– bleed	57.15.02
– test	57.15.01
Steering column	
– remove and refit	57.40.01
– overhaul	57.40.10
Steering column lock and ignition/starter switch	
– remove and refit	57.40.31
Steering damper	
– remove and refit	57.35.10
Steering wheel	
– remove and refit	57.60.01
Track rod	
– remove and refit	57.55.09
– linkage – remove and refit	57.55.10
Universal joints and coupling shaft	
– remove and refit	57.40.25

POWER STEERING BOX

– Remove and refit   57.10.01

Service tools:
    18G1063–Ball joint extractor
    18G75A–Drop arm extractor

NOTE:

It is important that whenever any part of the system, including the flexible piping, is removed or disconnected, that the utmost cleanliness is observed.

All ports and hose connections should be suitably sealed off to prevent ingress of dirt etc. If metallic sediment is found in any part of the system, the complete system should be checked, the cause rectified and the system thoroughly cleaned.

Under no circumstances must the engine be started until the reservoir has been filled. Failure to observe this rule will result in damage to the pump.

Removing

1. Park the vehicle on level ground.
2. Prop open the bonnet.
3. Remove the filler cap from the power steering fluid reservoir. Disconnect the pipes from the pump. Drain and discard the fluid. Replace the filler cap.
4. Disconnect the flexible hoses from the steering box.
5. Blank off all disconnected hose connections to prevent ingress of foreign matter.
6. Jack up and support the chassis front end. Alternatively, raise the vehicle on a ramp.

    WARNING: Whichever method is adopted, it is essential that the wheels are chocked and the handbrake applied.
7. Disconnect the lead from the oil pressure switch.
8. Uncouple the drag link and remove. Use ball joint extractor 18G1063.
9. Remove the drop arm. Use drop arm extractor 18G75A.
10. Remove the pinch bolt attaching the universal joint to the power steering box.
11. Slacken the nut securing the tie bar to the chassis.
12. Remove the bolts securing the tie bar to the steering box and move the tie bar aside.
13. Remove the fixings attaching the power steering box to the chassis side member.
14. Withdraw the power steering box.

Continued

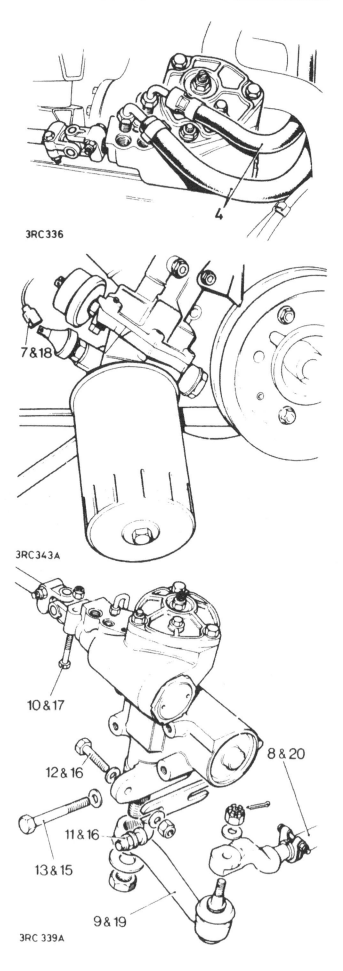

3RC336

7 & 18

3RC343A

10 & 17

12 & 16

11 & 16

8 & 20

13 & 15

9 & 19

3RC 339A

Refitting

15. Refit the steering box to the chassis side member and tighten the four Nyloc nuts.

16. Refit the tie bar to the steering box, and tighten the tie bar securing nut.

17. Reconnect the pinch bolt, attaching the universal joint to the power steering box.

18. Reconnect the lead to the oil pressure switch.

19. Refit the drop arm.

20. Refit the drag link and secure.

21. Lower the vehicle to ground level.

22. Remove the blanks and reconnect the flexible hoses to the steering box.

23. Remove the blank and refit the flexible hose to the power steering pump.

24. Ensure that the steering wheel spokes are horizontal when the wheels are in the straight-ahead position.

    NOTE: It may be necessary to remove the steering wheel and reposition on the splines to obtain this condition.

25. Remove the filler cap from the power steering fluid reservoir. Fill the reservoir to the mark on the dipstick with one of the recommended fluids (Division 09) and bleed the power steering system. 57.15.02.

26. Replace the reservoir filler cap.

27. Check, and if necessary, adjust the steering box. 57.10.13.

28. Test the steering system for leaks, with the engine running, by holding the steering hard on full lock in both directions.

    CAUTION: Do not maintain this pressure for more than 30 seconds in any one minute, to avoid causing the oil to overheat and possible damage to the seals.

29. Close the bonnet.

30. Road test the vehicle.

POWER STEERING BOX

– Overhaul   57.10.07

Service tools: 606600 'C' Spanner
606601 Peg Spanner
606602 Ring expander
606603 Ring compressor
606604 Seal saver, sector shaft
R01015 Seal saver, valve and worm
R01016 Torque setting tool

Dismantling:

1. Remove the steering box 57.10.01.
2. Rotate the retainer ring, as necessary, until one end is approximately 12 mm (0.500 in) from the extractor hole.
3. Lift the cover retaining ring from the groove in the cylinder bore, using a suitable pointed drift applied through the hole provided in the cylinder wall.
4. Complete the removal of the retainer ring, using a screwdriver.
5. Turn on left lock (L.H. stg.) until the piston pushes out the end cover (for R.H. stg. models, turn on right lock).
6. Slacken the grub screw retaining the rack pad adjuster.
7. Remove the rack pad adjuster.
8. Remove the sector shaft adjuster locknut.
9. Remove the sector shaft cover fixings.
10. Screw in the sector shaft adjuster until the cover is removed.
11. Slide out the sector shaft.
12. Withdraw the piston, using a suitable ½ in. U.N.C. bolt screwed into the tapped hole in the piston.

Continued

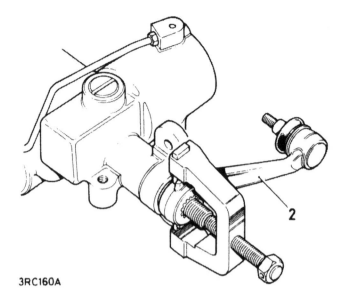

3RC160A

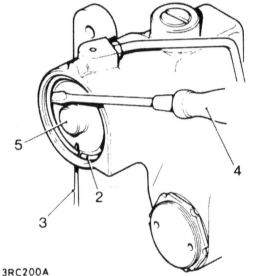

3RC200A

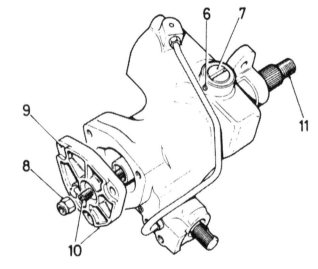

3RC149

57–4

13. Remove the worm adjusting screw locknut using 'C' spanner 606600.
14. Remove the worm adjusting screw using peg spanner 606601.
15. Tap the splined end of the spindle shaft to free the bearing.
16. Withdraw the bearing cup and caged ball bearing assembly.
17. Withdraw the valve and worm assembly.
18. Do not disturb the trim screw, otherwise the calibration will be adversely affected.
19. Withdraw the inner bearing ball race and shims. Retain the shims.

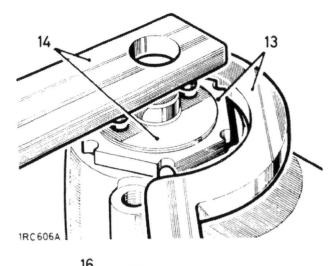

1RC606A

Steering box seals

20. Remove the circlip and seals from the sector shaft housing bore.

   NOTE: Do not remove the sector bush unless replacement is required. Refer to item 23.

21. Remove the circlip and seals from the input shaft housing bore.

   NOTE: Do not remove the input shaft needle bearing unless replacement is required.

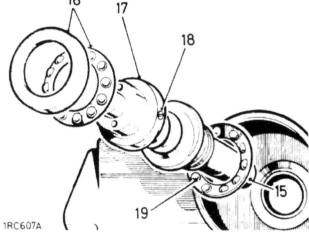

1RC607A

Inspecting

22. Discard all rubber seals and provide replacements.

   NOTE: A rubber seal is fitted behind the plastic ring on the rack piston. Discard the seal also the plastic ring and provide replacements.

   Continued

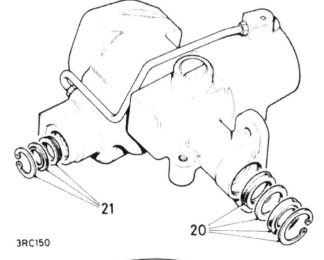

3RC150

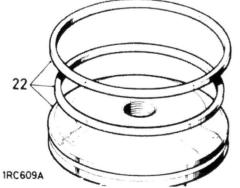

1RC609A

Steering box casing ·

23. If necessary, replace the sector shaft bush, using suitable tubing as a drift.
24. Examine the piston bore for traces of scoring and wear.
25. Examine the inlet tube seat for damage. If replacement is necessary this can be undertaken by using a suitable tap.
26. Examine the feed tube for signs of cracking.

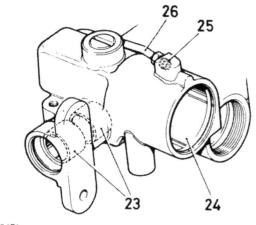

3RC151

Sector shaft assembly

27. Check that there is no side play on the rollers.
28. Check that the adjusting screw retainer is securely retained in the sector shaft by staking.
29. The adjusting screw end float must not exceed 0,12 mm (0.005 in.) if necessary, the end float may be decreased by turning in the threaded adjuster retainer then re-staking.

> WARNING: Re-staking of the adjuster retainer must be such that it cannot become loosened during service.

30. Examine the bearing areas on the shaft for excessive wear.
31. Examine the gear teeth for uneven or excessive wear.

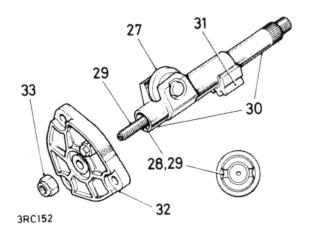

3RC152

Sector shaft cover assembly

32. The cover, bush and seat are supplied as a complete assembly for replacement purposes.

Sector shaft adjuster locknut

33. The locknut functions also as a fluid seal and must be replaced at overhaul.

Continued

Valve and worm assembly

34. Examine the valve rings which must be free from cuts, scratches and grooves. The valve rings should be a loose fit in the valve grooves.

35. Remove the damaged rings ensuring that no damage is done to the seal grooves.

36. If required, fit replacement rings, using the ring expander 606602. Both rings and tool may be warmed if found necessary. Use hot water for this purpose. Then insert into the ring compressor 606603 to cool.

    NOTE: The expander will not pass over rings already fitted. These rings must be discarded to allow access then renewed.

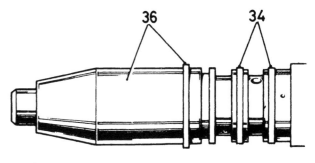

3RC153

37. Examine the bearing areas for wear. The areas must be smooth and not indented.

38. Examine the worm track which must be smooth and not indented.

39. Check for excessive end-float between the worm and valve sleeve. End float must not exceed 0,12 mm (0.005 in.).

40. Rotary movement between the components at the trim pin is permissible.

41. Check for wear on the torsion bar assembly pins; no free movement should exist between the input shaft and the worm.

    NOTE: Any sign of wear at locations 39, 40 and 41 above, make it essential that the complete valve and worm assembly are renewed.

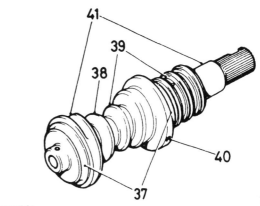

3RC201

Continued

Ball bearing and cage assemblies

42. Examine the ball races and cups for wear and general condition.
43. If the ball cage has worn against the bearing cup, fit replacements.
44. Bearing balls must be retained by the cage.
45. Bearings and cage repair are carried out by the complete replacement of the bearings and cage assembly. The bearing cup may be replaced separately only.
46. To remove the inner bearing cup and shim washers. jar the steering box on the work bench.

> NOTE: Should difficulty be experienced at this stage, warm the casing and the bearing assembly. Cool the bearing cup using a suitable mandrel and jar the steering box on the bench.

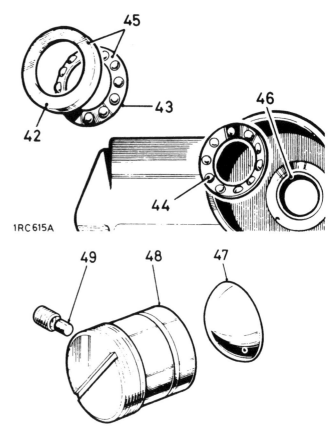

1RC615A

Rack thrust pad and adjuster

47. Examine the thrust pad for scores.
48. Examine the adjuster for wear in the pad seat.
49. Examine the nylon pad and adjuster grub screw assembly for wear.

Rack and piston

50. Examine for excessive wear on the rack teeth.
51. Ensure the thrust pad bearing surface is free of scores and wear.
52. Ensure that the piston outer diameters are free from burrs and damage.
53. Examine the seal and ring groove for scores and damage.
54. Fit a new rubber ring to the piston. Warm the white **nylon seal** and fit this to the piston. Slide the piston **assembly** into the cylinder with the rack tube outwards. Allow to cool.

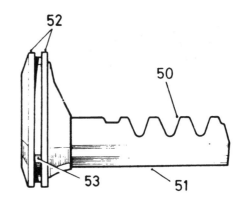

1RC616A

Continued

IRC617A

Input shaft needle bearing

55. If necessary, replace the bearing. The replacement must be fitted squarely in the bore (numbered face of the bearing uppermost). Then, carefully push the bearing in until it is flush with the top of the housing bore. Ideally, the bearing will be just clear of the bottom of the housing bore.

Reassembling

> **NOTE**: When fitting replacement oil seals, these must be lubricated with recommended fluid.

Input shaft oil seal

56. Fit the seal, lipped side first, into the housing. When correctly seated, the seal backing will lie flat on the bore shoulder.
57. Fit the extrusion washer and secure with the circlip.

Sector shaft seal

58. Fit the oil seal, upper side first.
59. Fit the extrusion washer.
60. Fit the dirt seal, lipped side last.
61. Fit the circlip.

Fitting the valve and worm assembly

62. If removed, refit the original shim washer(s) and the inner bearing cap. Only vaseline must be used as an aid to assembling the bearings.

> **NOTE**: If the original shims are not available, fit shim(s) of 0.76 mm (0.030 in) nominal thickness.

63. Fit the inner cage and bearings assembly.

Continued

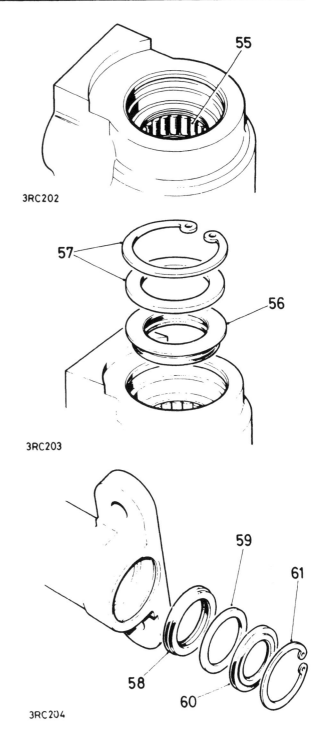

3RC202

3RC203

3RC204

64. Fit the valve and worm assembly, using seal saver RO1015 to protect the input shaft seal.
65. Fit the outer cage and bearings assembly.
66. Fit the outer bearing cup.
67. Renew the worm adjuster sealing ring and loosely screw the adjuster into the casing. Fit the locknut, but do not tighten.
68. Turn-in the worm adjuster until the end float at the input is almost eliminated.
69. Measure and record the maximum rolling distance of the valve and worm assembly, using a spring balance and cord coiled around the torque setting tool RO 1016.
70. Turn in the worm adjuster to increase the figure recorded in 69 by 1,8 to 2,2 kg (4 to 5 lb.) at 1¼ in. radius to settle the bearings, then back off the worm adjuster until the figure recorded in 69 is increased by 0,9 to 1,3 kg (2 to 3 lb.) only, with the locknut tight. Use peg spanner 606601 and 'C' spanner 606600.

Fitting the rack and piston

71. Screw a slave ½ in. U.N.C. bolt into the piston head for use as an assembly tool.
72. Fit the piston and rack assembly so that the piston is 2½ in. approximately from the outer end of the bore.
73. Feed in the sector shaft using seal saver 606604 aligning the centre gear pitch on the rack with the centre gear tooth on the sector shaft. Push in the sector shaft, and, at the same time rotate the input shaft about a small arc to allow the sector roller to engage the worm.

Continued

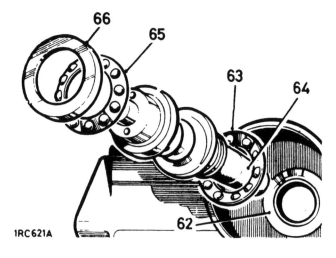

IRC621A

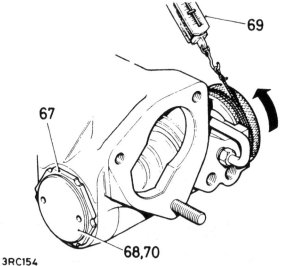

3RC154

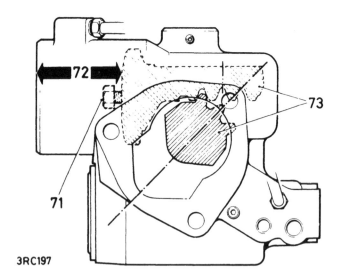

3RC197

Fitting the rack adjuster

74. Fit the sealing ring to the rack adjuster.
75. Fit the rack adjuster and thrust pad to engage the rack. Back off a half turn on the adjuster.
76. Loosely fit the nylon pad and adjuster grub screw assembly to engage the rack adjuster.

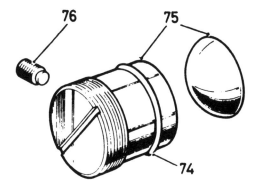

IRC625A

Fitting the sector shaft cover

77. Fit the sealing ring to the cover.
78. Screw the cover assembly fully on to the sector shaft adjuster screw.
79. Position the cover on to the casing.
80. Tap home the cover. If necessary back off on the sector shaft adjuster screw to allow the cover to joint fully with the casing.

    NOTE: Before tightening the fixings, rotate the input shaft about a small arc to ensure that the sector roller is free to move in the valve worm.

81. Fit the cover fixings and torque load to 2.2 to 2,8 kgf.m (16 to 20 lbf. ft.).

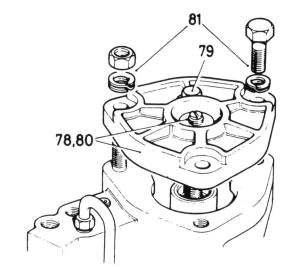

3RC157

Fitting the cylinder cover

82. Fit the square section seal to the cover.
83. Remove the slave bolt and press the cover into the cylinder just sufficient to clear the retainer ring groove.
84. Fit the retainer ring to the groove with one end of the ring positioned 12 mm (0.5 in.) approximately from the extractor hole.

Continued

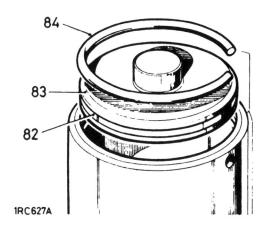

IRC627A

237

57–11

Adjusting the sector shaft

85. Set the worm on centre by rotating the input shaft 1½ turns approximately from either full lock position.

86. Rotate the sector shaft adjusting screw anti-clockwise to obtain backlash between the input shaft and the sector shaft.

87. Rotate the sector shaft adjusting screw clockwise until the backlash is just eliminated.

88. Measure and record the maximum rolling resistance at the input shaft, using a spring balance, cord and torque tool RO1016.

89. Hold still the sector shaft adjuster screw and loosely fit a new locknut.

90. Turn-in the sector shaft adjuster screw until the figure recorded in 88 is increased by 0,9 to 1,3 kg (2 to 3 lb.) with the locknut tightened.

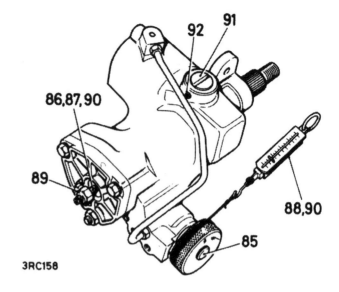

3RC158

Adjusting the rack adjuster

91. Turn-in the rack adjuster to increase the figure recorded in 90 by 0,9 to 1,3 kg (2 to 3 lb.). **The final figure may be less than but must not exceed 7,25 kg (16 lb.).**

92. Lock the rack adjuster in position with the grub screw.

Torque peak check

With the input shaft rotated from lock-to-lock, the rolling resistance torque figures should be greatest across the centre position (1½ turns approximately from full lock) and equally disposed about the centre position.

This condition depends on the value of shimming fitted between the valve and worm assembly inner bearing cup and the casing. The original shim washer value will give the correct torque peak position unless major components have been replaced.

> NOTE: During the following 'Procedure', the stated positioning and direction of the input shaft applies for both LH and RH boxes. However, the procedure for shim adjustment where necessary, differs between LH and RH steering boxes and is described under the applicable LH stg. and RH stg. headings.

Continued

Procedure

93. With the input coupling shaft toward the operator, turn the shaft fully anti-clockwise.

94. Check the torque figures obtained from lock-to-lock using a spring balance cord and torque tool R01016.

Adjustments

95. Note where the greatest figures are recorded relative to the steering gear position. If the greatest figures are not recorded across the centre of travel (i.e. steering straight ahead position), adjust as follows:

   L.H. stg. models. If the torque peak occurs **before** the centre position, add to the shim washer valve; if the torque peak occurs **after** the centre position, **subtract** from the shim washer valve.

   R.H. stg. models. If the torque peak occurs **before** the centre position, subtract from the shim washer valve; if the torque occurs **after** the centre position, **add** to the shim washer valve.

Shim washers are available as follows:

0,03 mm, 0,07 mm, 0,12 mm and 0,24 mm (0.0015 in., 0.003 in., 0.005 in. and 0.010 in.).

> **NOTE**: Adjustment of 0,07 mm (0.003 in) to the shim value will move the torque peak area by ¼ turn approximately on the shaft.

96. Refit the steering box 57.10.01.

97. Replenish the system with the correct grade of fluid. Refer to Recommended Lubricants—Division 09. Bleed the system, 57.15.02.

98. Test the system for leaks, with the engine running, by holding the steering hard on full lock in both directions.

> **NOTE**: Do not maintain this pressure for more than 30 seconds in any one minute to avoid overheating the fluid and possibly damaging the seals.

99. Road test the vehicle.

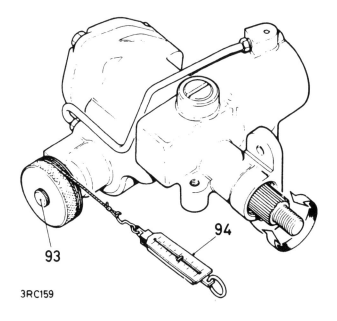

3RC159

POWER STEERING BOX

– Adjust   **57.10.13**

> **NOTE**: The condition of adjustment which must be checked is one of minimum backlash without overtightness when the wheels are in the straight-ahead position.

1. Jack up the front of the vehicle until the wheels are clear of the ground.
   WARNING: Wheels must be chocked in all circumstances.
2. Gently rock the steering wheel about the straight-ahead position to obtain the 'feel' of the backlash present. This backlash must not be more than 9,5 mm (0.375 in.).
3. Continue the rocking action whilst an assistant slowly tightens the steering box adjuster screw after slackening the locknut until the rim movement is reduced to 9,5 mm (0.375 in.) maximum.
4. Tighten the locknut, then turn the steering wheel from lock to lock and check that no excessive tightness exists at any point.
5. Lower the vehicle to ground level and remove the wheel chocks.
6. Road test the vehicle.

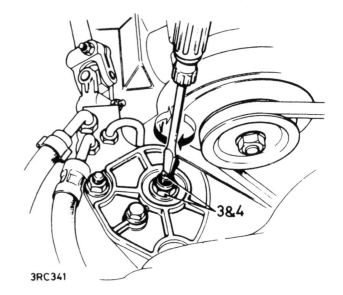

3RC341

POWER STEERING SYSTEM

−Test   57.15.01

Service tools:

JD10−Three-way adaptor, hose and pressure gauge for testing power steering
JD10-2−Adaptor pipe and tap for use with JD10

If lack of power assistance is evident, it is imperative that the pressure of the hydraulic pump, which is bolted to the front of the engine, is checked. It may be necessary to correct or fit a replacement unit prior to any action being taken to replace the complete steering unit.

This operation together with the fault finding chart following this procedure indicates the method of checking this point and also covers other faults on the power steering system which are best diagnosed before any removal or overhaul of the system is attempted.

Procedure

1. The hydraulic pressure test gauge is used in conjunction with the special adaptor (as illustrated) for testing the power steering system. This gauge is calibrated to read up to 140 kgf/cm^2 (2000 lbf/in^2) and the normal pressure which may be expected in the power steering system is 60 kgf/cm^2 (850 lbf/in^2).

2. Under certain fault conditions of the hydraulic pump it is possible to obtain pressures up to 105 kgf/cm (1500 lbf/in^2). Therefore, it is important to realise that the pressure upon the gauge is in direct proportion to the pressure being exerted upon the steering wheel. When testing, apply pressure to the steering wheel very gradually while carefully observing the pressure gauge.

3. Check, and if necessary replenish, the fluid reservoir.

4. Examine the power steering units and connections for leaks. All leaks must be rectified before attempting to test the system.

5. Check the steering pump drive belt for condition and tension, rectify as necessary.

Continued

6. Fit the test gauge JD10 and the adaptor JD10-2 to the steering pump outlet line.

7. Open the tap in the adaptor JD10-2.

8. Bleed the system. 57.15.02. Exercise extreme care when carrying out this operation so as not to overload the pressure gauge.

9. With the system in good condition, the pressures should be as follows:

    (a) Steering wheel held hard on full lock and engine running at 1,000 rev/min, the pressure should be 60 to 67 kgf/cm^2 (850 to 950 lbf/in^2).

    (b) With the engine idling and the steering wheel held hard on full lock, the pressure should be 28 kgf/cm^2 (400 lbf/in^2) minimum.

    These checks should be carried out first on one lock, then on the other.

    CAUTION: **Under no circumstances must the steering wheel be held on full lock for more than 30 seconds in any one minute, otherwise there will be a tendency for the oil to overheat and possible damage to the seals may result.**

    (c) Release the steering wheel and allow engine to idle; pressure should be 7 kgf/cm^2 (100 lbf/in^2) maximum.

10. If the pressures recorded during the foregoing test are outside the specified range, or pressure imbalance is recorded, a fault exists in the system. To determine if the fault is in the steering box or the pump, close the adaptor tap for a period not exceeding five seconds. If the gauge fails to register the specified pressures, the pump is inefficient and the pump relief valve should be examined and renewed as necessary.

11. Repeat the foregoing test after renewing the relief valve and bleeding the system. If the pump still fails to achieve the specified pressures, the pump should be overhauled or a new unit fitted.

12. If pump delivery is satisfactory and low pressure or marked imbalance exists, the fault must be in the steering box valve and worm assembly.

Continued

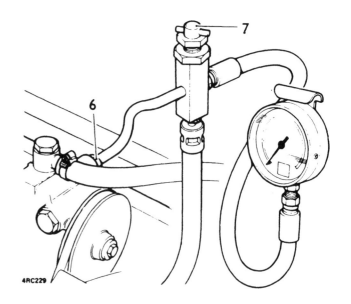

4RC229

FAULT DIAGNOSIS

SYMPTOM	CAUSE	TEST ACTION	CURE
INSUFFICIENT POWER ASSISTANCE WHEN PARKING	(1) Lack of fluid	Check hydraulic fluid tank level	If low, fill and bleed the system. 57.15.02
	(2) Engine idling speed too low	Try assistance at fast idle	If necessary, reset idle speed
	(3) Driving belt slipping	Check belt tension	Adjust the driving belt. 57.20.01
	(4) Defective hydraulic pump and/or pressure relief valve	(a) Fit pressure gauge between high pressure hose and steering pump, with steering held hard on full lock, see Note 1 below, and engine running at 1000 rev/min the pressure should be 60 to 67 kgf/cm^2 (850 to 950 lbf/in^2) with engine idling the pressure should be 28 kgf/cm^2 (400 lbf/in^2) minimum	If pressure is outside limits (high or low) after checking items 1 and 3, see Note 2 below
		(b) Release steering wheel and allow engine to idle, pressure should be 7 kgf/cm^2 (100 lbf/in^2) maximum	If pressure is greater, check steering box for freedom and self-centring action
POOR HANDLING WHEN CAR IS IN MOTION	Lack of castor action	This is caused by over-tightening the rocker shaft backlash adjusting screw on top of steering box	It is most important that this screw is correctly adjusted. Instructions governing adjustment are given in 57.10.13 and must be strictly adhered to.
	Steering too light and/or over-sensitive	Check for loose torsion bar fixings on steering box valve and worm assembly	Fit new valve and worm assembly
HYDRAULIC FLUID LEAKS	Damaged pipework, loose connecting unions, etc.	Check by visual inspection; leaks from the high pressure pipe lines are best found while holding the steering on full lock with engine running at fast idle speed (see Note 1 below). Leaks from the steering box tend to show up under low pressure conditions, that is, engine idling and no pressure on steering wheel	Tighten or renew as necessary
EXCESSIVE NOISE	(1) If the high pressure hose is allowed to come into contact with the body shell, or any component not insulated by the body mounting, noise will be transmitted to the car interior	Check the loose runs of the hoses	Alter hose route or insulate as necessary
	(2) Noise from hydraulic pump	Check oil level and bleed system. 57.15.02	If no cure, change hydraulic pump
CRACKED STEERING BOX	Excessive pressure due to faulty relief valve in hydraulic pump	Check by visual inspection	Fit new steering box and rectify hydraulic pump or replace as necessary

Note 1. Never hold the steering wheel on full lock for more than 30 seconds in any one minute, to avoid causing the oil to overheat and possible damage to the seals.

Note 2. High pressure – In general it may be assumed that excessive pressure is due to a faulty relief valve in the hydraulic pump.
Low pressure – Insufficient pressure may be caused by one of the following:
  1. Low fluid level in reservoir ⎫ Most usual cause of
  2. Pump belt slip   ⎬ insufficient pressure
  3. Leaks in the power steering system
  4. Faulty relief valve in the hydraulic pump
  5. Fault in steering box valve and worm assembly
  6. Leak at piston sealing in steering box
  7. Worn components in either steering box or hydraulic pump

57–17

POWER STEERING SYSTEM

– Bleed 57.15.02

Procedure

1. Fill the steering fluid reservoir to the mark on the dipstick with one of the recommended fluids (Section 09).
2. Start and run the engine until it attains normal operating temperature.
3. Run the engine at idle speed.

   NOTE: During the carrying out of items 4, 5 and 6, ensure that the steering reservoir is kept full. Do not increase the engine speed or move the steering wheel.

4. Slacken the bleed screw. When fluid seepage past the bleed screw is observed, retighten the screw.
5. Ensure that the fluid level is in alignment with the mark on the reservoir dipstick.
6. Wipe off all fluid released during bleeding.
7. Check all hose joints, pump and steering box for fluid leaks under pressure by holding the steering hard on full lock in both directions.

   CAUTION: Do not maintain this pressure for more than 30 seconds in any one minute, to avoid causing the oil to overheat and possible damage to the seals. The steering should be smooth lock-to-lock in both directions, that is, no heavy or light spots when changing direction when the vehicle is stationary.

8. Carry out a short road test. If necessary, repeat the complete foregoing procedure.

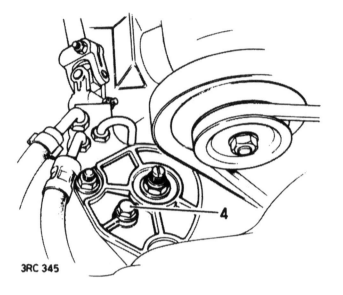

3RC 345

POWER STEERING FLUID RESERVOIR

– Remove and refit   **57.15.08**

Removing

1. Prop open the bonnet.
2. Remove the reservoir filler cap.
3. Disconnect the return hose from the steering box. Drain the fluid completely from the reservoir.

    CAUTION: It is most important that this fluid is not re-used.

4. Refit the return hose to the steering box.
5. Remove the fixings attaching the reservoir to the wing valance. Disconnect the flexible hoses and withdraw the reservoir.

    NOTE: If the reservoir is not to be refitted immediately, the hoses must be sealed to prevent the ingress of foreign matter.

6. Remove the top cover of the reservoir also the filter element.

Refitting

7. Fit a new filter element ensuring that the cover sealing washer is fitted correctly.
8. Refit the top cover of the reservoir.
9. Refit and tighten the fixings attaching the reservoir to the wing valance.
10. Re-connect the flexible hoses to the reservoir. Tighten the clips.
11. Fill the reservoir to the prescribed level with one of the recommended fluids (Section 09) and bleed the power steering system 57.15.02.
12. Fit the reservoir filler cap.
13. Close the bonnet.

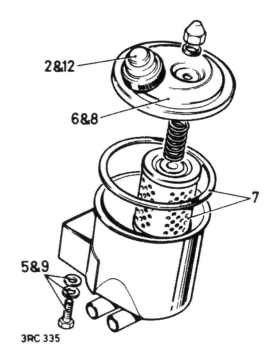

3RC 335

POWER STEERING PUMP DRIVE BELT

— Adjust   57.20.01

Procedure

1. Prop open the bonnet.
2. Check, by thumb pressure, the belt tension between the crankshaft and the pump pulley.
   There should be a free movement of between 11 to 14 mm (0.437 to 0.562 in.).
   If adjustment is necessary, proceed as follows:
3. Slacken the nut on the pivot bolt securing the pump mounting bracket to the cylinder head.
4. Slacken the bolt securing the pump lower bracket to the slotted adjustment link.
5. Slacken the bolt securing the slotted adjustment link to the support bracket mounted on the water pump cover.
6. Pivot the pump as necessary and adjust until the correct belt tension is obtained.
7. Maintaining the tension, tighten the pump adjusting bolts and pivot bolt nut.

   NOTE: Whenever a new belt is fitted, it is most important that its adjustment is re-checked after approximately 1.500 km (1,000 miles) running.

8. Close the bonnet.

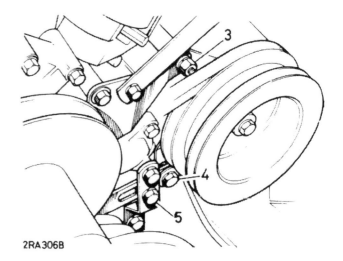

2RA306B

POWER STEERING PUMP DRIVE BELT

— Remove and refit   57.20.02

Removing or preparing for the fitting of a new belt

1. Prop open the bonnet.
2. Slacken the alternator fixings and remove the fan belt.
3. Slacken the power steering pump fixings. Move the pump clear of its mounting bracket.
4. Release the driving belt from the pump and crankshaft pulleys.

Continued

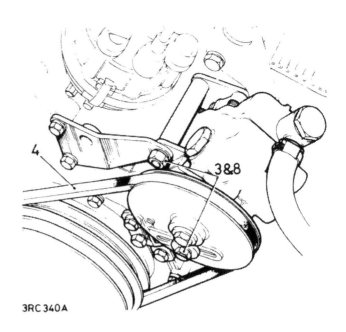

3RC 340A

Refitting

5. Locate the driving belt over the crankshaft and pump pulleys.

6. Refit the pump in position, leaving the fixings finger tight only.

7. Adjust the position of the pump to give a driving belt tension of 11 to 14 mm (0.437 to 0.562 in) movement when checked by thumb pressure midway between the crankshaft and pump pulleys.

8. Secure the pump adjusting and pivot bolts.

9. Refit the fan belt and adjust the tension to give 11 to 14 mm (0.437 to 0.562 in.) movement when checked by thumb pressure midway between the crankshaft and alternator pulleys.

10. Tighten the alternator fixings.

11. Close the bonnet.

POWER STEERING PUMP

– Remove and refit   57.20.14

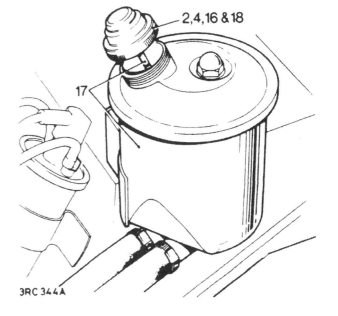

Removing

1. Prop open the bonnet.

2. Remove the filler cap from the power steering fluid reservoir.

3. Disconnect the inlet hose from the pump and drain the fluid into a suitable container. Blank off the hose orifice to prevent the ingress of foreign matter.

   CAUTION: Under no circumstances must the fluid be re-used.

4. Replace the reservoir filler cap.

5. Disconnect the outlet hose from the pump. Blank off the hose orifice to prevent ingress of foreign matter.

6. Slacken and remove the adjuster bolt below the pulley.

7. Slacken and remove the nut on the front mounting bracket.

8. Remove the two bolts securing the front mounting bracket to the water pump.

9. Slide off the power steering pump.

Continued

57–21

Refitting

10. Offer up the power steering pump, locating the driving belt over the pulley.
11. Refit the two bolts securing the front mounting bracket to the water pump. Tighten.
12. Refit the nut to the front mounting bracket.
13. Adjust the position of the pump to give a driving belt tension of 11 to 14 mm (0.437 to 0.562 in) movement when checked by thumb pressure midway between the crankshaft and pump pulleys. Fit and tighten the pump adjuster bolt below the pulley, also, tighten the nut at the front mounting bracket.
14. Remove the blank and reconnect the inlet hose to the pump.
15. Remove the blank and reconnect the outlet hose to the pump.
16. Remove the filler cap from the steering fluid reservoir.
17. Fill the steering fluid reservoir to the mark on the dipstick with one of the recommended fluids (Division 09).
18. Replace the filler cap.
19. Bleed the power steering system. 57.15.02.
20. Test the steering system for leaks with the engine running, by holding the steering hard on full lock in both directions.

    CAUTION: Do not maintain this pressure for more than 30 seconds in any one minute, to avoid causing the oil to overheat and possible damage to the seals.

21. Close the bonnet.
22. Road test the vehicle.

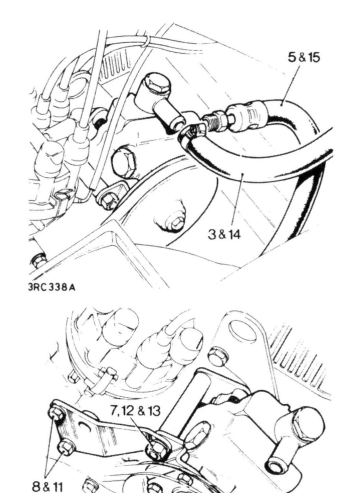

3RC338A

3RC337A

POWER STEERING PUMP

– Overhaul   57.20.20

Dismantling

1. Remove the steering pump 57.20.14.
2. Drain the pump of any oil and clean the exterior of the pump.
3. Remove the bolt, spring washer and large plain washer securing the pulley to the pump shaft.
4. Using a suitable puller, withdraw the pulley. Do not attempt to hammer the shaft from the pulley, or lever the pulley from the shaft, as this may cause internal damage to the pump.
5. Withdraw the square key from the shaft.
6. Remove the four bolts and spring washers securing the bearing retainer plate and front mounting plate to the pump body. Remove the plates.
7. Remove the three bolts and spring washers securing the rear mounting plate to the pump body. Remove the plate.
8. Clamp the pump body in a vice, ensuring that the jaws are protected.
9. Remove the union bolt and withdraw the fibre washer, inlet adaptor and rubber gasket.

NOTE: **The tubular steel venturi flow director under the inlet adaptor is pressed into the cover and should not be removed.**

10. Remove the six Allen screws securing the cover to the pump body. Separate the cover from the body vertically to prevent the parts falling out.
11. Remove the pump from the vice.
12. Remove the 'O' ring seals from the grooves in the pump body.
13. Carefully tilt the pump body, and remove the six rollers.
14. Draw the carrier off the shaft, and remove the drive pin.
15. Remove the shaft from the body.
16. Remove the cam and the cam lock peg from the pump body.
17. If necessary withdraw the sealed bearing from the shaft.
18. Remove the shaft seal from the body, ensuring that no damage is caused to the shaft bushing.
19. Remove the valve cap, 'O' ring, valve and valve spring from the body. Place all parts where they will not be damaged, or subject to contamination.

Inspection

20. Wash all parts in a suitable solvent, air dry, or wipe clean with a lint-free cloth if air is not available.
21. Check the pump body and cover for wear. Replace either part, if faces or bushes are worn.
22. Check the pump shaft around the drive pin slot. Remove any burrs.

NOTE: **Ensure that the aluminium restrictor in the output port is thoroughly cleaned but not dislodged.**

Continued

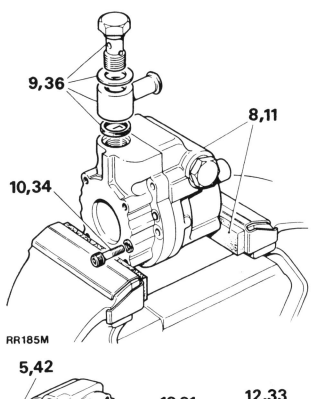

RR185M

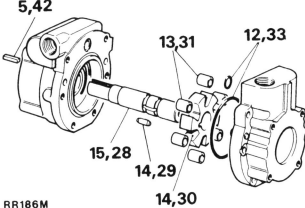

RR186M

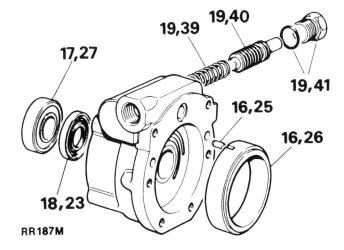

RR187M

Reassembling

23. Carefully examine a new shaft seal to ensure that it is clean and undamaged. Smear the sealing lips with grease and apply a fine smear of 'Wellseal' to the pump body where the outside diameter of the oil seal locates (applies to metal cased seals). Place the seal square to the housing recess with the lip towards the inside of the housing.

24. Press the seal into position approximately 1/32in. below the seal housing face, ensuring that it does not tilt.

25. Replace the cam lock peg into the location in the body.

26. Renew the cam if worn or damaged. Refit the cam, ensuring that it seats correctly in the body and that the slot locates over the locking peg.

27. Replace the sealed bearing onto the shaft.

28. Insert the shaft and bearing assembly into the seal side of the body.

29. Refit the carrier drive pin in the shaft.

30. Inspect the carrier and replace in position, ensuring that the greater angle on the carrier teeth is in the leading position as shown on illustration.

31. Inspect the rollers, paying particular attention to the finish on the end. Replace the rollers if scored, damaged or oval. Refit the rollers to the carrier.

32. Using a straight edge across the cam surface, and a feeler gauge, check the end clearance of the carrier and rollers in the pump body. If the end clearance is more than 0,05 mm (0.002 in.) replace the carrier and rollers.

33. Smear a fine trace of Loctite 275 to the pump body in a 'figure of 8' outside the 'O' rings and inside the bolt holes. Install new 'O' rings to the body of the pump.

34. Refit the cover on the pump body and secure with six Allen screws and spring washers.

35. Tighten the Allen screws, in diagonal sequence, checking that the shaft rotates freely and does not bind. Final torque to be 15 to 17 lbf.ft (20 to 23 Nm).

36. Replace the square sectioned rubber gasket to the groove around the inlet port and replace the inlet adaptor, fibre washer and union bolt. Torque to 28 to 30 lbf.ft (38 to 41 Nm).

37. Refit the rear mounting plate to the pump body and secure with three bolts and spring washers.

38. Refit the front mounting plate and the bearing retainer plate to the pump body and secure with four bolts and spring washers.

39. Refit the flow control valve spring in the bore. The spring tension should be 3,6 to 4,8 kgf (8 to 9 lbf) at 21 mm (0.820 in.). If not, renew the spring.

40. Replace the valve in the bore, inserting the valve so that the exposed ball end enters last. Ensure that the valve is not sticking.

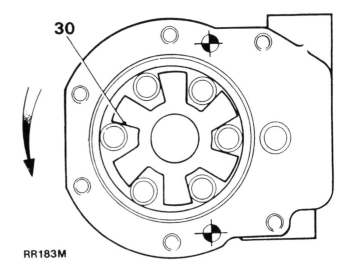

30

RR183M

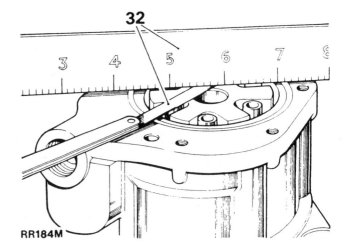

32

RR184M

41. Replace the 'O' ring on the valve cap and assemble in the pump. Tighten the cap to a torque figure of 4,0 to 4,9 kgf.m (30 to 35 lbf.ft).

42. Refit the pulley key.

43. Refit the pulley to the shaft and secure with the special washer, spring and washer and bolt. Tighten the bolt to a torque figure of 1,4 to 1,6 kgf.m (10 to 12 lbf.ft).

44. Refit the steering pump to the vehicle, see operation 57.20.14.

STEERING BOX

– Remove and refit   **57.30.01**

Service tool: **18G.1063 Ball joint extractor**

Removing.

1. Jack up the front of the vehicle and support securely.
2. Remove the road wheel (steering box side).
3. Using service tool 18G.1063 disconnect the drop arm ball joint.
4. Remove the pinchbolt securing the universal joint to the worm shaft.
5. (Tie bar type box). Slacken the nut securing the tie bar to the panhard rod mounting arm.
6. (Tie bar type box). Remove the two bolts, washers and nuts securing the tie bar to the steering box.
7. (Tie bar type box). Swing the tie bar clear of the steering box.
8. Remove the four nuts and bolts securing the steering box to the chassis.
9. Withdraw the steering box complete with drop arm.

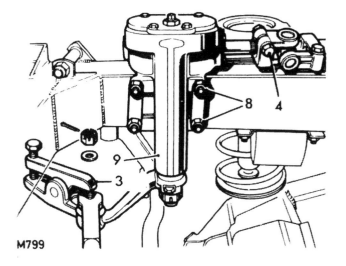

M799

Refitting.

10. Position the steering wheel with the centre spoke downward.
11. Set the steering box in its mid-way lock to lock position.
12. Locate the steering box in position on the vehicle and engage the universal joint and pinchbolt.
13. Secure the steering box to the chassis with four bolts and nuts.
14. Tighten the universal joint pinchbolt. Torque 3,5 kgf.m (25 lbf.ft).
15. (Tie bar type box). Secure the tie bar to the steering box with two bolts and nuts. Do not fully tighten the bolts. Tighten the nut securing the tie bar to the panhard rod. Slacken the tie bar bolt and then tighten fully.
16. Connect the drag link to the drop arm. Torque 4,0 kgf.m (30 lbf.ft).
17. Fit the road wheel and remove the chassis supports.

    Steering boxes incorporating a tie bar were introduced at the following Commission numbers:
    35503461 A. UK Market.
    35600097 A. Other Markets (Right hand steering).
    35800667 A. Other Markets (Left hand steering).

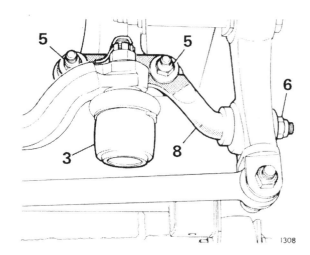

1308

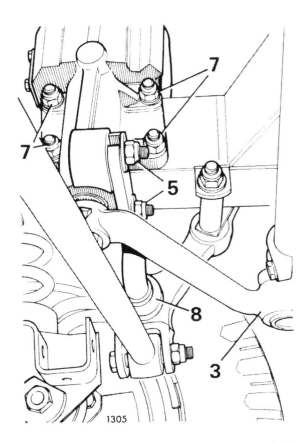

1305

57–25

STEERING BOX – MANUAL

– Overhaul   57.30.07

Service tool: 18G 75A Drop arm extractor

1. Remove the steering box from the vehicle 57.30.01.

Dismantling

2. Remove the drop arm 57.50.14.
3. Remove the oil filler plug and drain the oil.
4. (Tie bar type box). Remove the bolts securing the small triangular cover plate and remove the plate and shims. Remove the button and spring from the rocker shaft.
5. Remove the main cover plate bolts and lift off the triangular cover and gasket.
6. Lift out the rocker shaft and slide the roller from the main nut assembly.
7. Remove the end cover, seal plate, joint washer and shims.
8. Rotate and withdraw the worm shaft and lift out the main nut. Retain any dislodged bearing balls.
9. Withdraw the bearing balls and outer race.
10. Free the bearing inner race by tapping the housing. Withdraw the inner race.
11. Remove the 'O' ring seal from the housing.

Inspection

12. Thoroughly clean all components. Take care to retain all bearing balls.
13. Examine all bearing surfaces for wear and damage. Check that splines are not worn or dented.
14. Examine the main nut and worm shaft ball tracks for indentation and scaling. Check all balls for pitting. Worn or doubtful components must be renewed.
15. If required new the rocker shaft bush. It will be necessary to first remove the peened washer before pressing out the bush. Fit the washer and peen after installation.
16. Fit a new oil seal, lipped side first to the worm shaft end cover. Press into position until the seal outer lip is flush with, or very slightly below the cover inner face.

Assembling

17. Fit the worm shaft inner race to the housing and smear with general purpose grease. Install ten bearing balls in the inner race. The smaller balls are fitted to the worm shaft bearings; the larger balls belong to the main nut.
18. Assemble the larger bearing balls to the main nut and retain them in position with general purpose grease. On non tie bar type steering boxes twenty eight balls are accommodated in the main nut; on tie bar type boxes, twenty seven balls are accommodated.

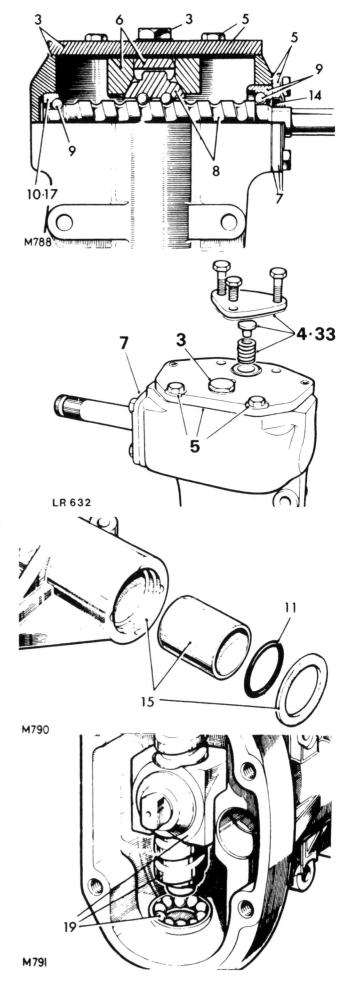

19. Locate the main nut in the steering box and carefully engage the worm shaft until it contacts the inner race. Ensure the bearing balls in both the main nut and inner race are not dislodged.

20. Smear the outer bearing race with general purpose grease and locate thirteen bearing balls in position. Carefully insert the outer race until the balls make contact with the worm shaft. Ensure no balls are dislodged.

21. Fit the end cover, joint washer and sufficient shims to provide worm shaft end-float.

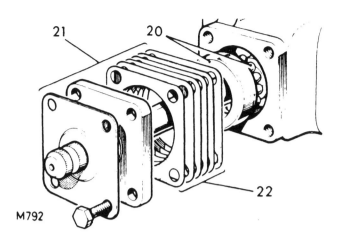

M792

22. Progressively remove shims until the worm shaft is free to rotate with an absence of end-float. Remove shims to the thickness of 0,07 mm (0.003 in.). This provides pre-load for the worm shaft bearings. A paper joint may be substituted for a metal shim to obtain fine adjustment.

23. With the end cover torqued to 2,1 to 2,5 kgf.m (15 to 18 lbf.ft) check that the worm shaft remains free to rotate. If satisfactory remove the end cover bolts and smear the threads with non-hardening sealing compound, such as 'Wellseal' or an equivalent. Fit the oil seal shield and retorque the bolts.

24. Fit a new 'O' ring to the rocker shaft bore. Lubricate the 'O' ring and rocker shaft bush.

25. Locate the main nut in the centre position of its travel. Ensure that it remains in this position until the drop arm is fitted.

26. Fit the roller to the main nut.

27. Insert the rocker shaft and engage the roller.

28. (Non tie bar type box). Fit the top cover and gasket. Using a non-hardening sealer on the gasket faces and bolt threads.

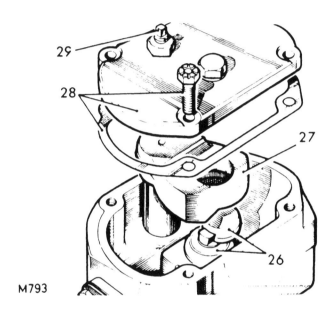

M793

29. (Non tie bar type box). Set the rocker shaft adjusting screw and locknut to eliminate rocker shaft end-movement ensuring that the worm shaft remains free to rotate. Centralise the main nut as necessary.

30. (Tie bar type box). Set the steering box in the straight ahead position. Temporarily fit the main cover using a new gasket. Use spacers or plain washers with the two long bolts which secure the triangular cover plate. The triangular cover plate should not be fitted at this stage.

31. (Tie bar type box). Place the triangular cover plate on the protruding rocker shaft and using feeler gauges check the clearance between the underside of the triangular cover plate and the steering box.

IMPORTANT: It is essential that there is NO pre-load on the rocker shaft in the bolted down position and that end float is minimal.

The shim pack for the triangular cover plate contains two paper washers, one at each end. To make up a shim pack include only one of the paper washers when measuring the total thickness required. Finally fit the other paper washer. Due to compression of the paper washers, when the triangular cover plate is bolted down, a minimal clearance should then be provided. The following shims are available:

Paper 0,13 mm (compresses 60% to 0,078 in.)

Steel 0,05 mm   Steel 0,13 mm   Steel 0,25 mm

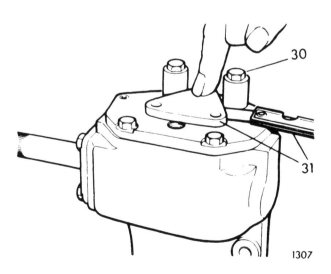

1307

continued

32. (Tie bar type box). Remove the main cover and gasket. Smear the gasket and all bolt threads with a non-hardening sealant. Fit the main cover and gasket leaving the bolts loose.
33. (Tie bar type box). Fit the spring and thrust button to the rocker shaft.
34. (Tie bar type box). Fit the triangular cover and the selected shim pack.
35. (Tie bar type box). Evenly tighten the main and triangular cover plate bolts to 2,1 to 2,5 kgf.m (15 to 18 lbf.ft).
36. (Tie bar type box). Check the rocker shaft for perceptible end float (with the steering box in the straight ahead position). If there is no end float or conversely there is end float exceeding 0,03 mm (0.001 in.) remove the triangular cover plate and adjust the thickness of the shim pack accordingly.
37. Fit the drop arm, lockwasher and nut. Tighten the nut to 17,3 to 18,0 kgf.m (125 to 130 lbf.ft). Secure the nut with the lockwasher. When tightening the drop arm nut the drop arm should be clamped in a vice to prevent strain being applied to the steering box components.
38. Fit the steering box with recommended lubricant.
39. Fit the oil filler plug.
40. Fit the steering box to the vehicle 57.30.01.

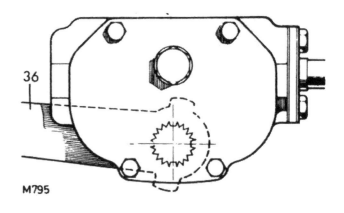

36

M795

STEERING BOX – MANUAL

– Adjust   57.35.01

1. Raise the front wheels of the vehicle clear of the ground.
2. Position the road wheels straight ahead.
   Non tie bar type box
3. Slacken the locknut securing the rocker shaft adjusting screw.
4. Slacken the adjusting screw and then tighten until rocker shaft end-float is reduced to the minimum.
5. Tighten the locknut.
6. Check steering lock to lock for tight spots. If oil seepage is evident at the adjusting screw, remove and clean the screw and coat the threads with Loctite Hydraulic Sealant.
   Tie bar type box
7. Remove the small triangular cover plate, shims, thrust button and spring from the steering box.
8. Refit two of the triangular cover plate bolts to the steering box using packing washers or spacers as necessary. This ensures that the steering box cover is held in position.
9. Carry out instructions 30 and 33 to 35, Operation 57.30.07.
10. Check steering movement lock to lock for tight spots.
11. Lower the vehicle and remove the jacks.

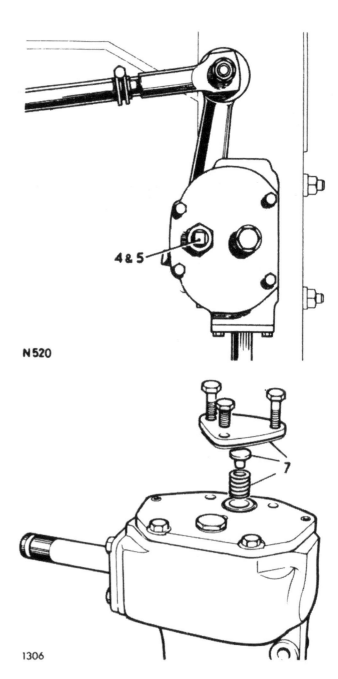

4 & 5

N 520

7

1306

STEERING DAMPER

– Remove and refit 57.35.10

Removing

1. Remove the fixings at the differential case bracket.
2. Remove the fixings at the track rod bracket.
3. Withdraw the steering damper.

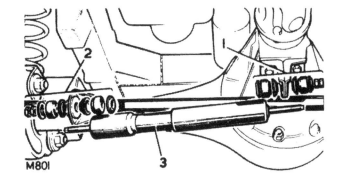

Refitting

4. Reverse 1 to 3.

STEERING COLUMN

– Remove and refit 57.40.01

Service tool: RO1002 Extractor for steering wheel

IMPORTANT: The steering column is of a 'safety' type and incorporates shear pins. Therefore do not impart shock loads to the steering column at any time.

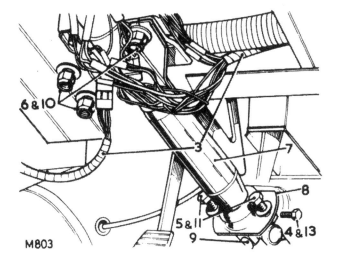

Removing

1. Remove the steering wheel 57.60.01. Extractor RO1002.
2. Remove the lower facia panel, 76.46.03 driver's side.
3. Disconnect the electrical leads from the steering column switches.
4. Remove the pinch bolt, universal joint to steering box.
5. Remove the fixings, steering column to toe board.
6. Remove the fixings, steering column to dash bracket.
7. Withdraw the steering column assembly.

Refitting

8. Position the sealing gasket on the end of the column assembly.
9. Feed the steering shaft through the toe board and engage the drive splines at the steering box shaft.
10. Loosely fit the column upper fixings.
11. Loosely fit the column lower fixings.
12. Tighten upper and lower fixings.
13. Fit universal joint pinch bolt. Torque loading 3,5 kgf.m (25 lbf.ft).
14. Reverse 1 to 3.

STEERING COLUMN ASSEMBLY

– Overhaul   57.40.10.

1.   Remove the steering column assembly.57.40.01

Dismantling.

2.   Remove the lighting switch from the lower shroud. 86.55.10.
3.   Remove the direction indicator switch from the column assembly. 86.65.55.
4.   Remove the wiper/washer switch from the column assembly 86.65.38.

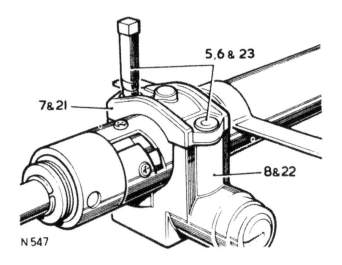

N 547

Removing the steering column lock assembly 5 to 8

5.   Drill a hole in each sheared bolt to accept an 'easy-out' extractor.
6.   Remove the sheared bolts.
7.   Detach the end cap.
8.   Withdraw the column lock assembly.
9.   Lift off the cam for the self-cancelling switch from the column.

Removing the top bearing assembly 10 to 15

10.   Remove the circlip.
11.   Withdraw the thrust washer and shim washer/s.
12.   Withdraw the wave washer.
13.   Remove the fixings, top bearing assembly to steering column.
14.   Withdraw the top bearing assembly.
15.   Remove the retaining ring from the groove in the steering column.
16.   If required, renew the mesh cover, using heat fusing to join together the replacement cover edges.

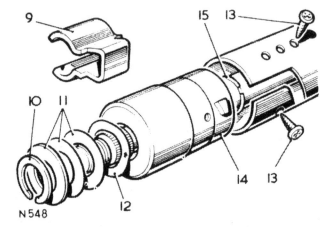

N 548

IMPORTANT. The steering column is now dismantled as far as is permitted. A replacement steering column comprises outer column, inner column and lower bearing assembly.

Assembling

17. Reverse 10 to 15.

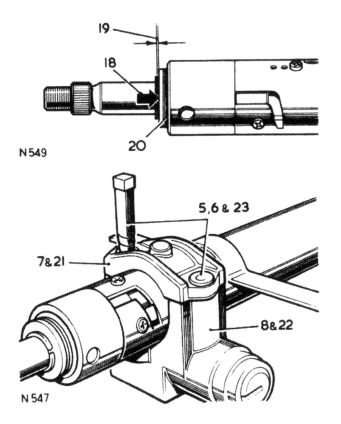

N549

Checking the top bearing end load, items 18 to 20.

18. Hold down the thrust washer and shims fully against the wave washer spring load.
19. Measure the clearance between the thrust washer and the circlip. The clearance must be 0,12 mm. (.005 in).
20. Adjust the clearance as necessary by fitting replacement shim washer/s which are available in the range 0,127 to 0,762 m.m.(.005 to .030 in) in 0,127 (.005) stages.

Fitting the steering column lock

21. Position the steering lock cap on the outer column, locating the spigot in the hole provided.
22. Offer the lock assembly to the column.
23. Fit the shear bolts to retain the cap and lock assembly.
24. Tighten the bolts sufficient to shear the heads.
25. Reverse 1 to 4.

N547

DATA

Top bearing end load   0,127 m.m. (.005 in) clearance between circlip and thrust washer with wave washer compressed

UNIVERSAL JOINTS AND COUPLING SHAFT

– Remove and refit   57.40.25

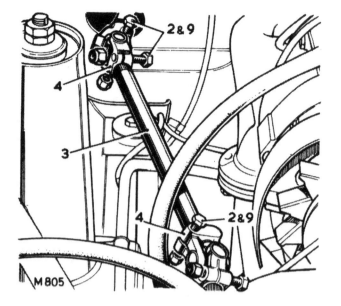

Removing

1. Note the position of the steering wheel spokes.
2. Remove the four pinch bolts.
3. Withdraw the shaft and universal joints
4. Withdraw the universal joints from the shaft

Refitting.

5. Position the universal joints on the shaft.
6. Offer the shaft assembly to the steering column, aligning the pinch bolt hole with the flat on the column.
7. Position the steering wheel as noted in 1.
8. Fit the front universal joint to the steering box shaft.
9. Fit the pinch bolts. Torque load 3.5 kgf.m (25 lbf.ft).

STEERING COLUMN LOCK AND IGNITION/STARTER SWITCH

– Remove and refit   57.40.31.

IMPORTANT. The steering column is of a 'safety' type and incorporates shear pins. Therefore do not impart shock loads to the steering column at any time.

Ignition / starter switch only remove and refit is described in operation 86.65.03

Removing

1. Remove the lower facia panel, driver's side. 76.46.03.
2. Remove the top shroud.
3. Remove the lighting switch from the lower shroud. 86.55.10.
4. Remove the steering column fixings at the toe board.
5. Remove the fixings, steering column to dash bracket.
6. Lower the steering column to gain access to the column lock fixings and (later models) remove the insulating cover.
7. Drill a hole in each sheared bolt to accept an 'Easi-out' extractor.
8. Remove the sheared bolts.
9. Detach the end cap.
10. Withdraw the column lock assembly.
11. Remove the ignition/starter switch.

Refitting

12. Position the steering lock cap on the outer column, locating the spigot in the hole provided.
13. Offer the lock to the column.
14. Fit the shear bolts to retain the cap and lock.
15. Tighten the bolts sufficient to shear off the heads.
16. Fit the ignition / starter switch.
17. Reverse 1 to 6.

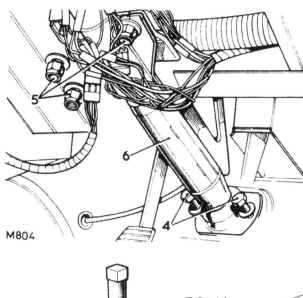

M804

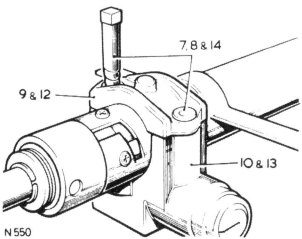

N550

DROP ARM

– Remove and refit   57.50.14

Service tools: 18G 1063 Balljoint extractor
18G 75A Drop arm extractor.

Removing

1. Disconnect the drag link from the drop arm ball joint. Extractor 18G1063.
2. Remove the drop arm from the steering box rocker shaft. Extractor 18G75A.

NOTE. **The drop arm ball joint is integral with the drop arm.**

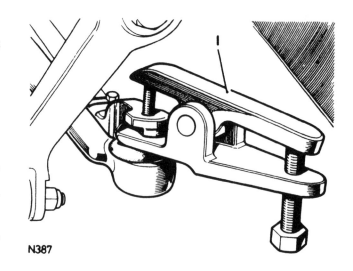

N387

Refitting

3. Set the steering box in the midway lock-to-lock position.
4. Fit the drop arm in the position as illustrated, aligning the dead splines.
5. Fit the drop arm fixings. Torque load 17,9 kgf.m (125 lbf. ft.).
6. Fit the drag link. Torque load for ball joint is 4,0 kgf.m (30 lbf. ft.).

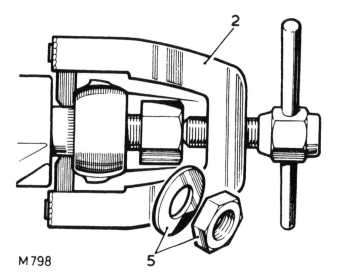

M798

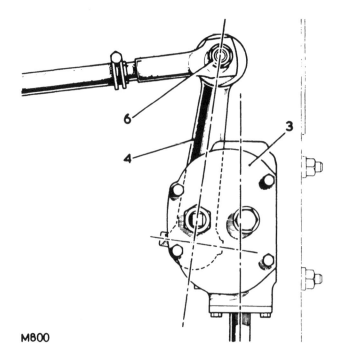

M800

TRACK ROD

– **Remove and refit**   **57.55.09.**

TRACK ROD LINKAGE

– **Remove and refit**   **57.55.10.**
Service tools: 18G1063 Ball joint extractor

TRACK ROD

Removing

1. Jack up and support chassis.
2. Disconnect the steering damper at the track rod.
3. Disconnect the track rod at the ball joints. Extractor 18G1063.
4. Withdraw the track rod complete.

LINKAGE.

Removing

5. Slacken the clamp bolts.
6. Unscrew the ball joints.
7. Unscrew the track rod adjuster, left-hand thread.

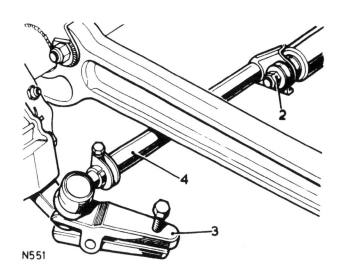

N551

Refitting

8. Fit the replacement parts. Do not tighten the clamp pinch bolts at this stage.
9. Screw in a ball joint to the full extent of the threads.
10. Set the adjuster dimensionally to the track rod as illustrated, to 8,9 mm (.350 in)
11. Set the adjuster end ball joint dimensionally, as illustrated, to 28,57 mm (1.125 in).
12. The track rod effective length of 1117,0 mm (48.4 in) is subject to adjustment during the subsequent wheel alignment check.

MC20

TRACK ROD

Refitting

13. Fit the track rod. Ball joint torque load 4,0 kgf.m (30 lbf.ft).
14. Check the front wheel alignment 57.65.01.
15. Reverse 1 and 2.

DRAG LINK

– Remove and refit   57.55.17.

DRAG LINK ENDS

–Remove and refit   57.55.16
Service tool 18G1063 Extractor for ball joint.

Removing

1. Jack up and support the chassis.
2. Remove the passenger's side road wheel.
3. Disconnect the drag link ball joint at the swivel housing arm. Extractor 18G1063
4. Disconnect the drag link end at the drop arm ball joint. Extractor 18G1063.
5. Withdraw the drag link.

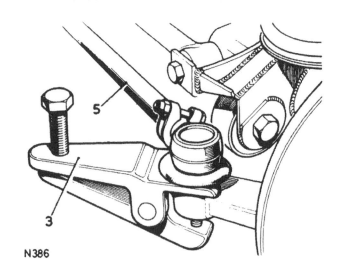

N386

DRAG LINK ENDS

Removing

6. Slacken the clamp bolts.
7. Unscrew the ball joint.
8. Unscrew the cranked end.

Refitting

9. Fit the replacement ends. Do not tighten the clamp bolts at this stage.
10. Set the ball joint dimensionally to the drag link, as illustrated, to 28,57 mm (1.125 in).
11. Adjust the cranked end to obtain the nominal overall length of 919,0 mm (36.2 in) . The length is finally adjusted during refitting.

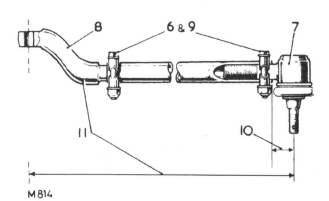

M814

DRAG LINK

Refitting.

12. Fit the drag link. Torque load for ball joints is 4,0 kgf.m (30 lbf.ft).
13. Check and if necessary set the steering lock stops. 57.65.03.
14. Turn the steering and ensure that full travel is obtained between the lock stops. Adjust the drag link length to suit.
15. Using a mallet, tap the ball joints in the direction indicated so that both ball pins are in the same angular plane.
16. Tighten the clamp bolts. Torque load 1,4 kgf.m (10 lbf.ft).
17. Reverse 1 and 2.

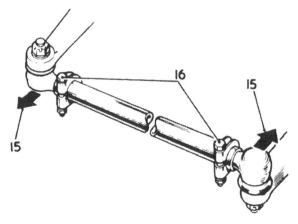

M815

STEERING WHEEL

– Remove and refit   **57.60.01**

Service tool- RO 1002 A Extractor for steering wheel.
               1862 (6312A) Puller.

IMPORTANT. The steering column is of a 'safety' type and incorporates shear pins. Therefore do not impart shock loads to the steering column during removing and refitting the steering wheel or at any time.

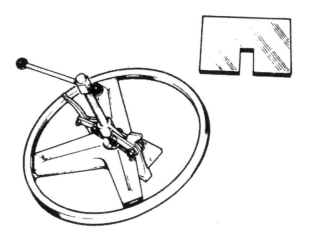

Removing

1. Withdraw the motif.
2. Remove the fixing nut and washer.
3. Extract the steering wheel. Extractor RO 1002.

Refitting.

4. Position the road wheels 'straight ahead'.
5. Position the steering wheel on the column splines with the centre spoke downwards.
6. Fit the nut and washer and secure the wheel to the column. Torque load 3,8 kgf.m (28 lbf.ft).
7. Fit the motif.

FRONT WHEEL ALIGNMENT

– Check and adjust   57.65.01.

Checking

Toe-out dimensions

NOTE. No adjustment is provided for castor, camber or swivel pin inclination.

1. Set the vehicle on level ground with the road wheels in the straight ahead position.
2. Push the vehicle back then forwards for a short distance to settle the linkage.
3. Measure the toe-out at the horizontal centre-line of the wheels.
4. Toe-out must be 1,2 to 2,4 m.m. (.04 to .09 in)
5. Check-tighten the clamp bolts fixings to 1,4 kgf.m (10 lbf.ft.).

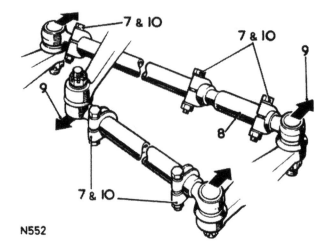

N552

Adjusting

6. Jack up and support the chassis.
7. Slacken the adjuster sleeve clamp bolts.
8. Rotate the adjuster to lengthen or shorten the track rod.
9. Check the toe-out setting as in 1 to 4. When the toe-out is correct, lightly tap the steering linkage ball joints, in the directions illustrated, to the maximum of their travel to ensure full unrestricted working travel.
10. Finally tighten the clamp bolts. Torque loading 1,4 kgf.m (10 lbf.ft.).

STEERING LOCK STOPS

– Check and adjust   57.65.03.

Checking

1. Measure the stop bolts protrusion as illustrated. This must be 40,5 m.m. (1.59 in)

Adjusting

2. Slacken the stop bolts locknuts.
3. Turn the stop bolt in or out as required.
4. Tighten the locknuts.
5. Check the wheels position at full lock.

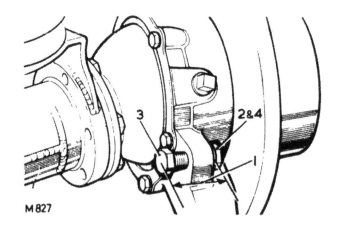

M 827

DATA

Toe-out setting   1,2 to 2,4 m.m. (.047 to .094 in)

FRONT SUSPENSION OPERATIONS

Operation No.

Bump stop
    – remove and refit   60.30.10.

Hub assembly
    – remove and refit   60.25.01.
    – overhaul   60.25.07.

Hub bearing end float
    – check and adjust   60.25.13.

Hub stub axle
    – remove and refit   60.25.22.
    – overhaul   60.25.24.

Panhard rod
    – remove and refit   60.10.10.

Radius arm
    – remove and refit   60.10.16.

Road spring
    – remove and refit   60.20.01.

Shock absorber
    – remove and refit   60.30.02.

Swivel pin housing
    – remove and refit   60.15.20.
    – overhaul   60.15.26.

PANHARD ROD

– Remove and refit   60.10.10

Removing

1. Remove the fixings at the mounting arm.
2. Remove the fixings at the axle bracket.
3. Withdraw the panhard rod.
4. If required press out the bushes.
5. Fit replacement bushes central in the rod.

Refitting

6. Reverse 1 to 4.

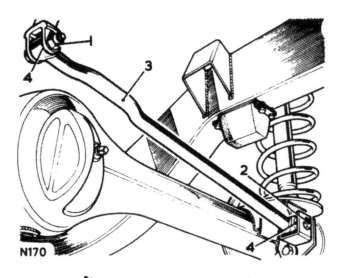

N170

RADIUS ARM

– Remove and refit   60.10.16.

Service tools: 18G 1063 Extractor for ball joint

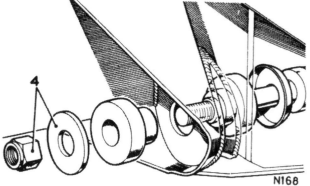

N168

Removing

1. Remove the road wheel.
2. Support the chassis.
3. Support the axle weight.
4. Remove the fixings, radius arm to chassis side member.
5. Disconnect the track rod at the ball joint. Extractor 18G 1063.
6. Remove the fixings, radius arm to axle.
7. Lower the radius arm front end to clear the axle and withdraw.
8. If required, press out the bush assemblies.
9. Fit the replacement bushes centrally in the arm.

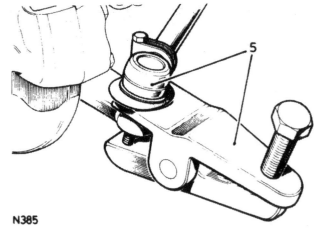

N385

Refitting

10. Reverse 1 to 7. Torque for ball joint is 4,0 kgf.m (30 lbf.ft).

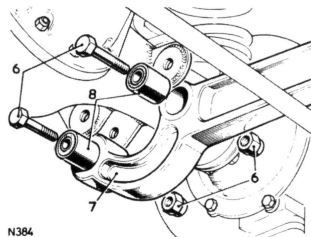

N384

60–2

SWIVEL PIN HOUSING ASSEMBLY

– Remove and refit 60.15.20.

Service tool: **18G 1063, Extractor for ball joint.**

Removing

1. Remove the front hub assembly. 60.25.01.
2. Remove the hub stub axle. 60.25.22.
3. LH side only. Disconnect the drag link at the ball joint. Extractor 18G 1063.
4. Disconnect the track rod at the ball joint. Extractor 18G 1063.
5. Remove the oil seal retainer fixings.
6. Move the retainer and oil seal away from the swivel housing.
7. Remove the top swivel pin fixings.
8. Withdraw the top swivel pin and shim washer/s.
9. Withdraw the swivel pin housing complete with bearings.
10. Lift out the constant velocity coupling.
11. Withdraw the axle shaft.
12. Remove the swivel bearing housing fixings.
13. Withdraw the bearing housing and joint washer.
14. Withdraw the oil seal and retainer.

Refitting

15. Pack the oil seal with 'Castrolease heavy' grease or suitable equivalent.
16. Offer the oil seal retainer to the swivel pin housing and the axle casing to determine its fitted position.
17. Position the oil seal and retainer on the bearing housing.
18. Reverse 11 and 12. Torque load for bearing housing fixings is 7,6 to 8,6 kgf.m (55 to 62 lbf.ft).
19. Reverse 6 to 10. Torque load for swivel pin fixings is 7,0 to 8,9 kgf.m (50 to 65 lbf.ft). Do not engage the lock plates at this stage.
20. Connect a spring balance to the swivel pin housing ball joint eye.
21. Measure the resistance to rotation of the swivel pin housing in a horizontal plane. This must be 1,2 to 1,3 kg (2.5 to 3 lb) after overcoming inertia.
22. Adjust as necessary by adding or subtracting shim washers under the top swivel pin. Shim range is 0,076; 0,127; 0,254; 0,762 mm (0.003; 0.005; 0.010; 0.030 in.).
23. Recheck after adjustment with the fixings torque loaded.
24. When satisfactory, engage the lock plates.
25. Reverse 5 and 6. Torque load for retainer fixings is 1,0 to 1,2 kgf.m (7 to 9 lbf.ft).
    Check that the oil seal wipes evenly over the surface of the bearing housing.
    Reverse 1 to 4. Torque load for ball joints is 4,0 kgf.m (30 lbf.ft).

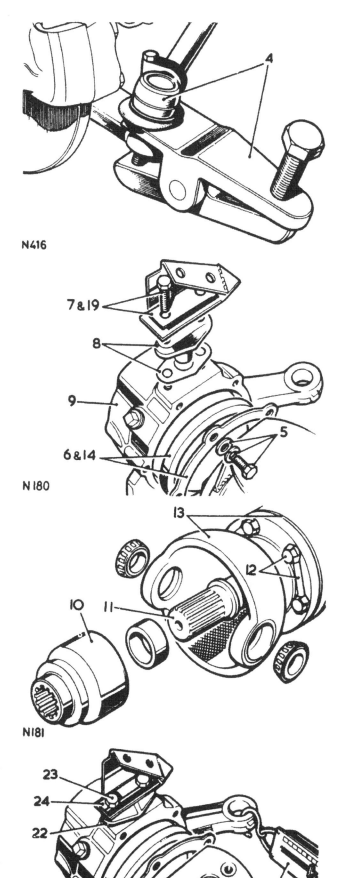

N416

N180

N181

N186

60–3

SWIVEL PIN HOUSING

– Overhaul 60.15.26.

1. Remove the swivel pin housing assembly 60.15.20.

Dismantling

2. Remove the lower swivel pin.
3. Withdraw the oil seal from the bearing housing.
4. Mark the relative positions of the constant velocity joint inner and outer race and the cage for correct reassembly.
5. Tilt and swivel the cage and inner race to remove the balls.
6. Swivel the cage into line with the axis of the joint and turn it until two opposite windows coincide with two lands of the joint housing.
7. Withdraw the cage.
8. Turn the inner track at right angles to the cage with two of the lands opposite the cage openings and withdraw the inner race.

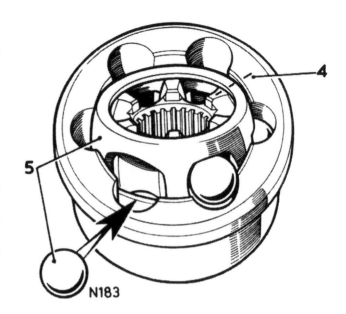

N183

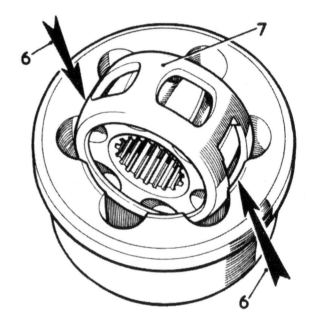

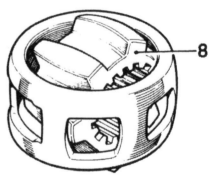

N184

Inspection

9. Examine all components for general condition.
10. The roller bearings must be a light push fit onto the swivel pins. If necessary, drive out the bearing outer tracks and press in replacements.
11. The surface of the bearing housing must be free of corrosion and damage.
12. Examine the inner and outer track, cage balls and bearing surfaces of the constant velocity joint for damage and excessive wear. Maximum acceptable end-float on the assembled joint 0,64 mm (0.025 in.).

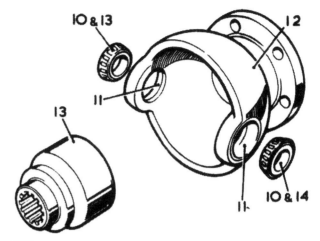

N417

Assembling

13. Reverse 2 to 8. Torque load for swivel pin fixings is 7,0 to 8,9 kgf.m (50 to 65 lbf.ft).
14. Position a roller bearing onto the lower swivel pin.
15. Position a roller bearing into the bearing housing upper track.
16. Fit the swivel pin housing assembly 60.15.20.
17. Check the front wheel alignment. 57.65.01.
18. Check the steering lock stops setting 57.65.03.

FRONT ROAD SPRING

– **Remove and refit**   60.20.01.

Removing

1. Remove the front shock absorber 60.30.02.

WARNING. During the following procedure avoid over stretching the brake hoses. If necessary, turn back the hose connectors locknuts to allow the hoses to follow the axle.

2. Lower the axle sufficient to free the road spring.
3. Withdraw the road spring.
4. Withdraw the shock absorber bracket securing ring.

Refitting

5. Reverse 4. Retain in position with a slave nut.
6. Reverse 2 and 3.
7. Remove the nut retaining the securing ring.
8. Fit the front shock absorber. 60.30.02.

60–6

FRONT HUB ASSEMBLY

– Remove and refit	**60.25.01.**

Service tool: **606435 - Spanner for hub nuts.**

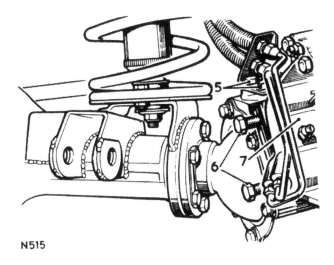

N515

Removing

1. Jack up under the front axle.
2. Remove the road wheel.
3. Unscrew the swivel pin housing drain plug, drain off the oil and refit the plug.
4. Unscrew the differential casing drain plug, drain off the oil and refit the plug.
5. Remove the fixings, brake pipe retaining bracket to swivel housing.
6. Remove the brake caliper fixings and lockplates.
7. Tie aside the brake caliper free of the front disc; avoid bending the brake pipes.
8. Remove the fixings at the driving shaft hub.
9. Withdraw the hub and integral drive shaft.
10. Remove the hub lock nut and locker. Spanner 606435.
11. Remove the hub nut and key washer. Spanner 606435.
12. Withdraw the hub assembly.

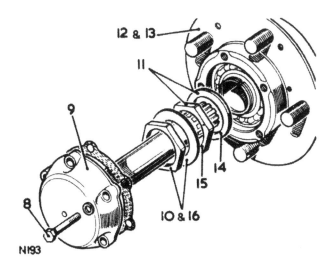

N193

Refitting

13. Position the hub assembly on the stub axle.
14. Locate the key washer onto the stub axle.
15. Fit the hub nut fully on then back off until the hub is just free to rotate.
16. Fit the locker and locknut.
17. Fit a dial gauge and bracket to record the end-float between the hub and the stub axle.

continued

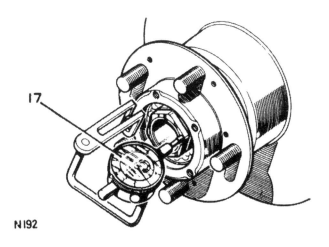

N192

18. Rotate the hub to settle the bearings.
19. Check the hub to stux axle end-float. End-float must be 0,05 to 0,10 mm (0.002 to 0.004 in.).
20. Adjust as necessary by rotating the hub nut.
21. Recheck end-float with the locknut tightened and locked.
22. Fit the hub cap and drive shaft. Torque is 4,1 to 5,1 kgf.m (30 to 38 lbf.ft).
23. Fit the brake caliper. Torque is 8,5 kgf.m (60 lbf.ft).
24. Fit the brake pipe bracket to the swivel pin housing. Torque 7,0 to 8,9 kgf.m (50 to 65 lbf.ft).
25. Reverse 1 and 2.
26. Replenish swivel housing oil and differential oil. Refer to Maintenance 10.

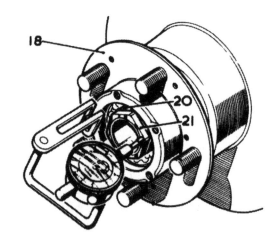

DATA

Hub to stub axle end float   0,05 to 0,10 mm (0.002 to 0.004 in.)

FRONT HUB ASSEMBLY

– Overhaul
60.25.07

1. Remove the front hub assembly 60.25.01.

Service tool: 18G 1349 Hub oil seal fitting tool
(dual-lipped seal)

Dismantling

2. Lift out the oil seal.
3. Withdraw the bearing inner members.
4. Press out the bearing outer members.
5. Wash out the casing, using clean kerosine or other suitable fluid.

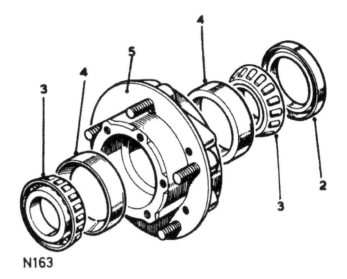

N163

Assembling

6. Reverse items 2 to 4, using a press to fit the bearing outer members.
    Early models Before fitting the oil seal, coat the seal outer diameter with Stag A jointing compound. Then press in the seal until the rear face is flush with the top face of the seal housing bore only, not fully into the bore.
    Later models with dual-lipped seal Use service tool 18G 1349 to press seal into correct position.
    NOTE: Replacement leather oil seals should be soaked in engine oil for one hour before fitting.
7. Lubricate the assembly by adding 14.2 gms (5 oz) approximately of grease to the hub casing and bearing assembly. Group 09 refers for recommended greases.
8. Refit the hub assembly 60.25.01.

FRONT HUB BEARING END FLOAT

– Check and adjust   60.25.13.

Service tool: **606435 - Spanner for hub nuts.**

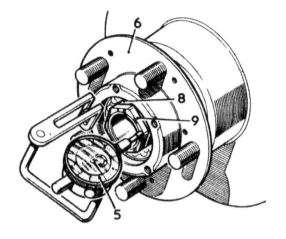

Checking

1. Jack up under the front axle.
2. Remove the road wheel.
3. Remove the hub and driving shaft.
4. Remove the brake pads from the caliper assembly 70.40.02.
5. Fit a dial gauge and bracket to measure end-float between the hub and the stub axle.
6. Rotate the hub to settle the bearings.
7. Check the hub to stub axle end-float. This must be 0,05 to 0,10 mm (0.002 to 0.004 in.).

Adjusting

8. To adjust, slacken the locknut and turn the hub nut in to decrease, out to increase, the end-float. Spanner 606435.
9. Re-check with locknut tightened and locked.
10. Fit the hub and driving shaft. Torque load 4,1 to 5,1 kgf.m (30 to 38 lbf.ft).
11. Reverse 1 and 2.

DATA

Hub to stub axle end float   0,05 to 0,10 mm (0.002 to 0.004 in.)

FRONT HUB STUB AXLE

— Remove and refit. 1 to 4, 11 to 13.   60.25.22

— Overhaul. 5 to 10.   60.25.24

Removing

1. Remove the front hub assembly 60.25.01.
2. Remove the fixing bolts and lockplates.
3. Position a waste oil can under the joint faces.
4. Withdraw the stub axle and joint washer.

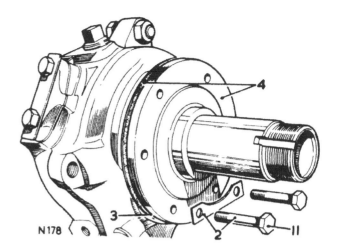

Overhaul

5. Withdraw the bush.
6. Press in the replacement.
7. Early axles — Remove the distance piece, using a chisel.
8. Apply Loctite Hydraulic Sealant or Bostik 777 or equivalent sealing compound to the stub axle distance piece seat.
9. Press on the distance piece.
   Later stub axles are in one piece.
10. Remove all traces of sealant from the exposed areas.

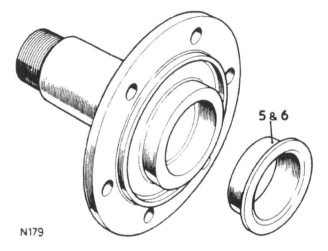

Refitting

11. Fit the stub axle and joint washer. Torque load for fixings is 6,0 kgf.m (44 lbf.ft).
12. Fit the front hub assembly 60.25.01.
13. With the road wheel replaced and the vehicle off jacks, check the swivel pin housing oil level 60.15.20.

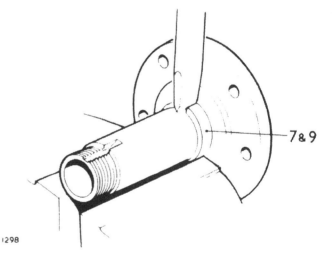

FRONT SHOCK ABSORBER

– Remove and refit   60.30.02

Removing

1.   Remove the road wheel.
2.   Support the axle weight.
3.   Remove the shock absorber lower fixing.
4.   Withdraw the cupwasher, rubber bush and seating washer.
5.   Remove the shock absorber bracket fixings.
6.   Withdraw the shock absorber and bracket complete.
7.   Withdraw the sealing washer, rubber bush and cupwasher.
8.   Remove the fixings, shock absorber to mounting bracket.
9.   Withdraw the mounting bracket.
10. Lift off the top seating washer, rubber bush and cupwasher.

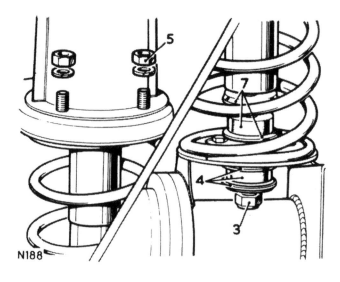

Refitting

11. Reverse 1 to 10.

BUMP STOP

Remove and refit   60.30.10

Removing

1.   Remove the fixings.
2.   Withdraw the bump stop rubber and carrier complete.

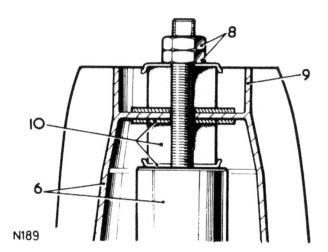

Refitting

3.   Position the fixing bolts in the slots in the chassis brackets.
4.   Fit the bump stop and carrier.

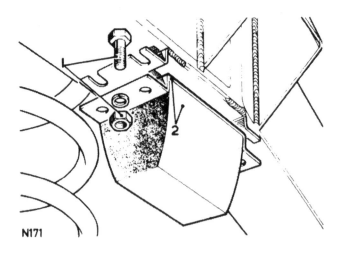

REAR SUSPENSION OPERATIONS

Operation No.

64–1

Bump stop
    — remove and refit   64.30.15.

Hub
    — remove and refit   64.15.01.
    — overhaul   64.15.07.

Hub bearing end-float
    — check and adjust   64.15.13.

Hub bearing sleeve (stub axle)
    — remove and refit   64.15.20.
    — overhaul   64.15.21.

Road spring
    — remove and refit   64.20.01.

Shock absorber
    — remove and refit   64.30.02.

Self-levelling unit
    — remove and refit   64.30.09.

Self-levelling unit ball joints
    — remove and refit   64.30.10.

Suspension link, lower
    — Bush — remove and refit   64.35.05
    — remove and refit   64.35.02

Suspension link, upper
    — Bush — remove and refit   64.35.04
    — remove and refit   64.35.01

REAR HUB ASSEMBLY

– Remove and refit   64.15.01.

Service tools: 606435   **Hub nut spanner.**

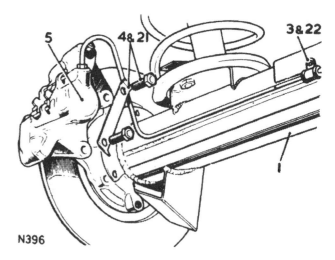

N396

Removing

1. Jack up under rear axle.
2. Remove road wheels.
3. Remove the drive screws from the brake pipe outer clips at the rear axle.
4. Remove the brake caliper fixings and lockplates.
5. Tie aside the brake calipers, free of the rear disc; avoid bending the brake pipes.
6. Remove the hub cap fixings.
7. Withdraw the hub cap and integral axle shaft complete.
8. Remove the hub locknut and locking plate.
9. Remove the hub nut and key washer.
10. Withdraw the hub and brake disc complete with the outer bearing inner member.

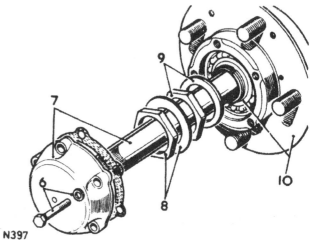

N397

Refitting

11. Position the hub assembly on the bearing sleeve.
12. Locate the key washer onto the sleeve.
13. Fit the hub nut fully on then back off until the hub is just free to rotate.
14. Fit the locker and locknut.
15. Fit a dial gauge and bracket to record the end-float between the hub and the bearing sleeve.
16. Rotate the hub to settle the bearings.
17. Check the hub to bearing sleeve end-float. This must be 0,05 to 0,10 mm (0.002 to 0.004 in.)
18. Adjust as necessary by rotating the hub nut.
19. Recheck end-float with the locknut tightened and locked.
20. Fit the hub cap and axle shaft. Torque loading is 4,1 to 5,1 kgf.m (30 to 38 lbf.ft).
21. Fit the brake caliper. Torque loading is 8,5 kgf.m (60 lbf.ft). To gain access to the RH caliper fixings, disconnect the shock absorber at the lower fixings.
22. Fit brake pipe clip to axle.
23. Refit shock absorber.
24. Reverse 1 and 2.

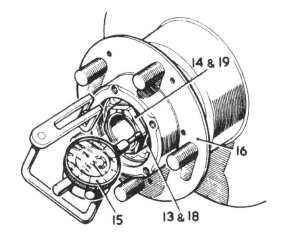

N389

DATA

Hub to bearing sleeve end-float   0,05 to 0,10 mm (0.002 to 0.004 in.)

REAR HUB ASSEMBLY

– Overhaul 64.15.07

1. Remove the rear hub assembly 64.15.01.

Service tool: 18G 1349 Hub oil seal fitting tool
(dual-lipped seal)

Dismantling

2. Lift out the oil seal.
3. Withdraw the bearing inner members.
4. Drive out the bearing outer members.
5. Wash out the casing, using clean kerosine or other suitable fluid.

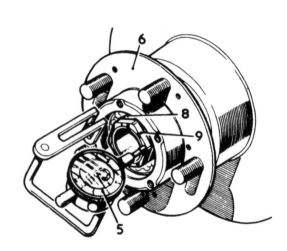

N163

Assembling

6. Reverse items 2 to 4, using a press to fit the bearing outer members.
   Early models Before fitting the oil seal, coat the seal outer diameter with Stag A jointing compound. Then press in the seal until the rear face is flush with the top face of the seal housing bore only, not fully into the bore.
   Later models with dual-lipped seal Use service tool 18G 1349 to press seal into correct position.
   NOTE: Replacement leather oil seals should be soaked in engine oil for one hour before fitting.
7. Lubricate the assembly by adding 142 gms (5 oz) approximately of grease to the hub casing and bearing assembly. Group 09 refers for recommended greases.
8. Refit the hub assembly 64.15.01.

REAR HUB BEARING END FLOAT

– Check and adjust 64.15.13

Service tool: 606435 Spanner for hub nuts

Checking

1. Jack up under the rear axle.
2. Remove the road wheel.
3. Remove the hub and driving shaft.
4. Remove the brake pads from the caliper assembly. 70.40.03.
5. Fit a dial gauge and bracket to measure end-float between the hub and the stub axle.
6. Rotate the hub to settle the bearings.
7. Check the hub to stub axle end-float. This must be 0,05 to 0,10 mm (0.002 to 0.004 in.).

N388

Adjusting

8. To adjust, slacken the locknut and turn the hub nut, in to decrease, out to increase, the end-float. Spanner 606435.
9. Re-check with locknut tightened and locked.
10. Fit the hub and axle shaft. Torque load 4,1 to 5,1 kgf.m (30 to 38 lbf.ft).
11. Reverse 1 and 2.

DATA

Hub to bearing sleeve end float . . . . 0,05 to 0,10 mm
(0.002 to 0.004 in.)

64–4

REAR HUB BEARING
SLEEVE (STUB AXLE)

– Remove and refit. 1 to 4, 9 to 11	64.15.20.
–Overhaul. 5 to 8	64.15.21

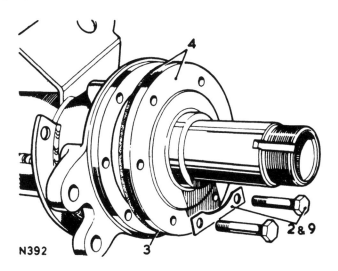

N392

Removing

1. Remove the rear hub assembly. 64.15.01.
2. Remove the fixing bolts and lockplates.
3. Position a waste oil can under the joint faces.
4. Withdraw the bearing sleeve and joint washer.

Overhaul

5. Remove the distance piece, using a chisel.
6. Apply Auto Gell No. 2 or equivalent sealing compound to the bearing sleeve distance piece seat.
7. Press on the distance piece.
8. Remove all traces of sealant from the exposed areas.

Refitting

9. Fit the bearing sleeve and joint washer. Torque load for fixings is 6,0 kgf.m (44 lbf. ft.).
10. Fit the rear hub assembly. 64.15.01.
11. With the road wheel replaced and the vehicle off jacks, check the rear axle oil level, refer to page 10–19.

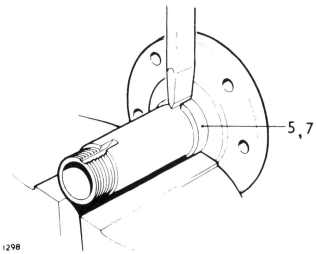

1298

REAR ROAD SPRING

– Remove and refit   64.20.01.

Service tool: RO1006   Extractor for ball joint.

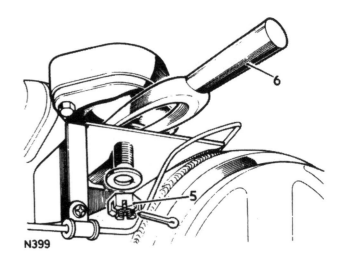

N399

Removing

1. Raise rear of vehicle and support chassis.
2. Remove road wheels.
3. Support the axle weight.
4. Disconnect the shock absorbers at one end.
5. Remove fixings, pivot bracket ball joint to axle.
6. Extract the ball joint pin from the tapered housing. Extractor RO 1006
7. Lower the axle further, sufficient to free the road spring from the upper seat.

WARNING. Avoid lowering the axle further than necessary otherwise the rear brake flexible hose will become stretched.

8. Remove the spring retainer plate.
9. Withdraw the road spring.
10. Lift off the spring seat.

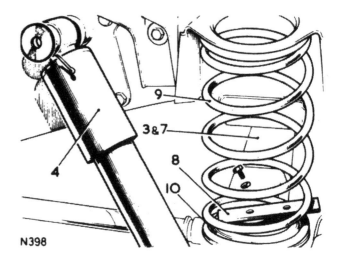

N398

Refitting

11. Reverse 1 to 10. Torque load for pivot bracket ball joint is 17,9 kgf.m (130 lbf.ft).

REAR SHOCK ABSORBER

– Remove and refit   64.30.02

Removing

1. Remove road wheel.
2. Support the rear axle.
3. Remove the fixings and withdraw the shock absorber from the axle bracket.
4. Remove upper fixings
5. Withdraw the shock absorber.
6. If required, remove the mounting bracket at the chassis side member.
7. If required, lift out the mounting rubbers at the upper end.

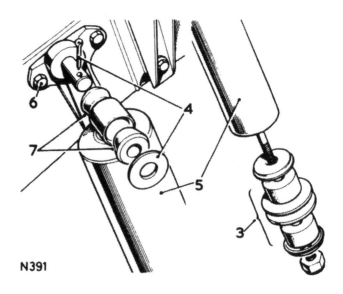

N391

Refitting

8. Reverse items 7 and 8 as applicable.
9. Reverse items 1 to 6.

LEVELLING UNIT

Description

A Boge Hydromat levelling unit is located in the centre of the rear axle.

When the vehicle is unladen the levelling unit has little effect. The unit is self-energising and hence the vehicle has to be driven before the unit becomes effective, the time taken for this to happen being dependant upon the vehicle load, the speed at which it is driven and the roughness of the terrain being crossed.

If the vehicle is overloaded the unit will fail to level fully and more frequent bump stop contact will be noticed.

Should the vehicle be left for a lengthy period e.g. overnight, in a laden condition, it may settle. This is due to internal leakage in the unit and is not detrimental to the unit performance.

LEVELLING UNIT

– Remove and refit   64.30.09

Removing

WARNING: The levelling unit contains pressurised gas and must not be dismantled nor the casing screws removed. Repair is by replacement of complete unit only.

1. Raise and support the chassis rear end.
2. Support the axle weight.
3. Disconnect the suspension upper links at the pivot bracket.
4. Ease up the lower gaiter.
5. Unscrew the lower ball joint at the levelling unit push rod, using thin jawed spanners.
6. Remove the top bracket fixings at the cross-member.
7. Withdraw the levelling unit and top bracket complete.
8. Ease back the upper gaiter.
9. Unscrew the upper ball joint at the levelling unit, using thin jawed spanners.
10. Withdraw the upper and lower gaiters and their retaining spring rings.

Refitting

11. Smear 'Loctite' grade CVX or suitable equivalent sealant on to ball pin threads.
12. Reverse items 1 to 10. Do not fully tighten the fixings until all items are in their fitted position. Torque loading for fitting ball pins to levelling unit is 3,5 kgf.m (25 lbf.ft).

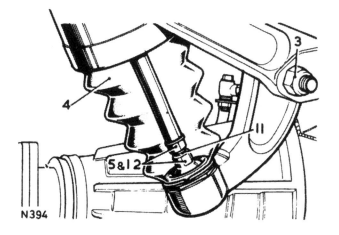

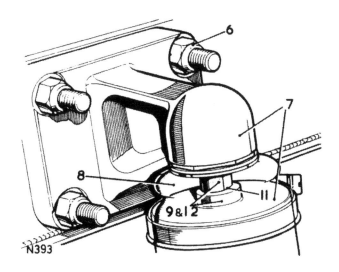

64–7

LEVELLING UNIT BALL JOINTS

– Remove and refit   64.30.10

Service tools: RO 1006 Extractor for axle bracket ball joint

Removing

1.   Remove the levelling unit 64.30.09.
2.   Remove the split pin and nut at the rear axle bracket.
3.   Extract the ball pin from the axle bracket. Extractor RO 1006
4.   Withdraw the pivot bracket complete with ball joints.
5.   Unscrew the ball joint assembly for the levelling unit.
6.   Remove the ball joint assembly for the axle bracket.
7.   Replacement ball joints are supplied as complete assemblies, less fixings, and are pre-packed with grease.
8.   The ball joint for the axle bracket must not be dismantled.
9.   The ball joints for the levelling unit may be dismantled and cleaned if required.
10. Pack the ball joint with Dextagrease GP or an equivalent grease when assembling.
11. Ensure that the ball seating is square in its housing before refitting.

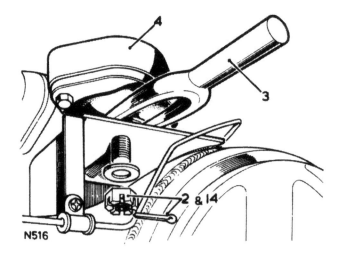

N516

Refitting

12. Press the knurled ball joint into the pivot bracket.
13. Screw the ball joints for the levelling unit into the mounting brackets. If the ball joints do not screw in easily and fully, remove and refit the assemblies ensuring that the plastic seats do not foul in the housings. Torque load 7,0 kgf.m (50 lbf.ft.).
14. Fit the pivot bracket complete with ball joints to the rear axle. Torque 17,9 kgf.m (130 lbf.ft.) for ball joint
15. Fit the levelling unit. 64.30.09.

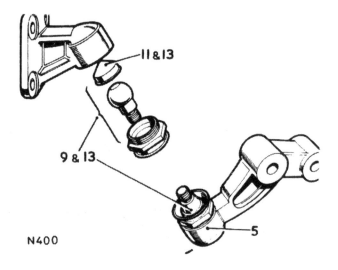

N400

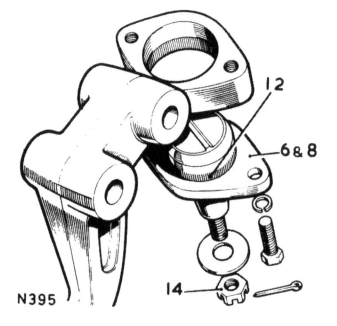

N395

BUMP STOP

– Remove and refit   **64.30.15.**

Removing

1. Remove the fixings.
2. Withdraw the bump stop rubber and carrier complete.

Refitting

3. Position the fixing bolts in the slots in the chassis brackets.
4. Fit the bump stop and carrier, position the shoulder on the carrier to suit the chassis configuration.

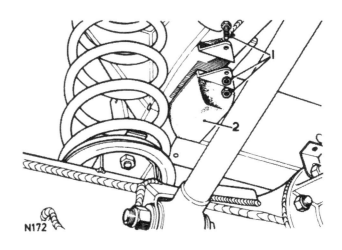

UPPER SUSPENSION LINK

– Remove and refit 1 to 6 and 9.   **64.35.01.**

BUSH

– Remove and refit 7 and 8   **64.35.04.**

Removing

1. Jack up under the chassis until the rear axle is freely suspended.
2. Remove the fixings, upper link bracket to frame.
3. Remove the fixings, upper links to pivot bracket.
4. Withdraw the upper link complete with frame bracket.
5. Remove the fixing bolt.
6. Separate link and bush assembly from bracket.

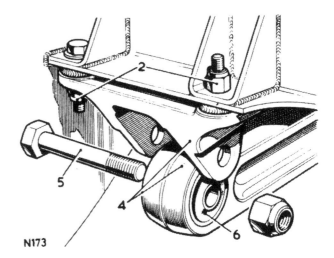

Replacing the bush

7. Press out the bush assembly.
8. Fit the replacement bush assembly central in the housing.

Refitting

9. Reverse 1 to 6. Do not fully tighten the fixings until all components are in position.

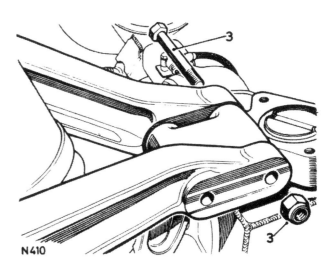

LOWER SUSPENSION LINK

– Remove and refit 1 to 7, 10 to 12   **64.35.02.**

BUSH

– Remove and refit 8 and 9   **64.35.05.**

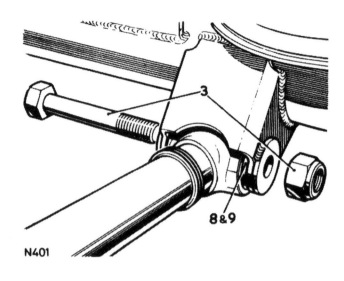

N401

Removing

1. Jack up the rear end or use a ramp for accessibility.
2. L.H. side only. Remove the shock absorber lower fixings and withdraw the shock absorber from the axle bracket. 64.30.02.
3. Remove the link rear fixings.
4. Remove the mounting bracket fixings at the side member bracket.
   a. Earlier model fixings illustrated.
   b. Later model fixings illustrated.
5. Withdraw lower link complete with mounting bracket.
6. Remove the locknut.
7. Withdraw the mounting bracket from the lower link.

Replacing the bush

8. Press out the bush assembly from the rear end.
9. Fit the replacement bush assembly central in the housing.

Refitting

10. Reverse items 6 and 7. Do not tighten the locknut at this stage.
11. Reverse items 3 to 5.
12. Lower the car, remove the jack and allow the axle to take up its static laden position.
13. Tighten the lock nut to a torque of 12,4 kgf.m (90 lbf.ft).

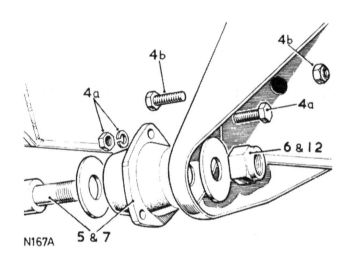

N167A

BRAKE OPERATIONS

BRAKE SYSTEM – Description of Unified threaded system used up to early 1981   70.00.00

1 – Primary section   2 – Secondary section

The hydraulic braking system fitted to the Range Rover is of the divided line type, incorporating primary and secondary sections.

The brake pedal is coupled directly to a mechanical servo which in turn operates a tandem master cylinder. The front disc brake calipers each house four pistons, the upper pistons are connected to the primary system, the lower pistons are connected to the secondary system. The rear disc brake calipers each house two pistons and these are connected to the primary system only. Both systems are connected to the brake failure switch, which causes illumination of an instrument panel warning light in the event of a pressure failure in either system.

The brake fluid reservoir is divided, the rear section feeds the master cylinder primary cylinder, and the front section feeds the master cylinder secondary cylinder. Under all normal conditions both the primary and secondary systems operate simultaneously on brake pedal application. In the event of a failure in the primary system, the secondary system will still function and operate the front calipers. Alternatively, if the secondary system fails, the primary system will still function and operate the front and rear calipers.

If the servo should fail, both systems will still function but would require greater pedal pressure.

The hand operated transmission brake is completely independant of the hydraulic system.

For details of the metric brake system introduced from Vehicle Identification Number (VIN) 112612 see page 70–4.

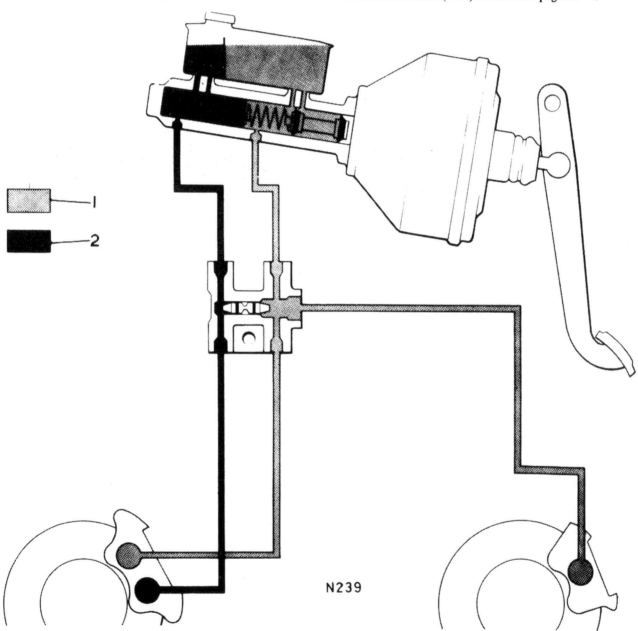

N239

70–2

FRONT DISCS

– Remove and refit   **70.10.10**

Removing

1. Remove the front hub assembly 60.25.01
2. Remove the front disc fixing bolts
3. Tap off the disc from the front hub

Refitting

4. Locate the disc onto the front hub
5. Fit the disc fixing bolts. Torque 5,0 kgf.m (38 lbf. ft.).
6. Using a dial test indicator, check the total disc run-out, this must not exceed 0,15 mm (0.006 in.). If necessary, reposition the disc.
7. Fit the front hub assembly 60.25.01

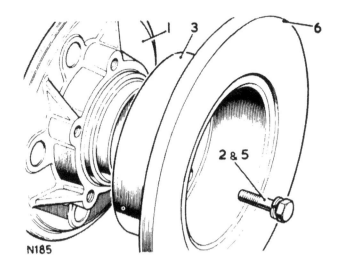

N185

REAR DISCS

– Remove and refit   **70.10.11**

Removing

1. Remove the rear hub assembly 64.15.01
2. Remove the rear disc fixing bolts
3. Tap off the disc from the rear hub.

Refitting

4. Locate the disc onto the rear hub.
5. Fit the disc fixing bolts. Torque 5.0 kgf.m (38 lbf ft).
6. Using a dial test indicator, check the total disc run out, this must not exceed 0.15mm (0.006 in.). If necessary reposition the disc.
7. Fit the rear hub assembly 64.15.01.

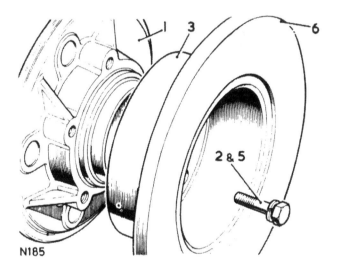

N185

Disc reclamation

Check the disc thickness marked on the disc boss – this dimension may be reduced to a minimum thickness of 0.510 in. (13mm) front and 0.460 in. (12mm) rear, by machining an equal amount off both sides.

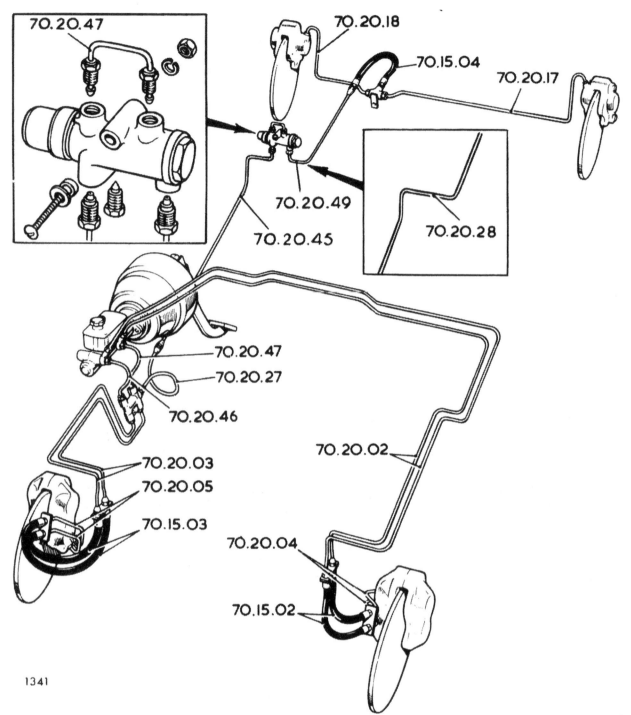

70.20.47

70.20.18

70.15.04

70.20.17

70.20.49

70.20.45

70.20.28

70.20.47

70.20.27

70.20.46

70.20.02

70.20.03
70.20.05
70.15.03

70.20.04

70.15.02

1341

NOTE: The above general layout of the braking system relates to the Unified threaded system fitted with a separate brake failure/pressure warning system.

To comply with EEC brake regulations, during early 1981 at V.I.N. 112612 a braking system was introduced using a new design master cylinder with an integral pressure failure warning switch. It should be noted that the Primary and Secondary sections in this master cylinder are the reversal of those described in the original Unified threaded braking system. Metric brake pipes and connections were also fitted throughout.

This type of system is identified by a yellow label displayed on the vacuum servo shell which states:

"ALL HYDRAULIC BRAKE PIPE FITTINGS HAVE METRIC THREADS"

IMPORTANT: For vehicles up to V.I.N. 112611 the original type of master cylinder and servo assembly, also the pressure reducing valve will still be available. However this is not the case with the front and rear brake calipers which will now only be supplied with metric threads. When fitting a new metric threaded caliper to an early vehicle with non metric (UNF) threaded calipers ensure that the original connections are replaced by compatible metric types.

HOSES

– Remove and refit

Front left hand	70.15.02
Front right hand	70.15.03
Intermediate	70.15.04

PIPES

– Remove and refit

Feed to front left hand hose connector	70.20.02
Feed to front right hand hose connector	70.20.03
Feed to front left hand caliper	70.20.04
Feed to front right hand caliper	70.20.05
Feed to rear left hand caliper	70.20.17
Feed to rear right hand caliper	70.20.18
Feed to two-way connector	70.20.27
Feed to intermediate hose	70.20.28
Feed to pressure reducing valve	70.20.45
Feed to brake failure switch, secondary system	70.20.46
Feed to brake failure switch, primary system	70.20.47
Feed – pressure reducing valve to rear brake hose	70.20.48
Transfer pipe – pressure reducing valve	70.20.49

Note.The operation numbers are included on the brake system illustration to facilitate indentification of the individual pipes.

Removing

1. Disconnect the hose or pipe at both connections.
2. Release the clipping.
3. Withdraw the hose or pipe.

Refitting

4. Reverse 1 to 3.
5. Bleed the brakes 70.25.02.

BRAKE FAILURE SWITCH

– Remove and refit   70.15.36

Removing

1. Disconnect the electrical leads from the brake failure switch.
2. Disconnect and blank off the five fluid pipes.
3. Remove the brake failure switch.

Refitting

4. Secure the brake failure switch in position.
5. Connect the fluid pipe from the master cylinder secondary system.
6. Connect the fluid pipe from the master cylinder primary system.
7. Connect the fluid pipe from the front brake secondary system.
8. Connect the fluid pipe from the front brake primary system.
9. Connect the fluid pipe from the rear brakes.
10. Connect the electrical lead to the brake failure switch.
11. Bleed the brakes 70.25.02.

    NOTE: On 1981 vehicles from V.I.N. 112612 the brake failure pressure warning switch is mounted directly into the master cylinder.

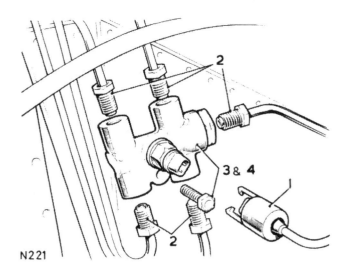

N221

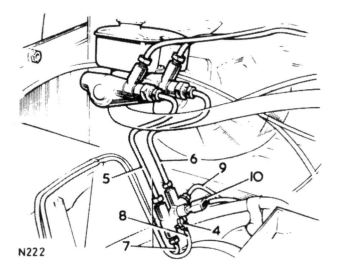

N222

70–6

BRAKE FAILURE SWITCH

– Overhaul   **70.15.41**

1. Remove the brake failure switch 70.15.36.

Dismantling

2. Remove the end plug.
3. Discard the copper washer.
4. Unscrew the switch.
5. Withdraw the shuttle valve, if necessary use a low pressure air line.
6. Remove and discard the seals from the shuttle valve.

Inspecting

7. Clean the shuttle valve, end plug and five-way connector, using new brake fluid or ethyl alcohol.
8. Examine the shuttle valve and its bore in the five-way connector, they must be in perfect condition with no signs of scratches or corrosion, otherwise fit a new switch complete.
9. To test the electrical switch, reconnect the leads and actuate the switch plunger by pressing it against an earthing point on the vehicle.

Assembling

10. Fit two new seals to the shuttle valve.
11. Coat the seals with Lockheed disc brake lubricant.
12. Insert the shuttle valve and seal assembly into the bore of the five-way connector.
13. Locate a NEW washer for the end plug in position.
14. Fit the end plug. Torque 2,2 kgf.m (16 lbf.ft).
15. Fit the electrical switch to the five-way connector.
16. Fit the brake failure switch 70.15.36.

    NOTE: On 1981 vehicles from V.I.N. 112612 the brake failure pressure warning switch is mounted directly into the master cylinder.

N223

BRAKES

– Bleed   70.25.02

The hydraulic system comprises two completely independent sections. The rear calipers and the upper pistons in the front calipers form the primary section, while the lower pistons in the front calipers form the secondary section. The following procedure covers bleeding the complete system, but it is permissible to bleed one section only if disconnections are limited to that section. When bleeding any part of the primary section, almost full brake pedal travel is available. When bleeding the secondary section only, then brake pedal travel will be restricted to approximately half.

On 1981 vehicles from V.I.N. 112612 where the brake failure/pressure warning switch is fitted directly into the master cylinder carry out the following procedure before bleeding the brakes.

1. Disconnect the leads from the switch.
2. Unscrew the switch four full turns before depressing the pedal, with either circuit open.
3. After completion of bleeding screw in the switch and tighten to a torque of 18 kg.cm (16 lbft.in.).

IMPORTANT: If bleeding the secondary section only, commence with the caliper furthest from the master cylinder and bleed from the screw on the same side as the fluid inlet pipes, then close the screw and bleed from the screw on the opposite side of the same caliper. Repeat for the other front caliper.

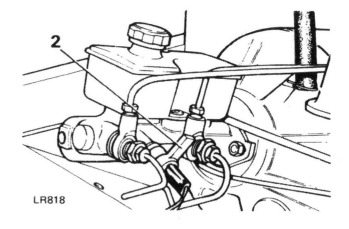

LR818

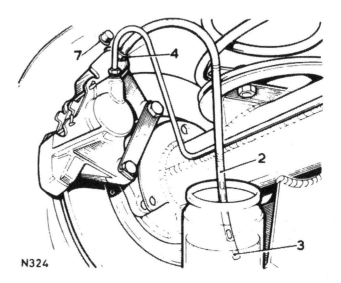

N324

Bleeding

1. Fill the fluid reservoir with the correct fluid, see group 09. On vehicles fitted with a pressure reducing valve in the rear brake fluid line, it is necessary to first bleed the pressure reducing valve.

IMPORTANT: The correct fluid level must be maintained throughout the procedure of bleeding.

2. Connect a bleed tube to the bleed screw on the rear caliper furthest from the master cylinder.
3. Submerge the free end of the bleed tube in a container of clean brake fluid.
4. Slacken the bleed screw.
5. Operate the brake pedal fully and allow to return.

IMPORTANT: Allow at least five seconds to elapse with the foot right off the pedal to ensure that the pistons fully return before operating the pedal again.

6. Repeat 5 until fluid clear of air bubbles appear in the container, then keeping the pedal fully depressed, tighten the bleed screw.
7. Remove the bleed tube and replace the dust cap on the bleed screw.
8. Repeat 1 to 7 for the other rear caliper.
9. Remove the front wheel on the side furthest from the master cylinder.
10. Connect a bleed tube to the primary bleed screw on the front caliper furthest from the master cylinder.
11. Connect a bleed tube to the secondary bleed screw on the same side of the caliper as the primary screw.

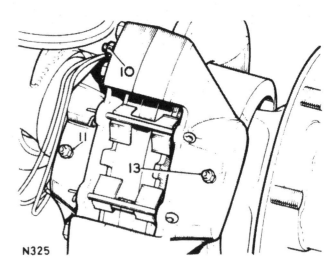

N325

12. Repeat 3 to 7 for the front caliper, bleeding from the two screws simultaneously.
13. Connect a bleed tube to the other screw on the front caliper furthest from the master cylinder.
14. Repeat 3 to 7 for the second secondary screw on the front caliper.
15. Refit the front wheel.
16. Repeat 9 to 15 for the front caliper nearest the master cylinder.

BRAKE PRESSURE REDUCING VALVE

– Remove and refit   70.25.21.

Removing.

1. Remove all dust, grime etc. from the vicinity of the pressure reducing valve fluid pipe unions.
2. Disconnect the inlet and outlet fluid pipes from the pressure reducing valve. Plug the pipes and reducing valve ports to prevent the ingress of foreign matter.
3. Remove the nut, spring washer, screw, plain washer and distance piece securing the reducing valve to the vehicle. Withdraw the reducing valve.

Refitting.

4. Reverse instructions 3 and 2.
5. Bleed the brake system commencing with the bleed nipple in the reducing valve.

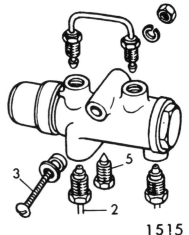

1515

MASTER CYLINDER – TANDEM

– Remove and refit   70.30.08

IMPORTANT. Two types of master cylinder and servo assemblies are in use and individual components must not be interchanged, but the latest type master cylinder together with the latest servo may be fitted in place of the earlier type.

Indentification – The early type master cylinder and servo assembly has an adaptor plate measuring approximately 90 x 100 mm (3.500 x 4in) fitted between the master cylinder and servo unit. The later type does not have an adaptor plate and the master cylinder is secured directly to the servo unit.

Only the latest type master cylinders and servo units are available as replacements, and these may be fitted individually when replacing components of the same type, but both must be fitted when replacing early type components.

Removing

1. Disconnect the pipes from the master cylinder.
2. Early type master cylinder – Remove the master cylinder complete with adaptor plate.
3. Latest type master cylinder – Remove the master cylinder.

Refitting

4. Early type master cylinder – Reverse 1 and 2.
5. Latest type master cylinder – Reverse 1 and 3.
6. Bleed the brakes 70.25.02.

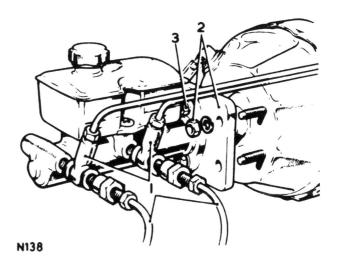

N138

MASTER CYLINDER – TANDEM

– Overhaul   70.30.09

1.   Remove the master cylinder 70.30.08.

Dismantling

2.   Remove the circlip from the end of the cylinder.
3.   Apply a high pressure air line to the rear outlet port
     to expell the primary piston.
4.   Withdraw the guide and spring.
5.   Remove the fluid reservoir.
6.   Withdraw the seals from the inlet ports.
7.   Depress the secondary piston against spring pressure.
8.   Withdraw the piston locating peg from the front inlet
     port.
9.   Temporarily seal both inlet ports and the rear outlet
     port.
10. Apply a high pressure air line to the front outlet port
     to expell the secondary piston.
11. Withdraw the guide and spring for the secondary
     piston.
12. Remove the seals and special washers from the
     pistons.

continued

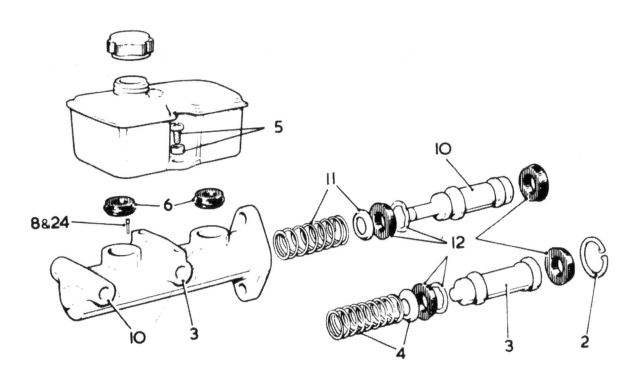

N160

Inspecting

13. Fit new seals, special washers, springs and retaining circlip.
14. Clean all components with new brake fluid. Dry with a lint free cloth.
15. All metal components must be in perfect condition. The pistons and the cylinder bore must not have any signs of scratches.

Assembling

16. Lubricate all components with clean brake fluid.
17. Place the special washer on the front end of the secondary piston with the concave side towards the seal.
18. Fit the front seal.
19. Fit the rear seal.
20. Place the special washer on he front end of the primary piston with the concave side towards the seal.
21. Fit the front seal.
22. Fit the rear seal.
23. Reverse 7,8 and 10 to 11.
24. Check that the secondary piston can move approximately 12mm (½in) against spring pressure, to ensure that the locating peg is correctly positioned.
25. Reverse 2 to 6.
26. Refit the master cylinder 70.30.08.

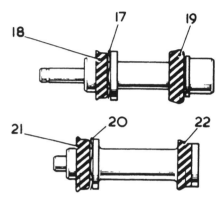

N159

FLUID RESERVOIR

– Remove and refit   70.30.15

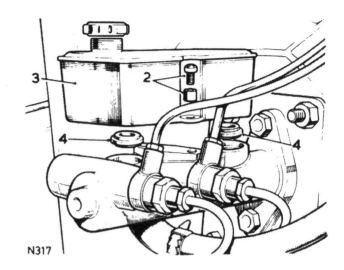

N317

Removing

1. Drain the fluid reservoir by bleeding through the front caliper nearest the reservoir.
2. Remove the fixings.
3. Withdraw the fluid reservoir.
4. If required, remove the reservoir seals from the master cylinder.

Refitting

5. Reverse 1 to 4.
6. Bleed the brakes 70.25.02.

PEDAL ASSEMBLY

– Remove and refit   70.35.01

Removing

1. Remove the lower facia panel 76.46.03.
2. Disconnect the servo operating rod from the brake pedal.
3. Remove the fixings securing the brake pedal and servo assemblies to the dash.
4. Withdraw the brake pedal assembly sufficient to disconnect the electrical leads from the stop light switch.
5. Lift the pedal assembly clear.

N224

Continued

Refitting

6. Reverse 3 to 5.
7. Connect the servo operating rod to the brake pedal with the pivot bolt eccentric in the forward position. Do not fully tighten the pivot bolt nut.
8. Early type servos only – Check that there is free play in the brake pedal action, if necessary turn the adjister sleeve on the servo operating rod to give slight clearance between the pedal and the rubber buffer.
9. Turn the pivot bolt to bring the brake pedal back until it just contacts the rubber buffer, then secure the pivot bolt nut.
10. Fit the lower fascia panel 76.46.03.

PEDAL ASSEMBLY

–Overhaul   **70.35.02**

1. Remove the pedal assembly 70.35.01.

Dismantling

2. Disconnect the pedal return spring.
3. Remove the circlip from the 'D' shaped end of the pedal shaft.
4. Withdraw the pedal shaft.
5. Withdraw the pedal from the box.
6. Remove the bushes from the pedal pivot tube.

Assembling.

7. Press the bushes into the pedal pivot tube.
8. If necessary, reamer the bushes to 15,87mm plus 0,05mm (.625 in plus .002in).
9. Lightly oil the bushes and pedal shaft.
10. Reverse 2 to 5.
11. Refit the pedal assembly 70.35.01.

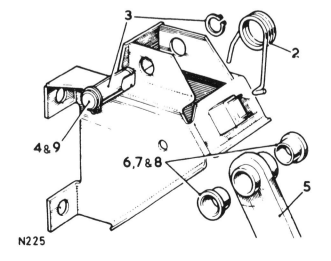

N225

FRONT BRAKE PADS

–Remove and refit   70.40.02.

Service tools: 18G672 Piston Clamp

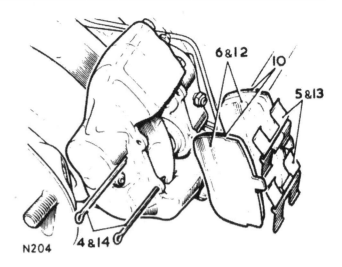

N204

Removing

1.   Slacken the front wheel fixings.
2.   Jack up the front of the car.
3.   Remove the front wheel.
4.   Remove the split pins from the front brake caliper
5.   Lift off the friction pad retaining springs.
6.   Withdraw the friction pads.

Refitting

7.   Clean the exposed parts of the pistons and the friction pad recesses. Use only new brake fluid to clean the pistons.

CAUTION. During the following item brake fluid will be displaced into the brake fluid reservoir. Care must be taken to prevent overflow by lowering the level in the reservoir.

8.   Using the piston clamp 18G 672, press each piston back into its bore.
9.   Lightly smear the faces of the pistons and the friction pad recesses with Lockheed Disc Brake Lubricant.
10.   Check the bearing edges of the new friction pads for blemishes. High spots on the steel pressure plates may be rectified by carefully filing.
11.   Lightly smear the metal to metal contact edges with Lockheed Disc Brake Lubricant. Keep the lubricant clear of the friction material.
12.   Insert the friction pads.
13.   Locate the friction pad retaining springs in position.
14.   Fit NEW split pins.
15.   Depress the foot brake pedal firmly several times to locate the friction pads correctly.
16.   Check and if necessary, replenish the brake fluid reservoir.
17.   Reverse 1 to 3.

IMPORTANT. Always renew the friction pads in both calipers of an axle, otherwise unbalanced braking will result.

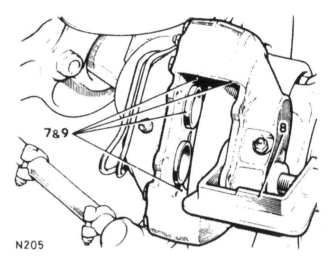

N205

REAR BRAKE PADS

– Remove and refit   70.40.03.

Service tool: 18G 672 Piston clamp

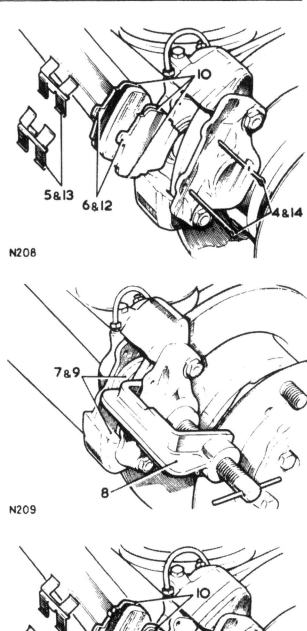

N208

Removing

1. Slacken the rear wheel fixings.
2. Jack up the rear of the car.
3. Remove the rear wheel.
4. Remove the split pins from the rear brake calipers.
5. Lift off the friction pad retaining springs.
6. Withdraw the friction pads and shims.

Refitting.

7. Clean the exposed parts of the pistons and the friction pad recesses. Use only new brake fluid to clean the pistons.

CAUTION. During the following item brake fluid will be displaced into the brake fluid reservoir. Care must be taken to prevent overflow by lowering the level in the reservoir.

N209

8. Using the piston clamp 18G672, press each piston back into its bore.
9. Lightly smear the faces of the pistons and the friction pad recesses with Lockheed Disc Brake Lubricant.
10. Check the bearing edges of the new friction pads for blemishes. High spots on the steel pressure plates may be recitified by carefully filing.
11. Lightly smear the metal to metal contact edges with Lockheed Disc Brake Lubricant. Keep the lubricant clear of the friction material.
12. Transfer the shims to the new friction pads and insert the pads with straight edge of the 'D'-shaped cut-out uppermost.
13. Locate the friction pad retaining springs in position.
14. Fit NEW split pins.
15. Depress the foot brake pedal firmly several times to locate the friction pads correctly.
16. Check and if necessary, replenish the brake fluid reservoir.
17. Reverse 1 to 3.

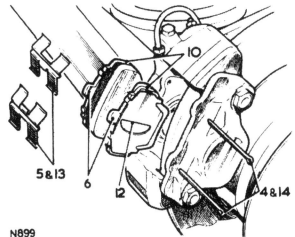

N899

IMPORTANT. Always renew the friction pads in both calipers of an axle, otherwise unbalanced braking will result.

TRANSMISSION BRAKE LEVER AND LINKAGE

– Remove and refit   70.45.01

Removing.

1. Slide the driver's seat fully rearward and remove the carpet from the driver's floor.
2. Pull back the carpet from the gearbox cover to expose the handbrake lever gaiter.
3. Remove the screws securing the handbrake gaiter securing plate and withdraw the gaiter and plate assembly.
4. Remove the circular plate located on the gearbox cover below the handbrake.
5. Disconnect the leads from the handbrake switch.
6. Remove the three nuts securing the handbrake mounting brackets to the gearbox and detach the wiring harness earth lead.
7. Remove the split pin, washers and clevis pin securing the handbrake linkage to the relay lever.
8. Withdraw the handbrake lever and bracket assembly.

Refitting.

9. Reverse instructions 1 to 8.

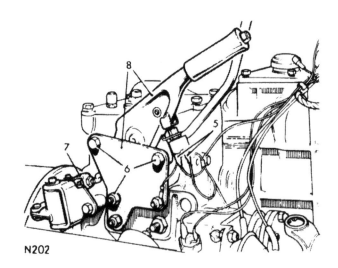

N202

TRANSMISSION BRAKE LEVER

– Overhaul   70.45.02

1. Release the handbrake.
2. Remove the handbrake lever grip by gently tapping it off the lever, using a mallet and a suitable drift.
3. Unscrew the release button from the end of the handbrake lever.
4. Withdraw the coil spring from inside the handbrake lever.
5. Withdraw the steel and felt washers (earlier models) or the rubber seating washer (later models).
6. Reverse instructions 1 to 5, using a clear Bostik adhesive to secure the handbrake grip to the lever.

TRANSMISSION BRAKE ASSEMBLY

— Adjust 37 to 41   70.45.09.

— Remove and refit — 1 to 41   70.45.16.

TRANSMISSION BRAKE SHOES

— Remove and refit 1 to 11, 26 to 41   70.45.18

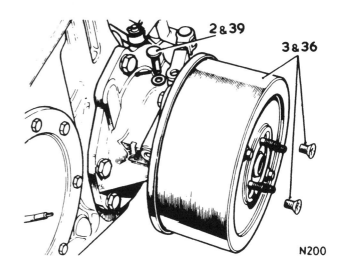

N200

Removing

1. Disconnect the rear propeller shaft.
2. Remove the clevis pin at the expander lever.
3. Remove the brake drum.
4. Remove the cup washers and springs.
5. Withdraw the steady posts.
6. Relieve the adjuster tension.
7. Spring apart the brake shoes and withdraw the adjuster.
8. Unhook and withdraw the tensioner spring.
9. Unhook and withdraw the pull off spring.
10. Withdraw the leading shoe.
11. Withdraw the trailing shoe, actuator lever and sealing shim.
12. Remove the output coupling flange.
13. Remove the pivot bolt.
14. Remove the brake backplate and oil catcher.

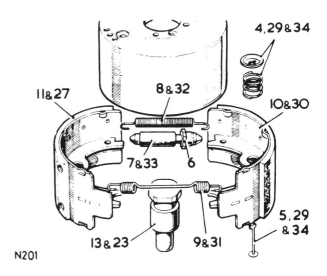

N201

Inspection

15. Clean all components in Girling cleaning fluid and allow to dry.
16. Examine all items for obvious signs of wear and replace with new components as necessary.
17. Examine the brake drum for scoring, ovality, and skim if required. Standard diameter: 184,15mm (7.250 in). Reclamation limit: 0,76 mm (0.030 in.) oversize on diameter.
18. If there is any sign of oil contamination on the brake linings, check, and if necessary, replace the rear output shaft oil seal, 37.23.01.
19. Brake shoes may be re-lined, following standard practices. Re-lined shoes are also available.

continued

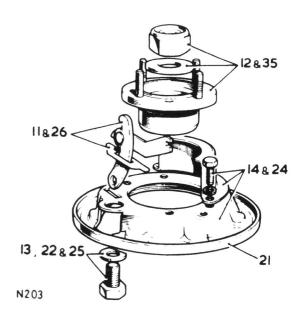

N203

70–17

Refitting

20. Lightly coat metal to metal contact points and the adjuster threads with molybdenum disulphide grease. Avoid depositing grease on the brake linings.
21. Position the back plate on the speedometer drive housing.
22. Apply Loctite CVX grade to the pivot bolt threads.
23. Loosely fit the pivot bolt then align the back plate and oil catcher fixing holes.
24. Fit the back plate fixings. Torque loading 3,5 kgf.m (25 lbf.ft).
25. Tighten the pivot bolt. Torque loading is 5,9 kgf.m (43 lbf.ft).
26. Fit the actuator lever and sealing shim.
27. Position the trailing shoe on the backplate.
28. Engage the actuator lever in the trailing shoe slot.
29. Fit a steady post, spring and cup washer.
30. Position the leading shoe on the backplate.
31. Fit the pull-off spring (the larger diameter spring) to the brake shoes with the hook ends toward the backplate.
32. Fit the tensioner spring with the hook ends toward the backplate.
33. Pull the shoes apart and fit the adjuster assembly; adjuster wheel nearest to the leading shoe.
34. Fit the steady post, spring and cupwasher to the leading shoe.
35. Fit the output coupling flange. Torque loading for fixing nut is 16.6 kgf.m (120 lbf.ft).
36. Fit the brake drum.
37. Turn the adjuster, using a screwdriver entered through the access hole until the brake shoes abut the drum.
38. Turn the adjuster back two divisions.
39. Connect the handbrake linkage.
40. Operate the handbrake several times to centralise the brake shoes.
41. Recheck for correct shoe adjustment.

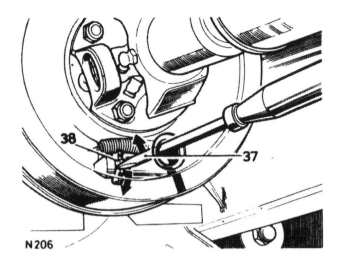

N 206

DATA

Brake drum standard diameter	184,15mm (7.250in)
reclamation limit	0,76mm (0.030in)

SERVO ASSEMBLY

– Remove and refit   70.50.01

IMPORTANT: Two types of servo assembly (with Unified thread) are in use and individual components must not be interchanged, but in the event of a fault the later type servo and master cylinder must be fitted in place of the earlier version which is not serviceable.

Identification – The early type servo assembly has an adaptor plate measuring approximately 90 x 100 mm (3.500 x 4in.) fitted between the master cylinder and the servo unit. The later type does not have an adaptor plate and the master cylinder is secured directly to the servo unit. The new metric type brake servo and master cylinder (introduced in 1981) has a reduced brake pedal travel which gives improved pedal 'feel'. To achieve this the master cylinder bore diameter has been increased from 22 mm to 23,8 mm. This master cylinder may also be used with earlier servo as a replacement if required.

Removing

1. Disconnect the vacuum pipe.
2. Disconnect the pipes from the master cylinder, the brake failure switch lead and the vacuum loss switch lead as applicable.
3. Remove the lower fascia panel 76.46.03.
4. Disconnect the servo operating rod from the brake pedal.
5. Remove the fixings securing the brake pedal and servo assemblies to the dash.
6. Withdraw the servo assembly from the engine compartment.
7. If required, remove the master cylinder.

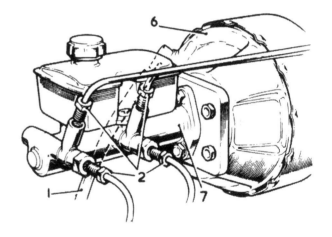

Refitting

8. Reverse 5 to 7.
9. Connect the servo operating rod to the brake pedal with the pivot bolt eccentric in the forward position. Do not fully tighten the pivot bolt nut.
10. Reverse 1 and 2.
11. Early type servo only – check that there is free play in the brake pedal action, if necessary turn the adjuster sleeve on the servo operating rod to give slight clearance between the pedal and the rubber buffer.
12. Turn the pivot bolt to bring the brake pedal back until it just contacts the rubber buffer, then secure the pivot bolt nut.
13. Fit the lower fascia panel 76.46.03.
14. Connect the brake pipes to the master cylinder, and fit the brake failure switch lead and the vacuum loss switch lead as applicable.
15. Bleed the brakes 70.25.02.

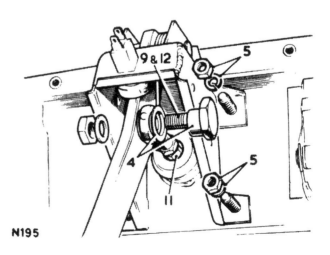

N195

BRAKE CALIPERS-FRONT

– Remove and refit   70.55.02

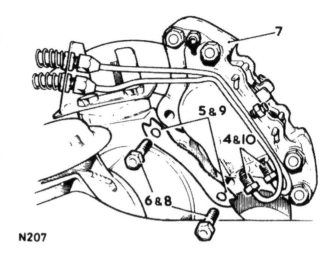

N207

Removing

1. Slacken the front wheel fixings.
2. Jack up the front of the car.
3. Remove the front wheel.
4. Disconnect the fluid pipes from the front brake caliper, taking precautions against fluid spillage.
5. Straighten the lockplates for the caliper securing bolts. (Early models only).
    NOTE: The bolts are fitted with spring washers on later models.
6. Remove the front brake caliper.

*Refitting

7. Locate the front brake caliper in position.
8. Fit the lockplates (earlier models only) and caliper securing bolts. Torque 8,3 kgf.m (60 lbf.ft).
9. Engage the lockplates (earlier models only).
10. Connect the fluid pipes to the front brake caliper.
11. Bleed the brakes 70.25.02
12. Reverse 1 to 3.

BRAKE CALIPERS-REAR

– Remove and refit   70.55.03.

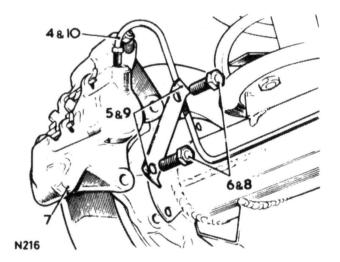

N216

Removing

1. Slacken the rear wheel fixings.
2. Jack up the rear of the car.
3. Remove the rear wheel.
4. Disconnect the fluid pipe from the rear brake caliper, taking precautions against fluid spillage.
5. Straighten the lockplate for the caliper securing bolts. (earlier models only).
    NOTE: The bolts are fitted with spring washers on later models.
6. Remove the rear brake caliper and (later models) the caliper splash shields.

*Refitting

7. Locate the rear brake caliper in position with (later models) the caliper splash shield correctly located.
8. Fit the lockplate (earlier models only) and caliper securing bolts. Torque 8,3 kgf.m (60 lbf.ft).
9. Engage the lockplate (earlier models only).
10. Connect the fluid pipe to the rear brake caliper.
11. Bleed the brakes 70.25.02.
12. Reverse 1 to 3.

NOTE: *In the event of fitting new front or rear calipers, see 'IMPORTANT' on page 7–04.

BRAKE CALIPERS-FRONT

– Overhaul   70.55.13

Service Tool: 18G 672 Piston clamp

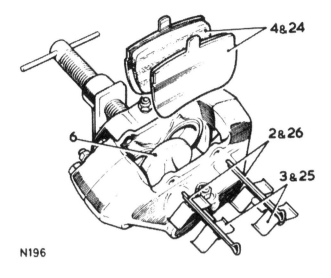

N196

1. Remove the front brake calipers 70.55.02.

Dismantling.

CAUTION. DO NOT separate the two halves of the caliper unecessarily, the piston seals can be changed without splitting the caliper. See 27 to 34.

2. Remove the split pins.
3. Lift off the friction pad retaining springs.
4. Withdraw the friction pads.
5. Thoroughly clean the outer surfaces of the caliper, using ethyl alcohol.
6. Fit the piston clamp 18G672, to retain both the pistons in the mounting flange, half of the caliper.
7. Using a high pressure air line, gently apply air pressure into the fluid inlet ports to expel the rim half pistons.

CAUTION. DO NOT scratch the pistons or cylinder bores, use extreme care when removing the seals.

8. Prise the retainers for the wiper seals from the mouths of the cylinder bores.
9. Withdraw the wiper seals.
10. Withdraw the fluid seals.

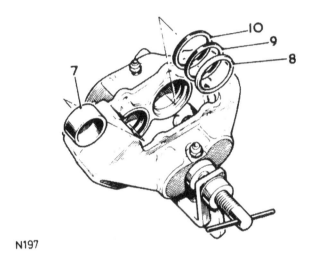

N197

Inspecting

11. Using clean brake fluid or ethyl alcohol only, clean the pistons, cylinder bores and the seal grooves.
12. Examine the cylinder bores and pistons, they must be in perfect condition with no signs of scratches or corrosion, otherwise fit new components.

continued

Assembling

13. Coat the new fluid seals with Lockheed Disc Brake Lubricant.
14. Dealing with each bore in turn, ease the fluid seal into the groove, using fingers only.

NOTE. The fluid seal groove and the seal are not the same in section, and the seal will feel proud to the touch at the edge furthest away from the mouth of the bore.

15. Slacken the bleed screw on the rim half of the caliper one complete turn.
16. Coat the piston with Lockheed Disc Brake Lubricant.
17. Insert the piston into the bore with the fingers, leaving approximately 8mm (5/16 in) of the piston projecting.
18. Coat the new wiper seal with Lockheed Disc Brake Lubricant.
19. Fit the wiper seal into its retainer.
20. Slide the wiper seal assembly, seal side first, into the mouth of the bore using the piston as a guide.
21. Use the piston clamp to press the seal assembly into place.
22. Tighten the bleed screw.
23. Overhaul the pistons, seals and retainers in the mounting flange half of the caliper as described for the rim half.
24. Insert the friction pads into the caliper recesses.
25. Locate the friction pad retaining springs in position.
26. Fit new split pins.
27. If it is necessary to separate the two halves of the caliper, carry out 28 to 34.
28. Remove the friction pads.
29. Remove the caliper half securing bolts.
30. Ensure that the jointing face of each half of the caliper are clean and that the threaded holes for the securing bolts are completely dry.
31. Locate new fluid channel seals in the appropriate caliper half.
32. Locate the two halves of the caliper together ensuring that the seals are not dislodged.
33. Fit NEW caliper half securing bolts. Torque 8,3 kgf.m (60 lbf.ft).
34. Refit the friction pads. 70.40.02.
35. Refit the front brake calipers 70.55.02.

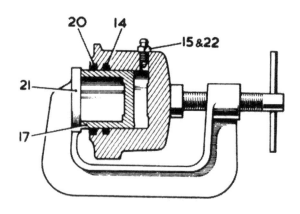

N198

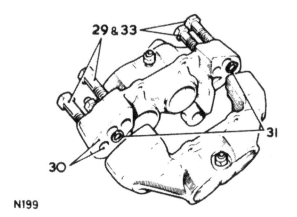

N199

BRAKE CALIPERS-REAR

– Overhaul   **70.55.14**

Service Tool: **18G 672 Piston clamp**

1. Remove the rear brake calipers 70.55.03.

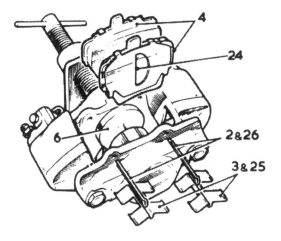

N941

Dismantling

CAUTION. DO NOT separate the two halves of the caliper unecessarily, the piston seals can be changed without splitting the caliper, see 27 to 34.

2. Remove the split pins.
3. Lift off the friction pad retaining springs.
4. Withdraw the friction pads and shims.
5. Thoroughly clean the outer surfaces of the caliper, using ethyl alcohol.
6. Fit the piston clamp 18G672, to retain the piston in the mounting flange half of the caliper.
7. Using a high pressure air line, gently apply air pressure to the fluid inlet port to expel the rim half piston.

CAUTION. DO NOT Scratch the piston or cylinder bore, use extreme care when removing the seals.

8. Prise the retainer for the wiper seal from the mouth of the cylinder bore.
9. Withdraw the wiper seal.
10. Withdraw the fluid seal.

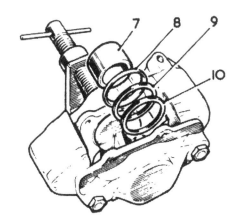

N218

Inspecting

11. Using clean brake fluid or ethyl alcohol only, clean the piston, cylinder bore and seal grooves.
12. Examine the cylinder bores and pistons, they must be in perfect condition with no signs of scratches or corrosion, otherwise fit new components.

Continued

Assembling

13. Coat the new fluid seals with Lockheed Disc Brake Lubricant.
14. Dealing with each bore in turn, ease the fluid seal into the groove, using fingers only.

NOTE. The fluid seal groove and the seal are not the same in section, and the seal will feel proud to the touch at the edge furthest away from the mouth of the bore.

15. Slacken the bleed screw on the rim half of the caliper one complete turn.
16. Coat the piston with Lockheed Disc Brake Lubricant.
17. Insert the piston into the bore with the fingers, leaving approximately 8 mm (0.0312 in.) of the piston projecting.
18. Coat the new wiper seal with Lockheed Disc Brake Lubricant.
19. Fit the wiper seal into its retainer.
20. Slide the wiper seal assembly, seal side first, into the mouth of the bore using the piston as a guide.
21. Use the piston clamp to press the seal assembly into place.
22. Tighten the bleed screw.
23. Overhaul the pistons, seals and retainers in the mounting flange half of the caliper as described for the rim half.
24. Fit the shims to the friction pads and insert the pads so that the straight edge of the 'D'-shaped cut-out will be uppermost when the caliper is fitted.
25. Locate the friction pad retaining springs in position.
26. Fit new split pins.
27. If it is necessary to separate the two halves of the caliper, carry out 28 to 34.
28. Remove the friction pads.
29. Remove the caliper half securing bolts.
30. Ensure that the jointing face of each half of the caliper are clean and that the threaded holes for the securing bolts are completely dry.
31. Locate new fluid channel seals in the appropriate caliper half.
32. Locate the two halves of the caliper together ensuring that the seals are not dislodged.
33. Fit NEW caliper half securing bolts. Torque 8,3 kgf.m (60 lbf.ft).
34. Refit the friction pads. 70.40.03.
35. Refit the brake calipers 70.55.03.

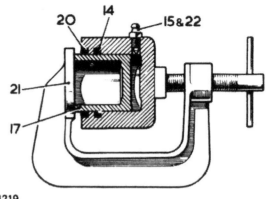

N219

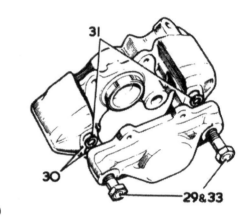

N220

WHEELS AND TYRES OPERATIONS

Description	Operation No.
	74–1
Tyres — Pressure check	74.00.00
— remove and refit	74.10.01
Wheels and Tyres — Balance	74.15.02
Wheels — remove and refit	74.20.01

Tyres   74.00.00
– Pressure Check

Tyre pressures should be checked at least every month for normal road use and at least weekly, preferably daily, if the vehicle is used off the road.

Normal on-off road use – All speeds and loads

Front	Rear
1,8 kg/cm^2	2,5 kg/cm^2
(25 lb/in^2)	(35 lb/in^2)
1.72 bars	2.4 bars

NOTE: For extra comfort rear tyre pressures can be reduced to 1,5 kg/cm^2, 25 lb/in^2, 1.72 bars, when the rear axle weight does not exceed 1250 kg (2755 lb).

These pressures may be increased for rough off-road usage where the risk of tyre cutting or penetration may be increased. Pressures may also be increased for high speed motoring near the vehicles maximum speed. Any such increase in pressures may be up to an absolute maximum pressure of 2,9 kg/cm^2 (42 lb/in^2) (2,94 bars). For off-road use in soft conditions where a maximum speed of 64 kph (40 mph) is used pressures may be reduced to obtain maximum traction as follows:

Front	Rear
1,1 kg/cm^2	1,8 kg/cm^2
(15 lb/in^2)	(25 lb/in^2)
(1,03 bars)	(1,72 bars)

Normal operating pressures should be restored as soon as reasonable road conditions or hard ground is reached.

After any usage off the road, tyres and wheels should be inspected for damage particularly if high cruising speeds are subsequently to be used.

Refer to Maintenance – page 10–25 for details of tyre wear inspection.

TYRES

– Remove and refit   74.10.01

Removing

1. Remove the road wheel 74.20.01.
2. Remove the valve cap and core and deflate the tyre.
3. Press each bead in turn off its seating.
4. Insert a tyre lever at the valve position and, while pulling on this lever, press the bead into the well diametrically opposite the valve.
5. Insert a second lever close to the first and prise the bead over the wheel rim.
6. Continue round the bead in small steps until it is completely off the rim.
7. Remove the inner tube and pull the second bead over the rim.

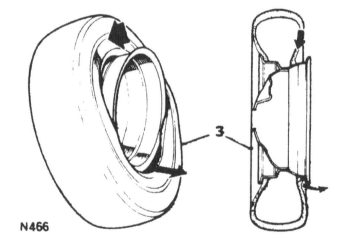

N466

74–2

Refitting.

8. Place the cover over the wheel and press the lower bead over the rim edge into the well.

9. Inflate the inner tube until it is just rounded out.

10. Dust the inner tube with French chalk and insert it in the cover, with the white spots near the cover bead coinciding with the black spots on the tube.

11. Press the upper bead into the well diametrically opposite the valve and lever the bead over the rim edge.

12. Push the valve inwards to ensure that the tube is not trapped under the bead, pull it back and inflate the tyre.

13. Visually check the concentricity of the fitting lines on the cover and the rim of the wheel flange.

14. Deflate the tube completely and re-inflate to the correct pressure, to relieve any strains in the tube.

15. Refit the road wheel. 74.20.01.

WHEELS AND TYRES

– Balance   74.15.02.

Balancing.

1. Wheel balance should always be checked whenever new tyres are fitted to ensure that the dynamic balance of the wheel and tyre is correct.

2. Ensure that the outer cover is fitted correctly.

3. Remove each wheel in turn and balance it complete with tyre dynamically, using standard garage equipment.

NOTE. Incorrect wheel balance is the most likely cause of front end vibration, steering shimmy and possibly uneven tread wear.

Usually during the life of a tyre, due to the many contributory factors involved in the suspension assembly, the tyre can wear slightly uneven on its total circumference and at a point in tyre life, although the wheels may have been balanced initially, a vibration or wheel shimmy may recur and this may purely be a case where rebalancing is necessary. It is also feasible that if a puncture occurs the tyre will not be refitted in the same position and in consequence rebalancing should be carried out.

Wheels

– Remove and refit	74.20.01

WARNING:

Always chock the wheels. If either rear wheel is to be jacked up it will be necessary to engage the gearbox differential lock. Because the engagement is vacuum operated it is preferably done prior to parking the vehicle and stopping the engine. A vehicle which has been parked for some time may have to be started and driven a few yards before engagement can be made.

Engagement of the differential lock is confirmed while the ignition is switched on by a warning light in the switch.

Application of the transmission handbrake and/or the selection of first and 'low' transfer gears will then effectively prevent the vehicle from rolling, once the limited back lash in the transmission is taken up.

It is unsafe to work under the vehicle with only the jack to support it. Always use stands or other suitable supports to provide adequate safety.

Removing

1. Select first and 'low' transfer gears and apply the hand brake.
2. Chock the wheels.
3. Slacken the five wheel nuts, using the hinged type wheel nut wrench from the vehicle tool kit in the fully extended position. This will provide additional leverage for removal of wheel nuts.
4. *Front wheels* Jack up the corner of the vehicle under the front axle casing immediately below the coil spring between the flange at the end of the axle casing and the front suspension attachment bracket.
   Rear wheels Jack up the corner of the vehicle under the rear axle casing immediately below the coil spring as close to the shock absorber bracket as possible.
5. Remove the wheel nuts and gently withdraw the wheel over the studs.

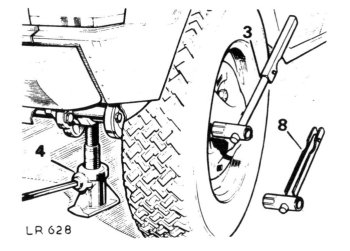

LR 628

Refitting

6. Apply a drop of oil or grease on the wheel studs to assist in replacement.
7. Refit wheel on to studs taking care not to damage the stud threads.
8. Replace nuts and tighten as much as possible using the hinged wheel nut wrench. The extended part will automatically fold to provide normal leverage for refitting.
9. Lower the vehicle to the ground and finally tighten the nuts to a torque figure of between 10,0 and 11,7 kgf.m (75 and 85 lbf.ft) for steel wheels. (See page 74–5) for fitting alloy road wheels.

IMPORTANT: DO NOT use foot pressure or an extension tube to tighten the nuts since this could cause over-stressing of the nuts and studs.

Alloy road wheels

Road wheels manufactured from an aluminium alloy are available as optional equipment on new Range Rovers through Unipart, for vehicles in service in certain markets. **WARNING: Do not fit alloy wheels in place of steel wheels unless the vehicle is fitted with the latest universal hubs. Failure to do this may result in stud failure and loss of road wheel.**
The correct hubs are identified by a groove across the end of each stud or by a triangle, as illustrated. DO NOT fit alloy wheels to hubs that are not marked in this manner. Do not fit alloy wheels that have not been supplied by Unipart specifically for Range Rover. Hub and stud assemblies, identified as above, suitable for fitting alloy wheels can be obtained through Unipart. Standard steel wheels can, however, be fitted to the latest universal hubs designed for alloy wheels.

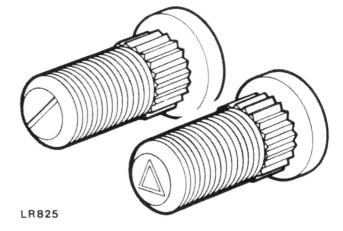

LR825

Fitting alloy wheels

Special care must be taken when fitting alloy wheels to prevent cross-threading, the retaining studs and nuts must be clean and screwed on to the studs by hand for at least three full threads before using any form of spanner or power wrench. Tighten the nuts evenly to a torque of between 12,5 and 13,35 kgf.m (90 and 95 lbf.ft).

IMPORTANT: DO NOT use foot pressure or an extension tube to tighten the nuts since this could cause over-stressing of the nuts and studs.

This page intentionally left blank

BODY OPERATIONS

Description		Operation No.
Body	— Rear corner panel, left hand — remove and refit	76.10.20
	— Rear corner panel, right hand — remove and refit	76.10.21
	— Rear quarter panel, left hand — remove and refit	76.10.22
	— Rear quarter panel, right hand — remove and refit	76.10.23
	— Side glass, front — remove and refit	76.81.17
	— Side glass, rear — remove and refit	76.81.18
	— Side, wheelarch and inner floor, left hand — remove and refit	76.10.08
	— Side, wheelarch and inner floor, right hand — remove and refit	76.10.09
Bonnet — remove and refit		76.16.01
Chassis frame — alignment check		76.10.02
Dash panel assembly — remove and refit		76.10.36
Door, front	— Door — remove and refit	76.28.01
	— Glass — remove and refit	76.31.01
	— Glass regulator — remove and refit	76.31.45
	— Lock — remove and refit	76.37.12
	— Private lock — remove and refit	76.37.39
	— Quarter vent — remove and refit	76.31.28
Door, rear	— Door — remove and refit	76.28.02
	— Check strap	76.40.80
	— Glass — remove and refit	76.31.02
	— Glass regulator — remove and refit	76.31.46
	— Hinges — remove and refit	76.28.68
	— Lock — remove and refit	76.37.13
Facia	— Console unit — remove and refit	76.25.01
	— Glove box — remove and refit	76.52.01
	— Lower panel — remove and refit	76.46.03
	— Top rail — remove and refit	76.46.04
Front decker panel — remove and refit		76.10.35
Front floor — remove and refit		76.10.12
Headlining	— Front section — remove and refit	76.64.02
	— Rear section — remove and refit	76.64.05
Roof panel — remove and refit		76.10.13
Seat	— Base — remove and refit	76.70.06
	— Front seats — remove and refit	76.70.01
Tailgate	— Frame — remove and refit	76.10.31
	— Glass — remove and refit	76.81.10
	— Lock, lower — adjust	76.37.18
	— Lock, lower — remove and refit	76.37.19
	— Lock, upper — remove and refit	76.37.17
	— Lower — remove and refit	76.28.30
	— Stays, upper — remove and refit	76.40.33
	— Upper — remove and refit	76.28.29
Windscreen glass — remove and refit		76.81.01
Wings	— Front — remove and refit	76.10.26
	— Front valance, left hand — remove and refit	76.79.01
	— Front valance, right hand — remove and refit	76.79.02
	— Rear, left hand — remove and refit	76.10.27
	— Rear, right hand — remove and refit	76.10.28

Body repairs, general information

1. The Range Rover body consists of a steel frame to which alloy outer panels are attached. The radiator grille, front deck panel, front wings, side door outer panels, body side outer panels, roof, rear floor and upper rear quarter panels are made from a special light magnesium-aluminium alloy known as 'Birmabright'.

2. 'Birmabright' was developed for aircraft use, and it is much stronger and tougher than pure aluminium. It melts at a slightly lower temperature than pure aluminium and will not rust nor corrode under any normal circumstances. It is work-hardening, and so becomes hard and brittle when hammered, but it is easily annealed. Exposed to the atmosphere, a hard oxide skin forms on the surface of it.

Panel beating 'Birmabright'

3. 'Birmabright' panels and wings can be beaten out after accidential damage in the same way as sheet steel. However, under protracted hammering the material will harden, and then it must be annealed to prevent the possibility of cracking. This is quite easily done by the application of heat, followed by slow air-cooling, but as the melting point is low, heat must be applied slowly and carefully. A rough but very useful temperature control is to apply oil to the cleaned surface to be annealed. Play the welding torch on the underside of the cleaned surface and watch for the oil to clear, which it will do quite quickly, leaving the surface clean and unmarked. Then allow to cool naturally in the air, when the area so treated will again be soft and workable. Do not quench with oil or water. Another method is to clean the surface to be annealed and then rub it with a piece of soap. Apply heat beneath the area, as described above, and watch for the soap stain to clear. Then allow to cool, as for the oil method. When applying the heat for annealing, always hold the torch some little distance from the metal, and move it about, so as to avoid any risk of melting it locally.

4. **Gas welding 'Birmabright'**

   A small jet must be used, one or two sizes smaller than would be used for welding sheet steel of comparable thickness. For instance, use a No. 2 nozzle for welding 18 swg (0.048 in.) sheet, and a No. 3 for 16 swg (0.064 in.) sheet.

5. The flame should be smooth quiet and neutral, have a brilliant inner core with a well defined rounded end. The hottest point of the flame is close to the jet, and the flame should have a blue to orange envelope becoming nearly colourless at the end.

6. A Slightly reducing flame may also be used, that is, there may be a slight excess of acetylene. Such a flame will have a brilliant inner core with a feathery white flame and a blue to orange envelope.

7. Do not use an oxydising flame, which has a short pointed inner core bluish white with a bluish envelope.

8. Use only 5 per cent magnesium/aluminium welding rod (5 Mg/A). Sifalumin No. 27 (MG.5 Alloy) (Use Sifbronze Special flux with this rod) or a thin strip cut of parent metal – that is to say, a strip cut from an old and otherwise useless 'Birmabright' panel or sheet. Do not use too wide or thick a strip, or trouble may be experienced in making it melt before the material which is being welded.

9. Clean off all grease and paint, dry thoroughly and then clean the edges to be welded, and an area at least half an inch on either side of the weld, with a stiff wire scratch-brush or wire wool. Cleanliness is essential. Also clean the welding rod or strip with wire wool.

10. A special acid flux must be used, and we recommend 'Hari-Kari' which is obtainable from:
    The Midland Welding Supply Co. Ltd.,
    105 Lakey Lane,
    Birmingham 28, England.

    or

    Sifbronze Special Flux, which is obtainable from:

    Suffolk Iron Foundry (1920) Ltd.,
    Sifbronze Works,
    Stowmarket, England.

11. A small quantity of 'Hari-Kari' may be made into a paste with water, following the directions on the tin, and the paste must be applied to both surfaces to be welded, and also to the rod. In the case of Sifbronze Special Flux use in powder form as directed. Remember that aluminium and its alloys do not show 'red-hot' before melting, and so there is nothing about the appearance of the metal to indicate that it has reached welding temperature. A little experience will enable the operator to gauge this point, but a useful guide is to sprinkle a little sawdust over the work; this will sparkle and char when the right temperature is approached; a piece of dry wood rubbed over the hot metal will sparkle at the point of contact.

12. As the flux used is highly acid, it is essential to wash if off thoroughly immediately after a weld is completed. The hottest possible water should be used, with wire wool or a scratch-brush. Very hot soapy water is good, because of the alkaline nature of the soap, which will tend to 'kill' the acid.

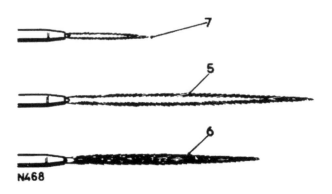

N468

13. It is strongly recommended that a few welds are made on scrap metal before the actual repair is undertaken if the operator is not already experienced in welding aluminium and its alloys.

14. The heat of welding will have softened the metal in the area of the repair, and it may be hardened again by peening with a light hammer. Many light blows are preferable to fewer heavy ones. Use a 'dolly' or anvil behind the work to avoid denting and deformation, and to make the hammering more effective. Filing of surplus metal from the weld will also help to harden the work again.

Welding tears and patching

15. If a tear extends to the edge of a panel, start the weld from the end away from the edge and also at this point drill a small hole to prevent the crack spreading, then work towards the edge.

16. When welding a long tear, or making a long welded joint, tack the edges to be welded at intervals of from 2 in. to 4 in. (50 to 100 mm) with spots. This is done by melting the metal at the starting end and fusing into it a small amount of the filler rod, repeating the process at the suggested intervals. After this, weld continuously along the joint from right to left, increasing the speed of the weld as the material heats up.

17. After the work has cooled, wash off all traces of flux as described above, and file off any excess of build-up metal.

18. When patching, cut the patch to the correct shape for the hole to be filled, but of such sizes as to leave a gap of 1/32 in. between it and the panel, and then weld as described above. Never apply an 'overlay' patch.

Electric welding

19. **CAUTION. The battery earth lead must be disconnected before commencing electric welding, otherwise the alternator will be damaged.**

20. At the Rover Factory the 'Argon-Arc' process is used, and this is very satisfactory, since all atmospheric oxygen is excluded from the weld by the Argon gas shield. For all body repair work normally undertaken by a Distributor's or Dealer's service department, the gas welding method is sufficient and quite satisfactory.

Spot-welding

21. Spot-welding is largely used in the manufacture of Range-Rover bodies, but this is a process which can only be carried out satisfactorily by the use of the proper apparatus. Aluminium and its alloys are very good conductors of heat and electricity, and thus it is most important to maintain the right conditions for successful spot-welding. The correct current density must be maintained, and so must the 'dwell' of the electrodes. Special spot-welding machines have been developed, but they are expensive, and though the actual work can be carried out by comparatively unskilled labour, supervision and machine maintenance must be in the hands of properly qualified persons.

Riveting

22. Where both sides of the metal are accessible and it is possible to use an anvil or 'dolly', solid aluminium rivets may be used, with a suitable punch or 'pop' to ensure clean rounded head on the work. For riveting blind holes, 'pop-rivets' must be used. These are inserted and closed by special 'Lazy-Tong' 'pop-rivet' pliers.

Painting 'Birmabright'

23. Refer to the procedure detailed in Paintwork Division 78.

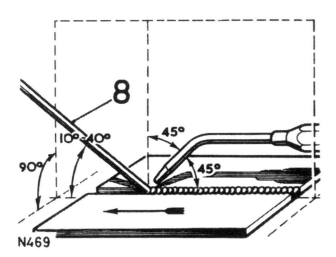

N469

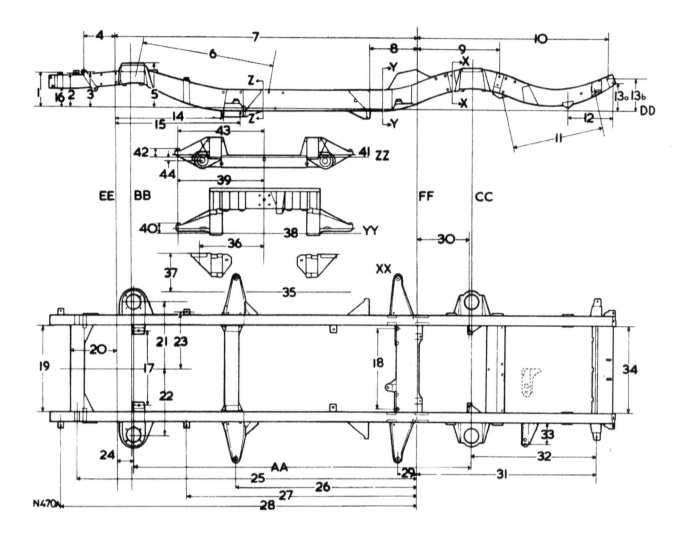

CHASSIS FRAME
— Alignment check   76.10.02

Diagram reference			millimetres	inches
AA	Wheelbase — Reference dimension		2540,00	100,000
BB	Centre line of front axle			
CC	Centre line of rear axle			
DD	Frame datum line			
EE	Side member datum line			
FF	Datum line			
1		(with mounting washers)	254,00 ± 0,63	10.000 ± .025
		(without washers)	264,525 $^{+1,91}_{-0,63}$	10.373 $^{+.075}_{-.025}$
2			261,11 ± 2,54	10.280 ± .100
3			266,70 ± 2,54	10.500 ± .100
4			237,74 ± 1,27	9.360 ± .050
5			327,81 ± 2,54	12.906 ± .050
6			979,93 ± 1,27	38.580 ± .100
7			2244,12 ± 2,54	88.375 ± .100
8			356,74 ± 2,54	14.045 ± .100
9			605,15 ± 2,54	23.825 ± .100
10			1405,38 ± 2,54	55.330 ± .100
11			694,44 ± 2,54	27.340 ± .100
12			338,83 ± 2,54	13.340 ± .100
13a			222,25 ± 5,08	8.750 ± .200
13b			240,54 ± 2,54	9.470 ± .100
14	Reference dimension		794,91	31.296
15	To face of boss		935,43 ± 2,54	36.828 ± .100
16	Frame datum to underside of crossmember		150,79	5.937
17			535,94 ± 2,54	21.100 ± .100
18			590,55 ± 0,64	23.250 ± .025
19	Check Figure		630,93 ± 1,27	24.840 ± .050
20			485,77 ± 1,27	13.550 ± .050
21			485,77 ± 2,54	19.125 ± .100
22			483,23 ± 2,54	19.125 ± .100
23			414,32 ± 2,54	16.312 ± .100
24			129,03 ± 2,54	5.080 ± .100
25			2544,44 ± 0,25	100.175 ± .010
26			1355,34 ± 0,38	53.360 ± .015
27			1722,04 ± 0,38	67.797 ± .015
28			2663,44 ± 0,38	104.860 ± .015
29			144,09 ± 0,38	5.673 ± .015
30			400,48 ± 2,54	15.767 ± .100
31			1333,88 ± 0,38	52.515 ± .015
32	Reference dimension		925,49	36.437
33	Reference dimension (also applicable to later lugs)		147,62	5.812
34	Reference dimension		635,00	25.000
SECTION XX				
35	Frame datum line DD			
36			488,95 ± 2,54	19.250 ± .100
37			295,27 ± 2,54	11.625 ± .100
SECTION YY				
38	Frame datum line DD			
39			660,40 ± 0,17	26.000 ± .007
40			80,95 $^{+1,91}_{-0,63}$	3.187 $^{+.075}_{-.025}$
SECTION ZZ				
41	Frame datum line DD			
42			80,95 $^{+1,91}_{-0,63}$	3.187 $^{+.075}_{-.025}$
43			660,4 ± 0,17	26.000 ± .007
44			9,525 ± 2,54	0.375 ± .100

NOTE: To provide a common chassis for two and four door bodies two additional rear seat belt anchorage brackets are welded to the front ends of the rear axle cross member on 1981 chassis frames.

BODY SIDE, WHEEL ARCH AND INNER FLOOR ASSEMBLY – 2 DOOR MODEL

– Left hand – remove and refit	76.10.08
– Right hand – remove and refit	76.10.09

Removing

1. Remove the bonnet 76.16.01.
2. Remove the air cleaner 19.10.01.
3. Remove the windscreen wiper arms 84.15.01.
4. Remove the front decker panel 76.10.35.
5. Remove the front parking and flasher lamp assembly 86.40.42.
6. Remove the front wing 76.10.26.
7. Remove the windscreen glass 76.81.01.
8. Remove the front door 76.28.01.
9. Remove the roof lamp assembly 86.45.02.
10. Remove the headlining 76.64.02/05.
11. Remove the front seat 76.70.01.
12. Remove the lower facia panel 76.46.03.
13. Remove the body rear quarter panel 76.10.22/23.
14. Remove the tail, stop and flasher lamp assembly 86.40.70/71.
15. Remove the body rear corner panel and rear wing 76.10.20/21 76.10.27/28.
16. Remove the roof panel 76.10.13.
17. Remove the trim panel from the side of the front foot well.
18. Remove the door post trim panels.
19. Remove the rear seat.
20. Remove the trim panels from the body side.
21. Remove the floor trim.
22. Disconnect the electrical harness from the body side.
23. Drill out all the pop-rivets securing the body side, wheelarch and inner floor assembly.
24. Remove the body side, wheelarch and inner floor assembly.

Refitting

25. Reverse 1 to 24, using a waterproof sealant between the body panel joints.

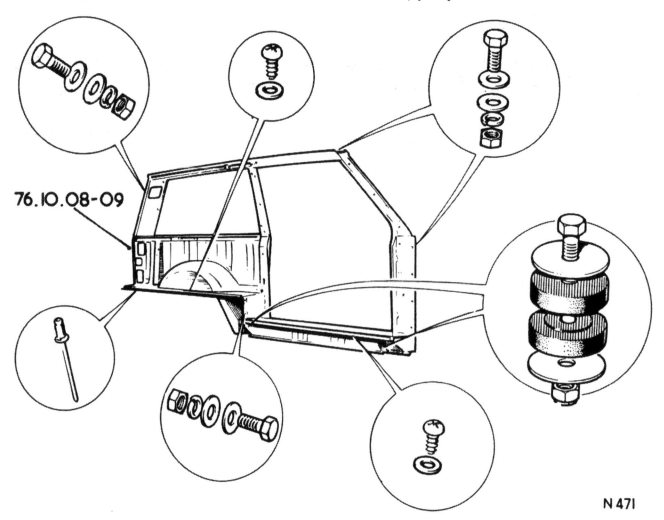

76.10.08-09

N 471

FRONT FLOOR

– Remove and refit   **76.10.12**

1. Remove the front seats 76.70.01.
2. Remove the seat bases outer fixings.
3. Remove the narrow floor panel.
4. Release the eyebolt fixings, seat bases to chassis.
5. Remove the hand brake lever grommet retainer.
6. Lift off the grommet.
7. Remove the transfer gear lever grommet.
8. Remove the maingear change lever.
9. Remove the differential lock vacuum control switch.
10. Disconnect the exhaust system heat shield at the floor.
11. Disconnect the handbrake linkage at a convenient clevis pin to allow full lever movements.
12. Remove the floor fixings. The two fixings concealed by the heater are accessible from the engine compartment.
13. Withdraw the front floor.

Refitting.

14. Reverse 1 to 13. Using a waterproof sealant around the floor joint flanges. Torque load for maingear change lever spherical seat fixings is   1,5 kgf.m (11 lbf. ft.).

ROOF PANEL

– Remove and refit   **76.10.13.**

Removing

1. Remove the roof lamp assembly 86.45.02.
2. Remove the rear view mirror.
3. Remove the sun visors.
4. Remove the headlining 76.64.02/05.
5. Remove the roof panel.

Refitting

6. Reverse 1 to 5, using a waterproof sealant between the roof to body joint.

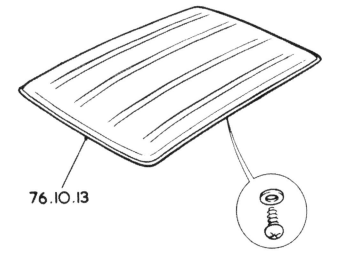

76.10.13

BODY REAR CORNER PANEL – 2 DOOR MODEL

– Left hand – remove and refit 3 to 8   76.10.20

– Right hand – remove and refit 1 , 2 and 4 to 8   76.10.21

Removing

1. Remove the fuel tank filler cap.
2. Remove the fuel tank filler neck.
3. Remove the spare wheel.
4. Remove the tail, stop and flasher lamp assembly 86.40.70/71.
5. Drill out all the pop-rivets securing the rear wing and body rear corner panel.
6. Remove the rear wing complete with the body rear corner panel.
7. Remove the body rear corner panel from the rear wing.

Refitting

8. Reverse 1 to 7, using a waterproof sealant between the body rear corner panel to rear wing joint.

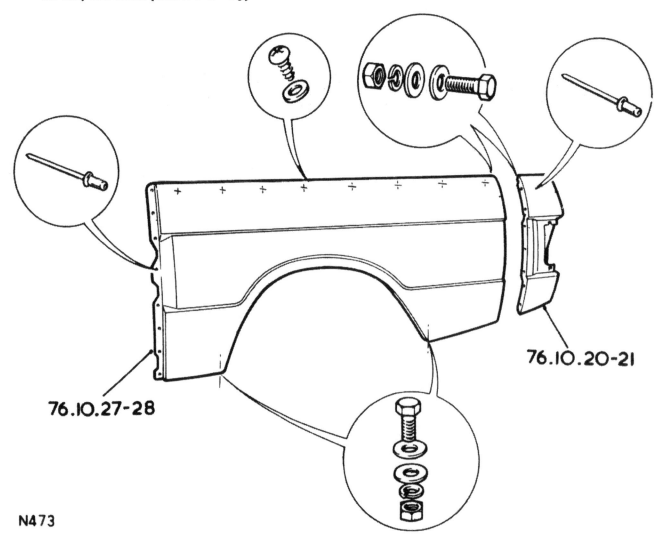

76.10.20-21

76.10.27-28

N473

BODY REAR QUARTER PANEL

– Left hand – remove and refit	76.10.22.
– Right hand – remove and refit	76.10.23.

Removing

1. Remove the headlining rear section 76.64.05.
NOTE. This instruction is not applicable for left hand panels on later models. These panels have plugs fitted for access to the panel fixing screws.
2. Remove the fixings from the rear quarter panel.
3. Withdraw the rear quarter panel.

Refitting.

4. Reverse 1 to 3.

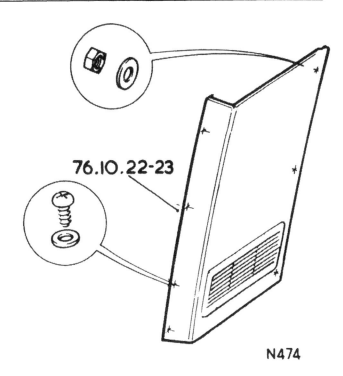

76.10.22-23

N474

N474A

FRONT WING

– Remove and refit   76.10.26.

Removing

1. Remove the bonnet 76.16.01.
2. Remove the air cleaner 19.10.01.
3. Remove the windscreen wiper arms 84.15.01.
4. Remove the front decker panel 76.10.35.
5. Remove the front parking and flasher lamp assembly
     86.40.42.
6. Remove the front wing.

Refitting

7.   Reverse 1 to 6.

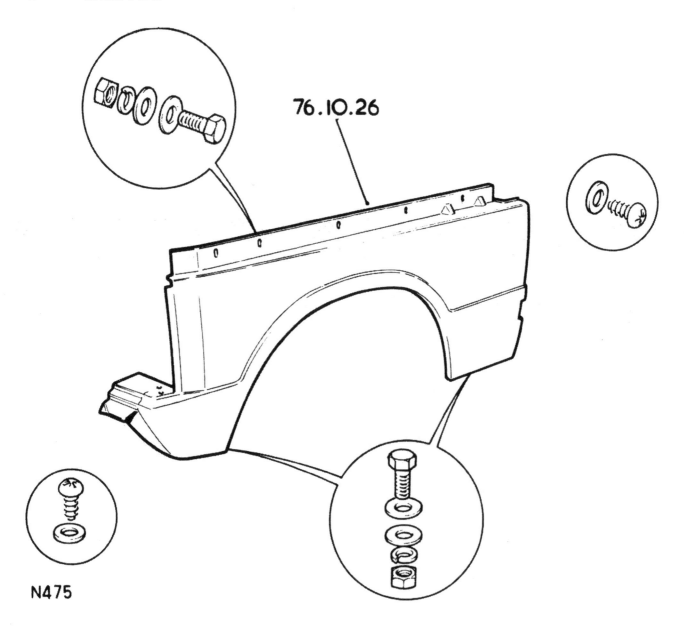

76.10.26

N475

REAR WING – 2 DOOR MODEL

– Left hand – remove and refit 3 to 8   76.10.27

– Right hand – remove and refit 1, 2 and 4 to 8   76.10.28

Removing

1. Remove the fuel tank filler cap.
2. Remove the fuel tank filler neck.
3. Remove the spare wheel.
4. Remove the tail, stop and flasher lamp assembly 86.40.70/71.
5. Drill out all the pop-rivets securing the rear wing and body rear corner panel.
6. Remove the rear wing complete with the body rear corner panel.
7. Remove the body rear corner panel from the rear wing.

Refitting

8. Reverse 1 to 7, using a waterproof sealant between the body rear corner panel to rear wing joint.

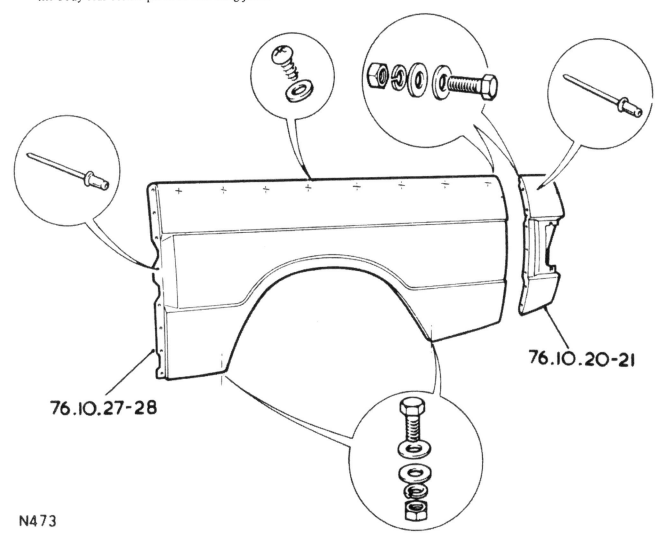

76.10.20-21

76.10.27-28

N473

TAILGATE FRAME

– Remove and refit   **76.10.31**

Removing.

1. Remove the upper tailgate 76.28.29.
2. Remove the lower tailgate 76.28.30.
3. Remove the roof lamp assembly 86.45.02.
4. Remove the headlining 76.64.02/05.
5. Remove the roof panel 76.10.13.
6. Remove the body rear quarter panels 76.10.22/23.
7. Remove the tail, stop and flasher lamp assemblies 86.40.70/71.
8. Remove the body rear corner panels and rear wings 76.10.20/21 76.10.27/28.
9. Drill out all the pop-rivets securing the tailgate frame.
10. Remove the tailgate frame

Refitting

11. Reverse 1 to 10 using a waterproof sealant between the tailgate frame and the body panels.

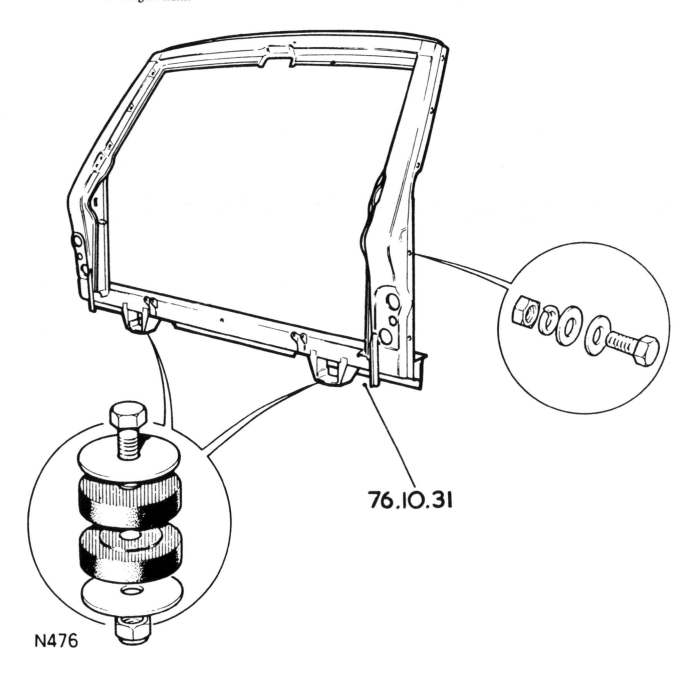

76.10.31

N476

FRONT DECKER PANEL.

– Remove and refit   **76.10.35.**

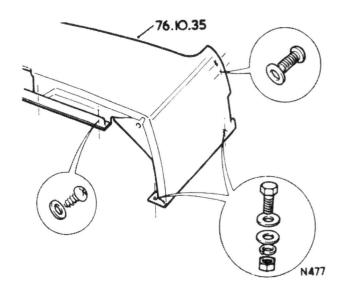

Removing

1. Remove the bonnet 76.16.01.
2. Remove the air cleaner 19.10.01.
3. Remove the windscreen wiper arms 84.15.01.
4. Remove the front decker panel.

Refitting

5. Locate the front decker panel in position under the lip of the windscreen rubber moulding.
6. Reverse 1 to 3.

76–13

DASH PANEL ASSEMBLY

– Remove and refit   76.10.36

Removing

1. Remove the bonnet 76.16.01.
2. Remove the air cleaner 19.10.01.
3. Remove the windscreen wiper arms 84.15.01.
4. Remove the front decker panel 76.10.35.
5. Remove the front parking and flasher lamp assemblies 86.40.42.
6. Remove the front wings 76.10.26.
7. Remove the front wing valances 76.79.01/02.
8. Remove the windscreen glass 76.81.01.
9. Remove the lower facial panel 76.46.03.
10. Remove the glove box 76.52.01.
11. Remove the console unit 76.25.01.
12. Remove the instrument housing 88.20.13.
13. Remove the facia top rail 76.46.04.
14. Remove the heater unit 80.20.01.
15. Remove the steering column 57.40.01.
16. Remove the front floor 76.10.12.
17. Remove or release all components fitted or attached to the dash panel assembly.
18. Remove the dash panel assembly.

Refitting

19. Reverse 1 to 18, using a waterproof sealant between the dash panel assembly to other panel joints.

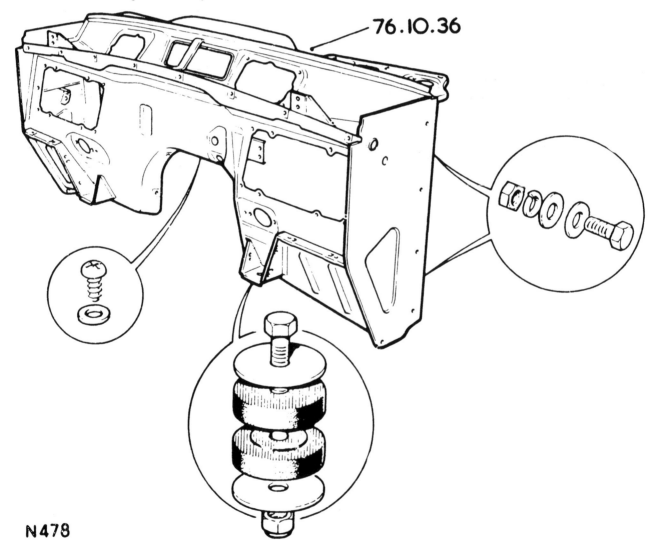

76.10.36

N478

BONNET

— Remove and refit   76.16.01

Removing

1. Prop the bonnet open.
2. Disconnect the pipe from the windscreen washer reservoir.
3. Remove the fixings, hinges to bonnet.
4. Lift the bonnet clear.

Refitting

5. Reverse 1 to 4.

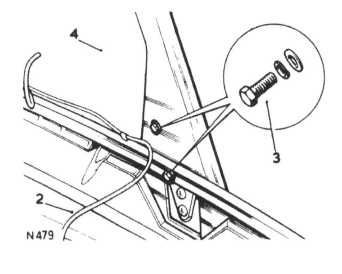

N479

CONSOLE UNIT

— **Remove and refit**   76.25.01

Removing

1. Remove the lower facia panel 76.46.03.
2. Open the glove box and release the check strap.
3. Remove the facia finisher.
4. Remove the heater control knobs.
5. Remove the heater control panel.
6. Remove the drive screws securing the console unit.
7. Withdraw the console unit sufficient to disconnect the harness from the clock and any other instruments that may be fitted.
8. Lift out the console unit.

Refitting

9. Reverse 1 to 8.

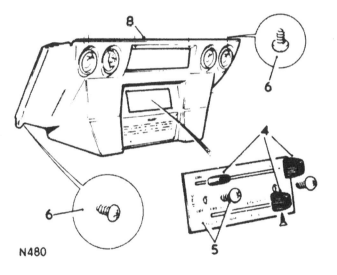

N480

FRONT DOOR

— **Remove and refit**   76.28.01

Removing

1. Drive out the retaining pin from the door check strap.
2. Support the door and remove the fixings, hinges to door.
3. Lift the door clear.

Refitting

4. Reverse 1 to 3.
5. Check the location of the door and the operation of the lock. If necessary, adjust the door and striker plate.
6. Door adjustment can be obtained in three separate ways, 7 to 9.

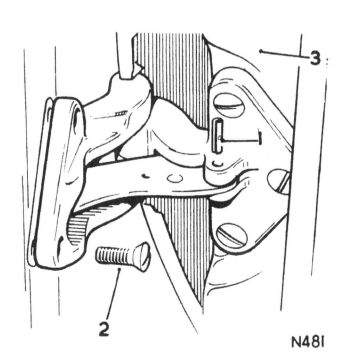

N481

7. By the addition or removal of shims between the door hinge and the door pillar, to bring the door either into or out of the aperture.

8. By the addition or removal of shims between the door and the hinge, to take the door forward or rearward in the aperture.
   NOTE: On earlier cars socket head or cross recess head drive screws may be fitted.

9. By slackening the six screws securing the hinges to the door and holding the door up or down while retightening, which will raise or lower the door in the aperture.

10. The door lock striker pin can be adjusted by adding or subtracting packing washers between the pin and the door pillar.

NOTE: This operation also applies to the front doors on 4 door vehicles which employ identical hinges, door restraints and striker fitments.

REAR DOOR – 4 DOOR MODEL

– Remove and refit   76.28.02

Removing

1. Open the bonnet and disconnect the battery negative earth lead.

2. Open the door and remove the two screws holding the arm rest to the inner door panel.

3. Wind the glass up into the fully closed position then remove the single screw retaining the window regulator handle on its shaft.

4. Carefully prise out the upper and lower halves of the inside door release handle bezel.

5. Unscrew the sill locking knob.

6. Remove the door trim pad by inserting a screwdriver between the trim pad and the inner door panel. gently prising out the eighteen plastic clips from their respective holes around the edges of the trim pad.

7. Remove the plastic weather sheet. This is held to the inner door panel by adhesive.

8. Remove the door check strap clevis pin circlip.

9. Remove the clevis pin and spacers.

10. Support door with a suitable block and jack.

11. Remove the upper and lower nuts (with spring washers) from both hinges and withdraw the bolts from the bracket inside the door.

12. Remove the central bolt (with spring and plain washers) from the **lower** hinge. This screws into a tapped hole in the bracket inside the door.

13. Repeat the above operation for the **upper** hinge and lower the door to the floor.

Refitting

14. Reverse the removal procedure, items 1 to 13.

15. Adjust door hinges/lock striker if necessary, as 'Refitting' front door 76.28.01.

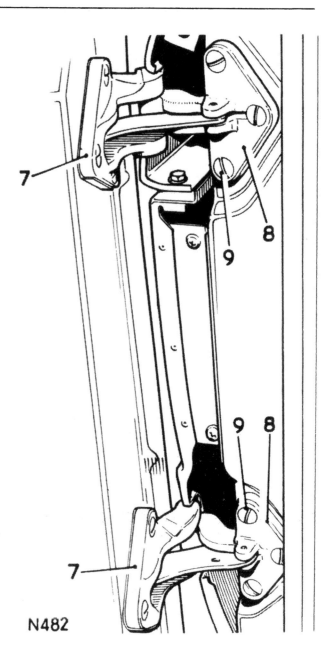

N482

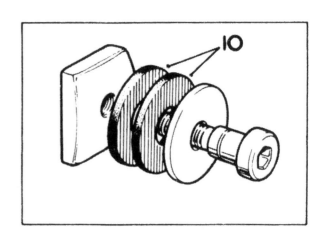

UPPER TAILGATE

– Remove and refit 76.28.29

Removing

1. Disconnect the stays from the upper tailgate. 76.40.33.
2. Remove the fixings, hinge to upper tailgate.
3. *Later Models.* Remove the tailgate glass wiper arm 84.15.01 and disconnect the heated backlight.
4. Lift the upper tailgate to clear.

Refitting

5. Reverse 1 to 4, ensuring that the hinge fixing sealing washers (later models) are correctly located.

LOWER TAILGATE

– Remove and refit 76.28.30

Removing

1. Disconnect the electrical leads from the rear number plate lamp.
2. Remove the fixings, tailgate to hinges.
3. Disconnect the check straps.
4. Withdraw the tailgate.

Refitting

5. Reverse 1 to 4.

REAR DOOR HINGES – 4 DOOR MODEL

– Remove and refit 76.28.68

Removing

1. Open the bonnet and disconnect the battery negative earth lead.
2. Open the door and remove the two screws holding the armrest to the inner door panel.
3. Wind the glass up into the fully closed position then remove the single screw retaining the window regulator handle on its shaft.
4. Carefully prise out the upper and lower halves of the inside door release handle bezel.
5. Unscrew the sill locking knob.
6. Remove the door trim pad by inserting a screwdriver between the trim pad and the inner door panel, gently prising out the eighteen plastic clips from their respective holes around the edges of the trim pad.
7. Remove the plastic weather sheet. This is held to the inner door panel by adhesive.
8. Ensure that the check strap clevis pin has been removed (see 76.40.30).
9. Support door with a suitable block and jack.
10. Remove the upper and lower nuts (with spring washers) from both hinges and withdraw the bolts from the bracket inside the door.
11. Remove the central bolt (with spring and plain washers) from the **lower** hinge. This screws into a tapped hole in the bracket inside the door.
12. Repeat the above operation for the **upper** hinge and lower the door to the floor.

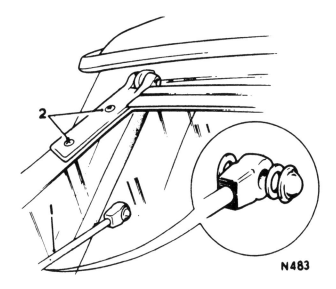

N483

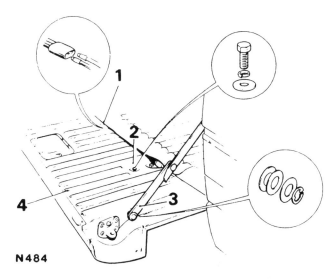

N484

Refitting

13. Reverse the removal procedure, items 1 to 12.

FRONT DOOR GLASS – 2 DOOR MODEL

– Remove and refit   76.31.01

Removing

1. Remove the armrest.
2. Remove both interior door handles.
3. Remove both door pull handles.
4. Remove the window winder.
5. Prise the upper trim panel from the door.
6. Prise the lower trim panel from the door.
7. Withdraw the seals from the top edge of the door.
8. Remove the door glass frame.
9. Withdraw the glass.

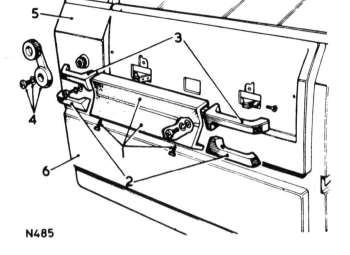

N485

Refitting

10. Reverse 1 to 9.

NOTE: On later models it may be found more convenient to remove the exterior door mounted mirror before commencing this operation.

FRONT DOOR GLASS – 4 DOOR MODEL

– Remove and refit   76.31.01

Removing

1. Remove items 1 to 8, 76.37.12.
2. Fit the window regulator handle temporarily and wind the glass down until the lifting arm stud is at the extreme end of the glass lifting channel.
3. Support the glass inside the door with a suitable length of wood. Remove the single bolt (spring and plain washers) from inside the door which secures the bottom of the front glass-run channel.
4. Remove the four window regulator retaining bolts (with shakeproof washers) from the inner door panel.
5. Disengage the lifting arm stud from the glass lifting channel, manoeuvre the window regulator past the loosened glass-run channel and remove from the lower centre cut-out in the **inner door panel**.
6. Remove the length of wood supporting the glass and lift the glass up into the closed position, taping it to the door frame.
7. Remove the single bolt (spring and plain washers) from inside the door which secures the bottom **rear** glass-run channel.
8. Remove the two screws (shakeproof and plain washers) from the hinge face of the door which secure the **front** door frame.
9. Remove the single screw (shakeproof and plain washers) from the shut face of the door which secures the **rear** door frame.
10. Remove the bolt (shakeproof and plain washers) from the recessed hole in the inner door panel under the front quarter light).

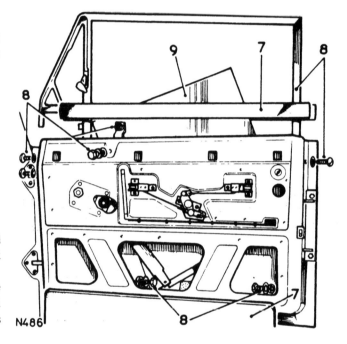

N486

11. Remove the waist rail finisher end mouldings. These are **each** retained by two self-tapping screws/pop rivets.
12. Remove the door glass weather strip secured to the **outer** door panel by six clips (the two front clips are different).
13. Remove the door glass inner seal secured to the **inner** door panel by four clips.
14. **Lift out the door frame with the glass in position.**

Refitting

15. Reverse removal procedure, items 1 to 14.

REAR DOOR GLASS – 4 DOOR MODEL

– Remove and refit   76.31.02

Removing

1. Open the bonnet and disconnect the battery negative earth lead.
2. Open the door and remove the two screws holding the arm rest to the inner door panel.
3. Wind the glass up into the fully closed position then remove the single screw retaining the window regulator handle on its shaft.
4. Carefully prise out the upper and lower halves of the inside door release handle bezel.
5. Unscrew the sill locking knob.
6. Remove the door trim pad by inserting a screwdriver between the trim pad and the inner door panel, gently prising out the eighteen plastic clips from their respective holes around the edges of the trim pad.
7. Remove the plastic weather sheet. This is held to the inner door panel by adhesive.
8. Fit window regulator handle temporarily and wind glass down until the lifting arm stud can be seen in the glass lifting channel through the cut-out in the inner door panel.
9. Support the glass inside the door with a suitable length of wood.
10. Remove the four window regulator retaining bolts (with shakeproof washers) from the inner door panel.
11. Carefully disengage the lifting arm stud from the glass lifting channel and remove the window regulator from the lower cut-out in the door panel.
12. Disconnect the control rod from the inside door release handle by pulling it out of the plastic connecting block.
13. Remove the length of wood supporting the glass and lift the glass up into the closed position, taping it to the top of the door frame.
14. Remove the single bolt (spring and plain washers) from inside the door which secures the bottom of the **short rear** glass run channel.

15. Remove the two screws (shakeproof and plain washers) from the hinge face of the door which secure the **front** door frame.
16. Remove the two screws (shakeproof and plain washers) from the **shut face** of the door which secure the **rear** door frame.
17. Remove the waist rail finisher end mouldings. These are each retained by two self-tapping screws/pop rivets.
18. Remove the door glass weather strip secured to the **outer** door panel by five clips.
19. Remove the door glass inner seals secured to the **inner** door panel by four clips.
20. Lift out the door frame with the glass in position.
21. Remove the tape and release the glass.

Refitting

22. Reverse the removal procedure, items 1 to 21.

FRONT QUARTER VENT

– Remove and refit   76.31.28

Removing

1. Remove the front door glass 76.31.01.
2. Remove the fixings from the quarter vent bottom pivot.
3. Withdraw the quarter vent.

Refitting

4. Reverse the removal procedure.

FRONT DOOR GLASS REGULATOR – 2 DOOR MODEL

– Remove and refit   76.31.45

Removing

1. Remove the front door glass 76.31.01.
2. Remove the fixings securing the regulator to the door.
3. Withdraw the regulator.

Refitting

4. Reverse 1 to 3.

FRONT DOOR GLASS REGULATOR – 4 DOOR MODEL

– Remove and refit   76.31.45

Removing

1. Remove the front door glass, items 1 to 5, 76.31.01.

Refitting

2. Reverse removal procedure.

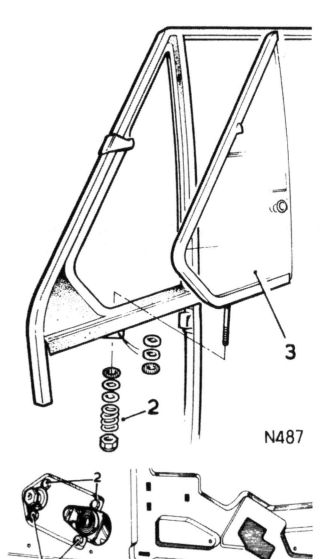

N487

N488

REAR DOOR GLASS REGULATOR – 4 DOOR MODEL

– Remove and refit   76.31.46

Removing

1. Open the bonnet and disconnect the battery negative earth lead.
2. Open the door and remove the two screws holding the arm rest to the inner door panel.
3. Wind the glass up into the fully closed position then remove the single screw retaining the window regulator handle on its shaft.
4. Carefully prise out the upper and lower halves of the inside door release handle bezel.
5. Unscrew the sill locking knob.
6. Remove the door trim pad by inserting a screwdriver between the trim pad and the inner door panel, gently prising out the eighteen plastic clips from their respective holes around the edges of the trim pad.
7. Remove the plastic weather sheet. This is held to the inner door panel by adhesive.
8. Fit window regulator handle temporarily and wind glass down until the lifting arm stud can be seen in the glass lifting channel through the cut-out in the inner door panel.
9. Support the glass inside the door with a suitable length of wood.
10. Remove the four window regulator retaining bolts (with shakeproof washers) from the inner door panel.
11. Carefully disengage the lifting arm stud from the glass lifting channel and remove the window regulator from the lower cut-out in the door panel.

Refitting

12. Reverse the removal procedure, items 1 to 11.

FRONT DOOR LOCK – 2 DOOR MODEL

– Remove and refit   76.37.12

Removing

1. Remove the front door glass 76.31.01.
2. Remove the external door handle.
3. Remove the operating rod from between the interior locking lever and the lock.
4. Disconnect the operating rod from the private key lock.
5. Disconnect the operating rod from the interior door handle relay, at the lock end.
6. Remove the front door lock.

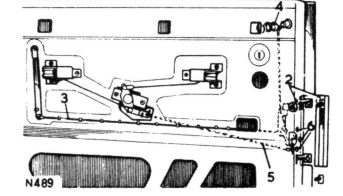

N489

Refitting

7. Reverse 1 to 6.

FRONT DOOR LOCK – 4 DOOR MODEL

– Remove and refit   76.37.12

Removing

1. Open the bonnet and disconnect the battery negative earth lead.
2. Open the door and remove the two screws holding the armrest to the inner door panel.
3. Wind the glass up into the fully closed position, then remove the single screw retaining the window regulator handle.
4. Carefully prise out the upper and lower halves of the inside door release handle bezel.
5. Unscrew the sill locking knob.
6. Remove the door trim pad by inserting a screwdriver between the trim pad and the inner door panel, gently prising out the nineteen plastic clips from their respective holes around the edges of the trim pad.
7. Unplug the two speaker connections inside the door and remove the door trim pad complete with speaker.
8. Remove the plastic weather sheet, held to the inner door panel by adhesive.
9. Remove the door glass (see 76.31.01).
10. Disconnect the control rod from the private key-operated lock by releasing the metal clip.
11. Disconnect the control rod from the **outside** door release handle by pulling it out of the plastic ferrule.
12. Disconnect the control rod from the **inside** door release handle by pulling it out of the plastic connecting block. (The control rod assembly also passes through a rubber padded guide bracket on the inside of the inner door panel.)
13. Release the door lock by removing the two countersunk screws from the door shut face and the single screw (with shakeproof washer) on the inside door panel.
14. Withdraw the door lock through the lower rear cut-out in the inner door.

NOTE: **If necessary the following items can also be removed.**

Outside door release handle, retained by two nuts (with shakeproof washers) from inside the door.

Inside door release handle, retained by four screws (with plain washers) on the inner door panel (the centre screw of the three front screws sets the door release position and must be re-fitted first).

Refitting

15. Reverse the removal procedure, items 1 to 14.

NOTE: When re-fitting the key-operated lock lever rod (item 10) the rod is entered from the back of the lever on left side doors and from the front of the lever on right side doors. The metal retaining clip is not affected.

REAR DOOR LOCK – 4 DOOR MODEL

– Remove and refit   76.37.13

Removing

1. Open the bonnet and disconnect the battery negative earth lead.
2. Open the door and remove the two screws holding the armrest to the inner door panel.
3. Wind the glass up into the fully closed position then remove the single screw retaining the window regulator handle on its shaft.
4. Carefully prise out the upper and lower halves of the inside door release handle bezel.
5. Unscrew the sill locking knob.
6. Remove the door trim pad by inserting a screwdriver between the trim pad and the inner door panel, gently prising out the eighteen plastic clips from their respective holes around the edges of the trim pad.
7. Remove the plastic weather sheet. This is held to the inner door panel by adhesive.
8. Disconnect the control rod from the **inside** door connecting block.
9. Disconnect the sill locking control rod from the door lock by releasing the metal clip.
10. Disconnect the control rod from the **outside** door release handle by pulling it out of the plastic ferrule.
11. Release the door lock by removing the two countersunk screws from the door shut face and the single screw (with shakeproof washers) on the inside door.
12. Withdraw door lock through rear cut-out in inner door panel with its two control rods attached.

NOTE. **If necessary the following items can also be removed.**

Outside door release handle, retained by two nuts (with shakeproof washers) from inside the door.

Inside door release handle, retained by four screws (with plain washers) on the inner door panel (the centre screw of the three front screws sets the door release position and must be re-fitted first).

Sill locking quadrant. Use a small screwdriver, or 1/8in. dia. (3.175mm) rod, to press the plastic locking pin through the square insert in the inner door panel until it can be retrieved from inside the door. The loosened insert can then be pushed inside the door releasing the quadrant assembly which can be withdrawn through the lower cut-out in the inner door panel with its control rods. (To refit, the locking pin is entered into the square insert from outside and pressed in flush.)

Refitting

13. Reverse the removal procedure, items 1 to 12.

If difficulty is experienced in reconnecting the control rod to the outside door handle **inside** the door (see item 10) the outside handle should be removed from the door sufficiently to enable the control rod to be attached from **outside**. Alternatively the outside handle can be fitted **after** the door lock.

ADJUSTMENT TO DOOR RELEASE CONTROLS

No adjustment is normally required to be made to outside or inside door release handle linkage which is pre-set during assembly. However, in the event of there being insufficient handle movement to release the door the following adjustments can be made:

Check that control rods are not damaged or incorrectly assembled, replace if necessary.

The bottom of the outside door release handle control rod is threaded at the door lock connection to enable its operating length to be varied as required.

The inside door release handle position is set during assembly by the centre of the three front securing screws. Remove the centre screw and slacken the remaining securing screws, the handle can then be moved as required and the slackened screws re-tightened.

If the adjustment is satisfactory drill a new lead hole and refit the self-tapping setting screw (with plain washer).

IMPORTANT: door release should be effective before the total handle movement is exhausted to provide a small over-throw movement.

DOOR LOCK-UPPER TAIL GATE

– Remove and refit   **76.37.17**

Removing

1. Remove the rear cover from the private lock (earlier models) or the side covers (later models).
2. Disconnect the bolt arms from the private lock (earlier models) or release the centre lock unit (later models).
3. Remove the fixings from the lock mechanisms each side of the upper tailgate.
4. Withdraw the lock mechanism complete with bolt arm(s).

Refitting

5. Reverse 1 to 4 as appropriate.

Adjusting (as required)

6. The bolt arms are threaded and can be adjusted in or out of the lock mechanisms.
7. The position of the striker plates can be adjusted by slackening the fixings, repositioning and retightening the fixings.

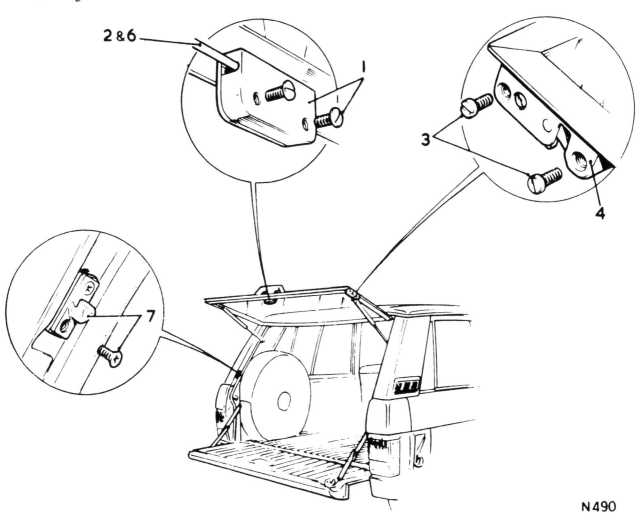

N 490

DOOR LOCK, LOWER TAILGATE

– Adjust 1 and 8 to 13   76.37.18

– Remove and refit 1 to 7 and 13   76.37.19

Removing

1. Remove the lock corner plate.
2. Remove the fixings from the lock mounting plate.
3. Disconnect either one of the bolt arms from the door lock.
4. Withdraw the lock complete with the fixed bolt arm.
5. Withdraw the loose bolt arm.

Refitting

6. Reverse 2 to 5.
7. Close the lower tailgate and check the operation of the lock, the bolts should engage automatically and release when the handle is moved fully right against spring pressure.

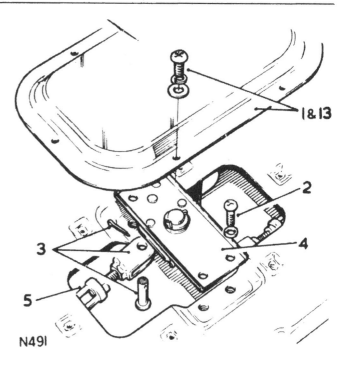

N491

Adjusting the lock 8 to 12

8. Slacken the locknuts at the lock end of the adjuster.
9. Slacken the locknuts at the bolt end of the adjuster, noting that they have a LEFT HAND thread.
10. Turn the adjuster as required to move the bolt in or out.
11. Secure the adjuster locknuts.
12. The eye brackets at each side of the tailgate can also be adjusted to align with their locating dowels, by slackening the fixings, slightly repositioning the brackets and retightening the fixings.
13. Fit the lock cover plate.

PRIVATE LOCK, FRONT DOOR – 2 DOOR MODEL

– Remove and refit   76.37.39

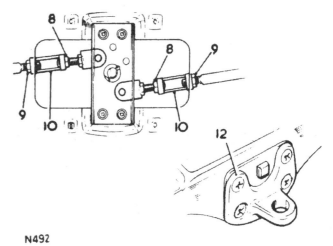

N492

Removing

1. Remove the front door glass 76.31.01.
2. Remove the circlip from the inner end of the private lock.
3. Withdraw the operating rod and lever.
4. Withdraw the special washer.
5. Withdraw the spring clip securing the private lock.
6. Withdraw the private lock.

Refitting

7. Reverse 1 to 6.

PRIVATE LOCK, FRONT DOOR – 4 DOOR MODEL

– Remove and refit   76.37.39

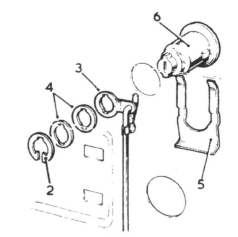

N493

Removing

1. Remove items 1 to 10, 76.37.12.
2. Remove the spring clip securing the private lock.
3. Withdraw the private lock.

Refitting

4. Reverse items 1 to 3.

76–26

REAR DOOR CHECK STRAP – 4 DOOR MODEL

– Remove and refit   76.40.30

Removing

1. Open the bonnet and disconnect the battery negative earth lead.
2. Open the door and remove the two screws holding the arm rest to the inner door panel.
3. Wind the glass up into the fully closed position then remove the single screw retaining the window regulator handle on its shaft.
4. Carefully prise out the upper and lower halves of the inside door release handle bezel.
5. Unscrew the sill locking knob.
6. Remove the door trim pad by inserting a screwdriver between the trim pad and the inner door panel, gently prising out the eighteen plastic clips from their respective holes around the edges of the trim pad.
7. Remove the plastic weather sheet. This is held to the inner door panel by adhesive.
8. Remove the door check strap clevis pin circlip.
9. Remove the clevis pin and spacers.
10. Remove the two bolts (with spring and plain washers) from the door hinge face.
11. Remove the check strap assembly from inside the door.
12. Remove the two bolts (with spring and plain washers) which retain the check strap bracket to the door pillar.

Refitting

13. Reverse the removal procedure, items 1 to 12.

TAILGATE STAYS, UPPER

– Remove and refit   76.40.33

Early Models

Removing

1. Prise off the retaining cap at each end of the stay.
2. Withdraw the large plain washer from each end of the stay.
3. Lift off the stay complete.

Refitting

4. Place a large plain washer onto the stay mounting pegs.
5. Reverse 1 to 3.

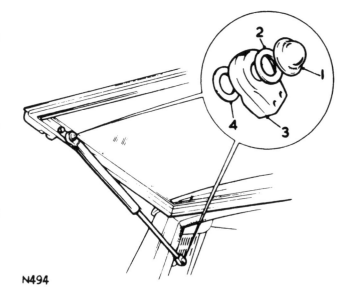

N494

Later Models

Removing

1. Prise the stay from the tailgate fixing and support the tailgate.
2. Prise the stay from the body fixing.

Refitting

3. Reverse instructions 1 and 2.

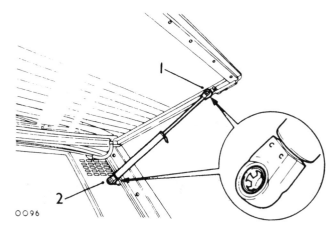

0096

LOWER FASCIA PANEL

Remove and refit   76.46.03

Removing

1. Remove the drive screws securing the lower facia panel.
2. Withdraw the lower fascia panel sufficient to disconnect the harness from the hazard warning switch.
3. Lift the lower fascia panel clear.

Refitting

4. Reverse 1 to 3.

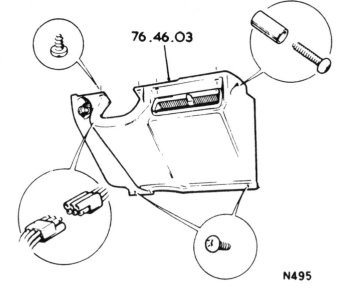

76.46.03

N495

FASCIA TOP RAIL

– Remove and refit   **76.46.04**

Removing

1. Remove the lower fascia panel 76.46.03.
2. Remove the instrument housing 88.20.13.
3. Remove the console unit 76.25.01.
4. Remove the fixings from the underside of the facia top rail on the passengers side.
5. Withdraw the fascia top rail.

Refitting

6. Locate the fascia top rail in position and connect the heater trunking and fresh air duct.
7. Reverse 1 to 4.

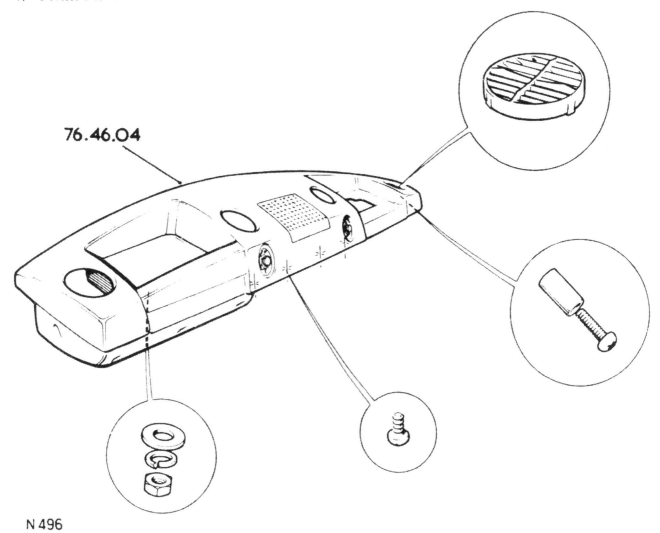

76.46.04

N 496

GLOVE BOX

— Remove and refit   76.52.01

Removing

1. Remove the drive screws securing the glove box hinge.
2. Release the glove box catch and withdraw the box sufficient to disconnect the check strap.
3. Lift the glove box clear.

Refitting

4. Reverse 1 to 3.

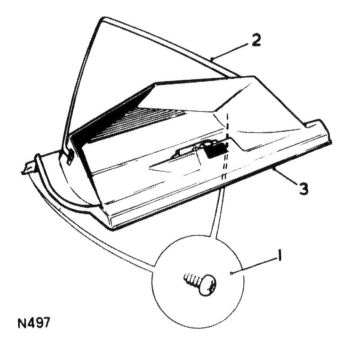

N497

HEADLINING, FRONT SECTION

— Remove and refit   76.64.02

Removing

1. Remove the roof lamp assembly 86.45.02.
2. Pull the rear view mirror from its flexible mounting.
3. Remove the flexible mounting.
4. Remove the sun visors.
5. Remove the grab handles by releasing the end covers and sliding them along the strap to gain access to the fixing screws.
6. Release the rear fixings (later models).
7. Withdraw the headlining rearwards.

NOTE: If any difficulty is experienced in withdrawing the headlining, it may be necessary to remove the rear section first, as some headlinings are fitted with connecting clips.

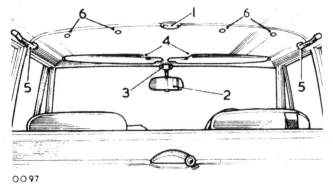

0097

Refitting

8. Reverse 1 to 7.

HEADLINING, REAR SECTION

— Remove and refit   76.64.05

Removing

1. Remove the roof lamp assembly 86.45.02.
2. Remove the rear quarter trim panels (earlier models as illustrated) or on later models the left hand rear quarter trim panel 76.10.22/23.
3. Remove the rear fixings from the grab handles and on later models the four centre fixings.
4. Ease the headlining clear of the rear location bracket and withdraw it rearwards.

Refitting

5. Reverse 1 to 4.

N499

FRONT SEAT – 2 DOOR MODEL

– Remove and refit   76.70.01

Removing

1. Remove the nut and bolt from the front of each seat slide.
2. Remove the six screws and the outside seat base cowling.
3. Remove the two screws and the plate from the outside rear of the seat base.
4. Remove the upper bolt from the rear retention bar and push the bar downward.
5. Release the retaining spring below the seat squab.
6. Working from the rear seat slide the front seat rearward and manoeuvre over the rear kick plate.
7. Withdraw the front seat. Note: It may be found necessary to remove the 'B' post trim if fouling is experienced.

Refitting

8. Reverse instructions 1 to 7.
9. Refit the 'B' post trim, if removed.

FRONT SEAT – 4 DOOR MODEL

– Remove and refit   76.70.01

Removing

1. Move front seat forward.
2. Remove seat runner stop bolts and nuts.
3. Unhook seat assisting spring from frame.
4. Move seat rearwards to clear runners.
5. Remove seat from vehicle.

Refitting

6. Reverse procedure, items 1 to 5.

FRONT SEAT BASE

– Remove and refit   76.70.06

Removing

1. Remove the seat 76.70.01.
2. Release carpet to provide access to seat base fixings.
3. Remove the seat base fixings.
4. Disengage the eyebolts, seat base to chassis.
5. Withdraw the seat base.

Refitting

6. Reverse procedure, items 1 to 5.

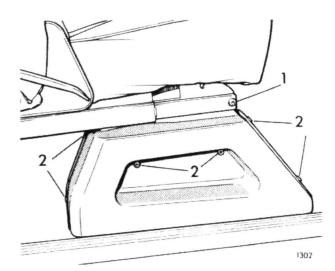

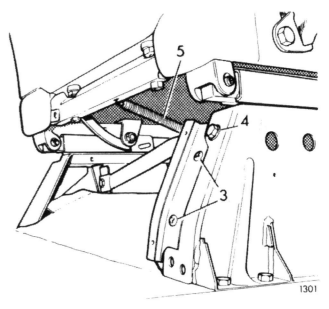

FRONT WING VALANCE

– **Left hand** – remove and refit	**76.79.01**
– **Right hand** – remove and refit	**76.79.02**

Removing

1. Remove the bonnet 76.16.01.
2. Remove the air cleaner 19.10.01.
3. Remove the windscreen wiper arms 84.15.01.
4. Remove the front decker panel 76.10.35.
5. Remove the front parking and flasher lamp assembly 86.40.26.
6. Remove the front wing 76.10.26.
7. Remove or release all components fitted or attached to the wing valance.
8. Remove the front wing valance.

Refitting

9. Reverse 1 to 8, using a waterproof sealant between the wing valance to dash panel joint.

76.79.01-02

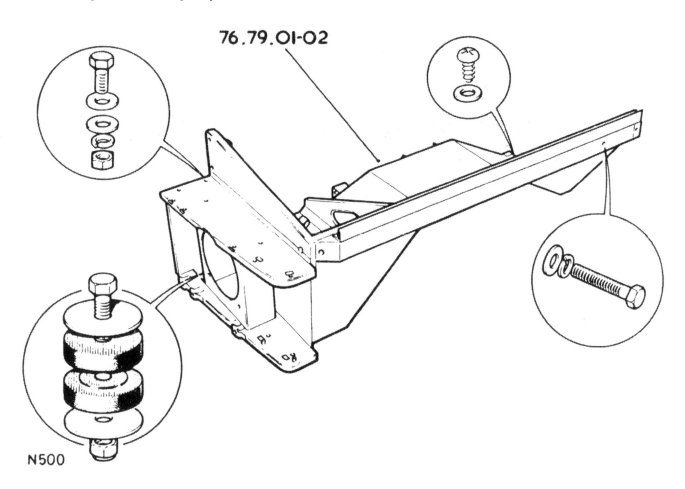

N500

WINDSCREEN GLASS

– Remove and refit   76.81.01

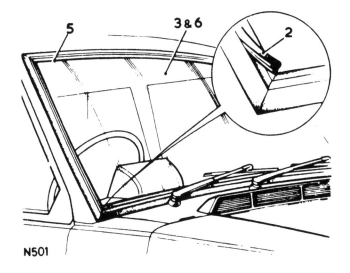

N501

Removing

1. Remove the windscreen wiper arms 84.15.01.
2. Remove the expander strip from the channel in the rubber moulding around the glass.
3. Ease the bottom edge of the windscreen glass from the rubber moulding.
4. Lift the windscreen glass clear.

Refitting

5. Smear soft soap around the windscreen glass location channel in the rubber moulding.
6. Locate the bottom edge of the windscreen glass into the rubber moulding.
7. Use wooden levers tapered to a thin end, to prise the rubber moulding over the windscreen glass all the way round.
8. Reverse 1 and 2.

TAILGATE GLASS

– Remove and refit   76.81.10

Removing

1. Remove the upper tailgate 76.28.29.
2. Remove the lock 76.37.17.
3. Remove the lift handle and trim.
4. The upper tailgate glass and frame are serviced as one unit.

Refitting

5. Reverse instructions 1 to 3.

BODY SIDE GLASS – 2 DOOR MODEL

– **Front** – remove and refit 3 to 7	76.81.17
– **Rear** – remove and refit 1, 2 and 6	76.81.18

Removing

1. Remove the headlining rear section 76.64.05.
2. Remove the expander strip from the channel in the rubber moulding around the glass.
3. Lift the tongue of the spring clips from each of the front glass runners.
4. Slide both runners clear of the glass.
5. Lift out the front glass.
6. Lift out the rear glass.

Refitting

7. Reverse 1 to 6.

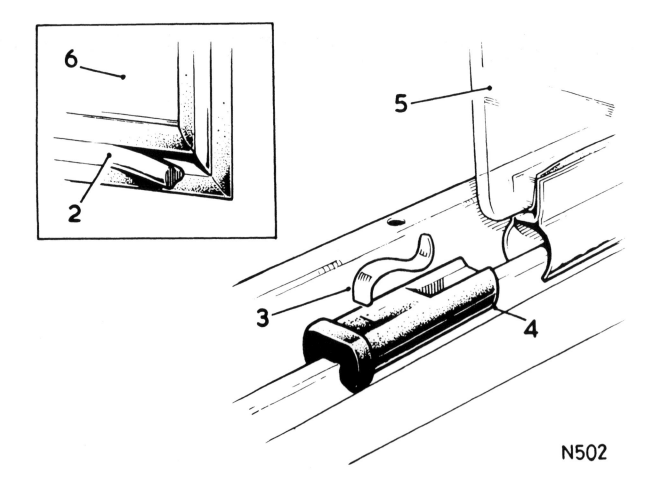

N502

PAINTWORK OPERATIONS

PAINTWORK OPERATIONS

PAINTWORK

General information 78.00.00

Body panels

1. Range Rover body panels are manufactured from a special aluminium-alloy known as 'Birmabright' and the following paintwork procedure should be followed on these panels.

Painting 'Birmabright'

2. The area to be painted must be flatted to remove the hard oxide skin which forms on the surface of the alloy when exposed to the atmosphere. Degrease and dry the area, then apply a suitable etch-primer. Unless an etch-primer is used, paint is liable to come away as it cannot 'key' into the hard oxide of an untreated alloy surface and the use of ICI Etching Primer P565-5002 is recommended. It is quick and easy to apply, and it prolongs the life of the paint film by ensuring excellent adhesion.

Application

3. The activated Etching Primer has a limited pot-life of about 8 hours at normal temperatures and should not be used after this time, as it may have inferior adhesion and corrosion resistance. Any Etching Primer which has been mixed for more than 8 hours must be thrown away, and not returned to the can.

4. Apply Etching Primer as soon as possible after cleaning, and paint as soon as the pre-treatment is completed. Undue delay may cause the surface to be contaminated again and thus nullify the treatment. Do not leave pre-treated work overnight before it is painted.

5. Etching Primer, when followed by a suitable paint system, gives a film which is very resistant to moisture, but the Etching Primer itself is water sensitive. It should therefore be coated with paint as soon as possible when it is dry.

6. Activate the Etching Primer by mixing it with an equal volume of Activator P273-5021 and allow to stand for 10 minutes.

7. Adjust the spraying viscosity of the mixture if necessary to 22-25 sec. BSB4 Cup by adding small quantities of Thinner 851-565; never add more Activator.

8. Apply by spray to a clean, dry surface in a thin uniform coat, rather than a thick heavy one which may impair adhesion.

9. Air dry for at least 15 minutes before applying undercoat by spray or for 2 hours before brush application. If required, these times can be shortened by force drying, this also gives increased hardness to the film.

10. Subsequent painting follows normal paintshop practice.

11. When wet flatting the subsequent paint layers take care not to rub through to the Etching Primer. If this does occur allow to dry out thoroughly, dry flat the area and spot in with Etching Primer.

HEATING AND VENTILATION OPERATIONS

Operation No.

Fan

    -Motor-remove and refit   80.20.15

    -Motor resistance unit-remove and refit   80.20.17

Heater

    -Controls-remove and refit   80.10.02

    -Radiator-remove and refit   80.20.29

    -Unit-remove and refit   80.20.01

HEATER CONTROLS

-Remove and refit 80.10.02

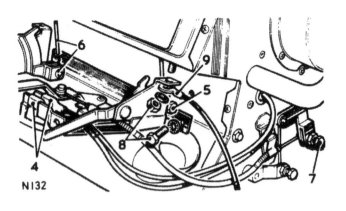

N132

Removing

1. Remove the lower fascia panel 76.46.03.
2. Remove the glove box 76.52.01
3. Remove the console unit 76.25.01
4. Disconnect the electrical leads from the control switch
5. Disconnect the relay rod from the 'SCREEN-CAR' lever
6. Disconnect the relay rod from the 'VENT' lever
7. Disconnect the control cables from each side of the heater unit.
8. Remove the heater controls assembly.

Refitting

9. Fit the heater controls assembly, incorporating the unit earth lead with R H upper fixing
10. Reverse 5 to 7
11. Check that the control levers give full movement of the flaps. If necessary, adjust at the relay rod or cable end fixings
12. Connect the electrical leads to the control switch, the Black/White lead connects to the front terminal.
13. Reverse 1 to 3

HEATER UNIT

-Remove and refit 80.20.01

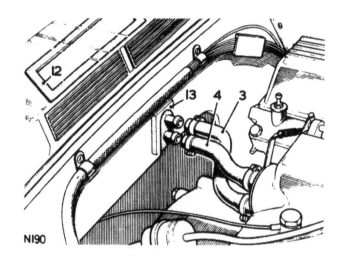

N190

Removing

1. Drain the cooling system 26.10.01
2. Remove the air cleaner 19.10.01
3. Disconnect the water inlet hose from the heater
4. Disconnect the water outlet hose from the heater
5. Remove the lower fascia panel 76.46.03.
6. Remove the console unit 76.25.01
7. Remove the glove box 76.52.01
8. Disconnect the four demister hoses from the heater unit
9. Lift the edge of the fascia rail if necessary release the fascia rail fixings, and withdraw the fresh air duct.
10. Disconnect the electrical leads from the heater unit
11. Remove the heater unit

Refitting

12. Check that the seal for the fresh air intake is in place on the back of the heater unit
13. Check that the seal for the heater radiator is in place on the radiator pipes.
14. Reverse 1 to 11

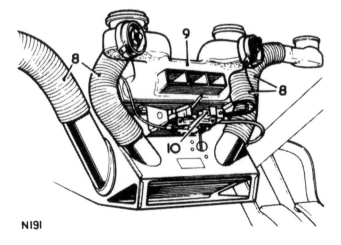

N191

80-2

HEATER FAN MOTOR

-Remove and refit   80.20.15

Early Models:

Removing

1. Remove the heater unit 80.20.01
2. Disconnect the electrical leads
3. Disconnect the air cooling hose
4. Remove the fan motor fixings
5. Partially withdraw the motor and fan assembly
6. Slacken the grub screws securing the fan to the motor
7. Withdraw the fan motor

Refitting

8. Engage the fan motor spindle into the boss on the fan and secure the grub screws
9. Locate the fan motor in position, engaging the fan spindle into the bearing
10. Hold the fan motor firmly in position and check that the fan rotates freely. If necessary, adjust the position of the fan on the motor spindle
11. Align the air cooling hose connection and secure the fan motor
12. Connect the air cooling hose
13. Connect the electrical leads, the green lead must be connected to the terminal nearest the air cooling hose
14. Refit the heater unit 80.20.01

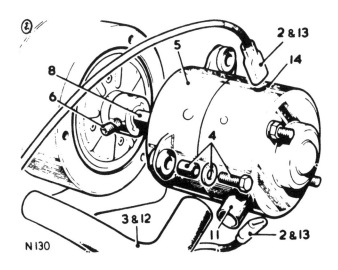

Later Models:

Removing

1. Withdraw the fan and motor from the unit 80.20.17 instructions 1 to 4.
2. Drill out the pop rivets.
3. Remove the fan from the motor.

Refitting

4. Reverse instructions 1 to 3.

FAN MOTOR RESISTANCE UNIT

-Remove and refit   80.20.17

Removing

1. Remove the heater unit 80.20.01
2. Disconnect the electrical leads from the fan motor
3. Disconnect the air cooling hose
4. Remove the fan motor and fan assembly.
5. Disconnect the electrical leads from the control switch
6. Drill out the two pop-rivets securing the resistance unit
7. Withdraw the resistance unit complete with leads.

Refitting

8. Reverse 1 to 7. When connecting the electrical leads, connect the Black/White resistance unit lead to the front terminal on the control switch, and connect the green fan motor lead to the terminal nearest the air cooling hose.

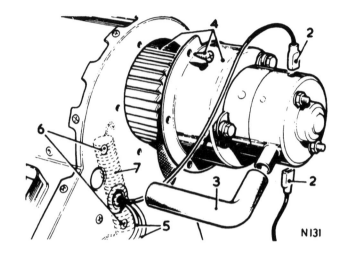

N131

HEATER RADIATOR

-Remove and refit   80.20.29

Removing

1. Remove the heater unit 80.20.01
2. Remove the fixings from the cam bracket for the fresh air flap, and move the cam and bracket assembly aside
3. Remove the lock-washers from the four flap spindles
4. Remove all the drive screws from the left hand side cover
5. Withdraw the left hand side cover complete with the fresh air flap.
6. Withdraw the heater radiator complete with seals.
7. Withdraw the seals from the heater radiator.

Refitting

8. Apply Bostik Sealing compound around the flange of the left hand side cover.
9. Reverse 1 to 7.

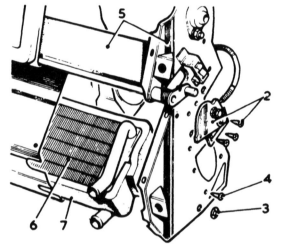

N133

80-4

AIR CONDITIONING OPERATIONS

AIR CONDITIONING – A.R.A. SYSTEM

Description   82.00.00

The A.R.A. air conditioning system comprises four units:
1. An engine-mounted compressor.
2. A condenser mounted in front of the radiator.
3. A receiver/drier unit located in the engine compartment.
4. An evaporator unit mounted behind the fascia.

The four units are interconnected by hoses carrying refrigerant, and the evaporator is linked into the vehicle ventilation system.

WARNING: Under no circumstances should refrigerant pipes be disconnected without first depressurising the system, see 82.30.35.

Cold refrigerant circuit

The function of the refrigeration circuit is to cool the evaporator.

Compressor
The compressor draws vaporized refrigerant from the evaporator. It is compressed, and thus heated, and passed on to the condenser as a hot, high pressure vapour.

Condenser
The condenser is mounted at the front of the car. Its function is to remove heat from the refrigerant and disperse it into the atmosphere.
The refrigerant is delivered as hot, high pressure vapour.

Air flow across the tubes, induced by vehicle movement and assisted by two electric condenser fans.
Early systems, however, used only one fan. The fans cool the vapour, causing it to condense into a high pressure liquid. As this change of state occurs a large amount of latent heat is released.

Receiver drier
This unit filters, removes moisture, and acts as a reservoir for the liquid. To prevent icing inside the system, extreme precautions are taken during servicing to exclude moisture. The receiver drier should be considered as a second stage insurance to prevent the serious consequences of ice obstructing the flow. A sight glass provided in the unit top enables a visual check to be made of the high pressure liquid flow.

Expansion valve and evaporator
High pressure liquid refrigerant is delivered to the expansion valve. A severe pressure drop occurs across the valve and as the refrigerant enters the evaporator space at a temperature of approximately $-6°C$ it boils and vaporizes. As this change of state occurs, a large amount of latent heat is absorbed. The evaporator is therefore cooled and as a result heat is extracted from the air flowing across the evaporator. The air flow is controlled by two evaporator fans regulated by the air conditioner fan control.

Second cycle
Vaporized refrigerant is then drawn from the evaporator by the compressor and a second cycle commences.

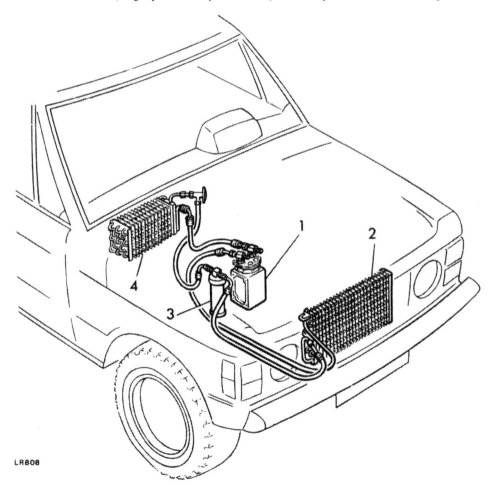

LR808

General service information

1. Introduction

Before any component of the air conditioning system is removed the system must be depressurised. When the component is replaced the system must be evacuated to remove all traces of old refrigerant and moisture. Then the system must be recharged with new refrigerant.

Any service work that requires loosening of a refrigerant line connection should be performed only by qualified service personnel. Refrigerant and/or oil will escape whenever a hose or pipe is disconnected.

All work involving the handling of refrigerant requires special equipment, a knowledge of its proper use and attention to safety measures.

2. Servicing equipment

The following equipment is required for full servicing of air conditioning.
Charging trolley.
Service valve adaptors.
Valve core removers.
Leak detector.
Tachometer.
Lock ring spanner.
Valve key.
Safety goggles.
Refrigerant charging line gaskets.
Compressor dip stick.
5/8 in. UNC bolt or Union nut (Part Number 534127) for extraction of the compressor pulley.
Thermometer $-20°C$ to $-60°C$ ($0°F$ to $-120°F$).

3. Servicing materials

Refrigerant: Freon R12.
Nominal charge weight:
    RHD vehicles – 1.43 kg (51 oz)
    LHD vehicles – 1.37 kg (48 oz)

CAUTION: Methychloride refrigerants must not be used.
Compressor oil: See Recommended Lubricants, Section 09, alternatives.

4. Precautions in handling refrigerant

Refrigerant 12 is transparent and colourless in both the gaseous and liquid state. It has a boiling point of $-30°C$ ($-22°F$) and at all normal pressures and temperatures it is a vapour. The vapour is heavier than air, non-flammable and non-explosive. It is non-poisonous except when in contact with an open flame, and non-corrosive until it comes into contact with water.

The following precautions in handling refrigerant 12 should be observed at all times:
a. Do not leave a drum of refrigerant without its heavy metal cap fitted.
b. Do not carry a drum in the passenger compartment of a car.
c. Do not subject drums to a high temperature.
d. Do not weld or steam clean near an air conditioning system.
e. Do not discharge refrigerant vapour into an area with an exposed flame, or into the engine air intake. Heavy concentrations of refrigerant in contact with a live flame will produce a toxic gas that will also attack metal.
f. Do not expose the eyes to liquid refrigerant. ALWAYS wear safety goggles.

5. Precautions in handling refrigerant lines

WARNING: Always wear safety goggles when opening refrigerant connections.

a. When disconnecting any pipe or flexible connection the system must be discharged of all pressure. Proceed cautiously, regardless of gauge readings. Open connections slowly, keeping hands and face well clear, so that no injury occurs if there is liquid in the line. If pressure is noticed allow it to bleed off slowly.
b. Lines, flexible end connections and components must be capped immediately they are opened to prevent the entrance of moisture and dirt.
c. Any dirt or grease on fittings must be wiped off with a clean alcohol dampened cloth. Do not use chlorinated solvents such as trichloroethylene. If dirt, grease or moisture cannot be removed from inside pipes, they must be replaced with new.
d. All replacement components and flexible end connections are sealed, and should only be opened immediately prior to making the connection. (They must be at room temperature before uncapping to prevent condensation of moisture from the air that enters.)
e. Components must not remain uncapped longer than 15 minutes. In the event of delay the caps must be replaced.
f. Receiver driers must never be left uncapped as they contain Silica Gel which will absorb moisture from the atmosphere. A receiver drier left uncapped must be replaced, and not used.
g. A new compressor contains an initial charge of 11 UK fluids ozs. (312.5 ml) of oil when received, part of which is distributed throughout the system when it has been run. The compressor contains a holding charge of gas when received which should be retained until the hoses are connected.
h. The compressor shaft must not be rotated until the system is entirely assembled and contains a charge of refrigerant.

j. The receiver-drier should be the last component connected to the system to ensure optimum dehydration and maximum moisture protection of the system.

k. All precautions must be taken to prevent damage to fittings and connections. Minute damage could cause a leak with the high pressures used in the system.

l. Always use two spanners of the correct size, one on each hexagon, when releasing and tightening refrigeration unions.

m. Joints should be coated with refrigeration oil to aid correct seating.

n. All lines must be free of kinks. The efficiency of the system is reduced by a single kink or restriction.

o. Flexible hoses should not be bent to a radius less than ten times the diameter of the hoses.

p. Flexible connections should not be within 50 mm (2.000 in.) of the exhaust manifold.

q. Completed assemblies must be checked for refrigerant lines touching sheet metal panels. Any direct contact of lines and sheet metal transmits noise and must be eliminated.

6. Periodic maintenance

The design of the system is such that routine servicing, apart from visual checks is not necessary.

These visual inspections are listed as follows.

a. **Condenser.** With a hose pipe or air line, clean the face of the condenser to remove flies, leaves, etc. Check pipe connections for signs of oil leakage.

b. **Compressor.** Check hose connections for signs of oil leakage. Check flexible hoses for swellings. Examine the compressor belt for tightness and condition. Checking compressor oil level and topping-up is only necessary after:

   i. Charging the system.

   ii. Any mal-function of the equipment. See 82.10.14.

Schrader service valve and valve core remover

1. Hose connection
2. Service valve
3. Core remover/adaptor
4. Schrader valve core
5. Plunger
6. Service hose
7. Compressor port

c. **Liquid receiver.** Examine the sight glass for bubbles with the system running. Check connections for leakage.

d. **Evaporator.** Examine refrigeration connections at the unit.

If the system should develop any fault, or if erratic operation is noticed, refer to the fault diagnosis charts.

7. Service valves (Schrader type)

These are secured to the head of the compressor, and the suction and discharge flexible end connections are secured to them by unions.

The service valves are identified as suction or low pressure, and discharge or high pressure. Whilst they are identical in operation they are not interchangeable, as the connections are of different sizes. When in position the valve can be identified by the letters 'DD DISCH' and 'SS SUCTION' on the compressor head. The valve with the larger connection fits the suction side.

As the name suggests, these valves are for service purposes, providing connections to external pressure/vacuum gauges for test purposes. In combination with charging and testing equipment they are used to charge the system with refrigerant.

The use of valve core removers (shown below) will facilitate servicing operations and should be used as follows:

1. Close all valves on the charging trolley.
2. Remove the service valve cap and seals from the valve core remover.
3. Withdraw the black plunger as far as possible and connect the core remover to the service valve.
4. Connect the hose to the core remover.
5. Depress the black plunger until it contacts the valve core. Unscrew the valve until it is free. Withdraw the plunger to its full extent.

Service valve caps must be replaced when service operations are completed. Failure to replace caps could result in refrigerant loss and system failure.

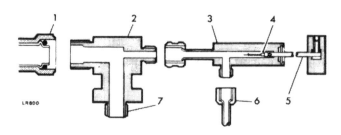

8.   Test procedure

Efficient testing of the air conditioning system requires the use of charging and testing equipment and an accurate thermometer, in addition to normal workshop equipment.

9.   Charging and testing equipment

This is standard equipment for the servicing of automotive air conditioners, and is used for testing, trouble shooting, evacuating and charging the system.

All evacuating and charging equipment is assembled into a compact portable unit. The use of this equipment reduces air conditioner servicing to a matter of connecting two hoses and manipulating clearly labelled valves. An instruction plate indicates the operations required for evacuating and charging the air conditioner system.

The compound or low pressure gauge is connected to the suction service valve to check pressure and vacuum on the low pressure side of the system. It is graduated in inches Hg of vacuum 0 to 30, and in pounds per square inch from 0 to 60.

The high pressure gauge is connected to the discharge service valve to check the pressures on the high pressure side of the system. It is graduated in pounds per square inch from 0 to 600.

The connections beneath the gauges are for attaching the hoses to the appropriate service valve.

The valves are opened and closed by means of hand-wheels at each end of the control panel. At no time is it possible to close off the gauges. The valves close off the centre section from each end connection and from one another. When the valves are turned fully clockwise, the centre section is closed to both sides of the system. As each valve is turned counter-clockwise, the appropriate service valve is put into communication with the centre section, gas flow being according to the pressure differential.

10.   Electrical supply, switches and fuses

The four main components of the air conditioning system draw current from three different sources. The evaporator blower motor and main condenser fan are supplied from the heated rear window relay. Power for the auxiliary condenser fan is taken direct from the battery while the compressor clutch circuit draws current from the starter relay. Each component in turn is energised and controlled by a series of relays and switches of various types as indicated by the circuit diagrams.

Four, in line, fuses in bayonet type holders protect the circuits and when fitting a replacement it is essential that a fuse of the same value is used. The blower motor and compressor clutch fuses are located on top of the inner right hand wing valance. The auxiliary condenser fan fuse is positioned adjacent to the battery connected into the lead from the positive terminal. The main condenser fan fuse is located on the bulkhead near the starter relay, behind the windscreen washer reservoir.

Circuit Diagram A

This circuit was used on LHD vehicles manufactured towards the end of 1978. The evaporator and compressor clutch circuits draw current through an ignition controlled air conditioning relay. This relay is energised from the heated rear window relay which is only live when the ignition switch is in the 'On' position. On vehicles with no heated rear window fitted, the green feed wire to W1 on the air conditioning relay is taken from three green wires in a double snap connector located close to the RH front corner of the heater unit. This system employs only one condenser fan.

Circuit Diagram B

From December 1978 onwards this circuit was used on all LHD Range Rovers. The layout is similar to diagram A except that all vehicles fitted with air conditioning have heated rear windows installed. The two feed wires to the heated rear window relay are taken from the same position adjacent to the heater unit. Also, only one condenser fan is used with this layout.

The compressor clutch circuit has been modified to relieve the pressure switch of excessive current load. Also the compressor clutch circuit now draws current from the existing starter motor relay. This provides a continuous feed to the compressor which is only interrupted when the starter motor is energised, thus reducing excessive drag during cranking.

Circuit Diagram C

This circuit was employed on the early twin condenser fan system. Two condenser fans were installed to provide additional cooling for the engine when the air conditioning is operating. The auxiliary condenser fan is operated, as before, by a pressure switch through a relay. The main condenser fan operates immediately the air conditioning system is switched on. The main fan is energised by a change-over relay which prevents the unnecessary use of the heated rear window, even if the switch is 'On' when the air conditioning system is in operation.

Circuit Diagram D

Functionally, this circuit is similar to diagram C. The main differences are changes in cable colours which are indicated by an asterisk. The feed pick-up point for the main fan, through the auxiliary pressure switch, is changed. The circuit between the starter relay and the compressor clutch relay is now protected by a 10 amp fuse. A new type, 26RA change-over relay is now used.

Circuit Symbols

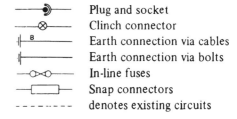

Plug and socket	
Clinch connector	
Earth connection via cables	
Earth connection via bolts	
In-line fuses	
Snap connectors	
denotes existing circuits	

Cable Colour Code

B	Black	G	Green	R	Red
U	Blue	O	Orange	W	White
N	Brown	P	Purple	Y	Yellow

The last letter of a colour code denotes the tracer.

Circuit Diagram A

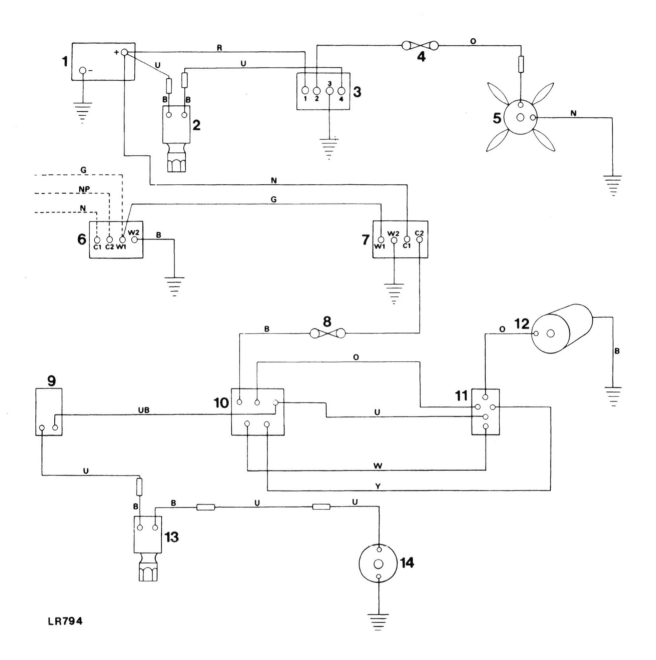

LR794

1. Battery
2. Condenser fan pressure switch
3. Condenser fan relay
4. 20 amp fuse
5. Condenser fan
6. Heated rear window relay
7. Air conditioning relay
8. 20 amp fuse
9. Thermostat switch
10. Blower motor switch
11. Resistor
12. Evaporator blower motor
13. Compressor clutch pressure switch
14. Compressor clutch

Circuit Diagram B

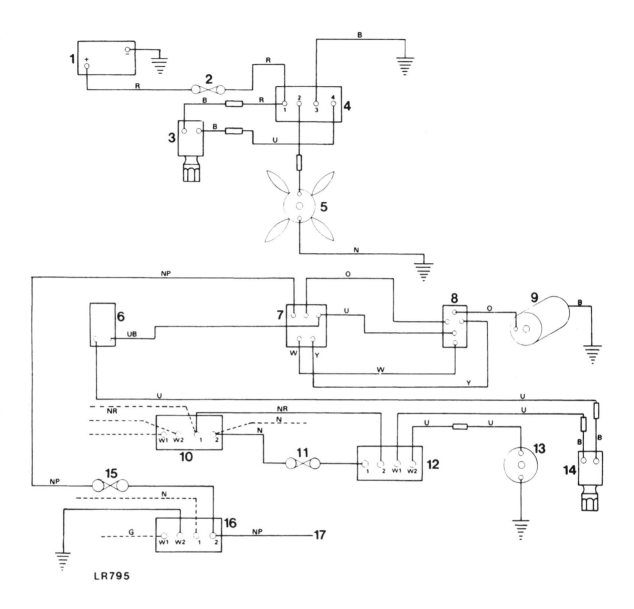

LR795

1. Battery	10. Starter relay
2. 20 amp fuse	11. 20 amp fuse
3. Pressure switch condenser fan	12. Compressor clutch relay
4. Condenser fan relay	13. Compressor clutch
5. Condenser fan	14. Compressor clutch pressure switch
6. Thermostat switch	15. 20 amp fuse
7. Blower motor switch	16. Heated rear window relay
8. Resistor	17. Existing lead to switch and heated rear window
9. Evaporator blower motor	

82-7

Circuit Diagram C

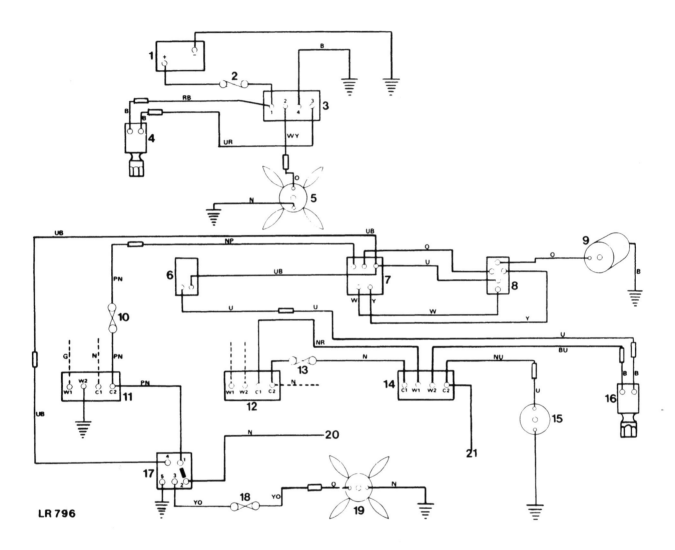

LR 796

1. Battery
2. 20 amp fuse
3. Auxiliary condenser fan relay
4. Condenser fan pressure switch
5. Auxiliary condenser fan
6. Thermostat switch
7. Blower motor switch
8. Resistor
9. Evaporator blower motor
10. 20 amp fuse
11. Heated rear window relay

12. Starter relay
13. 20 amp fuse
14. Compressor clutch relay
15. Compressor clutch
16. Compressor clutch pressure switch
17. Change-over relay
18. 20 amp fuse
19. Main condenser fan
20. Existing cable to heated rear window and switch
21. Throttle jack connection – if required

Circuit Diagram D

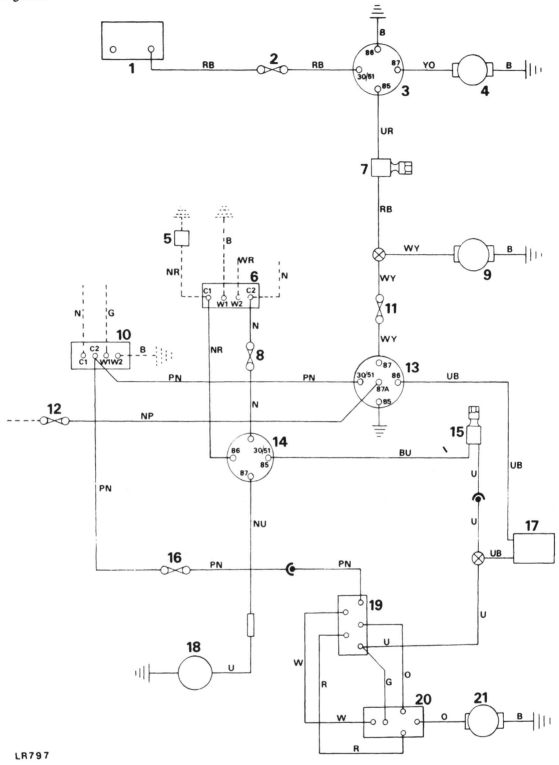

LR797

1. Battery	12. Existing cable to heated rear window and switch
2. 20 amp fuse	13. Change-over relay
3. Auxiliary fan relay	14. Compressor clutch relay
4. Auxiliary fan	15. Compressor clutch pressure switch
5. Starter solenoid	16. 20 amp fuse
6. Starter relay	17. Thermostat switch
7. Auxiliary fan pressure switch	18. Compressor clutch
8. 10 amp fuse	19. Blower motor switch
9. Main condenser fan	20. Resistor
10. Heated rear window relay	21. Blower motor
11. 20 amp fuse	

82-9

11. **Electrical System Fault Diagnosis**

FAULT	CAUSE	REMEDY
A. MOTOR INOPERATIVE OR SLOW RUNNING.	1 Incorrect voltage. 2 Open or defective fuse or relay. 3 Loose wire connection, including ground. 4 Switch open or defective. 5 Tight worn, or burnt motor bearings. 6 Open rotor windings. 7 Worn motor brushes. 8 Shaft binding – blade misaligned. 9 Defective resistor board.	1 Check voltage. 2 Check and replace as necessary. 3 Check system wires; tighten all connections. 4 Replace switch. 5 Replace motor. 6 Replace motor. 7 Replace motor. 8 Check alignment. Repair or replace as necessary. 9 Rectify or replace.
B. CLUTCH INOPERATIVE	1 Incorrect voltage. 2 Open or defective fuse or relay. 3 Defective thermostat control or pressure switch. 4 Shorted or open field coil. 5 Bearing seized (clutch will not disengage). 6 Refrigeration circuit problem causing heavy load and excessive drive torque.	1 Check voltage. 2 Check and replace as necessary. 3 Replace thermostat or pressure switch. 4 Replace coil. 5 Replace bearing. 6 Check and rectify.
C. CLUTCH NOISY	1 Incorrect alignment. 2 Loose belt. 3 Compressor bracket and/or braces not mounted securely. 4 Bearing in clutch-pulley assembly not pressed in properly. 5 Low voltage to clutch. 6 Clutch will not spin freely. 7 Oil on clutch face. 8 Slipping clutch. 9 Overloaded or locked compressor. 10 Icing.	1 Check alignment; repair as necessary. 2 Adjust to proper tension. 3 Repair as necessary. 4 Remove clutch and replace bearing. 5 Check connections and voltage. 6 Refer to B5 above. 7 Check compressor seals for leaks. 8 Refer to C5 above. 9 Repair or replace compressor. 10 Check for suction line frosting. Replace expansion valve if necessary. Replace receiver/drier if necessary.
D. CONDENSER AND/OR EVAPORATOR VIBRATION	1 Motor and/or blades improperly mounted. 2 Blade corrosion or foreign matter build-up. 3 Excessive wear of motor bearings.	1 Check mountings, adjust as necessary. 2 Clean blades with solvent or other non-inflammable cleaner. 3 Replace motor.

12.   **Refrigeration System Fault Diagnosis**

For any refrigeration system to function properly all components must be in good working order. The unit cooling cycle and the relationship between air discharge temperature and ambient temperature and the pressures at the compressor can help to determine proper operation of the system.

The length of any cooling cycle is determined by such factors as ambient temperature and humidity, thermostat setting, compressor speed and air leakage into the cooled area, etc. With these factors constant, any sudden increase in the length of the cooling cycle would be indicative of abnormal operation of the air conditioner.

The low and high side pressures at the compressor will vary with changing ambient temperature, humidity, cab temperature and altitude.

The following conditions should be checked after operating the system for several minutes:—

1.   All high side lines and components should be warm to the touch.
2.   All low side lines should be cool to the touch.
3.   Inlet and outlet temperatures at the receiver drier should be at the same temperature (warm). Any very noticeable temperature difference indicates a blocked receiver drier.
4.   Heavy frost on the inlet to the expansion valve may indicate a defective valve or moisture in the system.
5.   With ambient humidity between 30% and 60%, compressor pressures and evaporator air discharge temperature should fall within the general limits given in the table below.

Type of Weather	Evaporator Air Temp °F (°C)	Low Side Pressure lb/in^2 (Kg/cm^2)	High Side Pressure lb/in^2 (Kg/cm^2)
Cool Day 70-80°F (21-27°C)	35-45°F (1.7-7.2°C)	15-20 (1.1-1.4)	160-200 (11.2-14)
Warm Day 80-90°F (27-32°C)	40-50°F (4.4-10°C)	20-25 (1.4-1.8)	190-240 (13.4-16.9)
Hot Day Over 90°F (Over 32°C)	45-60°F (7.2-15.6°C)	25-30 (1.8-2.1)	220-270 (15.5-19)

NOTE:
1. Low and high side pressures are guides not specific limits.
2. Evaporator air temperatures will be lower on dry days, higher on humid days.

FAULT	CAUSE	REMEDY
A. HIGH HEAD PRESSURE	1 Overcharge of refrigerant. 2 Air in system. 3 Condenser air passage clogged with dirt or other foreign matter. 4 Condenser fan motor defective	1 Purge with bleed hose until bubbles start to appear in sight glass; then, add sufficient refrigerant gas to clear sight glass. 2 Slowly blow charge to atmosphere. Install new drier; evacuate and charge system. 3 Clean condenser of debris. 4 Replace motor.
B. LOW HEAD PRESSURE	1 Undercharge of refrigerant; evident by bubbles in sight glass while system is operating. 2 Split compressor gasket or leaking valves. 3 Defective compressor.	1 Evacuate and recharge the system. Check for leakage. 2 Replace gasket and/or reed valve; Install new drier, evacuate, and charge the system. 3 Repair or replace compressor.
C. HIGH SUCTION PRESSURE	1 Slack compressor belt. 2 Refrigerant flooding through evaporator into suction line; evident by ice on suction line and suction service valve. 3 Expansion valve stuck open. 4 Compressor suction valve strainer restricted. 5 Leaking compressor valves, valve gaskets and/or service valves. 6 Receiver-drier stopped; evident by temperature difference between input and output lines.	1 Adjust belt tension. 2 Check thermobulb. Bulb should be securely clamped to clean horizontal section of copper suction pipe. 3 Replace expansion valve. 4 Remove and clean or replace strainer. 5 Replace valves and/or gaskets. Install new drier, evacuate, and charge the system. 6 Install new drier, evacuate and charge the system.
D. LOW SUCTION	1 Expansion valve thermobulb not operating. 2 Expansion valve sticking closed. 3 Moisture freezing in expansion valve orifice. Valve outlet tube will frost while inlet hose tube will have little or no frost. System operates periodically. 4 Dust, paper scraps, or other debris restricting evaporator blower grille. 5 Defective evaporator blower motor, wiring, or blower switch.	1 Warm thermobulb with hand. Suction should rise rapidly to 20 lbs. or more. If not, replace expansion valve. 2 Check inlet side screen. Clean if clogged. Refer to C-2 and C-3. 3 Install new drier, evacuate and charge the system. 4 Clean grilles as required. 5 Refer to Fault Diagnosis Chart for Electrical System.

FAULT	CAUSE	REMEDY
E. NOISY EXPANSION VALVE (steady hissing)	1 Low refrigerant charge; evident by bubbles in sight glass.	1 Leak test. Repair or replace components as required.
F. INSUFFICIENT COOLING	1 Expansion valve not operating properly. 2 Low refrigerant charge – evident by bubbles in sight glass. 3 Compressor not pumping.	1 Refer to C-2, C-3, D-1, and E. 2 Refer to B-1 and E. 3 Refer to B-2 and B-3.
G. COMPRESSOR BELT SLIPPING	1 Belt tension. 2 Excessive head pressure. 3 Incorrect alignment of pulleys or worn belt not riding properly. 4 Nicked or broken pulley. 5 Seized compressor.	1 With tension gauge adjust to 100 lbs. (45 kg); or tighten until depression of about ½ inch (1.25 cm) occurs across longest span. 2 Refer to A-1 through A-4 and C-6. 3 Repair as needed. 4 Replace as needed. 5 Replace compressor.
H. ENGINE NOISE AND/OR VIBRATION	1 Loose or missing mounting bolts 2 Broken mounting bracket. 3 Loose flywheel or clutch retaining bolt. 4 Rough idler pulley bearing. 5 Bent, loose, or improperly mounted engine drive pulley. 6 Incorrect installation of clutch. bearing seal. 7 Insecure mountings of accessories: generator, power steering, air filter, etc. 8 Excessive head pressure. 9 Incorrect compressor oil.	1 Repair as necessary. 2 Replace bracket. 3 Repair as necessary. 4 Replace bearing. 5 Repair as necessary. 6 Replace bearing. 7 Repair as necessary. 8 Refer to A-1, A-2, A-3, A-4 and C-6. 9 Refer to compressor Oil Level Check.

COMPRESSOR DRIVE BELT

– Adjust 1, 2, 4 and 9 to 11　　　　82.10.01

Procedure

1. Prop open the bonnet.
2. Slacken the compressor adjuster and pivot bolts.
3. Pivot the compressor inwards as far as possible.
4. Adjust the position of the compressor by means of its pivot and slotted fixing, to give the correct belt tension. The belt must be tight with 4 to 6 mm (0.187 to 0.250 in.) total deflection when checked midway between the compressor and idler pulleys, by hand.
5. Tighten all fixings and recheck tension.
6. Close the bonnet.

COMPRESSOR DRIVE CLUTCH

– Remove and refit　　　　82.10.08

Compressor installed 1 to 3 and 5 to 22

Compressor removed 4 to 10 and 12 to 18

Removing

1. Remove the radiator block.
2. Remove the compressor drive belt.
3. Switch on the ignition and the air conditioning air flow control.
4. Connect the clutch lead to the positive terminal of a twelve volt battery, and connect the compressor crankcase to the battery negative terminal.
5. Remove the bolt and washer securing the compressor pulley to the crankshaft.
6. Screw a 5/8 in. UNC bolt into the thread provided in the pulley bore to extract the pulley from the crankshaft and remove the extractor bolt.

NOTE: If difficulty is experienced in removing the pulley, an alternative method is available, using a union nut, Rover Part No. 534127, and a 6 mm (0.250 in.) diameter bolt by 32 mm (1.250 in.) long. The bolt must be a clearance fit in the bore of the union nut. Proceed with items 7 to 10.

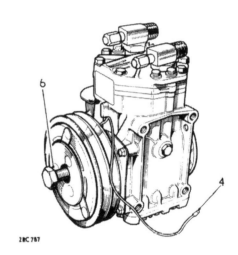

continued

82–14

7. Assemble the bolt and union nut as illustrated.
8. Screw the assembly into the tapping in the pulley centre.
9. Tighten the union nut to load the bolt head on to the compressor crankshaft end face.
10. Alternately tap with a hammer on the bolt end, screw in the union nut to free the pulley from the tapered seating.
11. Switch off the ignition and air flow control or disconnect the battery from the compressor, as applicable.
12. Disconnect the electrical lead from the compressor clutch at the snap connector.
13. Remove the four bolts and washers securing the drive clutch to the compressor crankcase.
14. Withdraw the compressor drive clutch and base plate assembly.

Refitting

15. Secure the clutch and base plate assembly to the compressor crankcase.
16. Connect the clutch lead to the positive terminal of a 12 volt battery, and connect the compressor crankcase to the battery negative terminal. If the compressor is installed in the car switch on the ignition and air conditioning air flow control.
17. With the clutch energised, fit the pulley and tighten the securing bolt to a torque of 2,2 kgf.m (16 lbf ft).
18. Disconnect the battery, or switch off the ignition and air-flow control, as applicable.
19. Reconnect the electrical lead for the compressor clutch at the snap connector.
20. Refit the compressor drive belt.
21. Refit the radiator block.
22. Refill the engine cooling system, using the correct mixture.

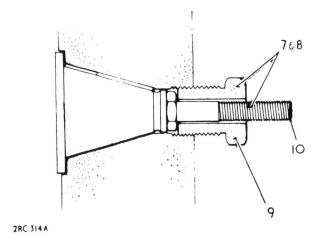

2RC 314A

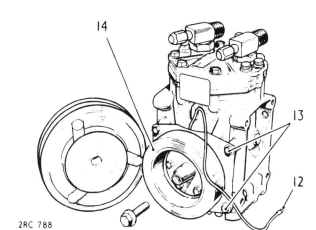

2RC 788

82-15

COMPRESSOR OIL LEVEL

– Check   **82.10.14**

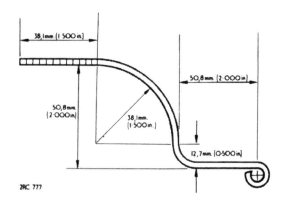

2RC 777

NOTE: The compressor oil level should be checked whenever any components, including the compressor are
removed and refitted, or when a pipe or hose has been
removed and reconnected or, if a refrigerant leak is
suspected. All compressors are factory charged with 11 UK
fluid ozs (312.5 ml) of oil. When the air conditioning
equipment is operated some of the oil circulates throughout
the system with the refrigerant, the amount varying with
engine speed. When the system is switched off the oil
remains in the pipe lines and components, so the level of
oil in the compressor is reduced, by approximately 2 UK
fluid ozs (56.8 ml). The compressor oil level must finally
be checked after the system has been fully charged with
refrigerant and operated to obtain a refrigerated temperature of the car interior. This ensures the correct oil balance
throughout the system.

The compressor is not fitted with an oil level dipstick, and
a suitable dipstick must be made locally from 3mm (0.125
in.) diameter soft wire in accordance with the accompany-
ing illustration. After shaping, mark the end of the dipstick with twelve notches, 3mm (0.125 in.) apart.

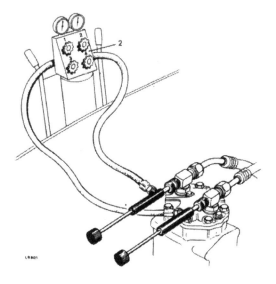

Procedure

1. Open the bonnet.
2. Fit the charging and testing equipment 82.30.01.
3. Start the engine and turn the temperature control to
    maximum cooling position, and the air flow control
    to HIGH speed. Operate the system for five minutes
    at 1,200 – 1,500 rev/min.

NOTE. It is important to open the valve slowly during the
following item to avoid a sudden pressure reduction in the
compressor crankcase that could cause a large amount of
oil to leave the compressor.

4. Reduce the engine speed to idling, and SLOWLY
    open the suction side valve on the test equipment
    until the compound gauge reads 0 or a little below.
5. Stop the engine at this point and quickly open the
    suction valve and discharge valve.
6. Loosen the oil level plug and unscrew it slowly by
    5 turns to bleed off crankcase pressure.
7. Remove the oil level plug, wipe the dipstick and
    insert it as near vertical as possible and to the lowest
    point of the crankcase. It may be necessary to turn
    the compressor crank to obtain clearance.

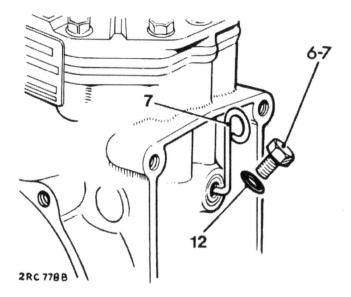

2RC 778B

continued

8. Withdraw the dipstick and determine the depth of oil. The depth should be 25 mm (1.00 in.) maximum, with a minimum of 22 mm (0.875 in.).

NOTE: 25 mm (1.00 in.) depth corresponds to 9 UK fluid ozs (255.6 ml). 22 mm (0.875 in.) depth corresponds to 6 UK fluid ozs (170.4 ml).

9. If required, top up the compressor crankcase to the correct level, using special compressor oil. See Recommended Lubricants, Section 09.

10. Lubricate a new 'O' ring with compressor oil, fit it over the threads of the level plug without twisting, and install the level plug loosely.

11. Evacuate the air from the compressor crankcase, using the vacuum pump on the charging and testing equipment, as follows:

    a. Open valve numbers 1 and 2.
    b. Ensure that valve number 4 is closed.
    c. Start the vacuum pump and open the vacuum pump valve.
    d. Slowly open valve number 3.
    e. Tighten the crankcase level plug securely.
    f. Close valve number 3.
    g. Switch off the vacuum pump.

12. Close fully the suction and discharge valves.

13. Start and run the engine at 1,200 rev/min and check for leak at the compressor level plug. Do not over-tighten to correct a leak. In the event of a leak isolate the compressor as previously described in items 4 to 6, and check the 'O' ring seats for dirt, etc.

14. Stop the engine.

15. Close all valves on the charging and testing equipment.

16. Disconnect the charging lines from the compressor.

17. Refit the blanking caps to the compressor valve stems and gauge connections, and to the charging lines.

18. Close the bonnet.

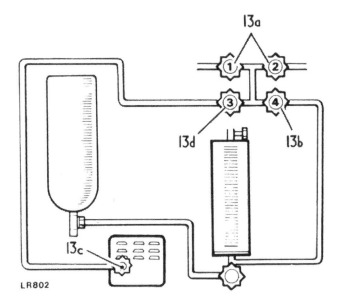

LR802

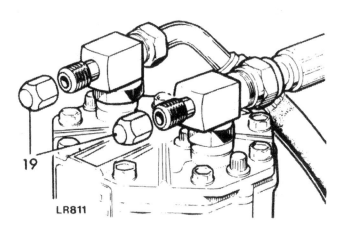

LR811

COMPRESSOR

– Remove and refit   82.10.20

Removing

1. Depressurise the air conditioning system 82.30.05.
2. Using goggles to protect the eyes and wearing gloves, disconnect the suction and discharge unions from the service valves on top of the compressor. Cap the flexible end connections and service valves immediately the joints are opened.
3. Disconnect the lead to the compressor magnetic clutch at the connector.
4. Remove the top bracket and lifting eye complete with radiator hose leaving the top bracket connected to the hose by the 'P' clip.
5. Slacken the compressor adjuster and pivot bolts, pivot the compressor inwards and release the driving belt.
6. Remove the compressor adjuster and pivot bolts and lift compressor clear.
7. If required, remove the mounting bracket from the compressor.

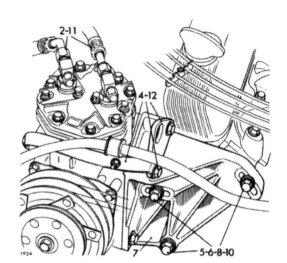

Refitting

NOTE: Before fitting the compressor to the engine, it must be complete with its side mounting/adjuster bracket.

8. If a new compressor is being fitted, check that it has the correct quantity of oil by making and using a dipstick as described in operation 82.10.14. Then drain off 2 UK fluid ozs (56.8 ml) to compensate for the oil already in the system.
9. If the original compressor is being refitted, drain off all the oil and refill with 9 UK fluid ozs only of the correct oil. See General Service Information.
10. Locate the compressor in position and fit the pivot and adjuster bolts, finger tight only.
11. Fit the compressor driving belt and adjust the compressor by means of its pivot and slotted fixing, to give the correct belt tension. The belt must be tight with 4 to 6mm (0.187 to 0.250 in.) total deflection when checked midway between the compressor and idler pulleys, by hand.
12. Tighten the adjuster and pivot bolts and recheck the driving belt tension.
13. Refit the suction and discharge flexible end connections to the service valves, lubricating the flares and threads of the unions with refrigeration oil.
14. Refit the top bracket and lifting eye, with hose.
15. Connect the lead to the compressor magnetic clutch at the snap connector.
16. Evacuate the air conditioning system 82.30.06. Maintain the vacuum for ten minutes.
17. Charge the air conditioning system 82.30.08.

82–18

CONDENSER FANS AND MOTORS
(TWIN FAN SYSTEM)

– Remove and refit   82.15.01

Removing fans and motors

1. Open the bonnet and disconnect the battery.
2. Remove the six screws and withdraw the grille panel.
3. Remove the insulation tape from the wiring harness to expose snap connectors. Make a note of the wiring colours to facilitate reconnection and disconnect snap connector.
4. Disconnect earth wiring retaining bolt.
5. Remove wiring securing clip.
6. Slacken the two upper bolts securing the left hand and right hand bonnet striker support stays.
7. Remove the lower bolts securing the lower ends of the stays and pivot both stays forward.
8. Remove the two upper bolts securing the fans (one for each fan).
9. Remove the four lower bolts securing the fans (two for each fan).
10. Turn each fan and motor assembly in an anti-clockwise direction and carefully remove from the vehicle.

NOTE: Later models have fan motor assemblies mounted on two bars across the front of the condenser.
Follow instructions 1 to 7 above then remove two nuts and washers securing each from motor and withdraw the assembly.

To dismantle fan motor and cowl assembly

NOTE: The fan cowl is deleted on later models.

11. Slacken the fan blade grub screw and withdraw the fan blades from the motor drive shaft. Make a note of the exact location of the fan blades on the shaft to facilitate reassembly.
12. Remove the wiring securing clip.
13. Slacken the fan cowl clamp screws.
14. Slacken the stay bracket clamp screws and remove the fan cowl and stay bracket.

To reassemble fan motor and cowl assembly

15. Fit stay bracket to fan motor.
16. Fit the fan cowl.
17. Correctly position fan motor and tighten clamp screws.
18. Locate fan in correct position on drive shaft and secure with grub screw. Check that the fan blade rotates freely.
19. Fit cable securing clip.

To fit fans and motors to vehicle

20. Offer up each assembly into its mounting position.
21. Fit the upper securing bolts (one for each assembly) but leave slack for the time being.
22. Fit the four lower securing bolts (two for each assembly).
23. Tighten all the securing bolts.
24. Secure earth wire.
25. Fit and tighten the bonnet striker support stays.

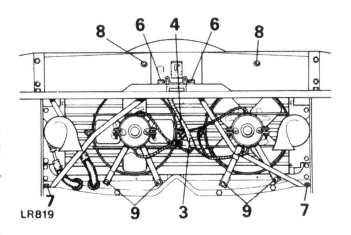

LR819

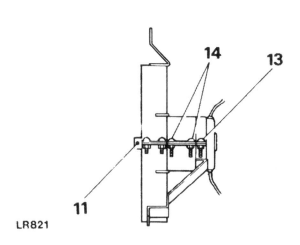

LR821

26. Fasten snap connector and apply insulation tape to the harness.
27. Fit the front grille and secure with the six screws.

82–19

**CONDENSER FAN MOTOR
(SINGLE FAN SYSTEM)**

– Remove and refit   82.15.01

Removing

1. Open the bonnet and isolate the battery.
2. Remove the front grille securing screws and remove the grille panel.
3. Disconnect the electrical leads from the condenser fan motor.
4. Unclip the electrical leads from the diagonal stays.
5. Remove the two bolts, nuts and washers securing the stays and the bonnet catch to the upper front panel.
6. Remove the two screws securing the fan assembly to the vehicle.
7. Slacken the two bolts securing the lower ends of the diagonal stays and pivot the stays forward.
8. Remove the fan assembly.
9. Slacken the hexagon socket screw and withdraw the fan from the motor shaft.
10. Slacken the motor clamp screws and remove the motor.

Refitting

11. Refit the motor into the support frame and tighten the clamp screws.
12. Refit the fan onto the motor shaft and tighten the hexagon socket screw. Ensure that the fan rotates without fouling the support frame.
13. Fit the fan assembly in position and secure with the two lower screws.
14. Refit the bonnet catch and secure the fan assembly and diagonal stays.
15. Connect the electrical leads and clip them to the diagonal stay.
16. Connect the battery.
17. Refit the front grille panel.

5–14

6–13

1853

82–20

CONDENSER (TWIN FAN SYSTEM)

– Remove and refit 82.15.07

Removing

NOTE: On later models it is not possible to withdraw the condenser through the grille aperture.

Ignore instructions 4 to 9 below. Remove radiator 26.40.04. Remove six condenser mounting bolts and withdraw condenser complete with fan motor assemblies.

1. Open the bonnet and disconnect the battery.
2. Depressurise the air conditioning system – refer to operation 82.30.05.
3. Remove six screws and withdraw the front grille panel.
4. Disconnect the left-hand and right-hand horn electrical leads.
5. Remove the horn bracket securing bolts and remove both horns.
6. Mark the position of the bonnet striker plate.
7. Remove the two bolts securing the striker plate and striker plate diagonal support stays.
8. Move the support stays aside to gain access to the condenser.
9. Remove the fan and motor assemblies, see operation 82.15.01 instructions 8 to 10.

CAUTION: Before carrying out instruction 10 protect the eyes with safety goggles and wear protective gloves.

10. Using two spanners on each union, carefully disconnect the pipes at the condenser end. Fit blanks to the exposed ends of the pipes.
11. Remove the six bolts retaining the condenser and withdraw the condenser, from the vehicle, through the grille aperture.

Refitting

12. Place the condenser into position and secure with the six bolts.
13. Remove the blanks from the pipes and fit new 'O' rings to the pipes.
14. Apply refrigerant oil to the pipe threads to aid sealing.
15. Connect the pipes to the condenser and tighten to the following torque: –
    Compressor hose 3.4 to 3.9 kgf.m (24-29 lbf ft)
    Receiver drier hose 1.4 to 2.1 kgf.m (10-15 lbf ft)
16. Fit the fan and motor assemblies.
17. Locate the bonnet striker support stays and align striker to previously made marks and tighten the retaining bolts.
18. Tighten the striker plate diagonal support stay lower bolts.
19. Refit the two horns, and connect the electrical leads.
20. Connect the battery and test the horns.
21. To compensate for oil loss, add 2 UK fluid ozs (56.8 ml) of the correct oil to the compressor.
22. Evacuate the system, operation 82.30.06.
23. Charge the system, operation 82.30.08.

24. Carry out a leak test on the disturbed joints, see operation 82.30.09.
25. Check the complete system as described in operation 82.30.16.
26. Fit the grille panel.

LR820

CONDENSER AND SINGLE FAN ASSEMBLY

– Remove and refit   82.15.17

Removing

1. Open the bonnet and connect the gauge set, 82.30.01.
2. Depressurise the system, refer to 82.30.05.
3. Isolate the battery.
4. Remove the front grille securing screws and remove the grille panel.
5. Disconnect the electrical leads from the condenser fan motor.
6. Unclip the electrical leads from the diagonal stays.
7. Remove the two bolts, nuts and washers securing the stays and the bonnet catch to the upper front panel.
8. Remove the two screws securing the fan assembly to the vehicle.
9. Slacken the two bolts securing the lower ends of the diagonal stays and pivot the stays forward.
10. Disconnect the electrical leads from the two horns and remove the horns from their mounting brackets.
11. Protect the eyes with safety goggles and wear gloves, during operation 12.
12. Carefully disconnect the two hose connections at the condenser. Use two spanners on each union to protect the delicate condenser pipe joints.
13. Remove the screws securing the condenser mounting brackets to the condenser.
14. Remove the screws securing the condenser brackets to the vehicle.
15. Withdraw the condenser forwards through the grille aperture.

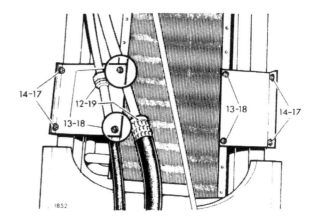

Refitting

16. Place the condenser in position in front of the radiator.
17. Secure the condenser mounting brackets to the vehicle.
18. Secure the brackets to the condenser.
19. Connect the two hose connections. Use refrigerant compressor oil on all mating surfaces to facilitate leakage prevention.
    Tighten the connections to the following torques:
    Compressor hose 3.4-3.9 kgf.m (24-29 lbf.ft)
    Receiver drier hose 1.4-2.1 kgf.m (10-15 lbf.ft)
20. Refit the horns and connect the electrical leads.
21. Fit the fan assembly in position and secure with the two lower screws.
22. Refit the bonnet catch and secure the fan assembly and diagonal stays.
23. Connect the battery.
24. To compensate for oil loss, add 2 UK fluid ozs (56.8 ml) to the compressor.
25. Evacuate the refrigeration system, 82.30.06.
26. Charge the complete system, 82.30.08.
27. Perform a leak test on any disturbed joints, 82.30.09.
28. Carry out a functional check, 82.30.16.
29. Disconnect the gauge set, 82.30.01.
30. Refit the front grille panel.

RECEIVER DRIER

– Remove and refit   82.17.01

CAUTION: Immediate blanking of the receiver drier is important. Exposed life of the unit is only 15 minutes.

Removing

1. Connect the gauge set, 82.30.01.
2. Discharge the complete system, 82.30.05.
3. Protect the eyes with safety goggles and wear gloves during operations 4 and 5.
4. Disconnect the electrical leads at the snap connectors and carefully unscrew the pressure switches from the receiver drier. Blank the exposed connections immediately.
5. Carefully disconnect the two hose connections. Use a second spanner to support the hose adaptor. Blank the exposed connections immediately.
6. Remove one bolt, nut and washers securing the mounting bracket to the wing valance.
7. Remove the clamp bolts, washers and nuts.
8. Withdraw the receiver drier from the mounting bracket.

Refitting

9. Insert the receiver drier into the mounting bracket with the inlet and outlet connections correct to the refrigerant circuit as shown.
10. Connect the two hose connections finger tight. Use refrigerant compressor oil on all mating surfaces to assist leakage prevention.
11. Fit the clamp bolts, washers and nuts.
12. Secure the mounting bracket to the wing valance.
13. Tighten the two hose connections. Use a second spanner to support the hose adaptor.
    Tighten the hose connections to a torque of 1.4-2.1 kgf.m (10-15 lbf.ft).
14. Carefully refit the pressure switches to the receiver drier. Use refrigerant compressor oil on all mating surfaces to assist leakage prevention and tighten the switches to a torque of 2.1-2.6 kgf.m (15-19 lbf.ft). Reconnect the electrical leads.
15. To compensate for oil loss, add 1 UK fluid oz (28.4 ml) of the correct oil to the compressor.
16. Evacuate the complete system, 82.30.06.
17. Charge the complete system, 82.30.08.
18. Perform a leak test on any disturbed joints, 82.30.09.
19. Carry out a functional check, 82.30.16.
20. Disconnect the gauge set, 82.30.01.

Illustration A early type.

Illustration B latest type

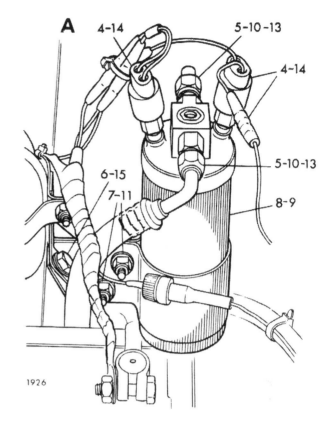

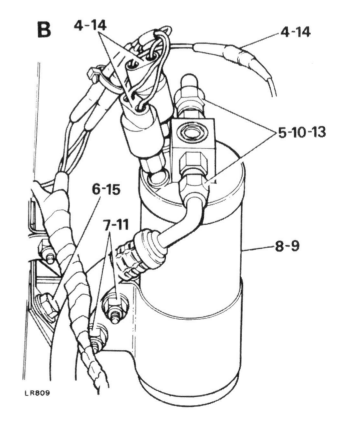

82–23

DASHBOARD UNIT – ARA

– EVAPORATOR

– Remove and refit	82.25.20

Expansion valve
– Remove and refit 1 to 22 and 34 to 56	82.25.01

Hose – compressor to evaporator
– Remove and refit 1 to 5 and 50 to 56	82.25.02

Hose – receiver drier to evaporator
– Remove and refit 1 to 5 and 50 to 56	82.25.03

Blower assembly – left hand
– Remove and refit 1 to 24 and 32 to 56	82.25.33

Removing

1. Open the bonnet and connect the gauge set, 82.30.01.
2. Depressurise the system, 82.30.05.
3. Isolate the battery.
4. Protect the eyes with safety goggles and wear gloves during instruction 5.
5. Disconnect the evaporator hoses from the compressor and the receiver drier. Use a second spanner to support the hose adaptors and blank all the exposed connections immediately.
6. Disconnect the dash unit electrical harness at the underbonnet relays and release the cable from the clips.
7. Working inside the vehicle, remove the three screws (two at right-hand, one at left-hand) securing the lower edge of the centre console.
8. Remove the six screws securing the lower edge of the grille panel to the console and evaporator case.
9. Remove the heater control panel and knobs from the centre console.
10. Withdraw slightly the centre console and remove the instruments.
11. Remove the centre console.
12. Remove the screws securing the evaporator plenum and grille panel to the dash top panel.
13. Withdraw the thermostat sensor from the evaporator.
14. Withdraw the grille panel clear of the dash.
15. Depress the left end of the plenum and remove the air hoses from the upper panel.
16. Remove the screws securing the lower right evaporator bracket.
17. Support the evaporator case and remove the two nuts securing the case and reinforcing strip to the upper mounting bracket.
18. Carefully withdraw the refrigerant hoses, electrical harness and evaporator condensate tubes through the bulkhead and remove the rear left-hand air hose from the plenum. Remove the evaporator and plenum assembly from the vehicle.

continued

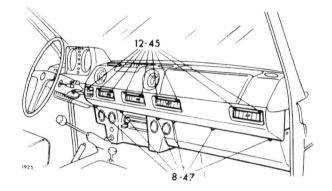

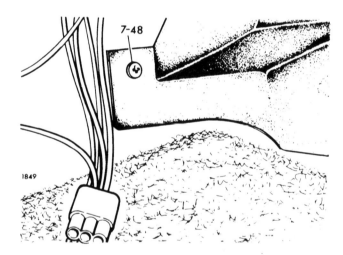

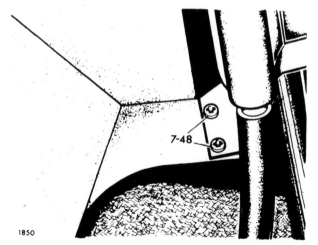

82–24

Dismantling

19. Remove the insulation from the evaporator and expansion valve hose connections.
20. Disconnect the hoses from the expansion valve and evaporator. Use a second spanner to support the hose adaptors and blank all the exposed connections immediately.
21. Unclip the sensor coil from the evaporator outlet pipe.
22. Carefully unscrew the expansion valve from the evaporator. Blank the exposed connections immediately.
23. Unplug the electrical harness at the connector on the left-hand blower casing.
24. Remove the eight securing screws and detach the blower units from the evaporator case.
25. Remove the screws securing the upper evaporator/plenum casing to the evaporator and lower casing.
26. Remove the heater seal and lift off the upper casing.
27. Remove the insulation pad and the two screws securing the evaporator to the lower casing.
28. Withdraw the evaporator from the casing.

Assembling

29. Secure the evaporator to the lower casing.
30. Fit the insulation pad and run the electrical leads under the evaporator outlet pipe.
31. Secure the casings together with the screws and refit the heater seal.
32. Refit and secure the blower units to the evaporator casing.
33. Connect the electrical lead to the connector on the left-hand blower casing.
34. Assemble the expansion valve to the evaporator with the inlet facing downwards. Use refrigerant compressor oil on all mating surfaces to assist leakage prevention. Tighten the connection to a torque of 4.5-5.3 kgf.m (33-39 lbf.ft).
35. Clip the sensor coil to the evaporator outlet pipe.
36. Connect the hoses to the evaporator and expansion valve. Use new 'O' rings and refrigerant compressor oil on all mating surfaces to assist leakage prevention. Tighten the connections to the following torques:
Compressor hose 3.6-4.2 kgf.m (26-31 lbf.ft)
Receiver drier hose 1.4-2.1 kgf.m (10-15 lbf.ft)
37. Wrap all exposed metal at the hose connections with 'prestite' tape.

continued

82–25

Refitting

38. Place the evaporator assembly on the floor of the vehicle and route the electrical harness and the refrigerant hoses through the bulkhead.
39. Lift the unit into the mounting position and connect the rear left-hand air hose. Fit the reinforcement strip and secure the casing to the upper bracket with two nuts.
40. Secure the lower right mounting bracket to the vehicle.
41. Feed the hoses, electrical harness and evaporator condensate tubes through the bulkhead. Ensure that the apertures and grommets are adequately sealed against ingress of dust and moisture.
42. Depress the left end of the plenum and connect the two upper air hoses.
43. Position the left-hand of the plenum so that the opening is centered over the fresh air outlet of the heater.
44. Carefully push the thermostat pipe into the evaporator fins.
45. Refit the grille panel and secure the plenum casing and grille panel to the dash top panel with screws.

46. Refit the instruments to the centre console and fit the console in position.
47. Refit the six screws securing the grille panel to the centre console and evaporator case.
48. Secure the lower edge of the console with three screws (two at right-hand, one at left-hand).
49. Working under the bonnet, connect the dash harness to the two relays.
50. Connect the two refrigerant hoses to the compressor and receiver drier. Use refrigerant compressor oil on all mating surfaces to assist leakage prevention. Tighten the connections to the following torques:
Compressor hose 3.6-4.2 kgf.m (26-31 lbf.ft)
Receiver drier hose 1.4-2.1 kgf.m (10-15 lbf.ft)
51. To compensate for oil loss, add 3 UK fluid ozs (85.24 ml) of the correct oil to the compressor.
52. Evacuate the system, 82.30.06.
53. Charge the complete system, 82.30.08.
54. Perform a leak test on any accessible disturbed joints, 82.30.09.
55. Perform a functional check, 82.30.16.
56. Disconnect the gauge set, 82.30.01.

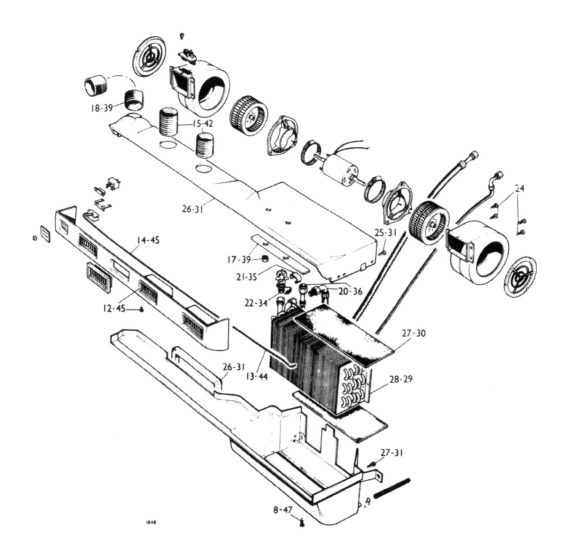

CHARGING AND TESTING EQUIPMENT

– Fit and remove   **82.30.01**

For evacuating or charging with liquid refrigerant
1 to 5 and 7 to 23.

For sweeping or charging with gaseous refrigerant
1 to 4, 6 to 13 and 18 to 23.

NOTE: There are two methods of connecting the charging and testing equipment, depending on the Operation to be carried out. The method described for 'evacuating or charging with liquid refrigerant' also applies to 'Pressure test' and 'Compressor oil level check' operations.

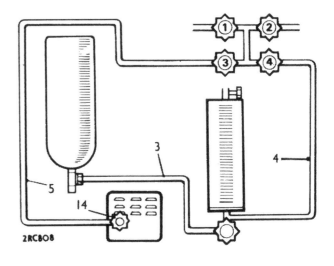

Fitting

1. Ensure that all the valves on the charging and testing equipment are closed. Control valves on the particular equipment selected are numbered 1 to 4 as illustrated. The sequence may vary on other proprietary equipment.
2. Mount a 11,3 kg (25 lb) drum of refrigerant upside-down on the support at the rear of the charging equipment, and secure with the web strap.
3. Connect the hose from the bottom of the charging cylinder to the refrigerant drum valve.
4. Connect the hose between the bottom of the charging cylinder and the refrigerant control valve (No. 4).
5. **For evacuating or charging with refrigerant in a liquid state** – Connect the hose between the vacuum pump valve and the vacuum control valve (No. 3).
6. **For sweeping or charging with refrigerant in gaseous state** – Connect the hose between the top of the charging cylinder, and the vacuum control valve (No. 3).
7. Prop open the car bonnet.
8. Remove the caps from the compressor service valves.
9. Remove the caps from the gauge connections on both services valves.

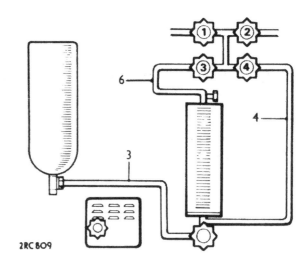

continued

10. Remove the blanking caps from the equipment charging lines and coat the threads and flares with refrigerant oil.
11. Connect the low pressure charging line (blue) from valve No. 1, to the compressor suction service valve.
12. Connect the high pressure charging line (red) from valve No. 2, to the compressor discharge service valve.
13. Start the vacuum pump and open the vacuum pump valve.
14. Open valve numbers 1, 2 and 3, to evacuate the charging lines.
15. Close valve numbers 1, 2 and 3 and stop the vacuum pump.

NOTE: The charging and testing equipment is now connected and ready for proceeding with the required Operations.

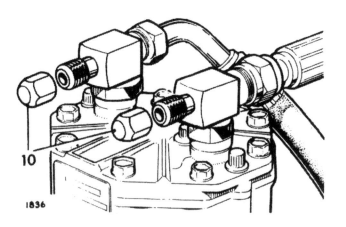

1836

Removing

16. If the engine has been run, it must be stopped prior to disconnecting the charging and testing equipment.
17. Close all valves on the charging and testing equipment.
18. Disconnect the charging lines from the compressor.
19. Refit the blanking caps to the compressor valve stems and gauge connections, and to the charging lines.
20. Close the bonnet.

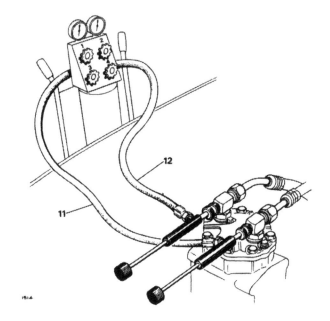

1814

82–28

AIR CONDITIONING SYSTEM

– Depressurise   82.30.05

NOTE: The air conditioning refrigeration system contains 'Refrigerant 12' under pressure, and before any component is disconnected or removed, the system must be discharged of all pressure.

Refrigerant 12 is transparent and colourless in both the gaseous and liquid state. It has a boiling point of $-30°C$ $(-21.7°F)$ and at all normal pressures and temperatures it is a vapour. The vapour is heavier than air, non-flammable and non-explosive. It is non-poisonous except when in contact with an open flame, and non-corrosive until it comes into contact with water.

Proceed cautiously, regardless of gauge readings.

WARNING: Open connections slowly, keeping the hands and face well clear, so that no injury occurs if there is liquid in the line. If pressure is noticed allow it to bleed off slowly.

Always wear safety goggles when opening refrigerant connections.

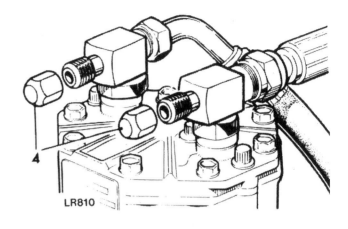

Depressurising

1. Place the car in a ventilated area away from open flames and heat sources.
2. Stop the engine.
3. Prop open the bonnet.
4. Remove the caps from the compressor service valves.
5. Close all valves on the charging and testing equipment.
6. Put on safety goggles.
7. Connect the high pressure charging line (red) from valve No. 2, to the compressor discharge service valve.
8. Slowly open valve No. 2, one turn.
9. Hold the end of the low pressure charging line (blue) in an absorbent rag.

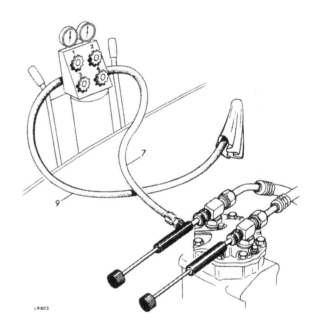

continued

10. Slowly open valve No. 1, and discharge the refrigerant vapour into the rag. If oil is discharged, reduce the valve opening.

11. When the pressure has been reduced, and the hissing sound ceases, close the valves Nos. 1 and 2 on the charging and testing equipment.

12. Disconnect the high pressure charging line from the compressor service valve.
   Any component of the refrigerant system should be capped immediately when disconnected.

13. Open the refrigeration drum valve.

14. Open the valve at the base of the charging cylinder and allow approximately 0,25 kg (½ lb) of refrigerant to enter the cylinder.

15. Close the refrigeration drum valve and the valve at the base of the charging cylinder.

16. Open valve No. 4 (refrigerant control) and flush out the high and low pressure lines by opening valve numbers 1 and 2 momentarily until a white stream of refrigerant is observed.

17. Close all valves on the charging and testing equipment, and fit the blanking caps.

18. Remove safety goggles.

19. The air conditioning system is now depressurised.

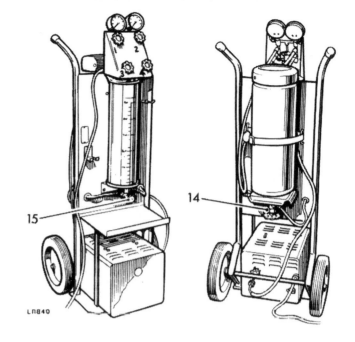

LN840

AIR CONDITIONING SYSTEM

– Evacuate 82.30.06

NOTE: Evacuation of the system is an essential preliminary to charging the system with refrigerant 12. The operation (a) removes air from the system so that it can be fully charged with refrigerant, (b) helps to remove the moisture that is harmful to the system, (c) provides a check for leaks due to faulty connections.

Where a system has been open for some time, or is known to have excessive moisture, it is recommended that the additional operation of 'sweeping' be carried out to eliminate as much of the accumulated moisture as possible. This is done after evacuation, and is detailed in Operation 82.30.07. If sweeping is intended, the receiver-drier unit must be replaced before commencing evacuation. It is recommended that vacuum of the system is maintained for 20 minutes, following sweeping of the system.

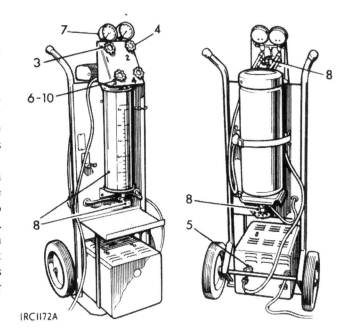

IRC1172A

Evacuating

1. Depressurise the air conditioning system, 82.30.05.
2. Connect the charging and testing equipment as for evacuating, 82.30.01.
3. Open the low pressure valve (No. 1).
4. Open the high pressure valve (No. 2).
5. Start the vacuum pump and check that the vacuum pump valve is open.
6. Slowly open the vacuum control valve (No. 1). If vacuum is applied to the system too quickly the residual oil may be drawn out.
7. A vacuum of 711 mm (28 in.) Hg at or near sea level should be reached. Allow 25 mm (1 in.) Hg reduced vacuum for each 300 metres (1,000 feet) of elevation.
8. While the system is evacuating fill the charging cylinder as required:
   a. Ensure that the refrigerant drum valve is opened.
   b. Open valve at base of charging cylinder and fill cylinder with required amount of refrigerant, i.e. 0,25 to 0,45 kg (½ to 1 lb) if sweeping the system or 0,9 kg (2 lb) if charging the system. Liquid refrigerant will be observed rising in the sight glass.
   c. As refrigerant stops filling the cylinder, open the valve at top of cylinder behind control panel intermittently to relieve head pressure and allow refrigerant to continue filling the cylinder.
   d. When refrigerant reaches desired level in the sight glass, close both the valve at base of cylinder and valve at bottom of refrigerant tank. Be certain top cylinder valve is fully closed. If bubbling is present in sight glass, reopen the cylinder base valve momentarily to equalise drum and cylinder pressures.

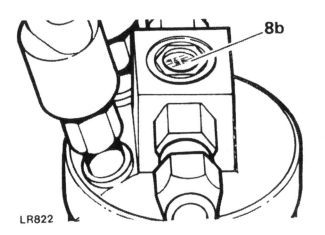

LR822

continued

9. If 711 mm (28 in.) Hg of vacuum cannot be obtained, close the vacuum control valve (No. 3), stop the vacuum pump and check the system for leaks.
10. Close the vacuum control valve (No. 3).
11. Close the vacuum pump valve, switch off the pump and allow the vacuum to hold for 15 minutes, then check that no pressure rise — loss or vacuum — is evident on the compound gauge. Any pressure rise denotes a leak that must be rectified before proceeding further. See leak detection, Operation 82.30.09. It is possible for residual liquid refrigerant in the compressor oil to vapourise and create a slight pressure rise. This can be eliminated by starting the engine and energising the magnetic clutch to rotate the compressor for about 30 seconds, then evacuate the system again.
12. With the system satisfactorily evacuated, the system is ready for sweeping or charging with refrigerant.

AIR CONDITIONING SYSTEM

— Sweep   82.30.07

NOTE: This operation is in addition to evacuating, and is to remove moisture from systems that have been open to atmosphere for a long period, or that are known to contain excessive moisture.

Sweeping

1. Fit a new liquid receiver/drier, 82.17.01.
2. Ensure that a full drum of refrigerant is fitted on the charging and testing equipment.
3. Fit the charging and testing equipment, 82.30.01, as described for evacuating.
4. Evacuate the air conditioning system, 82.30.06, allowing 0,25 to 0,45 kg (½ to 1 lb) of refrigerant to enter the charging cylinder.
5. Close all valves on the charging and testing equipment.
6. Disconnect the intake hose from the vacuum pump.

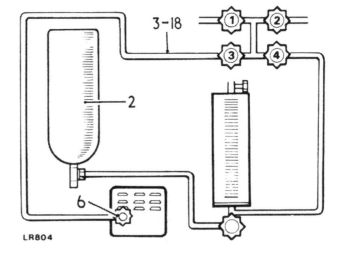

LR804

continued

7. Connect the intake hose to the valve at the top of the charging cylinder.

8. Open the valve at the top of the charging cylinder.

9. Put on safety goggles.

10. Crack open the hose connection at valve No. 3 and allow some refrigerant to purge the hose, then close the connection.

11. Open the high pressure valve (No. 2).

12. Slowly open valve No. 3, which is now connected to the top valve of the charging cylinder, and allow gas to flow into the system until the reading on the compound gauge remains steady. Between 0,25 and 0,45 kg (½ to 1 lb) of refrigerant will enter the system.

13. Allow the dry refrigerant introduced into the system to remain for 10 minutes.

14. Crack the suction valve charging line at the connection on the compressor to allow an escape of refrigerant, at the same time observing the sight glass in the charging cylinder. A slight drop in the level should be allowed before closing the connection at the compressor.

15. Close the high pressure valve (No. 2).

16. Close valve No. 3.

17. Close the valve at the top of the charging cylinder.

18. Reconnect the charging and testing equipment, 82.30.01, as described for evacuating.

19. Evacuate the air conditioning system, 82.30.06. Maintain the vacuum for twenty minutes.

20. The air conditioning system is now ready for charging with refrigerant, 82.30.08.

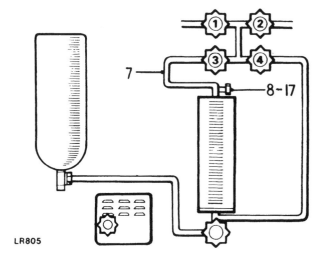

LR805

AIR CONDITIONING SYSTEM

– Charge 82.30.08

CAUTION: Do not charge liquid refrigerant into the compressor. Liquid cannot be compressed, and if liquid refrigerant enters the compressor inlet valve severe damage is possible. In addition, the oil charge may be absorbed, with consequent damage when the compressor is operated.

NOTE: NOMINAL CHARGE WEIGHT:
R.H.D. VEHICLES 1.25 kg (44 oz)
L.H.D. VEHICLES 1.08 kg (38 oz)

Charging

1. Fit the charging and testing equipment, 82.30.01, as described for evacuating.
2. Evacuate the air conditoning system, 82.30.06, allowing 1.37 kg (3 lb) of refrigerant to enter the charging cylinder.
3. Put on safety goggles.
4. Close the low pressure valve (No. 1).
5. Open the refrigerant control valve (No. 4) and release liquid refrigerant into the system through the compressor discharge valve. The pressure in the system will eventually balance.
6. If the full charge of liquid refrigerant will not enter the system, proceed with items 7 to 12.
7. Reconnect the charging and testing equipment as described for charging with gaseous refrigerant, 82.30.01.
8. Open the low pressure valve (No. 1).
9. Open valve No. 3.
10. Close the high pressure valve (No. 2).
11. Start and run the engine at 1000 to 1500 rev/min and allow refrigerant to be drawn through the low pressure valve (No. 1) until the full charge has been drawn into the system.
12. Close valve number 1 and 3.
13. Close valve No. 4.
14. Check that the air conditioning system is operating satisfactorily by carrying out a pressure test, 82.30.10.

CAUTION: Do not overcharge the air conditioning system as this will cause excessive head pressure.

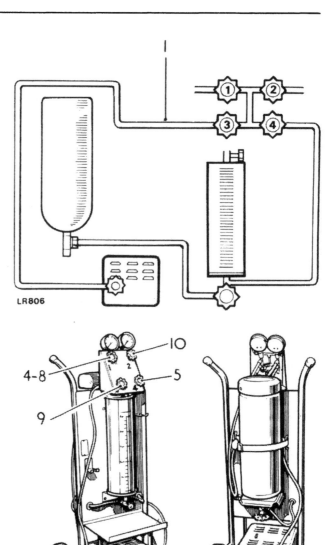

LR806

2RC807

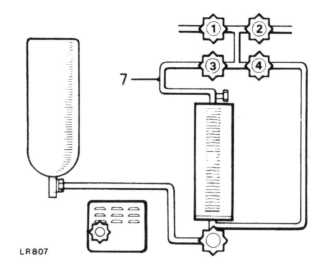

LR807

AIR CONDITIONING SYSTEM

– Leak test   82.30.09

NOTE: The following instructions employ an electronic type leak detector which is the safest, most sensitive and widely used.

1. Place the car in a well ventilated area but free from draughts, since a seepage from the system could be dissipated without detection.
2. Follow the instructions issued by the manufacturer of the particular leak detector being used. Some detectors have visual and audible indicators.
3. Commence searching for leaks by passing the detector probe round all joints and components, particularly on the underside, as the refrigerant gas is heavier than air.
4. Insert the probe into an air outlet of the evaporator. Switch the air conditioning blower on and off at intervals of ten seconds. Any leaking refrigerant will be gathered in by the blower and detected.
5. Insert the probe between the magnetic clutch and compressor to check the shaft seal for leaks.
6. Check all service valve connections, valve plate, head and base plate joints and back seal plate.
7. Check the condenser for leaks at the pipe connections.
8. If any leaks are found, the system must be depressurised before attempting rectification. If repairs by brazing are necessary the component must be removed from the vehicle and all traces of refrigerant expelled before heat is applied.
9. After repairs check the system again for leaks and evacuate prior to charging.

AIR CONDITIONING SYSTEM

– Pressure test   82.30.10

Procedure

1. Fit the charging and testing equipment, 82.30.01.
2. Start the engine.
3. Run the engine at 1000 to 1200 rev/min with the air conditioner air flow control at high speed and the temperature control at normal cooling operating position. (Half of cooling movement.)
4. Note the ambient air temperature in the immediate test area in front of the car, and check the high pressure gauge reading – discharge side – against the following table.

Ambient Temperature		Compound Gauge Readings		High Pressure Gauge Readings	
°C	°F	kgf/cm^2	lbf/in^2	kgf/cm^2	lbf/in^2
16	60	1,05-1,4	15-20	7,0-10,2	100-150
26,7	80	1,4-1,75	20-25	9,8-13,3	140-190
38	100	1,75-2,1	25-30	11,6-15,8	180-225
43,5	110	2,1-2,45	30-35	15,1-17,5	215-250

The pressure gauge readings will vary within the range quoted with the rate of flow of air over the condenser, the higher readings resulting from a low air flow. It is advisable to place a fan for additional air flow over the condenser if the system is to be operated for a long time. Always use a fan if temperatures are over 26.7°C (80°F) so that consistent analysis can be made of readings.

5. If the pressure readings are outside the limits quoted, refer to the fault diagnosis chart at the beginning of this Division.
6. Stop the engine.
7. Close all valves on the charging and testing equipment.
8. Disconnect the charging lines from the compressor.
9. Refit the blanking caps to the compressor valve stems and gauge connections, and to the charging lines.
10. Close the bonnet.

AIR CONDITIONING EQUIPMENT

– Test   82.30.16

Procedure

1. Place the car in a ventilated, shaded area free from excessive draught, with the car doors and windows open.
2. Check that the surface of the condenser is not restricted with dirt, leaves, flies, etc. Clean as necessary.
3. Switch on the ignition and the air conditioner air flow control. Check that the blower is operating efficiently at all speeds. Switch off the blower and the ignition.
4. Check that the evaporator condensate drain tubes are open and clear.
5. Check the tension of the compressor driving belt, and adjust if necessary, 82.10.01.
6. Inspect all connections for the presence of refrigerant oil. If oil is evident, check for leaks, 82.30.09, and rectify as necessary.

NOTE: The compressor oil is soluble in Refrigerant 12 and is deposited when the refrigerant evaporates from a leak.

7. Start the engine.
8. Set the temperature control switch to maximum cooling and switch the air conditioner blower control on and off several times, checking that the magnetic clutch on the compressor engages and releases each time.
9. With the temperature control at maximum cooling and the blower control at high speed, warm up the engine and fast idle at 1000 rev/min. Check the sight glass in the top of the receiver drier for bubbles or foam. The sight glass should be generally clear after five minutes running, occasional bubbles being acceptable. Continuous bubbles may appear in serviceable system on a cool day, or if there is insufficient air flow over the condenser at a high ambient temperature.
10. Repeat at 1800 rev/min.
11. Gradually increase the engine speed to the high range, and check the sight glass at intervals.
12. Check for frosting on the service valves and evaporator fins.
13. Check the high pressure pipes and connections by hand for varying temperature. Low temperature indicates a restriction or blockage at that point.
14. Switch off the air conditioning blower and stop the engine.
15. If the air conditioning equipment is still not satisfactory, proceed with the pressure test, 82.30.10.

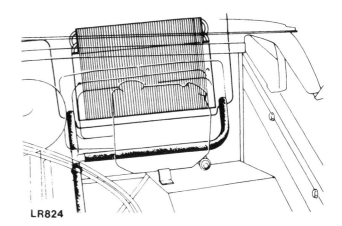

LR824

This page intentionally left blank

WIPERS AND WASHERS OPERATIONS

N571/B

Headlamp Wipers and Washers

For certain markets and as an optional extra, a headlamp wiper and washer facility can be fitted.

A separate switch actuated the headlamp wash and wipe on earlier cars. On later vehicles there are no extra controls, the headlamp wipers and washers being brought into action when the headlamps are illuminated and the windscreen wiper and washer controls are used.

Washer Reservoir

On later models a combined reservoir supplies the windscreen, tailgate glass and headlamp washers, three separate supply pipes being used with individual electric pumps for each facility.

Tailgate Glass Wiper

The wiper motor is mounted at the rear of the car and is controlled by a dashboard mounted switch combining with the control for the tailgate glass washer facility.

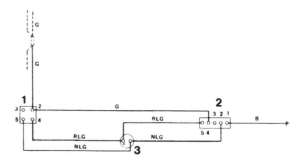

Circuit diagram – Headlamp Wipers and Washers – Earlier vehicles.

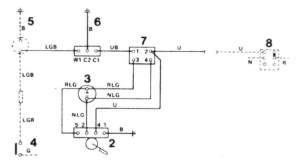

Circuit diagram – Headlamp Wipers and Washers – Later vehicles.

Key.

1. Headlamp Wiper/Washer Switch.
2. Headlamp Wiper Motor.
3. Headlamp Washer Pump.
4. Windscreen Washer Switch.
5. Windscreen Washer Pump.
6. Headlamp Wiper Relay.
7. Headlamp Wiper Delay Unit.
8. Vehicle Lighting Switch.

WASHER RESERVOIR

Earlier type windscreen washer

-Remove and refit.   84.10.01

Removing.

1. Disconnect the tubing from the reservoir.
2. Disconnect the electrical leads from the pump.
3. Slide the reservoir upwards out of its retaining bracket.

Refitting.

4. Reverse instructions 1 to 3.

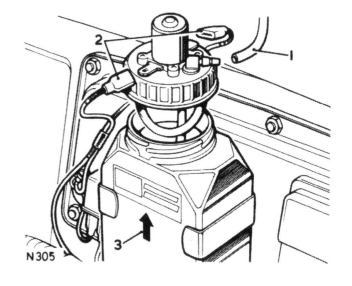

Earlier type headlamp washer

-Remove and refit.   84.10.01

Removing.

1. Remove the reservoir cap.
2. Lift the reservoir from the bracket.
3. Remove two screws to release the bracket.

Refitting.

4. Reverse instructions 1 to 3.

Earlier type tailgate glass washer.

-Remove and refit.   84.10.01

Removing.

1. Remove the right hand rear quarter trim panel. 76.10.23.
2. Remove the reservoir cap, wiping away any surplus water to avoid damaging the trim.
3. Lift the reservoir from the bracket.
4. Remove two screws to release the bracket.

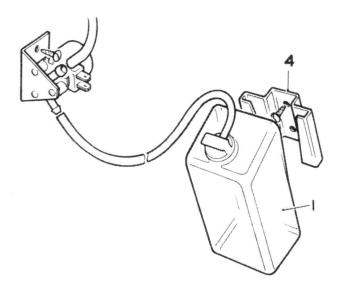

Refitting.

5. Reverse instructions 1 to 4.

Later type combined washer reservoir.

-Remove and refit   84.10.01

Removing.

1. Remove the two reservoir caps.
2. Lift the reservoir from the bracket.
3. Remove one bolt, nut and washers.
4. Drill out three rivets to release the bracket.

Refitting.

5. Reverse instructions 1 to 4.

WINDSCREEN WASHER JETS

-Remove and refit   84.10.09

Removing

1. Disconnect tubes from the jets.
2. Remove the locknut and washer securing jet to bonnet.
3. Remove jet.

Refitting

4. Reverse instructions 1 to 3.

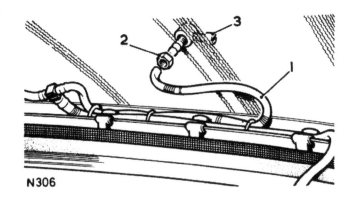

N306

WINDSCREEN WASHER TUBES

-Remove and refit   84.10.15

Removing

1. Disconnect tubing from reservoir pump.
2. Disconnect tubing from washer jets.
3. Disconnect tubing from three-way tee piece.
4. Release tubing from edge clips.

Refitting

5. Reverse the removal procedure.

NOTE. On later models a non return valve was fitted between the washer motors and jets. Ensure that this valve is correctly fitted when replacing the washer tubes.

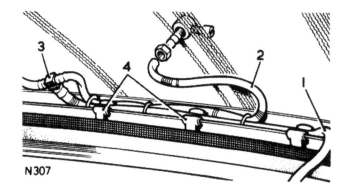

N307

WINDSCREEN WASHER PUMP

-Remove and refit   84.10.21

Early models

Removing

1. Disconnect tubing from reservoir pump.
2. Disconnect electrical leads from reservoir pump.
3. Unscrew pump unit from reservoir.

Refitting

4. Reverse the removal procedure.

Later Models.

A motor/pump unit separate to the reservoir was fitted, and was secured to the body by two screws.

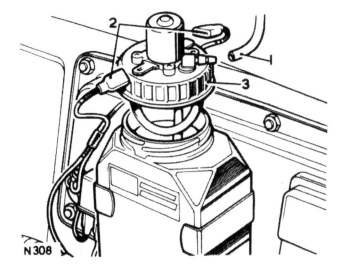

N308

WIPER ARMS

-Remove and refit   84.15.02

Removing

1. Hold back the small spring clip which retains the wiper arm on the spindle boss, by means of a suitable tool.
2. Gently prise off the wiper arm from the spindle boss.

Refitting.

3. Allow the motor to move to the 'park' position.
4. Push the arm on to the boss, locating it on the splines so that the wiper blades are just clear of the screen rail.
5. Ensure that the spring retaining clip is located in the retaining groove on the spindle boss.

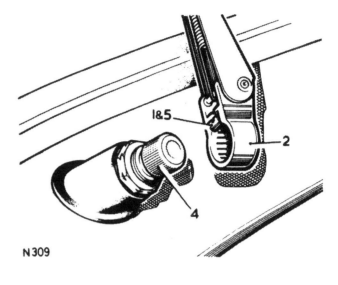

N 309

WIPER BLADES

-Remove and refit   84.15.06

Removing

1. Pull the wiper arm away from the glass.
2. Lift the spring clip and withdraw the blade from the arm.

Refitting.

3. Reverse instructions 1 and 2.

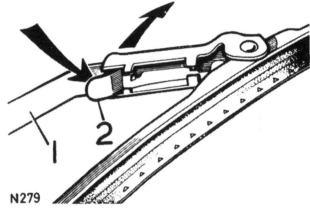

N279

WINDSCREEN WIPER MOTOR, LINKAGE AND WHEEL BOXES

-Remove and refit   84.15.10

Removing

1. Remove wiper arms 84.15.01.
2. Remove locknuts from wheel boxes.
3. Remove grommet from wheel boxes.
4. Remove bonnet 76.16.01.
5. Remove front decker panel 76.10.35.
6. Remove the spring clips securing the primary links to the wheelbox spindle links.
7. Remove the spring clips securing the primary links to the motor crank.
8. Remove the primary links.
9. Remove the lower grommet from the wheelboxes.
10. Remove the screws securing the motor and linkage assembly to the bulkhead.
11. Gently ease the unit out of its mounting aperture and disconnect the electrical leads at the plug and socket.
12. Withdraw the unit.
13. Remove the three bolts securing the motor to the mounting plate.
14. Separate the motor from the mounting plate by pulling the motor crank through the grommet.

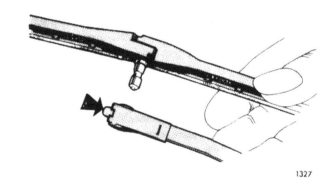

1327

continued

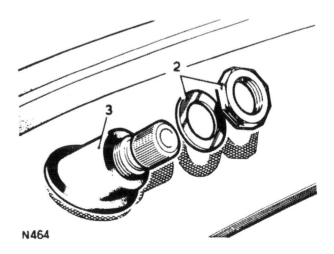

N464

Refitting.

15. Reverse the removal procedure.
16. When replacing the primary links ensure that they are mounted with the bushes on the inside, that is, towards the wiper motor. The shorter primary link is mounted on the drivers' side.
17. Replace the front decker panel 76.10.35.
18. Replace the bonnet 76.16.01.
19. Replace the wiper arms 84.15.01.

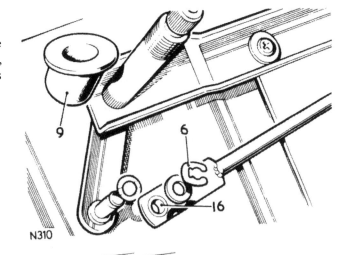

N310

WIPER MOTOR

Overhaul   84.15.18

Checking.

If unsatisfactory operation of the wiping equipment is experienced, a systematic check to determine the origin of the fault should be carried out as follows:

1. Check the blades for signs of excessive friction. Frictional blades will greatly reduce the wiping speed of the motor and cause increased current draw which may damage the armature. Check by substitution.
2. Check the motor light running current and speed with the motor coupling link disconnected from the wiper spindle transmission linkage. Connect a first grade moving coil ammeter in series with the motor supply and measure the current consumption when the motor is switched on. Check the operating speed by timing the speed of rotation of the motor coupling link. The results should compare with the figures given in the technical data.

NOTE. If the vehicle wiring connections are disconnected and an alternative supply source is applied it is essential that the correct polarity is observed. The Positive supply lead must be connected to the red/green cable and the Negative supply lead must be connected to the green cable. Failure to observe this will cause the motor to rotate in the reverse direction, which may result in the limit switch contacts being damaged.

3. If the motor does not run satisfactorily or takes higher than normal current, then a fault is apparent and should be investigated.
4. If the current consumption and speed of the motor are satisfactory, then a check should be carried out for proper functioning of the transmission linkage and wiper arm spindles.

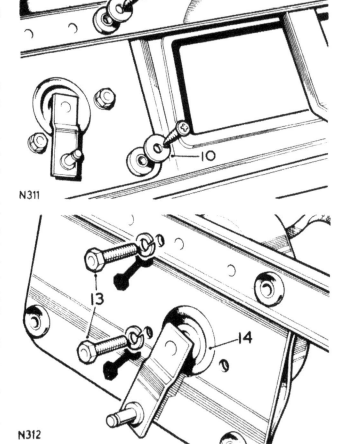

N311

N312

Continued

84-6

Dismantling.

Remove the wiper motor – 84.15.10.

5. Mark the gearbox cover adjacent to the arrow head on the limit switch cover. This will allow the original setting of the limit switch to be determined on reassembly.

6. Remove the two yoke fixing through bolts and spring washers.

7. Withdraw the yoke assembly from the gearbox.

8. Withdraw the armature.

NOTE. Ensure that the working area is clean. Protect the inside of the yoke from foreign matter that would normally be attracted by the exposed field magnets.

9. Remove the brushgear fixing screws.

10. Remove the gearbox cover fixing screws.

11. Remove the limit switch complete with connecting cables and brushgear plate.

12. Remove the main gearwheel lock-nut.

13. Remove the gearwheel and driving plate.

14. Withdraw the shaft and link assembly from underneath the gearbox.

15. Remove the dished washer from the main gearshaft.

16. Remove the intermediate gearwheel from its pivot pin.

continued

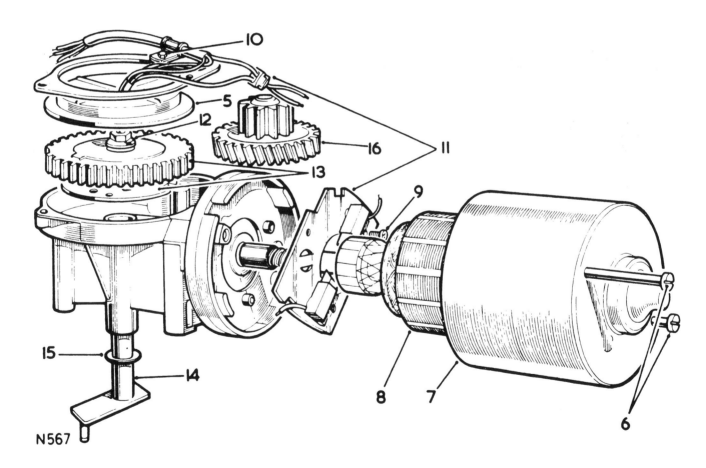

N567

Bench testing.

17. Examine all parts for signs of damage or wear.
18. Check that brush spring pressures and brush lengths are in accordance with the test data. Use push type spring gauge.
19. Test the insulation of the armature windings with a 15W mains test lamp. Lighting of the lamps indicates faulty insulation.
20. Check the armature windings for open and short circuits, using armature testing equipment.
21. Inspect the limit switch contact arms and ensure that they are firmly rivetted.
22. To ensure correct pressure on the slip ring the length of the contact arms from the base to the contact point should be approximately 7,143 mm (0.281 in).

N569

Reassembly.

23. Apply Molybdenum Di-sulphide oil to the intermediate gearwheel pivot pin.
25. Fit the dished washer to the main gearshaft, ensuring that the concave side (larger diameter) is towards the gearbox casting.
26. Refit the shaft and link assembly, lubricating with Shell Turbo 41 oil.
27. Apply Ragosine Listate grease to the gearwheel teeth, and refit the gearwheel and driving plate.
28. Re-fit the final gearwheel with the slip ring outer edge segment pointing in the same direction as the external coupling link. This is essential if correct parking is to be obtained.
29. Fit the final gearwheel fixing nut to a torque of 0,91 to 1,03 kgf.m (80-90 lbf. in.).
30. Refit the limit switch, connecting cables and brush-gear plate.
31. Refit the gearbox cover fixing screws.
32. Refit the brushgear fixing screws.
33. Replace the armature and yoke.

NOTE. If a replacement armature is to be fitted, first slacken the thrust screw in the gearbox before tightening the through bolts.

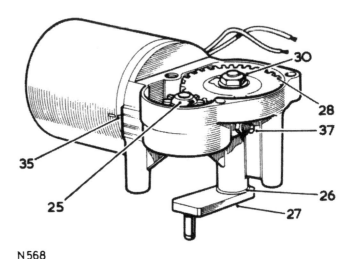

N568

continued

34. Ensure that the arrow head on the gearbox corrsponds with the mark on the yoke assembly.
35. Tighten yoke fixing through bolts to the specified torque.
36. Check the armature end float after re-assembly. Slacken the end float adjuster lock nut and check that the air-gap is within the limits 0,05 to 0,2 mm (0.002 to 0.008 in).
37. Tighten lock nut and recheck.
38. Refit the wiper motor — 84.15.10.

Technical test data.

Typical light running current (i.e. with link assembly disconnected) after 60 secs. from cold	1.2 amps. max.
Armature end plug (after tightening through bolts)	0,05 to 0,2 mm (0,002 to 0.008 in)
Yoke through bolt torque	0,138 to 0,184 kgf.m (12 to 16 lb. in)
Gearbox spacing ring torque	0,115 kgf.m (10 lbf. ft.).
Brush spring pressure with brush at bottom of slot in the brush box	150 to 210 gf (5 to 7 ozf)
Minimum brush length	4,8 mm (0.187 in)

HEADLAMP WASHER

JET

-Remove and refit   84.20.09

Removing

1. Release the headlamp and frame from the body. 86.40.02.
2. Withdraw the headlamp frame sufficient for access and remove the washer tube from the jet.
3. Remove the nut and washer and withdraw the jet from the headlamp frame.

Refitting.

4. Reverse instructions 1 to 3.

TUBE

-Remove and refit   84.20.15

Removing

1. Trace the run of the tube from the reservoir/motor pump to the jet.
2. Disconnect both ends of the tube, allowing any water in the tube to drain away.
3. Remove the tube.

Refitting.

4. Reverse instructions 1 to 3, noting that if a one way valve has been disturbed (later models) it must be fitted for water flow towards the jets.

PUMP

-Remove and refit   84.20.21

Early models

The pump was incorporated in the reservoir cap, refer to 84.10.01.

Later Models

A motor/pump unit separate to the reservoir was fitted, and was secured to the body by two screws.

SWITCH (Early models only).

-Remove and refit   84.20.27

Removing.

1. Depress the lugs at the back of the switch and push it forward from the panel.
2. Withdraw the switch sufficiently to remove the electrical connections noting their position for re-fitting.
3. Remove the switch.

Refitting.

4. Reverse instructions 1 to 3.

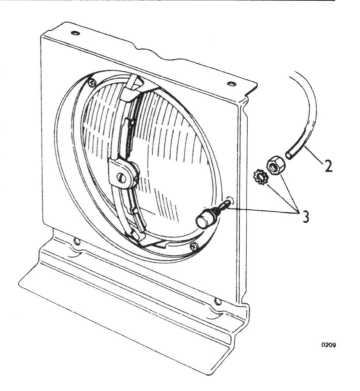

0209

84—10

HEADLAMP WIPER

ARM

-Remove and refit   84.25.02

Removing.

1. Using finger pressure to prevent the arm from turning, release the screw.
2. Ease the centre frame away from the headlamp and withdraw the wiper arm and blades.

Refitting.

3. Reverse instructions 1 and 2, ensuring that the arm is aligned with the centre frame.

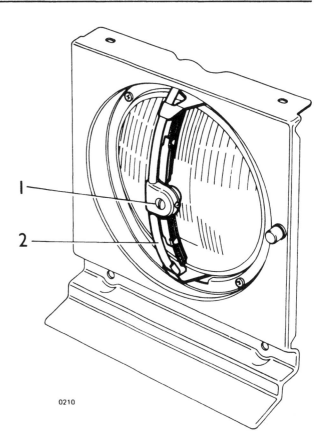

0210

BLADE

-Remove and refit   84.25.06

The wiper arm and blade are serviced as one unit. Refer to 84.25.02.

MOTOR

-Remove and refit   84.25.12

Removing

1. Disconnect the electrical leads to the motor.
2. Disconnect the flexible rack tubes from the motor.
3. Remove two nuts and washers, remove the clamp and support the wiper motor.
4. Using a second operator to gently turn the wiper blades at the headlamps, disengage the wiper racks.

NOTE. It will assist refitting the wiper rack if the blades are left in the position at which the rack disengaged.

5. Withdraw the racks from the flexible tubes and remove the wiper motor from the car.

Refitting.

6. Reverse instructions 1 to 5, ensuring that the head-lamp wiper blades are aligned with the centre frame when the rack has been fully engaged.

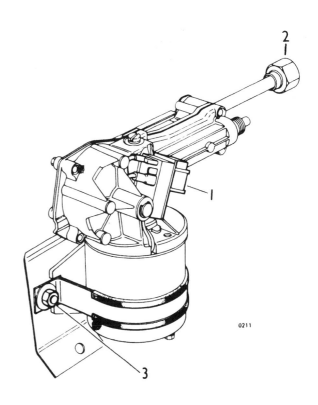

0211

HEADLAMP WIPER MOTOR

-Overhaul   84.25.18

1. Remove the wiper motor complete with racks 84.25.12.
2. Remove the racks from the motor 84.25.24.
3. Dismantle and overhaul the motor following the instructions in operation 84.35.18.

RACK

-Remove and refit   84.25.24

Removing

1. Remove the wiper motor complete with racks 84.25.12.
2. Remove six screws and lift clear the rack cover and plate.
3. Remove the circlip and washer to disconnect the racks from the driving gear.
4. Withdraw the racks from the motor.

Refitting

5. Reverse instructions 1 to 4.

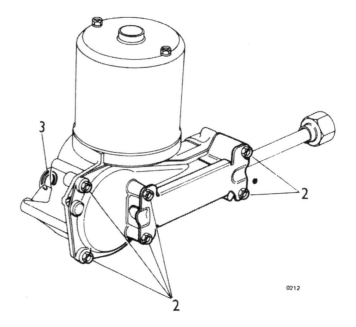

0212

HEADLAMP WIPER

WHEELBOX AND RACK TUBE

-Remove and refit   84.25.27

Removing

1. Remove the wiper motor complete with racks. 84.25.12.
2. Remove the headlamp and headlamp frame, carefully unclipping the wiper rack flexible tube and withdrawing it through the body grommets. Separate the headlamp rim from the headlamp 86.40.02.
3. The headlamp rim and flexible rack tube are serviced as one unit.

Refitting.

4. Reverse instructions 1 and 2.

TAILGATE GLASS WASHER

JET

-Remove and refit   84.30.09

Jet only – Instructions 1,2,6 and 7.
Jet and sleeve – Instructions 1 to 7 inclusive.

Removing.

1. Lower or remove the headlining rear section to gain access to the jet securing nut and washer tube connection. 76.64.05.

Early models

2. Disconnect the washer tube from the reservoir and drain away any water in the tube to avoid damaging the trim.
3. Hold the jet sleeve and remove the nut.
4. Remove the nut and distance piece from the end of the tube.
5. Withdraw the jet and sleeve from outside the car.
6. Unscrew the jet from the sleeve and disconnect the washer tube.

Refitting.

7. Reverse instructions 1 to 6 as apporpriate.

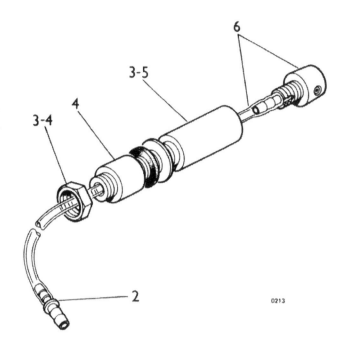

TAILGATE GLASS WASHER TUBE

-Remove and refit   84.30.15

Early models – washer motor located behind the right hand rear quarter trim panel.

Removing

1. Remove the right hand side rear quarter trim panel. 70.10.23.
2. Disconnect the washer tube from the motor and connector.
3. Remove the tube.

Refitting.

4. Reverse instructions 1 to 3.

-Remove and refit   84.30.15

Later models – washer motor located in the engine compartment.

Removing

1. Remove the rear headlining 76.64.05.
2. Disconnect the washer tube from the motor and connector.
3. Secure a length of strong cord to the rear end of the tube. On withdrawing the tube, the cord will follow the route of the tube through the body trim.
4. Remove the tube by feeding it in stages towards the engine compartment.

Refitting.

5. Reverse instructions 1 to 4 securing the cord to the tube and using it to draw the tube through the body trim.

TAILGATE GLASS WASHER PUMP

-Remove and refit   84.30.21

Early models – washer motor located behind the right hand rear quarter trim panel.

Removing

1. Remove the right hand side rear quarter trim panel. 70.10.23.
2. Disconnect the washer tube from the motor/pump.
3. Remove two screws and withdraw the motor/pump.

Refitting.
4. Reverse instructions 1 to 3.

-Remove and refit   84.30.21

Later models – washer motor located in the engine compartment.

Removing.

1. Raise the bonnet and locate the motor/pump adjacent to the washer reservoir.
2. Disconnect the washer tube from the motor/pump.
3. Remove two screws and withdraw the motor/pump.

Refitting.
4. Reverse instructions 1 to 3.

TAILGATE GLASS WASHER SWITCH

-Remove and refit   84.30.27

Early models

Removing.

1. Insert a suitable probe into the hole beneath the switch knob to depress the button and release the knob from the switch.
2. Holding the switch body remove the front nut and washer.
3. Withdraw the switch from behind the panel.
4. Remove the electrical connections from the switch noting their position for refitting.

Refitting.
5. Reverse instructions 1 to 4.

-Remove and refit   84.30.27

Later models

Removing

1. Remove the lower fascia panel 76.46.03.
2. Depress the lugs at the back of the switch and push it forward from the panel.
3. Withdraw the switch sufficiently to remove the electrical connections noting their position for refitting.
4. Remove the switch.

Refitting.

5. Reverse instructions 1 to 4.

TAILGATE GLASS WIPER MOTOR

-Remove and refit   84.35.12

On earlier models the tailgate glass wiper motor was situated behind the right hand rear quarter panel as illustrated below. It was secured to the body by a two bolt fixing.
On later models the motor is to be found behind the left hand rear quarter panel. This motor is secured to the body by a split bracket with a bolt and captive nut.

Removing.

1. Remove the appropriate rear quarter panel. 76.10.22/23.
2. Lower or remove the headlining rear section to gain access to the wiper wheelbox. 76.64.05.
3. Remove the wiper arm and blade.
4. Remove one nut and spacer to release the wiper wheelbox from the body.
5. Disconnect the electrical leads from the wiper motor and the rack tube clips to the body.
6. Support the wiper motor and remove the bolt(s) and bracket(s).
7. Withdraw the wiper motor from the body, complete with rack tube and wheelbox.

8. Slacken two bolts and nuts on the wheelbox to release it from the rack.
9. Slide the wheelbox and end rack tube off the rack.
10. Unscrew the rack tube nut at the wheelbox and slide the tube off the rack.

Refitting.

11. Reverse instructions 1 to 10, ensuring that the wheelbox is correctly located on the rack before tightening the two wheelbox bolts and nuts.

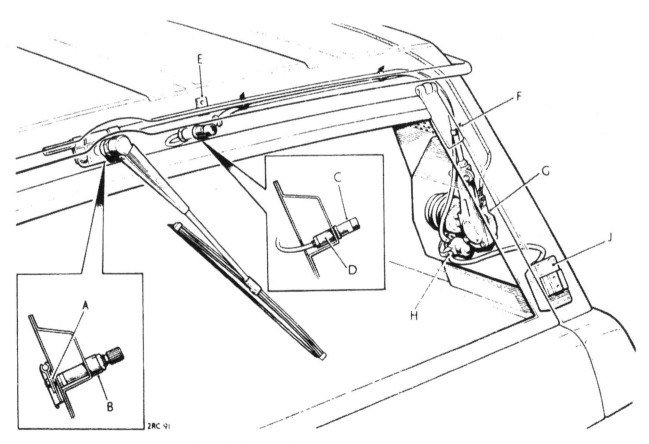

A.	Wiper Wheelbox.	F.	Wiper Rack.
B.	Nut and Spacer.	G.	Wiper Motor.
C.	Washer Jet.	H.	Washer Motor.
D.	Sleeve.	J.	Washer Bottle.
E.	Wiper Rack Clip.		

84-15

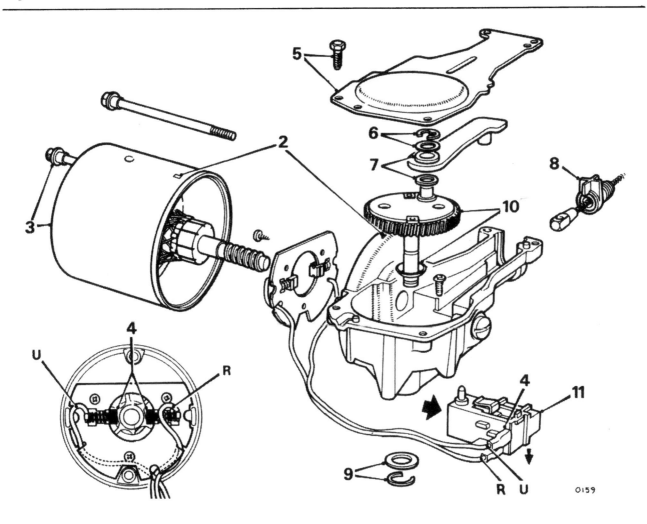

TAILGATE GLASS WIPER MOTOR

-Overhaul   84.35.18

Dismantling

NOTE: To change brushes only, follow procedures 1 to 4.

1. Remove the wiper motor and drive. 84.35.12.
2. Note the alignment marks on the yoke and gearbox. To ensure correct rotation of the motor, the mark on the yoke must be adjacent to the arrow-head marked on the gearbox case.
3. Unscrew the through-bolts and remove the yoke and armature assembly.
    CAUTION: The yoke must be kept clear of metallic particles which will be attracted to the pole-piece.
4. Note the colour and position of the leads. Withdraw the brushes from the insulating plate and disconnect the leads from the switch assembly.
5. Unscrew the four gearbox cover retaining screws and remove the cover.
6. Remove the circlip and flat washer securing the connecting rod to the crankpin.

7. Withdraw the connecting rod and the flat washer fitted under it.
8. Withdraw the cable rack with cross-head and outer casing ferrule.
9. Remove the circlip and washer securing the shaft and gear.
10. Clean any burrs from the gear shaft and withdraw the gear, taking care not to lose the dished washer fitted under it.
11. Switch assembly: Pull outwards and down to release the retaining clip.

Inspection

12. Examine the brushes. If the brushes are worn to 0.19 in (4.8 mm) they must be renewed. Renew the brush gear assembly if the springs are not satisfactory.

13. Test the armature for insulation and open- or short-circuits. Use a test lamp (110 volts, 15 watts). Renew the armature if faulty.

14. Examine the gear wheel for damage or excessive wear. Renew if necessary.

Reassembling

15. Reverse the procedure in 1 to 11.

16. Use Ragosine Listate Grease to lubricate the gear wheel teeth, armature shaft worm gear, connecting rod and pin, cross-head slide, cable rack, and wheelbox gear wheels.

17. Use Shell Turbo 41 oil to lubricate the bearing bushes, armature shaft bearing journals (sparingly), gear wheel shaft and crankpin, felt washer in the yoke bearing (thoroughly soak), and the wheelbox spindles.

18. Tighten the yoke fixing bolts to 14 lbf in (0.16 kgf m).

19. If a replacement armature is being fitted, slacken the thrust screw to provide end-float for fitting the yoke.

20. Fit the dished washer beneath the gear wheel with its concave side towards the gear wheel.

21. When fitting the gear wheel, ensure that the relationship of the crankpin and ramp is correct for the parking condition required. The park position of the motor is marked on the gearbox cover plate by the word 'PARK' and is immediately followed by an arrow pointing towards the word or away from it.

    a PARK   parking with cable rack retracted the cam should be fitted opposite the crankpin.

    b. PARK   parking with cable rack extended the cam should be fitted adjacent to the crankpin.

22. Ensure that the larger of the two washers is fitted to the crankpin beneath the connecting rod.

23. Armature end-float: Hold the yoke vertical with the adjuster screw uppermost. Carefully screw in the adjuster until resistance is felt, and screw back for a quarter-turn. This will give the required end-float.

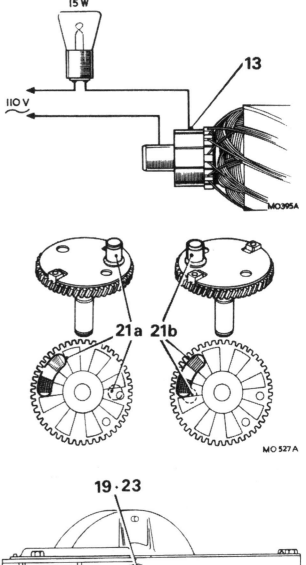

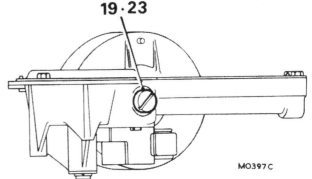

DATA

Wiper motor

Type	Lucas 14WA
Running current (rack disconnected)	1.5 A
Wiper speed (after 60 seconds)	45 to 52 rev/min
Armature end-float	0.002 to 0.008 in (0.05 to 0.2 mm)
Brush spring tension	5 to 7 ozf (140 to 200 gf)
Minimum brush length	0.18 in (4.7 mm)
Maximum pull to move rack in tubes	6 lbf (2.7 kgf)

TAILGATE GLASS WIPER RACK

-Remove and refit   84.35.24

Removing.

1. Remove the wiper motor complete with rack and wheelbox, Instructions 1 to 7, 84.35.12.
2. Remove the rack tube and wheelbox from the wiper motor, Instructions 8 to 10, 84.35.12.
3. Remove the wiper motor gearbox cover and withdraw the rack, Instructions 5 to 8, 84.35.18.

Refitting.
4. Reverse instructions 1 to 3.

TAILGATE GLASS WIPER WHEELBOX

-Remove and refit   84.35.28

Removing.

1. Remove the wiper motor complete with rack and wheelbox, Instructions 1 to 7, 84.35.12.
2. Slacken two bolts and nuts on the wheelbox to release it from the rack.
3. Slide the wheelbox and end rack tube off the rack.

Refitting.

4. Reverse instructions 1 to 3, ensuring that the wheelbox is correctly located on the rack before tightening the wheelbox bolts and nuts.

ELECTRICAL OPERATIONS

Description	Operation No.
Alternator	
— remove and refit	86.10.02
— overhaul	86.10.08
— test	86.10.14
Battery – remove and refit	86.15.11
Bulb chart	86.00.09
Cigar lighter – remove and refit	86.65.60
Distributor	
— remove and refit	86.35.20
- overhaul	86.35.26
Flasher unit — remove and refit	86.55.11
Fuel pump	*Refer to Section 19*
Fuse box — remove and refit	86.70.01
Hazard flasher unit — remove and refit	86.55.12
Horns — remove and refit	86.30.09
Ignition coil — remove and refit	86.35.32
Inspection sockets — remove and refit	86.45.33
Lamps -- remove and refit	
— differential lock illumination	86.45.42
— headlamp assembly	86.40.02
— number plate illumination	86.40.86
— panel illumination	86.45.32
— reverse	86.40.91
— roof	86.45.02
– side, front and flasher assembly.	86.40.42
— tail, stop and flasher assembly, LH	86.40.70
— tail, stop and flasher assembly, RH	86.40.71
— warning light cluster	86.45.62
Reflectors — remove and refit	86.40.65
Spark plugs — remove, clean, adjust and refit	86.35.01
Starter motor	
— remove and refit	86.60.01
— overhaul	86.60.13
— relay — remove and refit	86.55.05
— solenoid — remove and refit	86.60.08
Switches — remove and refit	
– brake failure	*Refer to 70.15.36*
— choke warning light	86.65.53
— combined direction indicator/headlight/horn	86.65.55
– differential lock warning	*Refer to 37.29.19*
— door pillar	86.65.15
— handbrake warning light	86.65.45
— hazard warning	86.65.50
– heated rear window	86.65.36
— ignition/starter	86.65.02
— lighting	86.65.10
— panel light	86.65.12
— reverse light	86.65.20
— roof lamp	86.65.21
— spot/fog light	86.65.19
— stop light	86.65.51
— windscreen wiper/washer	86.65.38

This page intentionally left blank

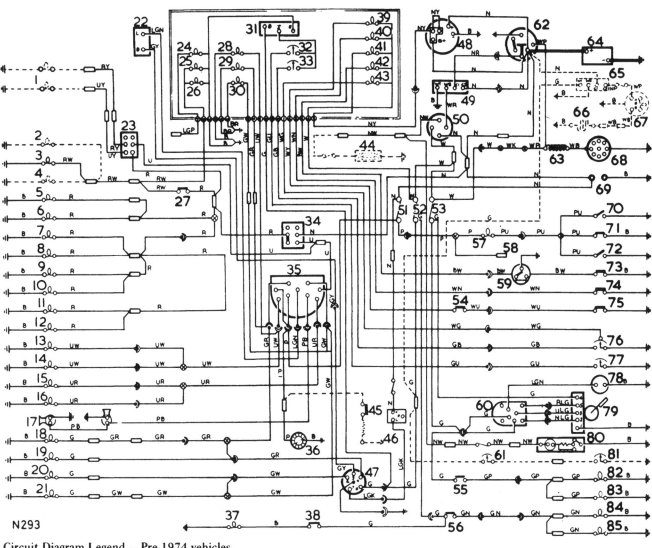

N293

Circuit Diagram Legend – Pre 1974 vehicles

1 Auxiliary driving lamps
2 Cigar illumination
3 Clock illumination
4 Auxiliary instrument illumination feed
5 L.H. front side lamp
6 R.H. front side lamp
7 L.H. rear marker lamp (NADA only)
8 L.H. rear tail lamp
9 Number plate illumination lamp
10 Number plate illumination lamp
11 R.H. rear tail lamp
12 R.H. rear marker lamp (NADA only)
13 R.H. headlamp, main
14 L.H. headlamp, main
15 R.H. headlamp, dip
16 L.H. headlamp, dip
17 Horns
18 L.H. rear indicator
19 L.H. front indicator
20 R.H. front indicator
21 R.H. rear indicator
22 Indicator unit
23 Auxiliary driving lamp switch
24 Trailer warning light
25 Instrument illumination
26 Instrument illumination
27 Panel light switch
28 Main beam warning light
29 L.H. indicator warning light

30 R.H. indicator warning light
31 Voltage stabiliser
32 Water temperature gauge
33 Fuel gauge
34 Lighting switch
35 Indicator, headlamp dip & horn switch
36 Clock
37 Differential lock warning light
38 Differential lock warning light switch
39 Choke warning light switch
40 Oil pressure warning light
41 Ignition warning light
42 Brake warning light
43 Fuel level warning light
44 Radio
45 Cigar lighter
46 Hazard warning unit
47 Hazard warning switch
48 Alternator, type 16 ACR
49 Starting relay
50 Ignition switch and steering lock
51 Fuse (1)
52 Fuse (2)
53 Fuse (3)
54 Choke switch
55 Stop lamp switch
56 Reverse light switch
57 Interior light

58 Trailer socket connection
59 Shuttle valve
60 Windscreen wiper and washer switch
61 Oil pressure gauge
62 Pre-engaged starter, type M45
63 Coil
64 Battery, type C9
65 Relay
66 Heated rear screen
67 Heated rear screen switch
68 Distributor
69 Inspection sockets
70 Courtesy light switch
71 Interior light switch
72 Courtesy light switch
73 Handbrake switch
74 Oil pressure switch
75 Choke thermostat
76 Fuel gauge unit
77 Water temperature transmitter
78 Screen washer motor
79 Windscreen wiper motor, type 17W 2-speed
80 Heater motor
81 Oil pressure transmitter
82 Stop light, R.H.
83 Stop light, L.H.
84 Reversing light, R.H.
85 Reversing light, L.H.

Symbol	Description
▭	Snap Connectors
─))─	Connections via Plug & Socket
─○─	Permanent In-line Connections
·‖─	Earth Connections via Fixing Bolts
·‖─B─	Earth Connections via Cables
○───○	Eureka Resistance Wire

CABLE COLOUR CODE

G	=	Green	U	=	Blue
P	=	Purple	N	=	Brown
L	=	Light	Y	=	Yellow
O	=	Orange	B	=	Black
S	=	Slate	R	=	Red
K	=	Pink	W	=	White

NOTE. All circuits shown dotted are additional extras. When cables have two colour code letters, the first denotes the main and the latter the tracer.

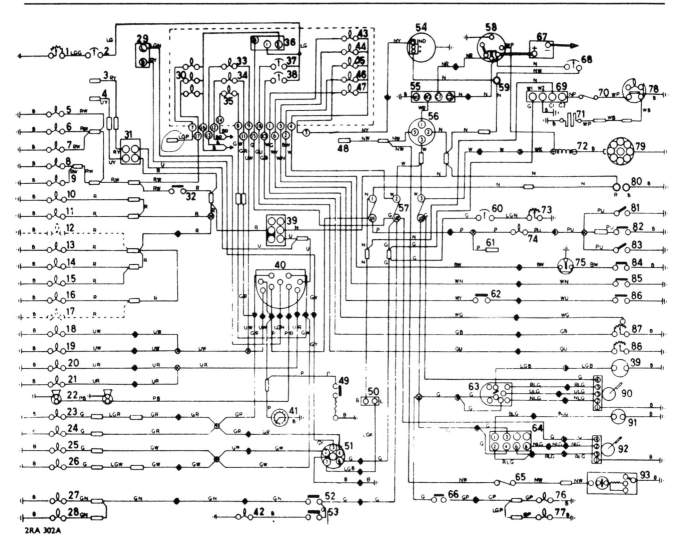

2RA 302A

Circuit Diagram Legend – Pre 1975 vehicles with auxiliary instruments

1 Oil temperature transmitter
2 Oil temperature gauge
3 Pick-up point for auxiliary driving lamps
4 Pick-up point for auxiliary driving lamps
5 Ammeter illumination
6 Oil temperature gauge illumination
7 Cigar lighter illumination
8 Oil pressure gauge illumination
9 Clock illumination
10 Side lamp, LH
11 Side lamp, RH
12 Side marker lamp, tail, LH, as applicable
13 Tail lamp, LH
14 Number plate illumination
15 Number plate illumination
16 Tail lamp, RH
17 Side marker lamp, tail, RH, as applicable
18 Headlamp main beam, RH
19 Headlamp main beam, LH
20 Headlamp dip, RH
21 Headlamp dip, LH
22 Horns
23 Indicator lamp, rear LH
24 Indicator lamp, front LH
25 Indicator lamp, front RH
26 Indicator lamp, rear RH
27 Reverse lamp
28 Reverse lamp
29 Indicator unit, 8 FL
30 Trailer illumination
31 Switch, auxiliary driving lamps

32 Switch panel lights
33 Warning light, headlamp main beam
34 Warning light, indicator, LH
35 Warning light, indicator, RH
36 Voltage stabiliser
37 Water temperature gauge
38 Fuel gauge
39 Switch, main light
40 Switch, headlamps, direction indicators & horn
41 Clock
42 Warning light, differential lock switch
43 Warning light cold start control
44 Warning light, oil pressure
45 Warning light, ignition
46 Warning light, relay
47 Warning light, fuel level
48 Pick-up point for radio
49 Cigar lighter
50 Hazard warning unit
51 Switch, hazard warning
52 Switch, reverse lights
53 Switch, differential lock
54 Alternator
55 Relay for starter motor
56 Switch ignition/starter
57 Fuses
58 Starter motor
59 Terminal post
60 Oil pressure gauge
61 Pick-up point for seven pin trailer socket
62 Switch, cold start

63 Switch, front wiper and washer
64 Switch, rear wiper and washer
65 In-line fuse for heater
66 Switch, stop lamps
67 Battery
68 Ammeter
69 Relay for heated rear screen
70 In-line fuse for heated rear screen
71 Heated rear screen
72 Ignition coil
73 Oil pressure transmitter
74 Interior light
75 Shuttle valve for brake switch
76 Stop lamp, LH
77 Stop lamp, RH
78 Switch, heated rear screen
79 Distributor
80 Inspection light sockets
81 Switch, courtesy light
82 Switch, interior light
83 Switch, courtesy light
84 Switch, handbrake
85 Switch, oil pressure
86 Switch, cold start control pick-up point
87 Fuel gauge, tank unit
88 Water temperature transmitter
89 Windscreen washer motor
90 Windscreen wiper motor 2-speed
91 Rear screen washer motor
92 Rear screen wiper motor, single speed
93 Heater motor, two-speed

		Snap Connectors
		Connections via Plug & Socket
		Permanent In-line Connections
		Earth Connections via Fixing Bolts
	B	Earth Connections via Cables
		Eureka Resistance Wire

CABLE COLOUR CODE

G	=	Green	U	=	Blue
P	=	Purple	N	=	Brown
L	=	Light	Y	=	Yellow
O	=	Orange	B	=	Black
S	=	Slate	R	=	Red
K	=	Pink	W	=	White

NOTE: All circuits shown dotted are additional extras. When cables have two colour code letters, the first denotes the main and the latter the tracer.

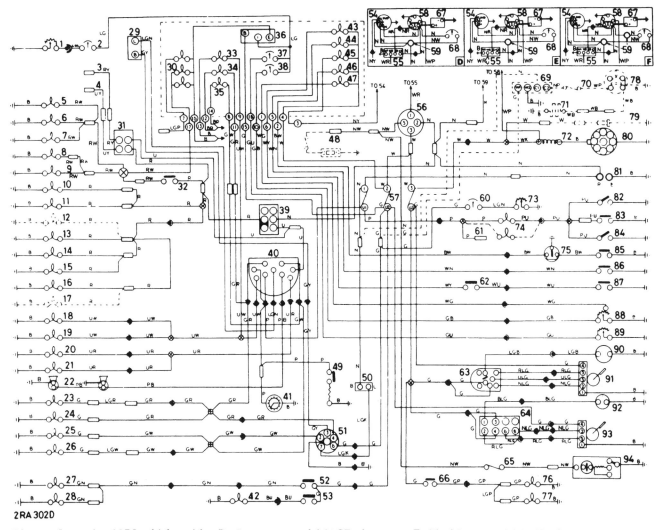

2RA 302D

Diagram Legend – 1975 vehicles with : D–Battery sensed 16ACR alternator, E–Machine sensed 16ACR alterator, F–Battery sensed 18ACR alternator

1 Oil temperature transmitter	33 Warning light, headlamp main beam	64 Switch, rear wiper and washer
2 Oil temperature gauge	34 Warning light, indicator, LH	65 In-line fuse for heater
3 Pick-up point for auxiliary driving lamps	35 Warning light, indicator, RH	66 Switch, stop lamps
4 Pick-up point for auxiliary driving lamps	36 Voltage stabiliser	67 Battery
5 Ammeter illumination	37 Water temperature gauge	68 Ammeter
6 Oil temperature gauge illumination	38 Fuel gauge	69 Relay for heated rear screen
7 Cigar lighter illumination	39 Switch, main light	70 In-line fuse for heated rear screen
8 Oil pressure gauge illumination	40 Switch, headlamps, direction indicators & horn	71 Heated rear screen
9 Clock illumination	41 Clock	72 Ignition coil
10 Side lamp, LH	42 Warning light, differential lock switch	73 Oil pressure transmitter
11 Side lamp, RH	43 Warning light, cold start control (when fitted)	74 Interior light
12 Side marker lamp, tail, LH, as applicable	44 Warning light, oil pressure	75 Shuttle valve for brake switch
13 Tail lamp, LH	45 Warning light, ignition	76 Stop lamp, LH
14 Number plate illumination	46 Warning light, brake	77 Stop lamp, RH
15 Number plate illumination	47 Warning light, fuel level	78 Switch, heated rear screen
16 Tail lamp, RH	48 Pick-up point for radio	79 Fuel pump
17 Side marker lamp, tail, RH, as applicable	49 Cigar lighter	80 Distributor
18 Headlamp main beam, RH	50 Hazard warning unit	81 Inspection light sockets
19 Headlamp main beam, LH	51 Switch, hazard warning	82 Switch, courtesy light
20 Headlamp dip, RH	52 Switch, reverse lights	83 Switch, interior light
21 Headlamp dip, LH	53 Switch, differential lock	84 Switch, courtesy light
22 Horns	54 Alternator	85 Switch, handbrake
23 Indicator lamp, rear LH	55 Relay for starter motor	86 Switch, oil pressure
24 Indicator lamp, front LH	56 Switch, ignition	87 Switch, cold start control pick-up point
25 Indicator lamp, front RH	57 Fuses	88 Fuel gauge, tank unit
26 Indicator lamp, rear RH	58 Starter motor	89 Water temperature transmitter
27 Reverse lamp	59 Terminal post	90 Windscreen washer motor
28 Reverse lamp	60 Oil pressure gauge	91 Windscreen wiper motor, two-speed
29 Indicator unit, 8FL	61 Pick-up point for seven pin trailer socket	92 Rear screen washer motor
30 Trailer illumination	62 Switch, cold start (when fitted)	93 Rear screen wiper motor, single speed
31 Switch, auxiliary driving lamps	63 Switch, front wiper and washer	94 Heater motor, two-speed
32 Switch panel lights		

Symbol	Description
▭	Snap Connectors
─)─	Connections via Plug & Socket
─○─	Permanent In-line Connections
·‖─	Earth Connections via Fixing Bolts
·‖─ B	Earth Connections via Cables
⊗──⊗	Eureka Resistance Wire

CABLE COLOUR CODE

G	=	Green	U	=	Blue
P	=	Purple	N	=	Brown
L	=	Light	Y	=	Yellow
O	=	Orange	B	=	Black
S	=	Slate	R	=	Red
K	=	Pink	W	=	White

NOTE: When cables have two colour code letters, the first denotes the main and the latter the tracer.

86–5

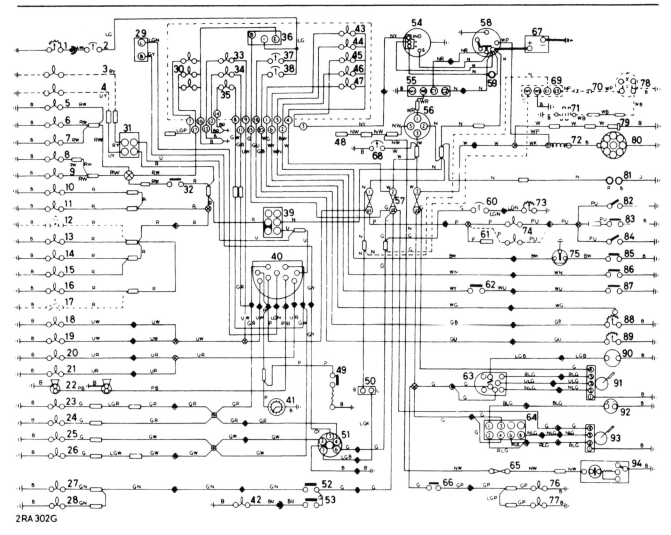

2RA 302G

Circuit Diagram Legend – 1978 Vehicles with Battery Voltmeter Gauge

1 Oil temperature transmitter	33 Warning light, headlamp main beam	64 Switch, rear wiper and washer
2 Oil temperature gauge	34 Warning light, indicator, LH	65 In-line fuse for heater
3 Pick-up point for auxiliary driving lamps	35 Warning light, indicator, RH	66 Switch, stop lamps
4 Pick-up point for auxiliary driving lamps	36 Voltage stabiliser	67 Battery
5 Battery voltmeter illumination	37 Water temperature gauge	68 Battery voltmeter
6 Oil temperature gauge illumination	38 Fuel gauge	69 Relay for heated rear screen
7 Cigar lighter illumination	39 Switch, main light	70 In-line fuse for heated rear screen
8 Oil pressure gauge illumination	40 Switch, headlamps, direction indicators & horn	71 Heated rear screen
9 Clock illumination	41 Clock	72 Ignition coil
10 Side lamp, LH	42 Warning light, differential lock switch	73 Oil pressure transmitter
11 Side lamp, RH	43 Warning light, cold start control (when fitted)	74 Interior light
12 Side marker lamp, tail, LH, as applicable	44 Warning light, oil pressure	75 Shuttle valve for brake switch
13 Tail lamp, LH	45 Warning light, ignition	76 Stop lamp, LH
14 Number plate illumination	46 Warning light, brake	77 Stop lamp, RH
15 Number plate illumination	47 Warning light, fuel level	78 Switch, heated rear screen
16 Tail lamp, RH	48 Pick-up point for radio	79 Fuel pump
17 Side marker lamp, tail, RH, as applicable	49 Cigar lighter	80 Distributor
18 Headlamp main beam, RH	50 Hazard warning unit	81 Inspection light sockets
19 Headlamp main beam, LH	51 Switch, hazard warning	82 Switch, courtesy light
20 Headlamp dip, RH	52 Switch, reverse lights	83 Switch, interior light
21 Headlamp dip, LH	53 Switch, differential lock	84 Switch, courtesy light
22 Horns	54 Alternator	85 Switch, handbrake
23 Indicator lamp, rear LH	55 Relay for starter motor	86 Switch, oil pressure
24 Indicator lamp, front LH	56 Switch, ignition	87 Switch, cold start cold pick-up point
25 Indicator lamp, front RH	57 Fuses	88 Fuel gauge, tank unit
26 Indicator lamp, rear RH	58 Starter motor	89 Water temperature transmitter
27 Reverse lamp	59 Terminal post (where fitted)	90 Windscreen washer motor
28 Reverse lamp	60 Oil pressure gauge	91 Windscreen wiper motor, two-speed
29 Indicator unit, 8FL	61 Pick-up point for seven pin trailer socket	92 Rear screen washer motor
30 Trailer illumination	62 Switch, cold start (when fitted)	93 Rear screen wiper motor, single speed
31 Switch, auxiliary driving lamps	63 Switch, front wiper and washer	94 Heater motor, two-speed
32 Switch panel lights		

NOTE: Later vehicles may be fitted with air cord 12 volt oil pressure gauges not wired via the voltage stabiliser or 10 volt gauges in circuit with it.

—▭—	Snap Connectors
—)—	Connections via Plug & Socket
—O—	Permanent In-line Connections
·‖—	Earth Connections via Fixing Bolts
·‖—B—	Earth Connections via Cables
O—O	Eureka Resistance Wire

CABLE COLOUR CODE

G	=	Green	U =	Blue
P	=	Purple	N =	Brown
L	=	Light	Y =	Yellow
O	=	Orange	B =	Black
S	=	Slate	R =	Red
K	=	Pink	W =	White

86–6

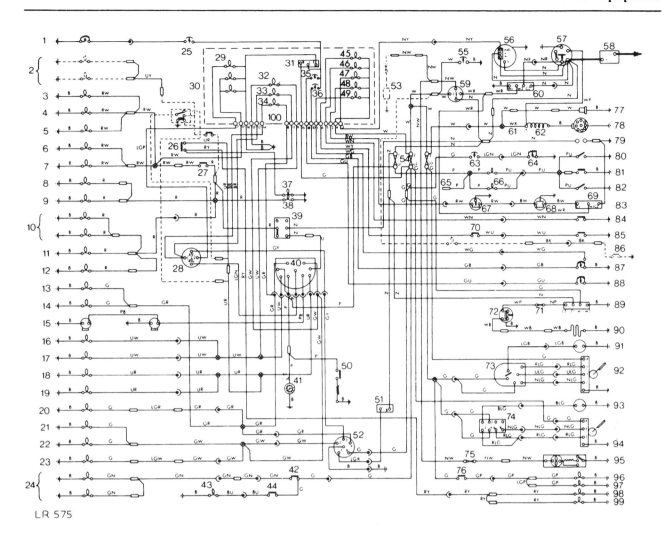

LR 575

Circuit Diagram Legend – 1979 Vehicles with Rear Fog Guard Lamps

1 Oil temperature transmitter
2 Pick-up point for front fog lamps
3 Battery voltmeter illumination
4 Oil temperature gauge illumination
5 Cigar lighter illumination
6 Oil pressure gauge illumination
7 Clock illumination
8 Side lamp, LH
9 Side lamp, RH
10 Number plate illumination
11 Tail lamp, LH
12 Tail lamp, RH
13 Side repeater lamp, LH
14 Indicator lamp, front LH
15 Horns
16 Headlamp main beam, RH
17 Headlamp main beam, LH
18 Headlamp dip, RH
19 Headlamp dip, LH
20 Indicator lamp, rear LH
21 Side repeater lamp, RH
22 Indicator lamp, front RH
23 Indicator lamp, rear RH
24 Reverse lamp
25 Oil temperature gauge
26 Switch, rear fog guard lamps
27 Switch, panel illumination
28 Indicator flasher unit
29 Warning light, trailer indicator
30 Panel illumination (two lamps)
31 Voltage stabiliser
32 Warning light, headlamp main beam
33 Warning light, indicator LH
34 Warning light, indicator RH

35 Water temperature gauge
36 Fuel gauge
37 Warning light, rear fog guard lamps
38 Warning light, side lamps
39 Switch, main, vehicle lighting
40 Switch, headlamps, direction indicators & horns
41 Clock
42 Switch, reverse lights
43 Switch, differential lock
44 Warning light, differential lock control
45 Warning light, cold start control
46 Warning light, oil pressure
47 Warning light, ignition
48 Warning light, brake circuit check
49 Warning light, fuel level
50 Cigar lighter
51 Hazard warning flasher unit
52 Switch, hazard warning
53 Pick-up point for radio
54 Fuses
55 Battery voltmeter
56 Alternator
57 Starter motor
58 Battery
59 Switch, ignition
60 Relay, starter motor
61 Ballast resistance wire, coil
62 Ignition coil
63 Oil pressure gauge
64 Oil pressure transmitter
65 Pick-up point for trailer socket
66 Interior illumination (two lamps)
67 Switch, brake fluid pressure

68 Switch, brake servo vacuum loss
69 Relay, brake check
70 Switch, cold start
71 In-line fuse, heated rear screen
72 Switch with warning light, heated rear screen
73 Switch, front wipers and washer
74 Switch, rear wiper and washer
75 In-line fuse, vehicle heater
76 Switch, stop lamps
77 Fuel pump
78 Distributor
79 Inspection lamp sockets
80 Switch, courtesy light
81 Switch, interior lights (two lamps)
82 Switch, courtesy light
83 Switch, brake circuit check
84 Switch, oil pressure
85 Switch, cold start thermostat
86 'Park' brake (option)
87 Fuel gauge, tank unit
88 Water temperature transmitter
89 Relay, heated rear screen
90 Heated rear screen
91 Windscreen washer motor
92 Windscreen wiper motor, two-speed
93 Rear screen washer motor
94 Rear screen wiper motor, single-speed
95 Heater motor, two-speed
96 Stop lamp, LH
97 Stop lamp, RH
98 Rear fog guard lamp, LH
99 Rear fog guard lamp, RH
100 Printed circuit connector pins

— ▭ — Snap Connectors

— ∘) — Connections via Plug & Socket

— ∘ — Permanent In-line Connections

⊣▮— Earth Connections via Fixing Bolts

⊣▮—B— Earth Connections via Cables

CABLE COLOUR CODE

G	=	Green	U =	Blue
P	=	Purple	N =	Brown
L	=	Light	Y =	Yellow
O	=	Orange	B =	Black
S	=	Slate	R =	Red
K	=	Pink	W =	White

The last letter of a colour code denotes the tracer colour.

86–7

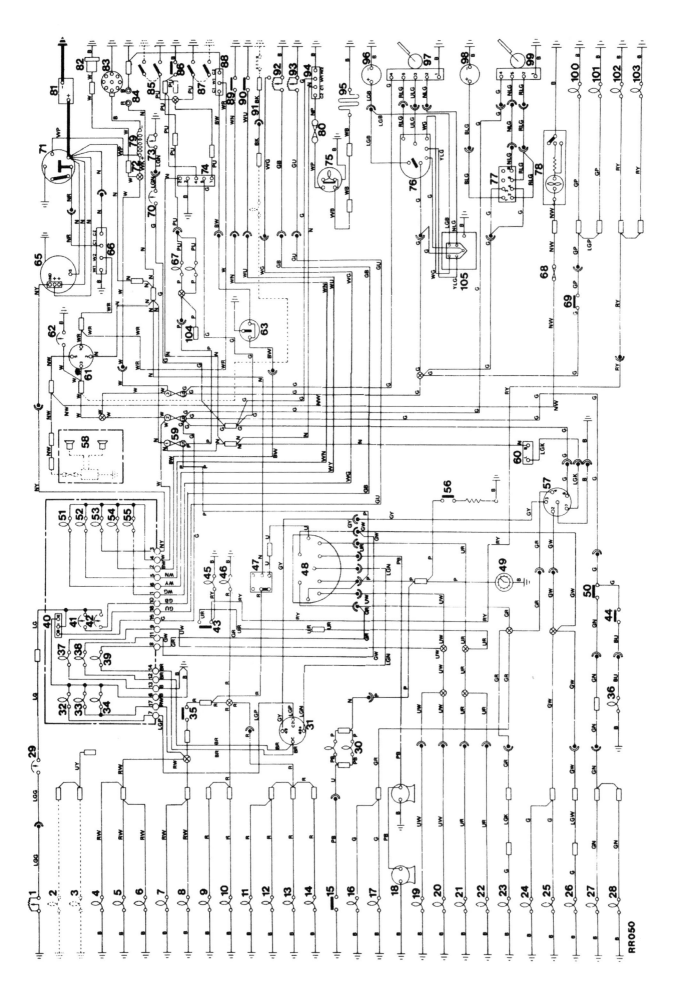

RR050

Circuit Diagram Legend – 1981 Vehicles with Programmed Wash/Wipe and Interior Light Delay Function

1 Oil temperature transmitter
2 Front fog lamp pick-up point
3 Front fog lamp pick-up point
4 Battery voltmeter illumination
5 Oil temperature gauge illumination
6 Cigar lighter illumination
7 Oil pressure gauge
8 Clock illumination
9 LH Front side lamp
10 RH Front side lamp
11 Number plate illumination
12 Number plate illumination
13 LH Rear tail lamp
14 RH Rear tail lamp
15 Under bonnet illumination
16 Direction indicator side repeater
17 LH Front indicator lamp
18 Horns
19 RH Headlamp main beam
20 LH Headlamp main beam
21 RH Headlamp dipped beam
22 LH Headlamp dipped beam
23 LH Rear indicator
24 Direction indicator side repeater
25 RH Front indicator lamp
26 RH Rear indicator lamp
27 Reverse lamp LH
28 Reverse lamp RH
29 Oil temperature gauge
30 Under-bonnet lamps
31 Direction indicator switch
32 Trailer warning light
33 Panel illumination
34 Panel illumination
35 Panel lighting switch

36 Differential lock warning light
37 Main beam warning light
38 LH Indicator warning light
39 RH Indicator warning light
40 Voltage stabiliser
41 Water temperature gauge
42 Fuel gauge
43 Rear fog lighting switch
44 Differential lock switch
45 Rear fog warning light
46 Sidelight warning light
47 Lighting switch
48 Indicator, headlamps, dipped beam & horn switch
49 Clock
50 Reverse lamp switch
51 Choke warning lamp
52 Oil warning lamp
53 Ignition warning lamp
54 Brake warning lamp
55 Fuel warning lamp
56 Cigar lighter
57 Hazard warning switch
58 Radio speakers
59 Fuse unit
60 Hazard unit
61 Ignition switch
62 Battery voltmeter
63 Brake pressure warning switch
64 Choke switch
65 Alternator
66 Starter relay
67 Interior lights
68 Heater fuse
69 Stop lamp switch
70 Oil pressure gauge

71 Starter motor
72 Resistive wire
73 Oil pressure transmitter
74 Courtesy lighting delay unit
75 Heated rear screen switch
76 Front wiper washer switch
77 Rear wiper washer switch
78 Heater motor
79 Coil
80 Fuse
81 Battery
82 Electric fuel pump
83 Distributor
84 Inspection sockets
85 Courtesy lighting switches
86 Interior lighting switch
87 Courtesy lighting switches
88 Brake circuit check relay
89 Oil pressure switch
90 Choke thermocoupling switch
91 Park brake warning light (optional)
92 Fuel gauge unit
93 Water temperature transmitter
94 Relay (heated rear screen)
95 Heated rear screen
96 Front screen washer motor
97 Two speed front wiper motor
98 Rear screen washer motor
99 Single speed rear wiper motor
100 LH stop lamp
101 RH stop lamp
102 LH rear fog lamp
103 RH rear fog lamp
104 Trailer socket connection
105 Programmed wash/wipe unit

Snap Connectors

Connections via Plug & Socket

Permanent In-line Connections

Earth Connections via Fixing Bolts

Earth Connections via Cables

CABLE COLOUR CODE

G	=	Green	U =	Blue
P	=	Purple	N =	Brown
L	=	Light	Y =	Yellow
O	=	Orange	B =	Black
S	=	Slate	R =	Red
K	=	Pink	W =	White

The last letter of a colour code denotes the tracer colour.

This page intentionally left blank

ELECTRICAL EQUIPMENT

IMPORTANT

The electrical system is Negative earth, and it is most important to ensure correct polarity of the electrical connections at all times. Any incorrect connections made when reconnecting cables may cause irreparable damage to the semiconductor devices used in the alternator and regulator. Incorrect polarity would also seriously damage any transistorised equipment such as radio and tachometer etc.

Before carrying out any repairs or maintenance to an electrical component, always disconnect the battery.

The V-drive fan belt used with alternators is not the same as that used with d.c. machines. Only use the correct Rover replacement fan belt. Occasionally check that the engine and alternator pulleys are accurately aligned.

It is essential that good electrical connections are maintained at all times. Of particular importance are those in the charging circuit (including those at the battery) which should be occasionally inspected to see that they are clean and tight. In this way any significant increase in circuit resistance can be prevented.

Do not disconnect battery cables while the engine is running or damage to the semi-conductor devices may occur. It is also inadvisable to break or make any connections in the alternator charging and control circuits while the engine is running.

The Model 8TR electronic voltage regulator employs micro-circuit techniques resulting in improved performance under difficult service conditions. The whole assembly is encapsulated in silicone rubber and housed in an aluminium heat sink, ensuring complete protection against the adverse affects of temperature, dust, and moisture etc. Battery voltage is sensed directly by a permanent connection (B+).

The regulating voltage is set during manufacture to give the required regulating voltage range of 14.1 to 14.5 volts, and no adjustment is necessary. The only maintenance needed is the occasional check on terminal connections and wiping with a clean dry cloth.

The alternator system provides for direct connection of a charge (ignition) indicator warning light, and eliminates the need for a field switching relay or warning light control unit. As the warning lamp is connected in the charging circuit, lamp failure will cause loss of charge. Lamp should be checked regularly and a spare carried.

When using rapid charge equipment to re-charge the battery, the battery must be disconnnected from the vehicle.

FAULT DIAGNOSIS

SYMPTOM	POSSIBLE CAUSE	CURE
A— Battery in low state of charge	1. Broken or loose connection in alternator circuit 2. Current voltage regulator not functioning correctly 3. Slip rings greasy or dirty 4. Brushes worn not fitted correctly or wrong type	1. Examine the charging and field circuit wiring. Tighten any loose connections and renew any broken leads. Examine the battery connection 2. Adjust or renew 3. Clean 4. Renew
B— Battery overcharging, leading to burnt-out bulbs and frequent need for topping-up	1 Current voltage regulator not functioning correctly	1. Renew
C— Lamps giving insufficient illumination	1. Battery discharged 2. Bulbs discoloured through prolonged use	1. Charge the battery from an independent supply or by a long period of daylight running 2. Renew
D— Lamps light when switched on but gradually fade out	1. Battery discharged	1. Charge the battery from an independent supply or by a long period of daylight runnings
E— Lights flicker	1. Loose connection	1. Tighten
F— Failure of lights	1. Battery discharged 2. Loose or broken connection	1. Charge the battery from an independent supply or by a long period of daylight running 2. Locate and rectify
G— Starter motor lacks power or fails to turn engine	1. Stiff engine 2. Battery discharged 3. Broken or loose connection in starter circuit 4. Greasy or dirty slip rings 5. Brushes worn not fitted correctly or wrong type 6. Brushes sticking in holders or incorrectly tensioned 7. Starter pinion jammed in mesh with flywheel	1. Locate cause and remedy 2. Charge the battery either by a long period of daytime running or from independent electrical supply 3. Check and tighten all battery, starter and starter switch connections and check the cables connecting these units for damage 4. Clean 5. Renew 6. Rectify 7. Remove starter motor and investigate
H— Starter noisy	1. Starter pinion or flywheel teeth chipped or damaged 2. Starter motor loose on engine 3. Armature shaft bearing	1. Renew 2. Rectify, checking pinion and the flywheel for damage 3. Renew
J— Starter operates but does not crank the engine	1. Pinion of starter does not engage with the flywheel	1. Check operation of starter solenoid. If correct, remove starter motor and investigate
K— Starter pinion will not disengage from the flywheel when the engine is running	1. Starter pinion jammed in mesh with the flywheel	1. Remove starter motor and investigate

86–12

FAULT DIAGNOSIS

SYMPTOM	POSSIBLE CAUSE	CURE
L— Engine will not fire	1. The starter will not turn the engine due to a discharged battery 2. Sparking plugs faulty, dirty or incorrect plug gaps 3. Defective coil or distributor 4. A fault in the low tension wiring is indicated when no spark occurs between the contacts when separated quickly with an insulated screwdriver with the ignition on 5. Dirty or pitted contacts 6. Contact breaker out of adjustment 7. Controls not set correctly or trouble other than ignition	1. The battery should be recharged by running the car for a long period during daylight or from an independent electrical supply 2. Rectify or renew 3. Remove the lead from the centre distributor terminal and hold it approximately 6 mm. (¼ in.) from some metal part of the engine while the engine is being turned over. If the sparks jump the gap regularly, the coil and distributor are functioning correctly. Renew a defective coil or distributor 4. Examine all the ignition cables and check that the bottom terminals are secure and not corroded 5. Clean 6. Adjust 7. See Starting Procedure in the Owner's Instruction Manual
M— Engine misfires	1. Distributor points incorrectly set 2. Faulty coil or condenser 3. Faulty sparking plugs 4. Faulty carburetter	1. Adjust 2. Renew 3. Rectify 4. Check and rectify
N— Frequent recharging of the battery necessary	1. Alternator inoperative 2. Loose or corroded connections 3. Slipping fan belt 4. Voltage control out of adjustment 5. Excessive use of the starter motor 6. Vehicle operation confined largely to night driving 7. Abnormal accessory load 8. Internal discharge of the battery	1. Check the brushes, cables and connections or renew the alternator 2. Examine all connections, especially the battery terminals and earthing straps 3. Adjust 4. Renew 5. In the hands of the operator 6. In the hands of the operator 7. Superfluous electrical fittings such as extra lamps etc. 8. Renew
P— Alternator not charging correctly	1. Slipping fan belt 2. Voltage control not operating correctly 3. Greasy, charred or glazed slip rings 4. Brushes worn, sticking or oily 5. Shorted, open or burnt-out field coils	1. Adjust 2. Rectify or renews, 3. Clean 4. Rectify or renew 5. Renew
Q— Alternator noisy	1. Worn, damaged or defective bearings 2. Cracked or damaged pulley 3. Alternator out of alignment 4. Alternator loose in mounting 5. Excessive brush noise	1. Renew 2. Renew 3. Rectify 4. Rectify 5. Check for rough or dirty slip rings, badly seating brushes, incorrect brush tension, loose brushes and loose field magnets. Rectify or renew
R— Defective distributor	1. Contact breaker gap incorrect or points burned and pitted 2. Distributor cap cracked 3. Weak or broken contact breaker spring 4. Excessive wear in distributor shaft bushes, etc. 5. Rotor arm pitted or burned 6. If the engine lacks power, or misfires, it may be due to a faulty condenser	1. Clean and adjust 2. Renew 3. Renew 4. Renew 5. Clean or renew 6. Renew the condenser

86-13

FAULT DIAGNOSIS

SYMPTOM	POSSIBLE CAUSE	CURE
S— Mixture control warning light fails to appear when engine reaches running temperature	1. Mixture control already pushed in 2. Broken connection in warning light circuit 3. Blown bulb 4. Faulty thermostat switch (at cylinder head) 5. Faulty manual switch (at mixture control) 6. Broken operating mechanism at manual switch	1. In the hands of the operator 2. Rectify 3. Renew 4. Renew 5 Renew 6. Rectify
T— Mixture control warning light remains on with engine at running temperature	1. Mixture control out 2, Faulty manual switch 3. Broken operating mechanism at manual switch	1. Push control right in 2. Renew 3. Rectify
U— Poor performance of horns	1. Low voltage due to discharged battery 2. Bad connections in wiring 3. Loose fixing bolt 4. A faulty horn	1. Recharge 2. Carefully inspect all connections and horn push 3. Rectify 4. Adjust or renew

86—14

ALTERNATOR

– Remove and refit   86.10.02.

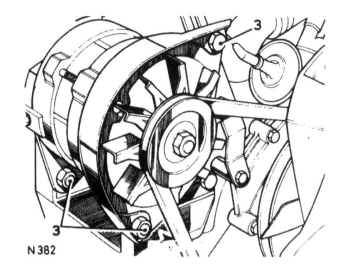

N 382

Removing

1. Disconnect battery earth lead.
2. Disconnect leads from alternator.
3. Slacken alternator fixings, pivot alternator inwards and remove fan belt.
4. Remove alternator.

Refitting

5. Attach the alternator lower fixing bolts and nuts.

NOTE. The fan guard is attached to the front fixing.

6. Slacken the alternator adjustment bracket and attach the alternator to the bracket.

NOTE. The fan guard is attached to the adjustment bracket bolt.

7. Fit the fan belt and adjust the belt tension 26.20.01.
8. Connect the wiring plug to the alternator.
9. Connect the battery.

NOTE. On vehicles fitted with air conditioning the alternator is located on the left hand side of the engine. Access to the adjustment bracket and fixing bolts is from beneath the vehicle.

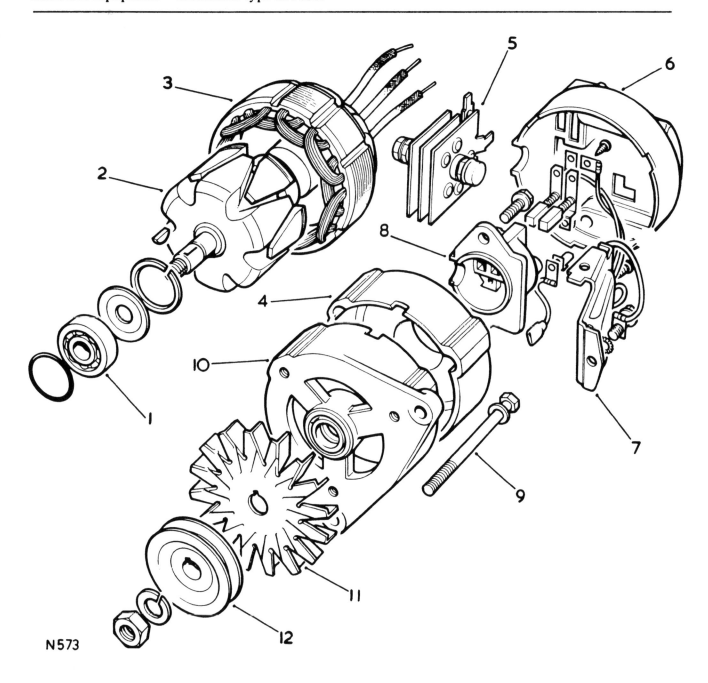

N573

ALTERNATOR

1– Drive end bearing
2– Rotor and slip ring
3– Stator
4– Slipping bracket
5– Rectifier
6– End cover
7– Regulator unit
8– Brush box
9– Through bolt
10– Drive end bracket
11– Fan
12– Pulley

ALTERNATOR

– Overhaul
86.10.08.

NOTE. Alternator charging circuit.

The ignition warning light is connected in series with the alternator field circuit. Bulb failure would prevent the alternator charging, except at very high engine speeds, therefore, the bulb should be checked before suspecting an alternator fault.

Precautions

Battery polarity is NEGATIVE EARTH, which must be maintained at all times.

No separate control unit is fitted; instead a voltage regulator of micro-circuit construction is incorporated on the slip-ring end bracket, inside the alternator cover.

The field connector block, which has three blades and is marked B+ and IND, has an offset moulded stop and must be removed before the main output connector block, which has two blades and is marked + and −. Since the B+ connector blade, although shrouded, is always live, disconnect the battery earth before removing the field connector block.

Testing in position

Output test

1. Check fan belt is correctly tensioned and all charging circuit connections are secure.
2. Run engine at fast idle speed until normal operating temperature is reached. Check battery is fully charged.
3. Disconnect both connector blocks from alternator.
4. Switch on ignition and connect voltmeter with negative lead to earth and positive lead to each cable connector blade of two connector blocks in turn. If battery voltage not available at any cable, locate and remedy fault.
5. Remove alternator end cover.
6. Bridge regulator field connector to earth.
7. Refit three-way connector block to alternator.
8. Do not refit two-way connector block; connect an ammeter in series with its positive blade and main positive output terminal of alternator. Do not make any connection to inner (negative) main terminal.
9. Start engine and run at 3,300 rpm. Ammeter should read 34 amps. If correct alternator output cannot be obtained repair or replace alternator.

Regulator test

10. Disconnect lead (item 6.) bridging regulator field connector to earth.
11. Connect voltmeter across battery terminals.
12. Start engine and run at 3,300 rpm. If ammeter (connected for item 8. of the output test) reads zero, regulator pack must be replaced.
13. Adjust engine speed until ammeter reading falls below 10 amps. Voltmeter should read between 13.6 and 14.4 volts; if not, either regulator is faulty or there is a high resistance in charging circuit cables.
14. Restore original connections to alternator and check charging circuit resistance.

Continued

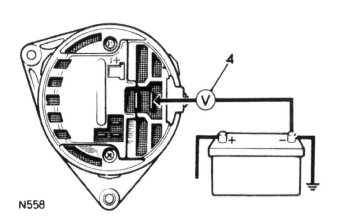

N558

Charging circuit resistance test

15. Connect voltmeter between positive terminal of alternator and positive terminal of battery.
16. Start engine and switch on headlamps; run engine at 3,300 rpm. Voltmeter reading should not exceed 0.5 volts.
17. Change voltmeter connections to negative terminals of alternator and battery. With engine running at 3,300 rpm voltmeter reading should not exceed 0.25 volts.
18. If voltmeter readings in 16. and 17. exceed voltage stated, charging circuit has high resistance fault which must be traced and rectified.

    If this test is satisfactory, then incorrect voltage obtained in 13. would have been caused by faulty regulator pack; then alternator must be either replaced or removed for overhaul.

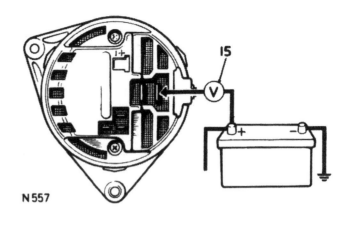

N557

Removing the regulator pack

19. Remove alternator (items 23. to 26.).
20. Remove moulded end cover.
21. Disconnect coloured tag lead connectors from brush box and detach black (earth) lead after removing lower mounting screw.
22. Remove remaining screw securing regulator pack.

Removing the alternator

23. Withdraw terminal block from alternator.
24. Remove adjusting link bolt from alternator.
25. Slacken alternator mounting bolts, lower alternator and slip fan belt from alternator pulley.
26. Unscrew alternator mounting bolts and remove alternator.

Testing – Alternator removed

27. Remove cover.
28. Unsolder stator connections from rectifier pack, noting connection positions.

IMPORTANT When soldering or unsoldering connections to diodes, great care must be taken not to overheat diodes or bend pins. During soldering operations diode pins should be gripped lightly with a pair of long nosed pliers which will act as a thermal shunt.

29. Unscrew brush moulding securing screws and if necessary, lower regulator pack securing screw.
30. Slacken rectifier pack retaining nuts and withdraw both brush moulding, with or without regulator pack, and rectifier pack.

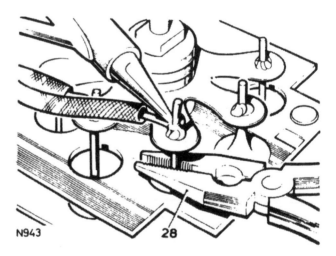

N943   28

Continued

86–18

Brushes

31. Check brushes for wear by measuring length of brush protruding beyond brush box moulding. If length is 5 mm (0.2 in) or less, brush must be renewed.
32. Check that brushes move freely in holders. If brush is sticking, clean with petrol moistened cloth or polish sides of brush with fine file.
33. Check brush spring pressure using push-type spring gauge. Gauge should register 198 to 283 gm. (7 to 10 oz) when brush is pushed back until face is flush with housing. If reading is outside these limits, renew brush assembly.

Slip rings

34. Clean surfaces of slip-rings using petrol moistened cloth.
35. Inspect slip-ring surfaces for signs of burning; remove burn marks using very fine sand-paper. On no account should emery-cloth or similar abrasives be used, or any attempt made to machine the slip-rings.

Rotor

36. Connect an ohmmeter or a 12-volt battery and an ammeter to the slip-rings. An ohmmeter reading of 4.3 ohms or an ammeter reading of 3 amps should be recorded.
37. Using a 110-volt a.c. supply and a 15-watt test lamp. test for insulation between one of the slip-rings and one of the rotor poles. If the test lamp lights, the rotor must be renewed.

Stator

38. Connect a 12-volt battery and a 36-watt test lamp to two of the stator connections. Repeat the test replacing one of the two stator connections with the third. If test lamp fails to light in either test, stator must be renewed.
39. Using a 110-volt a.c. supply and a 15-watt test lamp. test for insulation between any one of the three stator connections and stator laminations. If test lamp lights, stator must be renewed.

Diodes

40. Connect a 12-volt battery and a 1.5-watt test lamp in turn to each of the nine diode pins and its corresponding heat sink on the rectifier pack, then reverse the connections. Lamp should light with current flowing in one direction only. If lamp lights in both directions or fails to light in either, rectifier pack must be renewed.

Continued

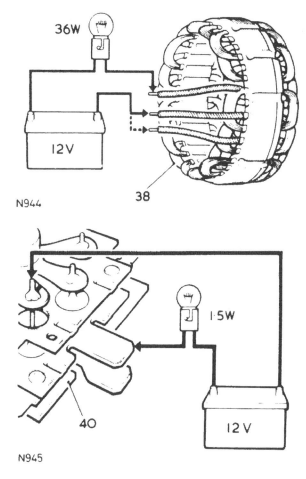

36W

12V

N944

38

1·5W

40

12 V

N945

Dismantling

41. Carry out operations detailed in items 27. - 30.
42. Remove the three through bolts.
43. Fit a tube over the slip-ring moulding so that it registers against outer track of slip-ring end bearing and carefully drive bearing from its housing.
44. Remove shaft nut, washer, pulley, fan and shaft key.
45. Press rotor from drive end bracket.
46. Remove circlip retaining drive end bearing and remove bearing.
47. Unsolder field connections from slip-ring assembly and withdraw assembly from rotor shaft.
48. Remove slip-ring end bearing.

Reassembling

49. Reverse the dismantling procedure, noting following points.
    (a) Use Shell Alvania 'RA' to lubricate bearings.
    (b) When refitting slip-ring end bearing, ensure it is fitted with open side facing rotor.
    (c) Use Fry's H.T.3. solder on slip-ring field connections.
    (d) When refitting rotor to drive end bracket, support inner track of bearing. Do not use drive end bracket to support bearing when fitting rotor.
    (e) Tighten through-bolts evenly.
    (f) Fit brushes into housings before fitting brush moulding.
    (g) Tighten shaft nut to correct torque figure (see page 86–30).
    (h) Refit regulator pack to brush moulding.

Refitting the alternator

50. Reverse removal procedure.

ALTERNATOR – Lucas 18ACR

–Overhaul	86.10.08
Including Test (Bench)	86.10.14

NOTE. Alternator charging circuit
The ignition warning light is connected in series with the alternator field circuit. Bulb failure would prevent the alternator charging, except at very high engine speeds, therefore, the bulb should be checked before suspecting an alternator failure.

Precautions

Battery polarity is NEGATIVE EARTH, which must be maintained at all times.
No separate control unit is fitted; instead a voltage regulator of micro-circuit construction is incorporated on the slip-ring end bracket, inside the alternator cover.
Battery voltage is applied to the alternator output cable even when the ignition is switched off, the battery must be disconnected before commencing any work on the alternator. The battery must also be disconnected when repairs to the body structure are being done by arc welding.

Surge protection device

Some protection of the alternator is provided by a surge protection device, connected across the 'IND' terminal, to absorb high transient voltages.

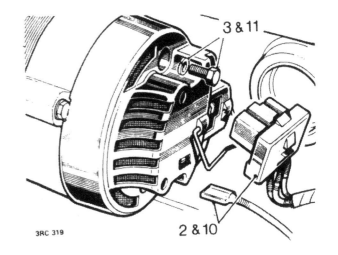

3RC 319

Testing in position
Surge protection device

If the alternator output falls to zero, the fault may be caused by the surge protection device failing safe short-circuit, or a fault in the alternator circuit. Check the surge protection device as follows:

1. Check that the fan belt is correctly tensioned.
2. Withdraw the connectors from the alternator.
3. Remove the alternator rear cover.
4. Remove the surge protection device from the 'IND' terminal.
5. Refit the alternator cover, and connectors. Ensure that all circuit connections are clean and tight.
6. Start and run the engine. If the alternator output is now normal, fit a new surge protection device.

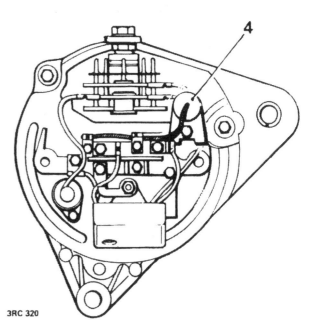

3RC 320

continued

433

86–21

Output test

7. Check that the fan belt is correctly tensioned and that all charging circuit connections are secure.
8. Run the engine at fast idle until normal operating temperature is attained.
9. Stop the engine.
10. Withdraw the connectors from the alternator.
11. Remove the alternator rear cover.
12. Connect the regulator case to the alternator frame.
13. Connect a 0-60 ammeter between the alternator and the battery.
14. Connect a 0-20 voltmeter across the battery terminals.
15. Connect a 15 ohm 35 amp, variable resistor across the battery terminals.

CAUTION. Do not leave the variable resistor connected across the battery terminals for longer than is necessary to carry out the following test, items 16 and 17.

16. Start the engine and run at 750 rev/min. The warning light bulb should be extinguished.
17. Increase the engine speed to 3000 rev/min, and adjust the variable resistance until the voltmeter reads 13.6 volts. The ammeter reading should then be approximately 43 amps. **Any appreciable deviation from this figure will necessitate removing and dismantling the alternator.** If the output test is satisfactory, proceed with the regulator test.

continued

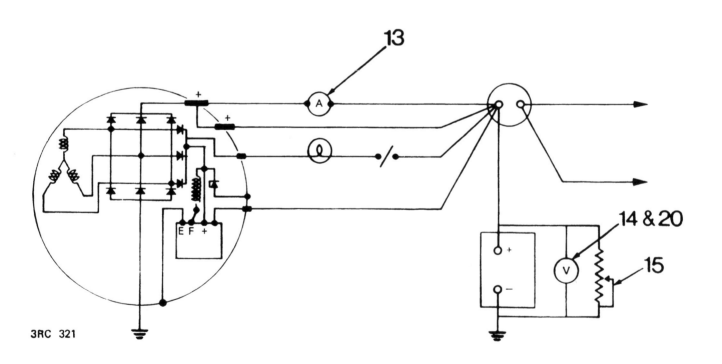

3RC 321

Regulator test

18. Disconnect the variable resistor and remove the connection between the regulator and the alternator frame.

19. With the remainder of the circuit connected as for the alternator output test, start the engine and run at 3,000 rev/min, until the ammeter shows an output current of less than 10 amperes.

20. The voltmeter should now give a reading of 13.6 to 14.4 volts. Any appreciable deviation from this (regulating) voltage indicates a faulty regulator which must be replaced.

21. If the foregoing output and regulator tests show the alternator and regulator to be performing satisfactorily, disconnect the test circuit, reconnect the alternator terminal connector and proceed with the charging circuit resistance test.

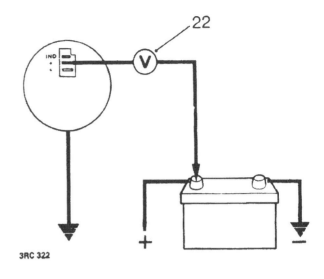

3RC 322

Charging circuit resistance test

22. Connect a low-range voltmeter between either of the alternator terminals marked + and the positive terminal of the battery.

23. Switch on the headlamps.

24. Start the engine and run at approximately 3,000 rev/min. Note the voltmeter reading.

25. Transfer the voltmeter connections to the frame of the alternator and the negative terminal of the battery, and again note the voltmeter reading.

26. If the reading exceeds 0.5 volt on the positive side or 0.25 volt on the negative side, there is a high resistance in the charging circuit which must be traced and remedied.

continued

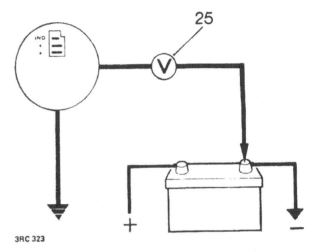

3RC 323

Testing – alternator removed

27. Remove the alternator. 86.10.02.
28. Remove the alternator rear cover.
29. Unsolder stator connections from rectifier pack, noting connection positions.

CAUTION. When soldering or unsoldering connections to diodes take care not to overheat the diodes or bend the pins. During soldering operations, diode pins should be gripped lightly with a pair of long nosed pliers which will act as a thermal shunt.

30. Unscrew brush moulding securing screws and if necessary, lower regulator pack securing scew.
31. Slacken rectifier pack retaining nuts and withdraw both brush moulding, with or without regulator pack, and rectifier pack.

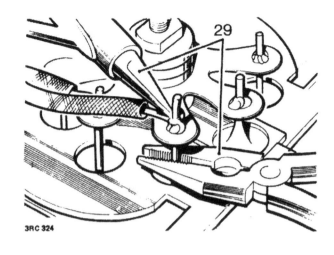

3RC 324

Brushes

32. Check brushes for wear by measuring length of brush protruding beyond brush box moulding. If length is 8 mm (0.3 in) or less, brush must be renewed.
33. Check that brushes move freely in holders. If brush is sticking, clean with petrol moistened cloth or polish sides of brush with fine file.
34. Check brush spring pressure using push-type spring gauge. Gauge should register 255 to 368g (9 to 13 oz) when brush is pushed back until face is flush with housing. If reading is outside these limits, renew brush assembly.

Slip-rings

35. Clean surfaces of slip-rings using petrol moistened cloth.
36. Inspect slip-ring surfaces for signs of burning; remove burn marks using very fine sandpaper. On no account should emery cloth or similar abrasives be used, or any attempt made to machine the slip-rings.

Rotor

37. Connect an ohmmeter or a 12-volt battery and an ammeter to the slip-rings. An ohmmeter reading of 3.2 ohms or an ammeter reading of 4 amps should be recorded.
38. Using a 110-volt a.c. supply and a 15-watt test lamp, test for insulation between one of the slip-rings and one of the rotor poles. If the test lamp lights, the rotor must be renewed.

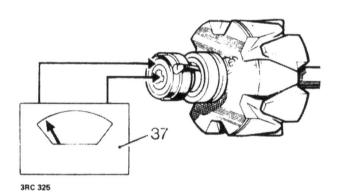

3RC 325

Stator

39. Connect a 12-volt battery and a 36-watt test lamp to two of the stator connections. Repeat the test replacing one of the two stator connections with the third. If test lamp fails to light in either test, stator must be renewed.
40. Using a 110-volt a.c. supply and a 15-watt test lamp, test for insulation between any one of the three stator connections and stator laminations. If test lamp lights, stator must be renewed.

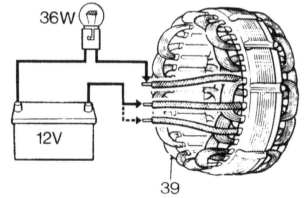

continued

3RC 326

Diodes

41. Connect a 12-volt battery and a 15-watt test lamp in turn to each of the nine diode pins and corresponding heat sink on the rectifier pack, then reverse the connections. Lamp should light with current flowing in one direction only. If lamp lights in both directions or fails to light in either, rectifier pack must be renewed.

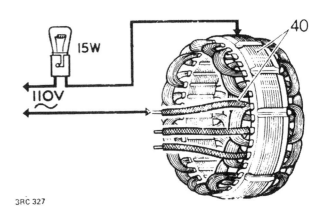

3RC 327

Dismantling

42. If not already completed, carry out items 27 to 31.
43. Remove the three through bolts.
44. Fit a tube over the slip-ring moulding so that it registers against outer track of slip-ring end bearing are carefully drive bearing from its housing.
45. Remove shaft nut, washer, pulley, fan and shaft key.
46. Press rotor from drive end bracket.
47. Remove circlip retaining drive end bearing and remove bearing.
48. Unsolder field connections from slip-ring assembly and withdraw assembly from rotor shaft.
49. Remove slip-ring end bearing.

Reassembling

50. Reverse the dismantling procedure, noting following points.
    a. Use Shell Alvania 'RA' to lubricate bearings.
    b. When refitting slip-ring end bearing, ensure it is fitted with open side facing rotor.
    c. Use Fry's H.T.3. solder on slip-ring field connections.
    d. When refitting rotor to drive end bracket, support inner track of bearing. Do not use drive end bracket to support bearing when fitting rotor.
    e. Tighten through-bolts evenly.
    f. Fit brushes into housings before fitting brush moulding.
    g. Tighten shaft nut to correct torque figure 3,5 to 4,2 kgf.m (25 to 30 lbf ft).
    h. Refit regulator pack to brush moulding.
51. Reconnect the leads between the regulator; brush box and rectifier, as illustrated.
    Lead colours B – Black
    　　　　　　　W – White
    　　　　　　　Y – Yellow
    　　　　　　　S – Sensing terminal
52. Refit the alternator 86.10.02.

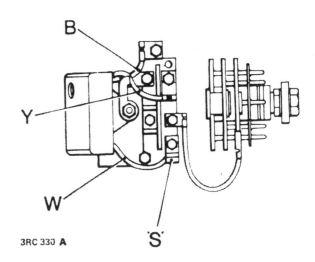

3RC 330 A

DATA

Alternator	Lucas 18ACR battery sensed
Nominal output	**43 amps at 6000 alternator rev/min.**
Field resistance	3.2 ohms
Brush spring pressure	255 to 368g (9 to 13 oz)
Brush minimum length	8 mm (0.312 in)
Regulating voltage	13.6 to 14.4 volts

Continued

437

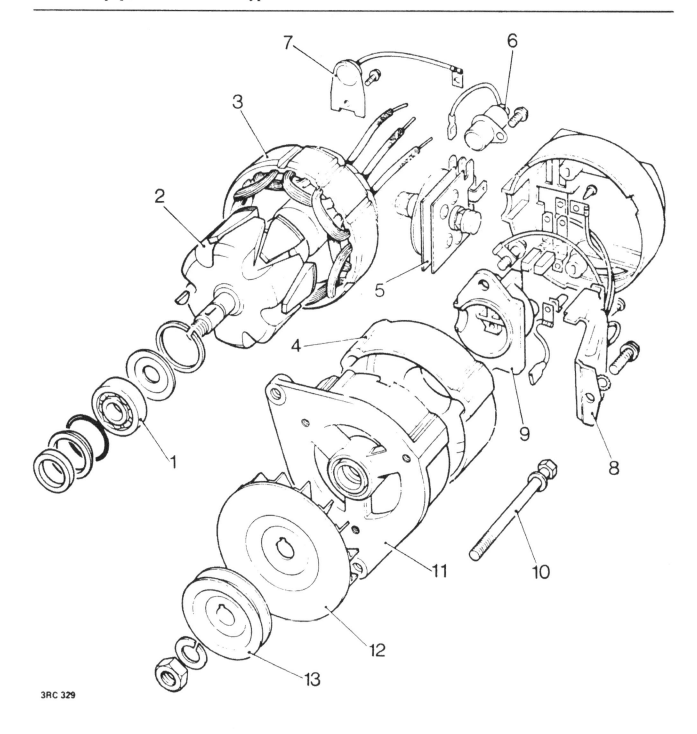

3RC 329

Alternator, type 18ACR

1.	Drive end bearing	6.	Suppressor	11.	Drive end bracket
2.	Rotor and slip ring	7.	Surge protection device	12.	Fan
3.	Stator	8.	Regulator unit	13.	Pulley
4.	Slip ring bracket	9.	Brush Box		
5.	Rectifier	10.	Through bolt		

ALTERNATOR

– Test (bench)   86.10.14.

1. Remove the cover retaining screws and detach the cover.
2. Unsolder the three stator connections from the rectifier pack, noting the connection positions.

NOTE. Use a pair of long nosed pliars as a thermal shunt when soldering or unsoldering connections to the diode pins.

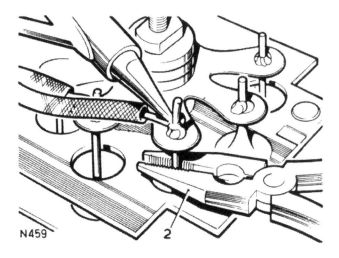

N459

3. Remove the two brush moulding securing screws, and the lower regulating pack securing screw, if necessary.
4. Slacken the rectifier pack retaining nuts and withdraw the brush moulding, rectifier pack and regulator pack as an assembly.
5. Renew the brushes and springs if they are not as specified. (See General Data page 86–30).
6. Make re-assembly marks on the drive-end bracket, stator lamination pack, and slip ring end bracket.
7. Remove the three through bolts.
8. Drive the slip ring bearing out of its housing using a tube of the correct dimensions, ensuring the tube registers against the outer track of the bearing.
9. Remove the retaining nut and detach the driving pulley, fan and shaft key.
10. Press the rotor from the drive-end bracket.
11. Remove the retaining circlip and remove the drive-end bearing.
12. Clean the slip rings with petrol (gasoline) or, if they are burned, with fine glass paper. Do not machine the slip rings.
13. Test the field windings.
    (a) Measure either resistance or current flow by connecting test equipment across the slip rings.
    (b) Test the insulation by connecting the test equipment between one of the rotor poles and each slip ring in turn.
14. Unsolder the field connections from the slip ring assembly to withdraw the assembly from the rotor shaft.

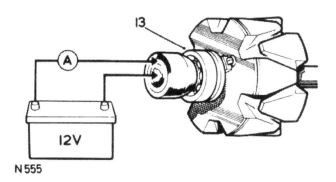

N555

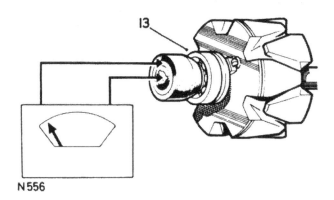

N556

continued

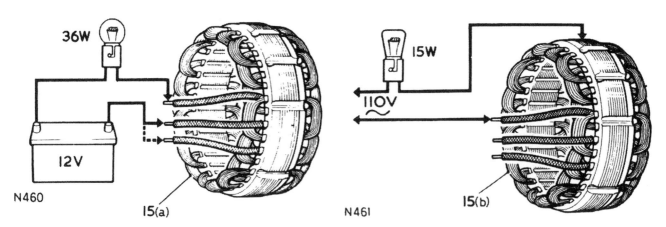

N460

15(a)

N461

15(b)

15. Test the stator windings.

    (a) Test for continuity by connecting any two of the cables in series with the test equipment, then repeat, using the third cable in place of one of the first two.

    (b) Test the insulation by connecting the test equipment between one of the three cables and the stator laminations.

continued

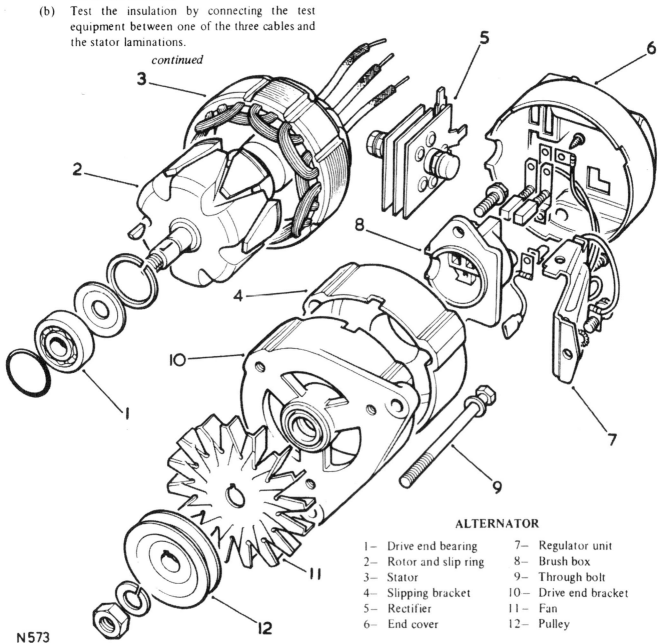

N573

ALTERNATOR

1–	Drive end bearing	7–	Regulator unit
2–	Rotor and slip ring	8–	Brush box
3–	Stator	9–	Through bolt
4–	Slipping bracket	10–	Drive end bracket
5–	Rectifier	11–	Fan
6–	End cover	12–	Pulley

86–28

16. Test each diode in turn by connecting it in series with the test equipment as shown, and then reversing the connections. Current should flow in one direction only.

17. Re-assemble reversing the procedure in 1. to 11. and 14., noting:

(a) Lubricate the bearings with Shell Aloenia 'RA'.

(b) Ensure the open side of the slip-ring end bearing faces towards the rotor and is fully home.

(c) Use Fry's H.T.3 solder to remake the connections to the slip-ring.

(d) Support the inner track of the bearing when re-fitting the rotor to the drive-end bracket. Do not use the bracket as a support.

(e) Tighten the through bolts evenly.

(f) Ensure that the brushes are entered in their housings before fitting the brush moulding.

(g) Tighten the shaft nut to a torque figure of 3,5 to 4,2 kgf.m (25 to 30 lbf. ft.).

continued

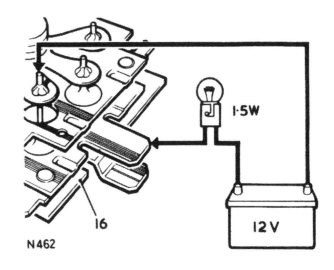

N462

441

Alternator – General Data

Output	34 amps at 6000 alternator RPM
Minimum brush length	5 mm (0.2 in) protruding beyond brush box moulding.
Brush spring pressure	198 to 283 gm. (7 to 10 oz.) when brush is pushed back flush with housing.

Field winding

Resistance	4.33 ohms ± 5%.
Current flow at 12 volts	3 amps
Insulation test equipment	110 volt A.C. supply and 15 watt test lamp.

Stator windings

Continuity test equipment	12 volt D.C. supply and 36 watt test lamp.
Insulation test equipment	110 volt A.C. supply and 15 watt test lamp.
Diode current flow test equipment	
12 volt D.C. supply and 1.5 watt test lamp.	

Alternator shaft nut	3,5 to 4.2 kgf.m (25 to 30 lbf. ft.).

1—Fixing nut
2—Rectifier pack
3—Yellow lead connecting field supply diodes to IND terminal on brush box
4—Terminals
5—B+ lead
6—8 TR regulator frame (heat sink)
7—Negative lead
8—Recess for suppression capacitor
9—F— terminal
10—Outer brush, connected to regulator +
11—Main negative terminal
12—Main positive terminal

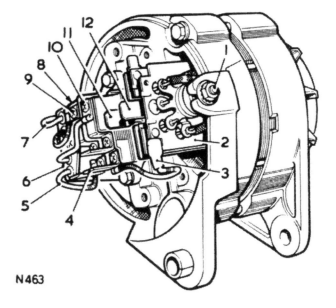

N463

Test	Conditions of test	Remarks
1. To ensure battery voltage is reaching the alternator.	Cables disconnected from alternator. Voltmeter with negative lead connected to earth and positive lead connected to each cable in turn.	Battery voltage should be available at each cable.
2. Output test.	Alternator end cover removed. Regulator field terminal bridged to earth. Ammeter connected in series with + blade. Negative (-) cable disconnected from its blade. All other cables connected to alternator blades. Engine running to give alternator speed of 6000 r.p.m.	The ammeter should read 34 amps approximately. See illustration for terminal connections.
3. Regulator test.	Voltmeter connected across battery terminals. Ammeter connected in series with positive blade and its cable. Negative cable disconnected from its blade. All other cables connected to alternator terminal blades. (a) Run engine to give alternator speed of 6000 r.p.m. (b) Run engine to give alternator speed of 6000 r.p.m. until the charging rate falls to below 10 amps.	If ammeter gives zero reading, renew the regulator. Voltmeter should read 14.0 to 14.4 volts. If outside these limits check resistance in charging circuit cables.
4. Charging circuit resistance test.	All cables connected to alternator terminals. Headlamps or equivalent load switched on. Engine running to give alternator speed of 6000 r.p.m. (a) Voltmeter connected between positive terminal of alternator and positive terminal of battery. (b) Voltmeter connections transferred to negative terminal of alternator and negative terminal of battery.	(a) Voltmeter reading should not exceed 0.5 volts. (b) Voltmeter reading should not exceed 0.25 volts. If readings are in excess of those stated, trace and remedy the high resistance fault in the charging circuit. If readings are satisfactory, renew the regulator.

86 – 31

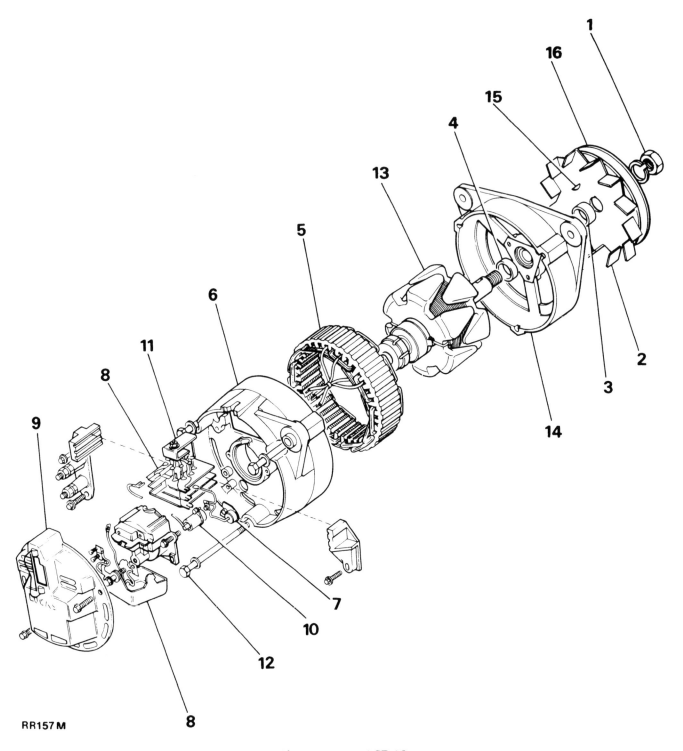

RR157M

Alternator, type ACR 25

1. Pulley retaining nut and spring washer
2. Fan
3. Outer drive end spacer
4. Inner spacer
5. Stator
6. Slip ring bracket
7. Surge protection diode
8. Regulator assembly
9. End cover
10. Suppressor
11. Brush box
12. Through bolt
13. Rotor and slip ring
14. Drive end bracket
15. Woodruff key
16. Pulley

86–32

ALTERNATOR – Lucas 25 ACR

– Overhaul   86.10.08

Including Test (Bench)   86.10.14

NOTE: Alternator charging circuit – The ignition warning light is connected in series with the alternator field circuit. Bulb failure would prevent the alternator charging, except at very high engine speeds, therefore, the bulb should be checked before suspecting an alternator failure.

Precautions

Battery polarity is NEGATIVE EARTH, which must be maintained at all times.
No separate control unit is fitted; instead a voltage regulator of micro-circuit construction is incorporated on the slip-ring end bracket, inside the alternator cover.
Battery voltage is applied to the alternator output cable even when the ignition is switched off, the battery must be disconnected before commencing any work on the alternator. The battery must also be disconnected when repairs to the body structure are being done by arc welding.

Surge protection device

Some protection of the alternator is provided by a surge protection device, connected across the 'IND' terminal, to absorb high transient voltages.

Testing in position
Surge protection device

If the alternator output falls to zero, the fault may be caused by the surge protection device failing safe short-circuit, or a fault in the alternator circuit. Check the surge protection device as follows:–
1. Check that the fan belt is correctly tensioned.
2. Withdraw the connectors from the alternator.
3. Remove the alternator rear cover.
4. Remove the surge protection device from the 'IND' terminal.
5. Refit the alternator cover, and connectors. Ensure that all circuit connections are clean and tight.
6. Start and run the engine. If the alternator output is now normal, fit a new surge protection device.

445

Output test

7. Check that the fan belt is correctly tensioned and that all charging circuit connections are secure.
8. Run the engine at fast idle until normal operating temperature is attained.
9. Stop the engine.
10. Withdraw the connectors from the alternator.
11. Remove the alternator rear cover.
12. Connect the regulator case to the alternator frame.
13. Connect a 0-60 ammeter between the alternator and the battery.
14. Connect a 0-20 voltmeter across the battery terminals.
15. Connect a 15 ohm 35 amp, variable resistor across the battery terminals.

CAUTION: Do not leave the variable resistor connected across the battery terminals for longer than is necessary to carry out the following test, items 16 and 17.

16. Start the engine and run at 750 rev/min. The warning light bulb should be extinguished.
17. Increase the engine speed to 3000 rev/min, and adjust the variable resistance until the voltmeter reads 13.6 volts. The ammeter reading should then be approximately 60 amps. Any appreciable deviation from this figure will necessitate removing and dismantling the alternator. If the output test is satisfactory, proceed with the regulator test.

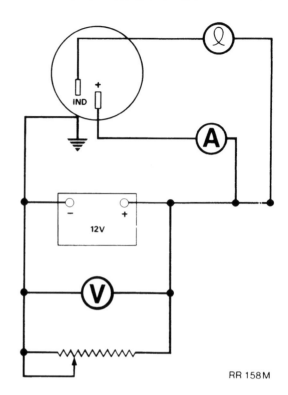

RR 158M

Regulator test

18. Disconnect the variable resistor and remove the connection between the regulator and the alternator frame.
19. With the remainder of the circuit connected as for the alternator output test, start the engine and run at 3000 rev/min, until the ammeter shows an output current of less than 10 amperes.
20. The voltmeter should now give a reading of 14.1 to 14.5 volts. Any appreciable deviation from this (regulating) voltage indicates a faulty regulator which must be replaced.
21. If the foregoing output and regulator tests show the alternator and regulator to be performing satisfactorily, disconnect the test circuit, reconnect the alternator terminal connector and proceed with the charging circuit resistance test.

Charging circuit resistance test

22. Connect a low-range voltmeter between either of the alternator terminals marked + and the positive terminal of the battery.
23. Switch on the headlamps.
24. Start the engine and run at approximately 3000 rev/min. Note the voltmeter reading.
25. Transfer the voltmeter connections to the frame of the alternator and the negative terminal of the battery, and again note the voltmeter reading.
26. If the reading exceeds 0.5 volt on the positive side or 0.25 volt on the negative side, there is a high resistance in the charging circuit which must be traced and remedied.

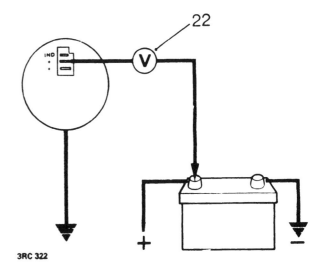

3RC 322

Testing – alternator removed

27. Remove the alternator, 86.10.02.
28. Remove the alternator rear cover.
29. Disconnect the terminal lead and connections associated with the suppression capacitor.
30. Detach the terminals and remove the surge diode.
31. Note the arrangement of regulator brush-box connector and disconnect. Remove regulator and fixings.
32. Unscrew brush moulding securing screws and rear brush-box.
33. Check brushes for wear by measuring length of brush protruding beyond brush box moulding. If length is 8 mm (0.3 in.) or less, brush must be renewed.
34. Check that brushes move freely in holders. If brush is sticking, clean with petrol moistened cloth or polish sides of brush with fine file.
35. Check brush spring pressure using push pull type spring gauge. Gauge should register 225g to 368g (9 to 13 oz) when brush is pushed back until face is flush with housing. If reading is outside these limits, renew brush assembly.

continued

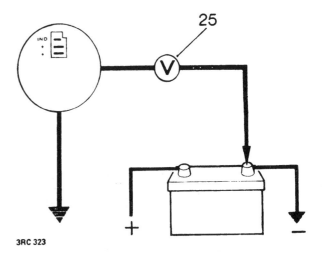

3RC 323

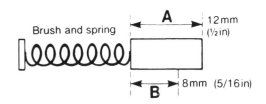

Brush and spring

A — 12mm (½in)

B — 8mm (5/16in)

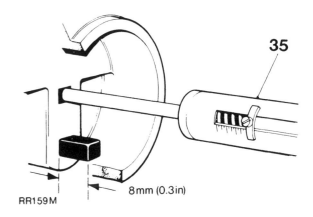

8mm (0.3in)

RR159M

447

36. Securely clamp alternator in vice and apply a soldering iron to the outer limbs of the rectifier terminals.
When the solder melts, prise out centre limb and
disconnect the stator winding cable ends.
CAUTION: When soldering or unsoldering connections to diodes take care not to overheat the diodes
or bend the pins. During soldering operations, diode
pins should be gripped lightly with a pair of long
nosed pliers which will act as a thermal shunt.

37. Connect a 12 volt battery and a 1.5 watt test lamp in
turn to each of the nine diode pins and corresponding heat sink on the rectifier pack, then reverse the
connections. Lamp should light with current flowing
in one direction only. If lamp lights in both directions or fails to light in either, rectifier pack must be
renewed.

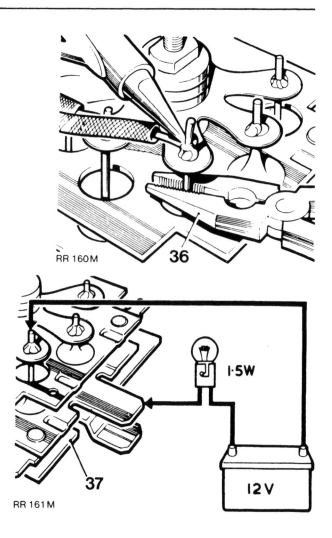

RR 160M   **36**

RR 161M

38. Clean surfaces of slip rings using petrol moistened
cloth.
Inspect slip ring surfaces for signs of burning; remove
burn marks using very fine sandpaper. On no account
should emery cloth or similar abrasives be used, or
any attempt made to machine the slip rings.

39. Connect an ohmmeter or a 12 volt battery and an
ammeter to the slip rings. An ohmmeter reading of
3.2 ohms or an ammeter reading of 4 amps should be
recorded.

40. Using a 110 volt a.c. supply and a 15 watt test lamp,
test for insulation between one of the slip rings and
one of the rotor poles. If the test lamp lights, the
rotor must be renewed.

continued

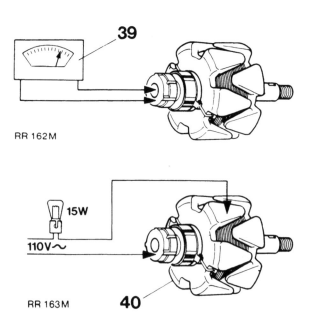

RR 162M

RR 163M   **40**

41. Remove the three through bolts.
42. Fit a tube over the slip ring moulding so that it registers against the outer track of the slip ring end bearing and carefully drive the bearing from its housing.
43. Remove the stator assembly noting its correct position to assist reassembly.

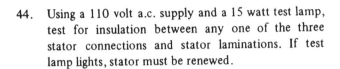
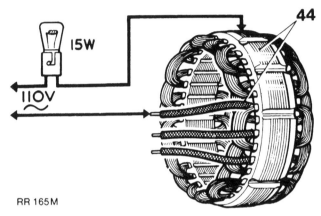

RR 164M

44. Using a 110 volt a.c. supply and a 15 watt test lamp, test for insulation between any one of the three stator connections and stator laminations. If test lamp lights, stator must be renewed.

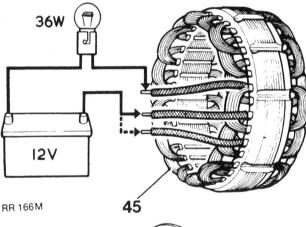

RR 165M

45. Connect a 12 volt battery and a 36 watt test lamp to two of the stator connections. Repeat the test replacing one of the two stator connections with the third. If test lamp fails to light in either test, stator must be renewed.

RR 166M

46. Remove the nut and washer; and detach the fan and pulley.

continued

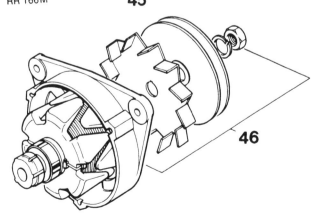

RR 167M

47. Remove the key by applying a flat ended engineers punch to the end of the woodruff key.
48. Remove the fan spacer.

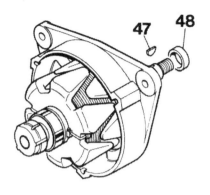

RR 168M

49. Using a suitable press, remove the rotor from the drive end bracket.

    CAUTION: Do not attempt to remove the rotor by applying hammer blows to the shaft end. Such action may burr over and damage the thread.

50. Remove the rotor spacing collar.

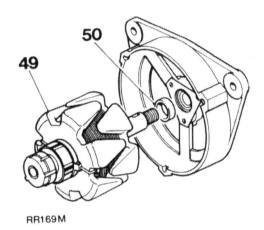

RR169M

51. Remove the three screws and remove the plate.
52. Press the bearing from the drive end bracket.
53. Collect up the metal plate and if fitted, the felt washer.

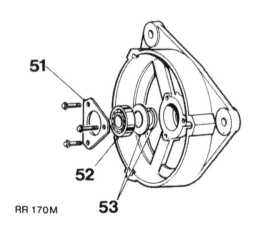

RR 170M

86-38

Reassembling

54.   Reverse the dismantling procedure, noting following points.
    a.   Use Shell Alvania 'RA' to lubricate bearings.
    b.   When refitting slip ring end bearing, ensure it is fitted with open side facing rotor.
    c.   Use Fry's H.T.3 solder on slip ring field connections.
    d.   When refitting rotor to drive end bracket, support inner track of bearing. Do not use drive end bracket to support bearing when fitting rotor.
    e.   Tighten through bolts evenly.
    f.   Fit brushes into housings before fitting brush moulding.
    g.   Tighten shaft nut to correct torque figure 3,5 to 4,2 kgf.m (25 to 30 lbf ft).
    h.   Refit regulator pack to brush moulding.

55.   Reconnect the leads between the regulator, brush box and rectifier.

56.   Refit the alternator 86.10.02.

DATA

Alternator . . . .   Lucas 25 ACR battery sensed
Nominal output . . . .   65 amps at 6000 alternator rev/min
Field resistance . . . . 3.2 ohms
Brush spring pressure . . . .   255 to 370 g (9 to 13 oz)
Brush minimum length . . . .   8 mm (0.312 in)
Regulating voltage . . . .   14.1 to 14.5 volts

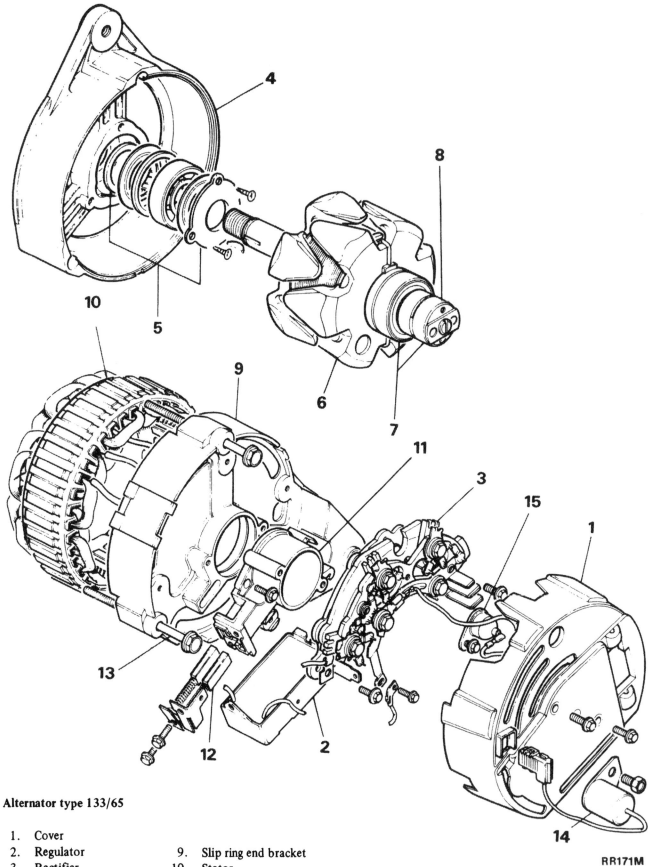

Alternator type 133/65

1. Cover
2. Regulator
3. Rectifier
4. Drive end bracket
5. Bearing assembly
6. Rotor
7. Slip ring end bearing
8. Slip rings
9. Slip ring end bracket
10. Stator
11. Brush box
12. Brushes
13. Through bolt
14. Suppressor
15. Surge protection diode

RR171M

ALTERNATOR − Lucas 113/65

− Overhaul	**86.10.08**
Including Test (Bench)	**86.10.14**

NOTE: Alternator charging circuit − The ignition warning light is connected in series with the alternator field circuit. Bulb failure would prevent the alternator charging, except at very high engine speeds, therefore, the bulb should be checked before suspecting an alternator failure.

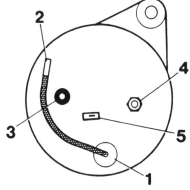

1. Suppression capacitor
2. Positive suppression terminal
3. IND terminal
4. + output terminal
5. Sensing terminal   RR 172 M

Precautions

Battery polarity is NEGATIVE EARTH, which must be maintained at all times.
No separate control unit is fitted; instead a voltage regulator of micro-circuit construction is incorporated on the slip ring end bracket, inside the alternator cover.
Battery voltage is applied to the alternator output cable even when the ignition is switched off, the battery must be disconnected before commencing any work on the alternator. The battery must also be disconnected when repairs to the body structure are being done by arc welding.

Surge protection device

Some protection of the alternator is provided by a surge protection device, to absorb high transient voltages.

Testing in position
Surge protection device

If the alternator output falls to zero, the fault may be caused by the surge protection device failing safe short-circuit, or a fault in the alternator circuit. Check the surge protection device as follows:−
1. Check that the fan belt is correctly tensioned.
2. Withdraw the connectors from the alternator.
3. Disconnect the suppressor and remove the alternator cover.
4. Remove the surge protection device.
5. Refit the alternator cover, and connectors. Ensure that all circuit connections are clean and tight.
6. Start and run the engine. If the alternator output is now normal, fit a new surge protection device.

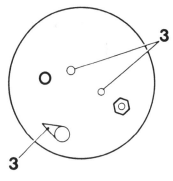

RR 173M

Output test

7. Check that the fan belt is correctly tensioned and that all charging circuit connections are secure.
8. Run the engine at fast idle until normal operating temperature is attained.
9. Stop the engine.
10. Withdraw the connectors from the alternator.

continued

86−41

11. Disconnect the suppressor and remove the alternator cover.
12. Connect the regulator case to the alternator frame.
13. Connect a 0-60 ammeter between the alternator and the battery.
14. Connect a 0-20 voltmeter across the battery terminals.
15. Connect a 15 ohm 35 amp variable resistor across the battery terminals.

CAUTION: Do not leave the variable resistor connected across the battery terminals for longer than is necessary to carry out the following test, items 16 and 17.

16. Start the engine and run at 750 rev/min. The warning light bulb should be extinguished.
17. Increase the engine speed to 3000 rev/min, and adjust the variable resistance until the voltmeter reads 13.6 volts. The ammeter reading should then be approximately 60 amps. Any appreciable deviation from this figure will necessitate removing and dismantling the alternator. If the output test is satisfactory, proceed with the regulator test.

Regulator test

18. Disconnect the variable resistor and remove the connection between the regulator and the alternator frame.
19. With the remainder of the circuit connected as for the alternator output test, start the engine and run at 3000 rev/min, until the ammeter shows an output current of less than 10 amperes.
20. The voltmeter should now give a reading of 13.6 to 14.4 volts. Any appreciable deviation from this (regulating) voltage indicates a faulty regulator which must be replaced.
21. If the foregoing output and regulator tests show the alternator and regulator to be performing satisfactorily, disconnect the test circuit, reconnect the alternator terminal connector and proceed with the charging circuit resistance test.

Charging circuit resistance test

22. Connect a low range voltmeter between either of the alternator terminals marked + and the positive terminal of the battery.
23. Switch on the headlamps and start the engine. Set the throttle to run at approximately 3000 rev/min. Note the voltmeter reading.
24. Transfer the voltmeter connections to the frame of the alternator and the negative terminal of the battery, and again note the voltmeter reading.
25. If the reading exceeds 0.5 volt on the positive side or 0.25 volt on the negative side, there is a high resistance in the charging circuit which must be traced and remedied.

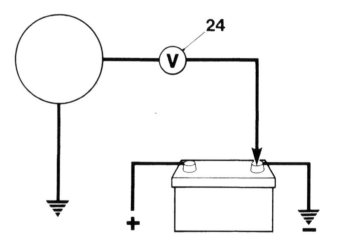

RR174M

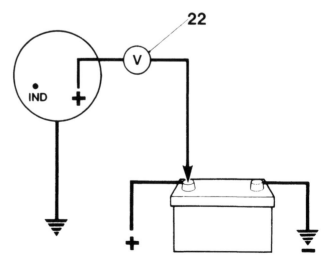

RR 175M

86–42

Testing – alternator removed

26. Withdraw the connectors from the alternator.
27. Remove the alternators, 86.10.02.
28. Disconnect the suppressor and remove the alternator cover.
29. Detach the surge protection device.
30. Disconnect the lead and remove the rectifier assembly.
31. Note the arrangement of the brush box connections and remove the screws securing the regulator to the brush box and withdraw. This screw also retains the inner brush mounting plate in position.
32. Remove the screw retaining the outer brush box in position and withdraw both brushes.
33. Check brushes for wear by measuring length of brush protruding beyond brush box moulding. If length is 10 mm (0.4 in.) or less, brush must be renewed.
34. Check that brushes move freely in holders. If brush is sticking, clean with petrol moistened cloth or polish sides of brush with fine file.
35. Check brush spring pressure using push-type spring gauge. Gauge should register 136 to 279g (5 to 10 oz) when brush is pushed back until face is flush with housing. If reading is outside these limits, renew brush assembly.
36. Remove the two screws securing the brush box to the slip ring end bracket and lift off the brush box assembly.
37. Securely clamp alternator in a vice and release the stator winding cable ends from the rectifier by applying a hot soldering iron to the terminal tags of the rectifier. Prise out the cable ends when the solder melts.

    CAUTION: When soldering or unsoldering connections to diodes take care not to overheat the diodes or bend the pins. During soldering operations, diode pins should be gripped lightly with a pair of long nosed pliers which will act as a thermal shunt.
38. Remove the two remaining screws securing the rectifier assembly to the slip ring end bracket and lift off the rectifier assembly. Further dismantling of the rectifier is not required.
39. Remove the slip ring end bracket bolts and lift off the bracket.
40. Connect a 12 volt battery and a 36 watt test lamp to two of the stator connections. Repeat the test replacing one of the two stator connections with the third. If test lamp fails to light in either test, stator must be renewed.

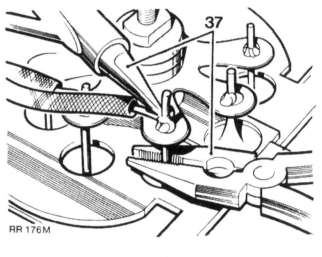

RR 176M

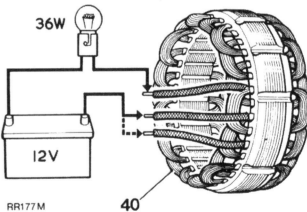

36W

12V

RR177M   40

continued

41. Using a 110 volt a.c. supply and a 15 watt test lamp, test for insulation between any one of the three stator connections and stator laminations. If test lamp lights, stator must be renewed.

42. Clean surfaces of slip rings using petrol moistened cloth.

43. Inspect slip ring surfaces for signs of burning; remove burn marks using very fine sandpaper. On no account should emery cloth or similar abrasives be used, or any attempt made to machine the slip rings.

44. Note the position of the stator output leads in relation to the alternator fixing lugs, and lift the stator from the drive end bracket.

45. Connect an ohmmeter or a 12 volt battery and an ammeter to the slip rings. An ohmmeter reading of 3.2 ohms or an ammeter reading of 4 amps should be recorded.

46. Using a 110 volt a.c. supply and a 15 watt test lamp, test for insulation between one of the slip rings and one of the rotor poles. If the test lamp lights, the rotor must be renewed.

47. To separate the drive end bracket and rotor, remove the shaft nut, washers, woodruff key and spacers from the shaft.

48. Remove bearing retaining plate by removing the three screws. Using a press, drive the rotor shaft from the drive end bearing.

49. If necessary, to remove the slip rings or the slip ring end bearing on the rotor shaft, unsolder the outer slip ring connection and gently prise the slip ring off the shaft, repeat the procedure for the inner slip ring connection. Using a suitable extraction tool, withdraw the slip ring bearing from the shaft.

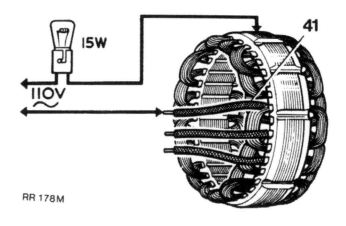

RR 178M

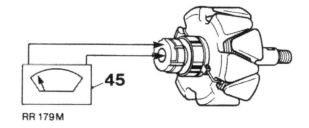

RR 179M

Reassembling

50. Reverse the dismantling procedure, noting following points.
    (a) Use Shell Alvania 'RA' to lubricate bearings.
    (b) When refitting slip ring end bearing, ensure it is fitted with open side facing rotor.
    (c) Use Fry's H.T.3 solder on slip ring field connections.
    (d) When refitting rotor to drive end bracket, support inner track of bearing. Do not use drive end bracket to support bearing when fitting rotor.
    (e) Tighten through-bolts evenly.
    (f) Fit brushes into housings before fitting brush moulding.
    (g) Tighten shaft nut to correct torque figure 3,5 to 4,2 kgf.m (25 to 30 lbf.ft).
    (h) Refit regulator pack to brush moulding.
51. Reconnect the leads between the regulator; brush box and rectifier, as illustrated.
52. Refit the alternator, 86.10.02.

DATA

Alternator . . . . Lucas A113/65 Battery sensed
Nominal output . . . . 65 Amps at 6000 alternator rev/min.
Field resistance . . . . 3.2 Ohms
Brush spring pressure . . . . 133 to 277 gm (4.7 to 9.8 oz)
Brush minimum length . . . . 10mm (0.312 in.)
Regulating voltage . . . . 13.6 to 14.4 Volts

BATTERY

– Remove and refit   86.15.11.

Removing

1. Disconnect the battery.
2. Remove wing nuts and battery retaining frame.
3. Remove battery.

Refitting

4. Reverse removal procedure, taking care when refitting the battery frame to avoid contact with the battery terminals.

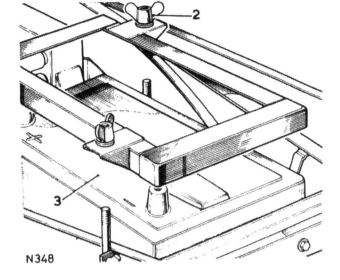

N348

HORNS

– Remove and refit   86.30.09.

Removing

1. Disconnect the battery.
2. Remove radiator grille.
3. Remove bolts, nuts and spring washers (nuts and washers only on later models) securing horn to bracket.
4. Disconnect wiring from horn.
5. Withdrawn horn.

NOTE. Twin horns are fitted. An identification letter is stamped on the front outer rim of the horn; 'H' – high note, 'L' – low note.

N349

Refitting

6. Reverse removal procedure.

SPARK PLUGS

– Remove, clean, adjust and refit   86.35.01

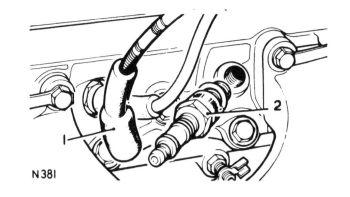

Removing

1. Withdraw leads by gripping end shrouds. DO NOT pull leads alone.
   NOTE: Remove the hot air pipe for access to the RH plugs as necessary.
2. Using special spanner and tommy bar supplied in vehicle tool kit, remove sparking plugs and washers.

Cleaning

3. Fit plug in plug cleaning machine.
4. Wobble plug with circular motion while operating abrasive blast **for a maximum of four seconds.**
   CAUTION: Excessive abrasive blasting will erode insulator nose.
5. Change to **air blast only** and continue to wobble plug for a minimum of thirty seconds to remove abrasive grit from plug cavity.
6. Wire brush plug threads, open gap slightly.
7. Using point file, square off electrode surfaces.
8. Set electrode gap: 0,60 mm (0.025 in.).
9. Test plugs in accordance with cleaning machine manufacturers instructions. If satisfactory, refit plugs in engine.
   IMPORTANT: If new plugs are necessary, refer to page 05−1.
10. Examine high tension leads, including coil to distributor lead, for insulation cracking or corrosion at end contacts. Fit new leads as necessary.
11. In addition to correct firing order, high tension leads must also be fitted in correct relation to each other to avoid cross firing. Figures in arrowed circles show plug lead numbers.
12. Leads at distributor cap must be connected as illustrated – Figures 1 to 8 inclusive – indicate plug lead numbers. RH – Right-hand side of engine when viewed from rear. LH – Left-hand side of engine when viewed from rear.
13. When pushing leads on plugs ensure ferrules within shrouds are firmly seated on plugs. A guide is that shroud ends are within 6 mm (0.250 in.) of metal body of plugs.

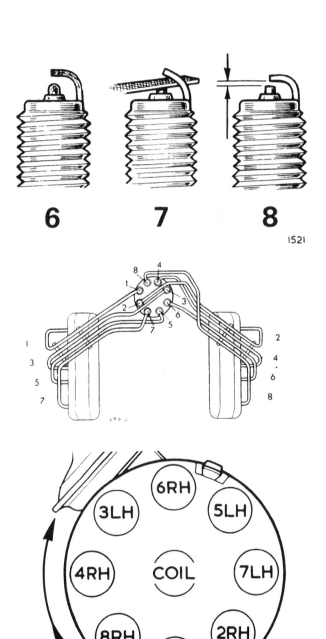

DISTRIBUTOR

– Remove and refit   86.35.20

Removing

1. Disconnect battery.
2. Disconnect vacuum pipe(s).
3. Remove distributor cap.
4. Disconnect low tension lead from coil.
5. Mark distributor body in relation to centre line of rotor arm.
6. Add alignment marks to distributor and front cover.
   NOTE: Marking distributor enables refitting in exact original position, but if engine is turned while distributor is removed, complete ignition timing procedure must be followed.
7. Release the distributor clamp and remove the distributor.

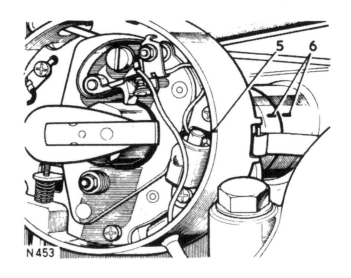

N 453

Refitting

NOTE: If a new distributor is being fitted, mark body in same relative position as distributor removed.

8. Lead for distributor caps should be connected as illustrated.

   Figures 1 to 8 inclusive indicate plug lead numbers.

   RH – Right hand side of engine, when viewed from the rear.

   LH – Left hand side of engine, when viewed from the rear.

9. If engine has not been turned whilst distributor has been removed, proceed as follows (items 10 to 17).
10. Fit new 'O' ring seal to distributor housing.
11. Turn distributor drive until centre line of rotor arm is 30° anti-clockwise from mark made on top edge of distributor body.
12. Fit distributor in accordance with alignment markings.
    NOTE: It may be necessary to align oil pump drive shaft to enable distributor drive shaft to engage in slot.
13. Fit clamp and bolt. Secure distributor in exact original position.
14. Connect vacuum pipe to distributor and low tension lead to coil.
15. Fit distributor cap.
16. Reconnect battery.
17. Using suitable electronic equipment, set dwell angle and ignition timing as follows.

continued

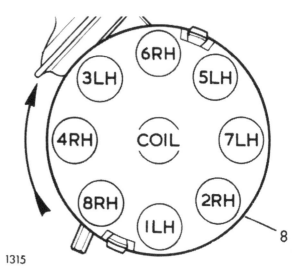

1315

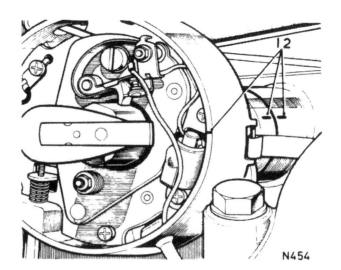

N454

86–48

18. If, with distributor removed, engine has been turned it will be necessary to carry out the following procedure.

19. Set engine – No. 1 piston to static ignition timing figure (see section 05) on compression stroke.

20. Turn distributor drive until rotor arm is approximately 30° anti-clockwise from number one sparking plug lead position on cap.

21. Fit distributor to engine.

22. Check that centre line of rotor arm is now in line with number one sparking plug lead on cap. Reposition distributor if necessary.

23. If distributor does not seat correctly in front cover, oil pump drive is not engaged. Engage by lightly pressing down distributor while turning engine.

24. Fit clamp and bolt leaving both loose at this stage.

25. Turn engine back until crankshaft pulley static ignition timing mark passes timing pointer on front cover, then turn engine forward until pointer aligns with the static ignition timing value.

26. Rotate distributor anti-clockwise until contact points just start to open.

27. Secure distributor in this position by tightening clamp bolt.

28. Connect vacuum pipe(s) to distributor and low tension lead to coil.

29. Fit distributor cap.

30. Reconnect battery.

31. Using suitable electronic equipment, set dwell angle and ignition timing as follows:–

Ignition timing and dwell angle settings

Ignition timing: Refer to Section 05.

Dwell angle: 26° to 28°

NOTE: It is essential that the following procedure is adhered to. Inaccurate timing can lead to serious engine damage.

32. Before attempting to start the engine, set the ignition timing statically as specified in the Engine Tuning Data, page 05–6. (This sequence is to give only an approximation in order that engine may be run. On no account should engine be started before this check is carried out.)

33. Start the engine and set to the correct idling speed according to the distributor type number.

NOTE: If the specification on page 05–6 calls for the distributor vacuum pipe to be disconnected, do this before setting the idle speed, and block the pipe to prevent an air leak into the manifold.

Set dwell angle as follows:–

34. Set selector knob to 'calibrate' position. Adjust calibration knob to give a zero reading on meter.
35. Couple up Tach-dwell meter to engine following manufacturers' instructions.
36. Set selector knob to 8-cylinder position and Tach-dwell selector knob to "dwell". Adjust distributor dwell angle by turning hexagon headed adjustment screw on distributor body until meter reads 26° to 28°.
37. If meter used does not have an 8-cylinder position, set selector knob to 4-cylinder position and adjust at distributor until meter reads 52° to 56°. Uncouple Tach-dwell meter.

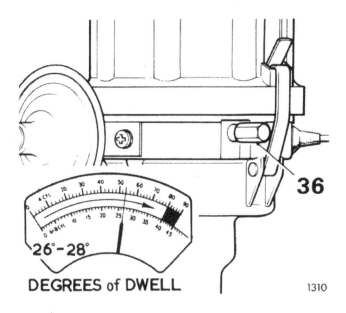

Set ignition timing as follows.

38. Couple Stroboscopic timing lamp to engine following manufacturers' instructions, with high tension lead attached into No. 1 cylinder plug lead.
39. Check distributor clamping bolt is slack and engine idle speed is correct.
40. Check ignition timing. Stroboscopic lamp must synchronise timing pointer and timing mark on crankshaft pulley as specified on page 05–6.
41. If necessary adjust timing. Turn distributor clockwise to retard or anti-clockwise to advance.
42. Tighten distributor clamping bolt.
43. Disconnect stroboscopic timing lamp.

NOTE: **Engine speed accuracy during ignition timing is of paramount importance. Any variation from the required idle speed particularly in an upwards direction will lead to wrongly set ignition timing.**

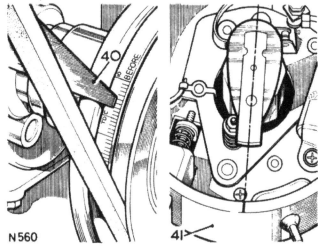

Automatic ignition advance mechanism

The distributor incorporates two automatic ignition advance mechanisms – a vacuum-controlled unit related to carburetter choke depression and a centrifugally-controlled unit related to engine speed. Both units are connected to the contact breaker assembly, and operate independently, progressively moving the contact breaker through a small arc about the cam.

A loss of engine performance, particularly a sudden loss, could be due to a malfunction of either of the automatic advance mechanisms, and where suitable electronic engine tuning and testing equipment is available, both units can be checked against the figures detailed in page 05–6.

The test should commence at maximum advance conditions and be checked during deceleration.

DISTRIBUTOR

– Overhaul   86.35.26

NOTE: A number of slightly varying designs of distributor may be fitted. The instructions below are generally applicable to all units unless otherwise stated.

For removal and replacement of the latest (Blue cap) type Distributor System see page 10–15.

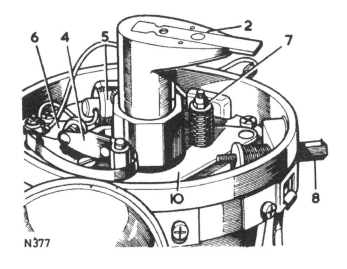

Dismantling

1. Unclip and remove distributor cap.
2. Withdraw rotor arm and felt lubricating pad.
3. Remove contact spring or remove the nut and lift off the insulating bush together with the low tension and capacitor leads.
4. *Early models* – Lift off the moving contact point and the insulating washer from the contact pivot and spring post.
   Later models – Remove the "Quickfit" contact set.
5. Remove the capacitor.
6. *Early models* – Remove the fixed contact.
7. *Early models* – Remove the nut, plain washer and spring from the contact breaker base plate pivot pin.
8. Remove the dwell angle adjuster screw and spring.
9. Remove the earth lead from the centrifugal advance cover plate.
10. Remove the contact breaker base plate.
11. Remove the vacuum unit and grommet.
    Reverse instructions 10 and 11 for later models.
12. Remove the centrifugal advance cover plate.
13. Carefully withdraw the two springs from the centrifugal advance unit.
14. Remove the car from inside the cam and lift off the cam and cam foot.
15. Remove the two weights.
16. Drive out the pin securing the driving gear and remove the gear and tab washer.
17. Check all parts for wear or damage and renew as necessary.
    NOTE: A replacement driving gear is undrilled. Drill as illustrated. The hole should align with the centre line of the driving slot and the centre of the gear teeth. A number 12 drill is suitable.

continued

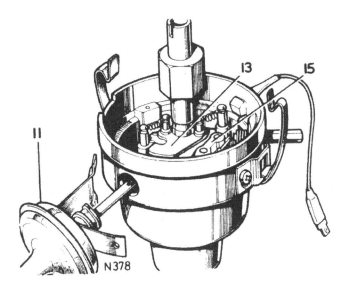

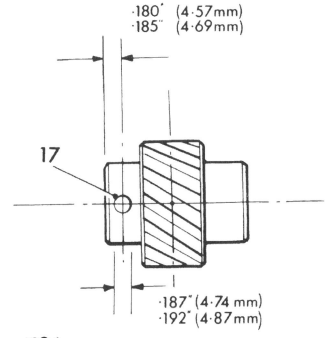

·180″ (4·57mm)
·185″ (4·69mm)

·187″ (4·74 mm)
·192″ (4·87mm)

1304

Assembling

18. Reassemble the distributor by reversing the dismant-
    ling instructions.
    NOTE: It will assist reassembly of later units if the
    vacuum advance lever is located to the base plate by
    the "Quickfit" contacts before the vacuum unit is
    secured to the distributor body.
19. When fitting the centrifugal governor springs, take
    care not to stretch them.
20. The points should be set to a clearance of 0,35 to
    0,40 mm (.014 to .016 in.).

NOTE: It is most important that the dwell angle is adjusted
to 26° to 28° using specialised equipment when the
distributor has been refitted.

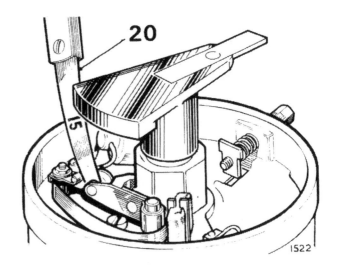

IGNITION COIL

– Remove and refit   86.35.32

Removing

1. Disconnect the battery.
2. Disconnect electrical leads from the coil.
3. Remove the two retaining bolts, nuts and washers
   securing the coil to the valance.

Refitting

4. Reverse the removal procedure.

NOTE: The electrical leads are fitted with male and female
connectors; ensure that they are fitted to the corresponding
blade on the ignition coil.

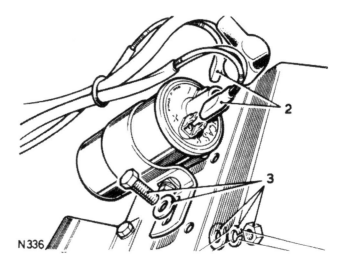

86–52

HEADLAMP ASSEMBLY

– Remove and refit (Tungsten)   86.40.02

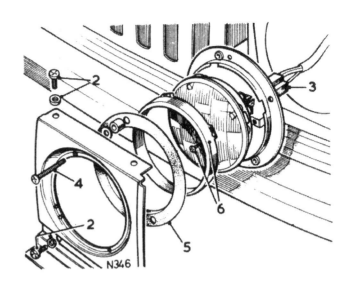

N346

Removing

1. Disconnect the battery.
2. Remove screws and washers securing the headlamp frame to the body.
3. Ease the headlamp assembly forward and disconnect.

NOTE: When headlamp washers and wipers are fitted, it will be necessary to disconnect the washer jet tube (84.20.09) and to disengage the wiper rack (84.25.12) before removing the headlamp and frame from the body.

4. Remove the two adjusting screws and (later models) one clamp to separate lamp unit from the frame.
5. Remove the rubber seal.
6. Separate the lamp unit from rim by loosening the three retaining screws and turning the lamp in an anti-clockwise direction.

Refitting

7. Reverse removal procedure, referring to 84.25.12, and 84.20.09 when headlamp washers and wipers are fitted.

– Remove and refit (Halogen)   86.40.02

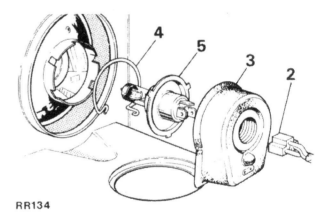

RR134

Removing

1. Prop open the bonnet. Two large clearance holes are provided, one on each side of the front valance, to give access to the respective bulb holders in the headlamp reflectors.

NOTE: To obtain access to the right-hand clearance hole it will be necessary to remove the battery from the vehicle.

2. Disconnect the multi-plug lead.
3. Remove the rubber dust cover.
4. Release the bulb retaining spring clip.
5. Remove the faulty bulb.

N948

Refitting

6. Fit the correct 'Halogen' type. The bulb holder is keyed to facilitate fitting.

   IMPORTANT: Do not touch the quartz envelope of the bulb with the fingers. If contact is accidentally made wipe gently with methylated spirits.

7. Refit the bulb retaining spring clip rubber dust cover and multi-plug lead.
8. In the case of right-hand bulb replacement, refit the battery.
9. Reverse remaining removal procedure.

86–53

REFLECTORS

– Remove and refit   86.40.65

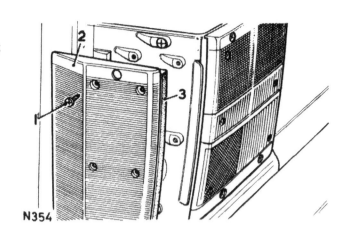

N354

Removing

1. Remove the five screws securing reflector.*
2. Remove reflector.
3. Remove rubber seal.

Refitting

4. Reverse the removal procedure.

*   On later version with additional fog guard lamps facility the reflector is secured with four screws.

TAIL, STOP AND FLASHER LAMP ASSEMBLY, LEFT HAND

– **Remove and refit**   86.40.70

TAIL, STOP AND FLASHER LAMP ASSEMBLY, RIGHT HAND

– Remove and refit   86.40.71

REVERSE LAMP ASSEMBLY

– Remove and refit   86.40.91

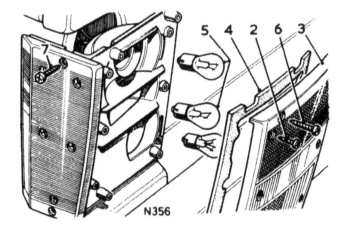

N356

Removing

1. Disconnect the battery.
2. Remove the six lens retaining screws.*
3. Remove lens.
4. Remove sealing rubber.
5. Remove the bulbs.
6. Remove the four screws securing the lamp unit to the body.
7. Remove the two through-screws from the reflector side, which also secure the lamp unit to the body.
8. Ease the lamp unit forward and disconnect leads at moulded connectors.

Refitting

9. Reverse the removal procedure.

*   On later versions with additional fog guard lamps facility the lens is secured by four screws.

NUMBER PLATE LAMP ASSEMBLY

– Remove and refit   86.40.86

Removing

1. Disconnect the battery.
2. Remove the two screws securing lens hood.
3. Remove lens hood.
4. Remove lens.
5. Remove rubber gasket.
6. Remove the bulb.
7. Remove the two screws, washers and nuts securing the lamp base.
8. Disconnect leads from base.

Refitting

9. Reverse the removal procedure.

ROOF LAMP ASSEMBLY

– Remove and refit   86.45.02

The courtesy lamps fitted to the interior are operated automatically by switches incorporated in both door pillars and also separately from a switch on the steering column nacelle. Alternative lamps having two bulbs (as illustrated) or a single bulb may be fitted.

Removing

1. Disconnect the battery.
2. Remove the lens from the courtesy lamp by pressing upward and turning it anti-clockwise.
3. Withdraw bulb(s) from spring clip holders.
4. Push out cable nipples from spring clip connectors.
5. Remove screws securing lamp base to roof panel.
6. Lower the lamp and release the cables.

Refitting

7. Reverse the removal procedure.

INTERIOR ROOF LAMPS CIRCUIT DELAY – 4-DOOR MODEL AND 1981 2-DOOR MODEL

The roof lamp circuit incorporates a delay function which is designed to allow the lamps to remain on for 12 to 18 seconds after the last door is closed. Note, switching on the ignition after the last door is closed will immediately override this feature, switching the interior lamps off.

MAINTENANCE

To gain access to the delay unit withdraw the lower fascia panel as described previously. The red delay unit is attached to the outer side of the steering column support bracket behind the fascia by a single nut and screw.
To remove the red delay unit pull off the five pin multiplug. The delay unit itself can also be pulled off leaving its mounting bracket still attached to the steering column support bracket.

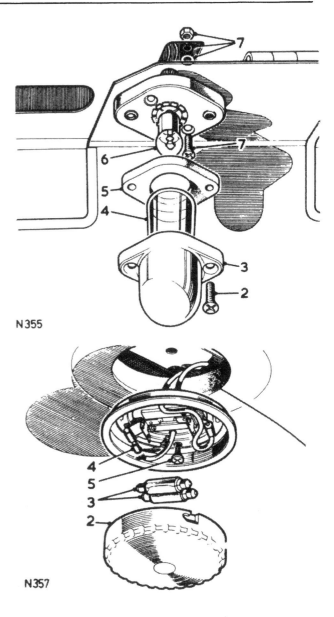

N355

N357

PANEL ILLUMINATION LAMP ASSEMBLY

– **Remove and refit**   86.45.32.

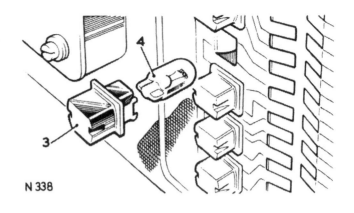

N 338

Removing

1. Disconnect the battery.
2. Release instrument binacle cowling by pressing in the cowling at the bottom rear and lifting it over the retaining clips.
3. Pull cowling to one side and remove appropriate bulb holder.
4. Remove bulb.

Refitting

5. Reverse removal procedure.

INSPECTION SOCKETS

– **Remove and refit**   86.45.33

N 339

Removing.

1. Disconnect the battery.
2. Remove the four drive screws securing bottom shroud to top shroud.
3. Remove single screw securing top shroud to switch housing bracket.
4. Remove top shroud and lower the bottom shroud.
5. Disconnect electrical leads from sockets.
6. Remove screw lock-rings spring and plain washers and insulation tab.

Refitting

7. Reverse the removal procedure, ensuring that the insulation tab is central between the two sockets.

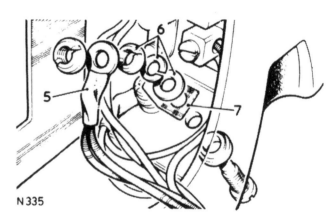

N 335

DIFFERENTIAL LOCK ILLUMINATION LAMP

— Remove and refit   86.45.42

Models with warning lamp by gear lever

Removing

1. Disconnect the battery.
2. Remove chrome locking ring on switch knob using a lock ring spanner.
3. Withdraw spring with bulb attached.
4. Unscrew bulb.

Refitting

5. Reverse the removal procedure.

Models with warning lamp in fascia

Removing

1. Gently ease the warning lamp body from the fascia.
2. Withdraw the bulb holder and remove the bayonet type bulb.

Refitting

3. Reverse instructions 1 and 2 ensuring that the locating lugs are correctly positioned.

WARNING LAMP CLUSTER

— Remove and refit   86.45.62

NOTE: Alternator charging circuit
The ignition warning light is connected in series with the alternator field circuit. Bulb failure would prevent the alternator charging, except at very high engine speeds, therefore, a failed bulb should be changed with the minimum delay, otherwise the battery will be discharged.

Removing

1. Disconnect the battery.
2. Release instrument binacle cowling by pressing in cowling at the bottom and lifting over the retaining clips.
3. Pull cowling to one side and remove appropriate bulb holder.
4. Remove bulb.

Refitting

5. Reverse the removal procedure.

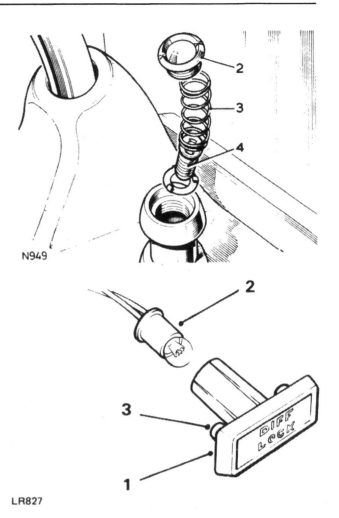

N949

LR827

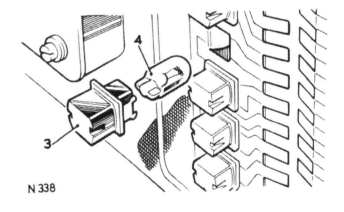

N 338

STARTER MOTOR RELAY

– Remove and refit   **86.55.05.**

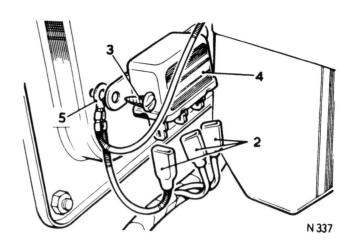

N 337

Removing

1. Disconnect the battery.
2. Disconnect electrical leads from the relay.
3. Remove the two self-tapping screws.
4. Remove relay.

Refitting

5. Reverse the removal procedure, securing the earth lead by one of the self-tapping screws, ensuring a good connection.

FLASHER UNIT,

– Remove and refit   **86.55.11.**

HAZARD FLASHER UNIT

– Remove and refit   **86.55.12.**

Removing

1. Disconnect the battery.
2. Remove the lower fascia panel.
3. Pull flasher/hazard unit from its spring clip, located on the steering column support.
4. Disconnect electrical leads from flasher unit.
5. Disconnect electrical socket from hazard unit.

N 334

Refitting

6. Reverse the removal procedure.

STARTER MOTOR

– Remove and refit 86.60.01

Removing

1. Place car on a suitable ramp.
2. Disconnect the battery.
3. Disconnect the leads from the solenoid and starter motor and (later models) remove the exhaust heat shield.
4. Remove the two bolts securing the starter motor to the flywheel housing.
5. Remove starter motor from underneath the vehicle.

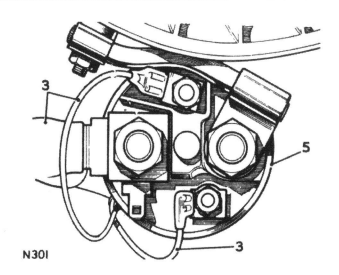

N301

Refitting

6. Reverse the removal procedure.
7. Tighten the bolts securing the starter motor to cylinder block to a torque of 4,0 to 4,9 kgf.m (30 to 35 lbf.ft).

STARTER SOLENOID

–Remove and refit 86.60.08

Removing.

1. Disconnect the battery.
2. Place car on a suitable ramp.
3. Disconnect the leads from the solenoid and starter motor.
4. Remove the two bolts securing the starter motor to the flywheel housing.
5. Remove the starter motor from underneath the vehicle.

NOTE. The starter solenoid is integral with the starter motor.

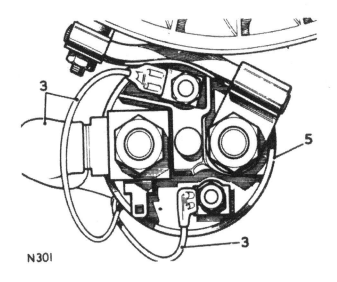

N301

Refitting.

6. Reverse the removal procedure.
7. Tighten the bolts securing the starter motor to cylinder block to a torque of 4,0 to 4,9 kgf.m (30 to 35 lbf ft).

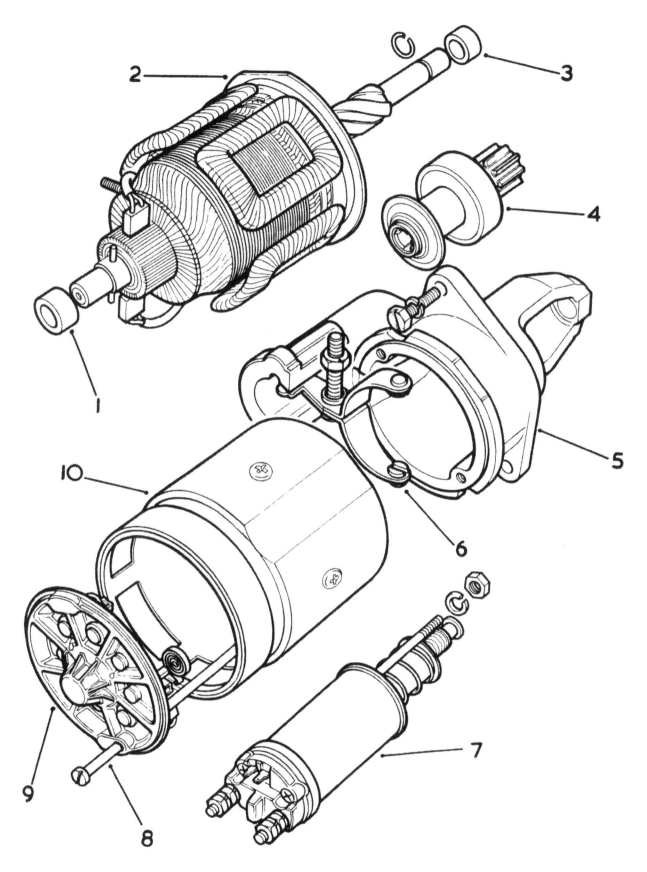

N574

STARTER MOTOR

1—	Bush, commutator end	6—	Pivot pin
2—	Field coil	7—	Starter solenoid
3—	Bush for bracket	8—	Bolt for starter
4—	Drive assembly	9—	Bracket, commutator end
5—	Bracket, drive end	10—	Cover band

STARTER MOTOR

– Overhaul   86.60.13.

Dismantling

1. Disconnect the copper link between lower solenoid terminal and starter motor casing.
2. Remove solenoid from drive end bracket.
3. Move starter pinion to end of its travel and disengage solenoid plunger from the engagement lever; withdraw plunger and spring.
4. Remove cover band.
5. Hold back brush springs and remove brushes.
6. Remove commutator end bracket from the starter yoke.
7. Separate the yoke from the drive end bracket.
8. Remove the eccentric pin and engagement lever from the drive end bracket.
9. Withdraw the armature, drive gear and intermediate bracket.
10. Using a suitable tube, remove the collar and jump ring from the armature shaft extension.
11. Remove the jump ring, collar, drive assembly, intermediate bracket and rubber seal.
12. Remove the brake ring, steel washer and tufnol washer from the commutator end bracket.

To overhaul

Clutch

13. Check that the clutch will:
    (a) Give instantaneous take up of the drive in one direction.
    (b) Rotate easily and smoothly in the other direction.
    (c) Be free to move round and along the shaft splines without any tendency to bind. All moving parts should be smeared with Shell Retinax 'A' grease or an equivalent.

NOTE. The roller clutch drive is sealed in a rolled steel outer cover and cannot be dismantled.

Continued

Brushes

14. Examine the brushes in the drive end bracket, intermediate and commutator end brackets; renew as necessary.

15. Check that brushes move freely in their holders while holding back the brush springs. If a brush is damaged or worn so that it does not make good contact with the commutator, all the brushes must be renewed. The new brushes are pre-formed; 'bedding' to the commutator is therefore unnecessary.

16. The flexible connectors are soldered to the terminal tabs, two to the field coils and two to the brush boxes. The flexible connectors must be unsoldered and the new brushes secured in position by re-soldering.

17. Using a spring balance, check that the tension of the brush springs is between 0.85 Kg. to 1.13 Kg. (30 oz. to 40 oz.). If tension is low, fit new springs.

18. Check the commutator, having first cleaned with a petrol (gasoline) moistened cloth. If necessary, rotate the armature and remove pits and burns using fine glass paper. If the commutator is badly worn, mount in a lathe and with a very sharp tool take a light cut, removing no more metal than is necessary.

IMPORTANT. Commutator segments must not be undercut.

Auxiliary contacts, to check

19. Disconnect all cables from the solenoid terminals and connectors.

20. Connect a test lamp between the connector marked 'IGN' and a good earth.

21. Connect the battery to the small unmarked connector.

22. Momentarily connect the large battery terminal to earth. The solenoid contacts should close fully and remain closed; the test lamp should emit a steady light.

23. Disconnect the battery from the small unmarked connector. The contacts should open, the solenoid will release and the test lamp extinguish. The period of energising should be as brief as possible to avoid overheating the windings.

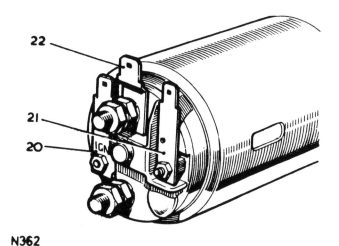

N362

Continued

Assembling

24. Fit the intermediate bracket onto the armature together with the drive assembly, stop collar, new jump ring and thrust washer.

25. Position the pre-engagement lever on to the drive assembly on the armature shaft and locate the lever and armature in the drive end bracket. The flatter edge on the pre-engagement lever must face toward the solenoid when fitted.

26. Fit the eccentric pivot pin to retain pre-engagement lever in drive end bracket, then temporarily turn pin in reverse direction until the eccentric cam on the pin allows the pre-engagement lever to move fully toward the armature shaft.

27. Fit the yoke over the armature and locate onto end bracket dowel.

28. Fit the tufnol washer, thrust washer and brake ring into the commutator end bracket, with the brake ring angled slots uppermost.

29. With the brushes positioned clear of the commutator, fit the bracket on to the yoke, locating the dowel and the drive pin into the brake ring.

30. Replace brushes.

31. Replace cover band.

32. With the drive assembly held in the forward position, locate the solenoid plunger over the engagement lever.

33. Fit the return spring over the plunger. Fit the solenoid on to the plunger and locate it in the drive end bracket. Secure with nuts and spring washers.

34. Refit the copper link between the lower solenoid terminal and the starter motor casing.

35. Before tightening the locknut on the eccentric bolt, check the pinion movement by connecting the small terminal on the solenoid to one side of a six volt battery, and the other side of the battery through a switch to a solenoid fixing stud.

36. Close the switch, thus throwing the drive assembly forward into the engaged position.

37. Measure the distance between the pinion and the collar on the armature shaft.
The setting should be 0.12 mm to 0.40 mm (0.005 in. to 0.015 in.) measured with the pinion pressed lightly towards the armature to take up any free play in the engagement linkage.

38. Adjust if necessary by rotating the eccentric bolt until the correct clearance is obtained, then secure with the lock nut.

STARTER MOTOR — Lucas 3M100PE

—Overhaul   86.60.13

Dismantling

1.   Remove the starter motor. 86.60.01.
2.   Remove the connecting link between the starter and the solenoid terminal 'STA'.
3.   Remove the solenoid from the drive end bracket.
4.   Grasp the solenoid plunger and lift the front end to release it from the top of the drive engagement lever.
5.   Remove the end cap seal.
6.   Using an engineer's chisel, cut through a number of the retaining ring claws until the grip on the armature shaft is sufficiently relieved to allow the retaining ring to be removed.
7.   Remove the two through bolts.
8.   Partially withdraw the commutator end cover and disengage the two field coil brushes from the brush box.
9.   Remove the commutator end cover.
10. Withdraw the yoke and field coil assembly.
11. Remove the retaining ring from the drive engagement lever pivot-pin, using the method previously described.
12. Withdraw the pivot pin.
13. Withdraw the armature.
14. Using a suitable tube, remove the collar and jump ring from the armature shaft.
15. Slide the thrust collar and the roller clutch drive and lever assembly off the shaft.

continued

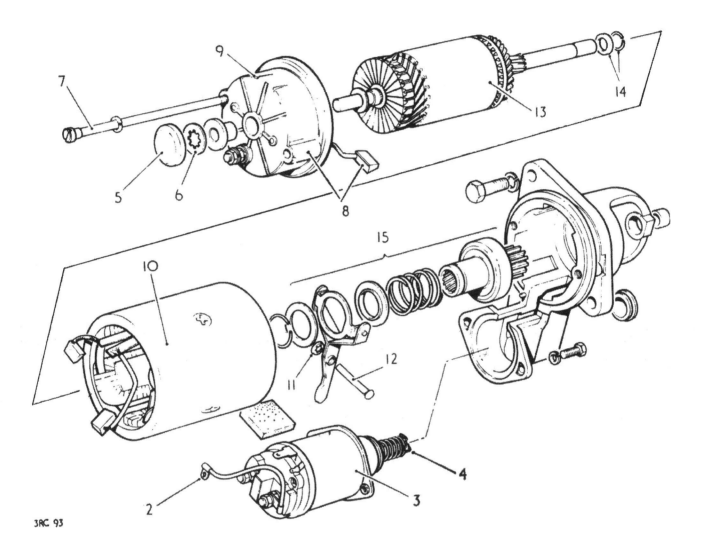

3RC 93

Inspecting

Clutch

16. Check that the clutch gives instantaneous take-up of the drive in one direction and rotates easily and smoothly in the other direction.
17. Ensure that the clutch is free to move round and along the shaft splines without any tendency to bind.
NOTE. The roller clutch drive is sealed in a rolled steel cover and cannot be dismantled.
18. Lubricate all clutch moving parts with Shell SB 2628 grease for cold and temperate climates or Shell Retinax 'A' for hot climates.

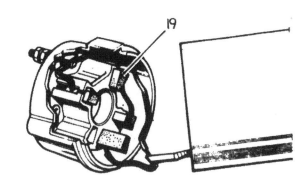

3RC94

Brushes

19. Check that the brushes move freely in the brush box moulding. Rectify sticking brushes by wiping with a petrol moistened cloth.
20. Fit new brushes if they are damaged or worn to approximately 9,5 mm (0.375 in).
21. Using a push-type spring gauge, check the brush spring pressure. With new brushes pushed in until the top of the brush protrudes about 1,5 mm (0.065 in) from the brush box moulding, the spring pressure reading should be 1,0 kgf (36 ozf.).
22. Check the insulation of the brush springs by connecting a 110V a.c 15W test lamp between a clean part of the commutator end cover and each of the springs in turn. The lamp should not light.

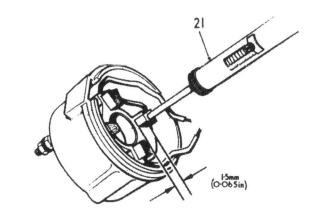

3RC95

Armature

23. Check the commutator. If cleaning only is necessary, use a flat surface of very fine glass paper, and then wipe the commutator surface with a petrol moistened cloth.
24. If necessary, the commutator may be machined providing a finished surface can be obtained without reducing the thickness of the commutator copper below 3,5 mm (0.140 in), otherwise a new armature must be fitted. Do not undercut the insulation slots.
25. Check the armature insulation by connecting 110V a.c. 15W test lamp between any one of the commutator segments and the shaft. The lamp should not light, if it does light fit a new armature.

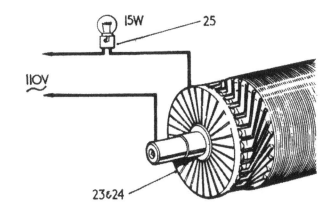

3RC96

continued

86–66

Field coil insulation

26. Disconnect the end of the field winding where it is riveted to the yoke, by filing away the riveted over end of the connecting-eyelet securing rivet, sufficient to enable the rivet to be tapped out of the yoke.
27. Connect a 110V a.c. 15W test lamp between the disconnected end of the winding and a clean part of the yoke.
28. Ensure that the brushes or bare parts of their flexibles are not touching the yoke during the test.
29. The lamp should not light, if it does light, fit a new field coil assembly.
30. Resecure the end of the field winding to the yoke.

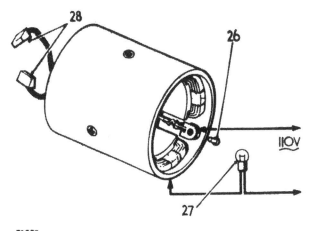

3RC97

Field coil continuity

31. Connect a 12V battery operated test lamp between each of the brushes in turn and a clean part of the yoke.
32. The lamp should light, if it does not light, fit a new field coil assembly.

Solenoid

33. Disconnect all cables from the solenoid terminals and connectors.
34. Connect a 12V battery and a 12V 60W test lamp between the solenoid main terminals. The lamp should not light, if it does light, fit new solenoid contacts or a new solenoid complete.
35. Leave the test lamp connected and, using the same 12V battery supply, energise the solenoid by connecting 12V between the small solenoid operating 'Lucar' terminal blade and a good earth point on the solenoid body.
36. The solenoid should be heard to operate and the test lamp should light with full brilliance, otherwise fit new solenoid contacts or a new solenoid complete.

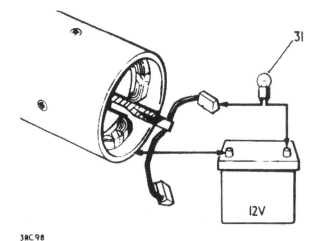

3RC98

Reassembling

37. Reverse 1 to 15, including the following:
38. Fit the commutator end cover before refitting the solenoid to facilitate assembly of the block shaped grommet which, when assembled, is compressed between the yoke, solenoid and fixing bracket.
39. Ensure that the internal thrust washer is fitted to the commutator end of the armature shaft.
40. Tightening torques:
    Through bolts 1,1 kgf.m (8.0 lbf ft).
    Solenoid fixing stud nuts 0,6 kgf.m (4.5 lbf ft).
    Solenoid upper terminal nuts 0,4 kgf.m (3.0 lbf ft).
41. Set the armature end float by driving the retaining ring on the armature shaft into a position that provides a maximum of 0,25 mm (0.010 in) clearance between the retaining ring and the bearing bush shoulder.

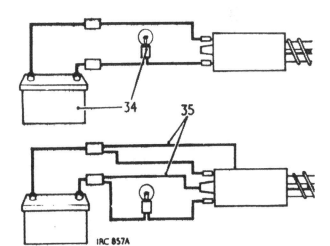

IRC 857A

IGNITION STARTER SWITCH

Early models

– Remove and refit   86.65.02

Removing

1. Disconnect the battery.
2. Remove the four drive screws securing bottom shroud to top shroud.
3. Remove the single screw securing top shroud to switch housing bracket.
4. Remove top shroud and lower the bottom shroud.
5. Disconnect the ignition switch cables at the connector socket.
6. Remove the two small screws securing ignition/starter switch to housing.
7. Withdraw the switch.

Refitting

8. Reverse the removal procedure.
9. Locate the lug on the side of the switch with the groove on the inside of the housing.

Later models Refer to 57.40.31.

The steering column switch layout has been standardised for left and right hand drive vehicles and is as follows:–

LEFT HAND CONTROLS

Upper switch – Main lighting switch
Lower switch – Main and dipped beam, direction indicators and horn.

RIGHT HAND CONTROLS

Upper switch – Rear fog guard
Lower switch – Windscreen programmed wash/wipe.

*LIGHTING SWITCH

– Remove and refit   86.65.10

Removing

1. Disconnect the battery.
2. Remove the four drive screws securing bottom shroud to top shroud.
3. Remove the single screw securing top shroud to switch housing bracket.
4. Remove top shroud, and lower the bottom shroud.
5. Disconnect cables at snap connectors.
6. Loosen the switch retaining lock-nut.
7. Slide switch unit away from its bracket.

Refitting

8. Reverse the removal procedure.

* 4-Door and latest 2-Door vehicles.

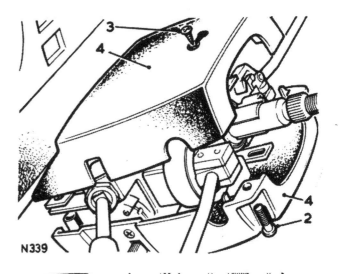

N339

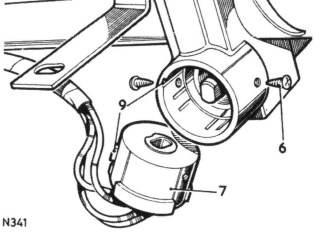

N341

N342

86–68

PANEL LIGHT SWITCH

– Remove and refit   86.65.12.

N339

Removing

1. Disconnect the battery.
2. Remove the four drive screws securing bottom shroud to top shroud.
3. Remove single screw securing top shroud to switch housing bracket.
4. Remove top shroud and lower the bottom shroud.
5. Disconnect electrical leads from switch.
6. Remove switch securing screws and cup washers.
7. Withdraw switch.

Refitting

8. Reverse the removal procedure.

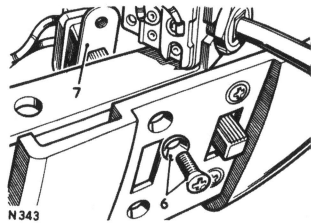

N343

DOOR PILLAR SWITCH

– Remove and refit   86.65.15.

Removing

1. Disconnect the battery.
2. Remove screw securing switch to door pillar.
3. Withdraw switch.
4. Disconnect electrical lead from connector blade.

Refitting

5. Reverse removal procedure.

N347

86–69

*SPOT/FOG/AUXILIARY LIGHT SWITCH

– **Remove and refit**   86.65.19

Removing

1. Disconnect the battery.
2. Remove the four drive screws securing bottom shroud to top shroud.
3. Remove the single screw securing top shroud to switch housing bracket.
4. Remove top shroud, and lower the bottom shroud.
5. Disconnect cables at snap connectors.
6. Loosen the switch retaining lock-nut.
7. Slide switch unit away from its bracket.

N339

Refitting

8. Reverse the removal procedure.

NOTE: From vehicle VIN number 100783 onwards the above switch facility was incorporated into the rear fog guard circuit. See circuit diagram (1979) on page 86–7.

*   See page 86–68.

N342

REVERSE LIGHT SWITCH

– **Remove and refit**   86.65.20

Removing

1. Disconnect the battery.
2. Remove gear lever rubber cover to give access to the switch.
3. Disconnect electrical leads from the switch.
4. Unscrew switch from gearbox.

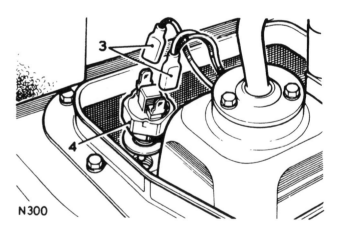

N300

Refitting

5. Reverse the removal procedure, tightening the switch to a torque of 1,4 to 2,0 kgf.m (15 to 20 lbf.ft).
6. Check the switch operation, adjusting the shims as necessary.

ROOF LIGHT PANEL SWITCH

– Remove and refit   86.65.21

Removing

1. Disconnect the battery.
2. Remove the four drive screws securing bottom shroud to top shroud.
3. Remove single screw securing top shroud to switch housing bracket.
4. Remove top shroud and lower the bottom shroud.
5. Disconnect electrical leads from switch.
6. Remove switch securing screws and cup washers.
7. Withdraw switch.

NOTE: See page 86–55 for latest circuit delay.

Refitting

8. Reverse the removal procedure.

HEATED REAR WINDOW SWITCH

– Remove and refit   86.65.36

Removing

1. Remove the consul unit 76.25.01.
2. Unscrew the switch knob and remove the bulb.
3. Remove the locking ring from the front of the switch and withdraw the nameplate.
4. Remove the switch from the back of the consul unit.

Refitting

5. Reverse the removal procedure.

*WINDSCREEN WIPER/WASHER SWITCH

– Remove and refit   86.65.38

Removing

1. Disconnect the battery.
2. Remove the four drive screws securing bottom shroud to top shroud.
3. Remove the single screw securing top shroud to switch housing bracket.
4. Remove top shroud and lower the bottom shroud.
5. Disconnect cables at snap connectors.
6. Remove the two screws securing switch to column switch bracket.
7. Remove switch.

Refitting

8. Reverse the removal procedure.

*   See page 86–68.

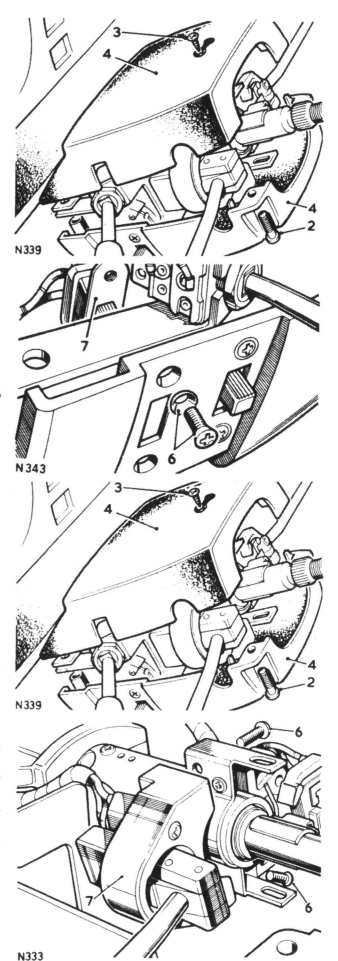

N339

N343

N339

N333

PROGRAMMED WASH/WIPE –
4-Door and 1981 2-Door vehicles

The windscreen wiper switch has an extra delay position which is designed to wipe once every 4 to 6 seconds until switched off.

Water spray for the windscreen is operated by pressing the wiper switch knob. In the 'OFF' position pressing the wiper switch knob automatically switches on the wipers which are designed to wipe for some 2.5 to 4.5 seconds after the switch knob is released.

MAINTENANCE

Failure of the wipers to park after use indicates a possible fault in the delay unit circuit.

To gain access to the delay unit control box withdraw the six screws and remove the lower fascia panel from the steering column. The black control box is located on the driver's side attached to the top side of the bulkhead behind the fascia.

To remove the control box lift the spring clip and pull off the five pin multi-plug. Remove the two fixing screws.

HANDBRAKE WARNING SWITCH

– Remove and refit   86.65.45

Removing

1. Disconnect the battery.
2. Remove handbrake cover.
3. Apply the handbrake.
4. Remove the rubber protector from switch.
5. Remove the hexagon nut.
6. Withdraw the switch from its mounting bracket.
7. Disconnect the electrical lead(s).

Refitting

8. Reverse the removal procedure.

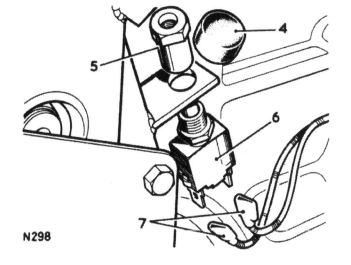

N298

HAZARD WARNING SWITCH

– Remove and refit   86.65.50

Special tools: Spanner for lock-ring, Part No. 601952

Removing

1. Disconnect the battery.
2. Unscrew the hazard warning switch knob assembly from the fascia.
3. Retain the bulb spring which is released when the knob is removed.
4. Pull out the warning lamp bulb.
5. Loosen the lock-ring using special spanner, Part No. 601952, and withdraw the plain washer.
6. Remove the lower fascia, 76.46.03.
7. Remove the locking ring and withdraw the switch.

Refitting

8. Reverse the removal procedure, referring to the circuit diagram when connecting electrical leads.

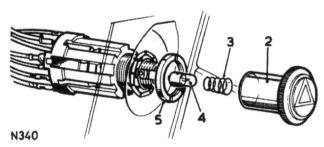

N340

86–72

STOP LIGHT SWITCH

—Remove and refit   **86.65.51**

Removing.

1. Disconnect the battery.
2. Remove the lower fascia panel 76.46.03
3. Depress the foot brake.
4. Remove the rubber protector from switch (where fitted).
5. Remove the hexagon nut.
6. Withdraw the switch.
7. Disconnect the electrical leads.

Refitting.

8. Reverse the removal procedure.

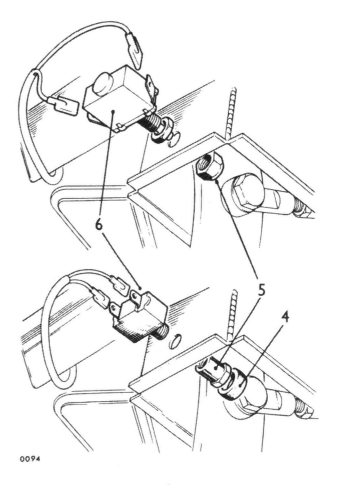

0094

CHOKE WARNING LIGHT SWITCH

– Remove and refit   **86.65.53.**

Removing

1. Disconnect the battery and (later models) remove one screw to release the switch cover.
2. Disconnect electrical leads from the switch.
3. Remove screw and clip securing switch to choke cable.
4. Remove switch.

Refitting

5. Reverse the removal procedure.

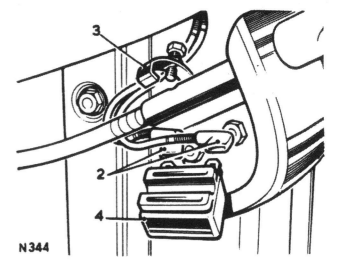

N344

86–73

*COMBINED DIRECTION INDICATOR HEADLIGHT HORN SWITCH

— Remove and refit   86.65.55

Removing

1. Disconnect the battery.
2. Remove windscreen wiper switch — 86.60.42.
3. Remove the two screws securing combined switch to column switch bracket.

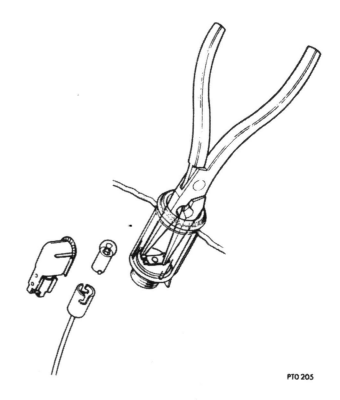

N345

Refitting

4. Reverse the removal procedure. Ensure that the lug on the switch locates in the recess in the steering column.

* See page 86—68.

CIGAR LIGHTER

— Remove and refit   86.65.60

Removing

1. Remove the consul unit 76.25.01.
2. Holding the back of the lighter, unscrew the centre barrel and withdraw it complete with chrome ring from the front of the consul unit.

Refitting

3. Reverse the removal procedure.

PTO 205

FUSE BOX

— Remove and refit   86.70.01

Removing

1. Disconnect the battery.
2. The fuse box cover is plastic and is a push-on fit. The fused circuits are illustrated on the cover. Two spare fuses are included.
3. To gain access to the cable entry into the fuse box, remove the glove box.
4. Push fuse box forward towards engine to release from spring clip retainers.

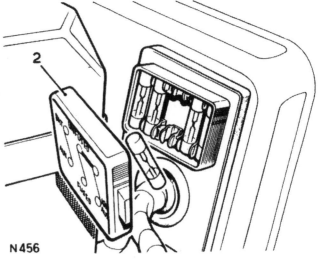

N456

Refitting

5. Reverse the removal procedure.

86—74

INSTRUMENT OPERATIONS

	Operation No
Ammeter – Remove and refit	88.10.01
Battery Voltmeter – Remove and refit	*Refer to* 88.10.01
Clock – Remove and refit	*Refer to* 88.10.01
Coolant temperature gauge – Remove and refit	88.25.14
Coolant temperature transmitter – Remove and refit	88.25.20
Fuel gauge – Remove and refit	88.25.26
Fuel gauge tank unit – Remove and refit	88.25.32
Instrument housing – Remove and refit	88.20.13
Instrument printed circuit – Remove and refit	88.20.19
Oil Pressure gauge – Remove and refit	*Refer to* 88.10.01
Oil pressure warning switch – Remove and refit	88.25.08
Oil temperature gauge – Remove and refit	*Refer to* 88.10.01
Speedometer – Remove and refit	88.30.01
Speedometer angle drive – instrument end – Remove and refit	88.30.15
Speedometer cable assembly – Remove and refit	88.30.06
Speedometer trip reset – Remove and refit	88.30.02
Voltage stabilizer – Remove and refit	88.20.26

AMMETER

Also applicable to Battery Voltmeter, Clock, Oil Pressure Gauge and Oil Temperature Gauge.

Remove and refit   88.10.01

Removing

1. Disconnect the battery.
2. Remove the grub screws securing the four heater control knobs to levers.
3. Remove the knobs.
4. Remove the screws securing the heater escutcheon plate.
5. Remove the escutcheon plate.
6. Remove the four upper retaining screws securing the heater console to the facia panel.
7. Remove the centre face level louvre.
8. Lower the passenger glove box lid.
9. Remove the lower screw securing the heater console.
10. Ease the console forward to gain access to the instrument.
11. Disconnect the electrical lead from the instrument.
12. Remove the illumination lead complete with holder and bulb from the instrument.
13. Remove the one or two, knurled nuts and support arms securing the instrument to the console.
14. Withdraw the instrument.

Refitting

15. Reverse the removal procedure.

INSTRUMENT HOUSING

Remove and refit   88.20.13

Removing

1. Disconnect the battery earth lead.
2. Remove the lower facia panel. 76.46.03
3. Remove the fixings securing the instrument housing to the facia top rail.
4. Withdraw the instrument housing sufficient to lift the rear cover.
5. Disconnect the speedometer cable.
6. Withdraw the electrical plug connector.
7. Lift the instrument housing clear.

Refitting

8. Reverse the removal procedure.

INSTRUMENT PRINTED CIRCUIT

Remove and refit   88.20.19

Removing

1. Remove instrument housing 88.20.13
2. Remove coolant temperature gauge 88.25.14
3. Remove voltage stabilizer 88.20.26
4. Remove fuel gauge 88.25.26
5. Remove all bulbholders.
6. Remove instrument printed circuit.

Refitting

7. Reverse the removal procedure.

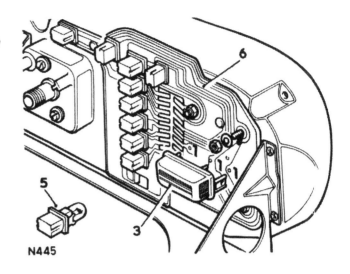

N445

VOLTAGE STABILIZER

Remove and refit   88.20.26

Removing

1. Disconnect the battery.
2. Release back cover of instrument housing and carefully move to one side.
3. Withdraw the voltage stabilizer from its socket connections.

Refitting

4. Reverse the removal procedure.

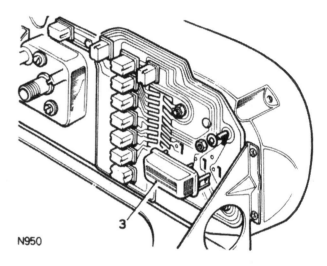

N950

OIL PRESSURE WARNING SWITCH

Remove and refit   88.25.08

Removing

1. Disconnect the battery.
2. Disconnect the electrical lead from the switch.
3. Unscrew the switch unit.
4. Remove switch and sealing washer.

Refitting

5. Reverse the removal procedure, using a new sealing washer.

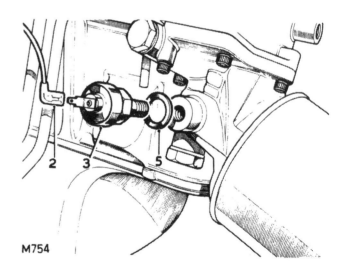

M754

COOLANT TEMPERATURE GAUGE

Remove and refit   88.25.14

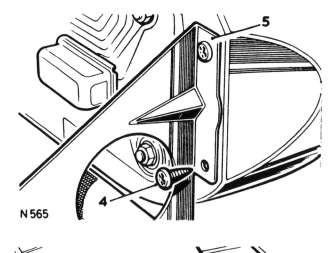

N 565

Removing

1. Remove the instrument housing 88.20.13
2. Remove the knurled nut securing the speedometer trip reset control.
3. Remove the instrument housing rear cover.
4. Remove the four screws securing the housing mounting bracket.
5. Remove the mounting bracket.
6. Remove the drive screws securing the body to the instrument housing front cover.
7. Separate the instrument housing front cover from the instrument mounting body.
8. Whilst holding the coolant temperature gauge from inside the mounting body, remove the two securing nuts and washers.
9. Withdraw the coolant temperature gauge.

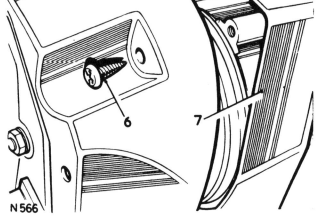

N 566

Refitting

10. Reverse the removal procedure.

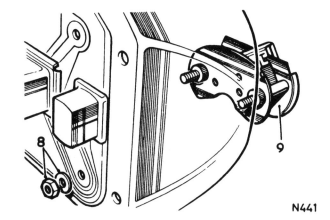

N441

COOLANT TEMPERATURE TRANSMITTER

Remove and refit   88.25.20

Removing

1. Disconnect the battery.
2. Disconnect the electrical lead from the transmitter and (later models) the air cleaner hose.
3. Remove the transmitter from the inlet manifold.

Refitting

4. Reverse the removal procedure, using a new joint washer.

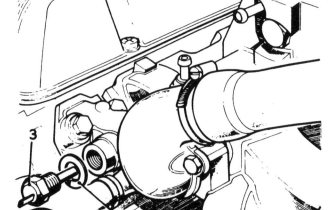

M901

88–4

FUEL GAUGE

Remove and refit   88.25.26

Removing

1. Remove the instrument housing 88.20.13
2. Remove the knurled nut securing the speedometer trip reset control.
3. Remove the instrument housing rear cover.
4. Remove the four screws securing the housing mounting bracket.
5. Remove the mounting bracket.
6. Remove the drive screws securing the body to the instrument housing front cover.
7. Separate the instrument housing front cover from the instrument mounting body.
8. Whilst holding the fuel gauge from inside the mounting body, remove the two securing nuts and washers.
9. Withdraw the fuel gauge.

Refitting

10. Reverse the removal procedure.

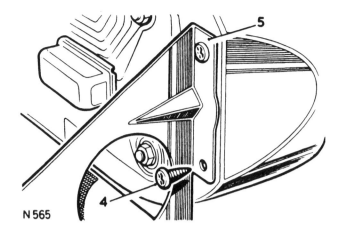

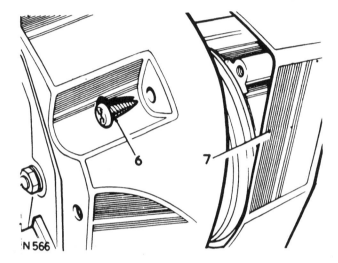

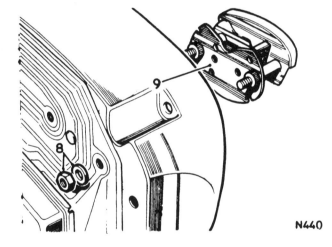

FUEL GAUGE TANK UNIT

Remove and refit   88.25.32

Service tool: 600964 Locking spanner

Removing

1. Disconnect the battery.
2. Disconnect the fuel pipe at the fuel gauge tank unit and drain the fuel from the tank.
3. Note electrical leads position and disconnect them from the unit.
4. Turn the gauge unit locking ring anti-clockwise until the indents in the locking ring align with the lugs on the tank. Tool 600964.
5. Withdraw the locking ring.
6. Withdraw the complete fuel gauge unit.
7. Remove the rubber seating ring.

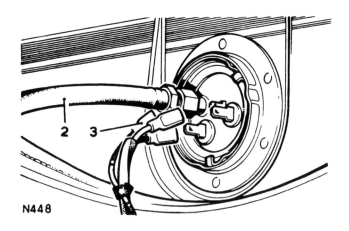

N448

Refitting

8. Slightly smear the joint faces on the gauge unit, seating washer and fuel tank with Bostik adhesive No. 772 or a suitable alternative.
9. Insert gauge unit and seating washer into tank and engage the two lugs in the gauge unit base plate into the cut-outs in the fuel tank flange.
10. Fit locking ring and turn ring clockwise until locked. Tool 600964.
11. Connect fuel pipe to unit.
12. Reconnect electrical leads.
13. Refill the fuel tank and check for leaks.
14. Check operation of fuel level gauge.

N564

88–6

SPEEDOMETER

Remove and refit   **88.30.01**

1. Remove the instrument housing 88.20.13
2. Remove the knurled nut securing the speedometer trip reset control.
3. Remove the instrument housing rear cover.
4. Remove the four screws securing the housing mounting bracket.
5. Remove the mounting brackets.
6. Remove the drive screws securing the body to the instrument housing front cover.
7. Separate the instrument housing front cover from the instrument mounting body.
8. Remove the speedometer angle drive from the rear of the instrument 88.30.15
9. Whilst holding the speedometer from inside the mounting body, remove the two securing screws, bush and washer.
10. Withdraw the speedometer.

Refitting

11. Reverse the removal procedure.

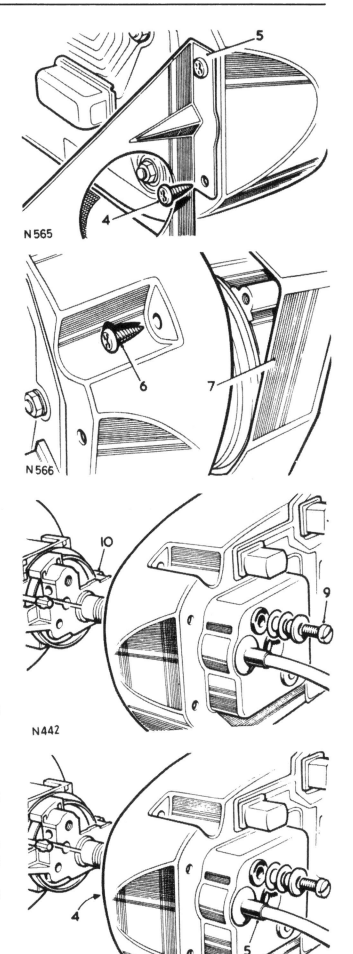

N 565

N 566

N 442

N 443

SPEEDOMETER TRIP RESET

Remove and refit   **88.30.02**

Removing

1. Remove the instrument housing 88.20.13
2. Remove the speedometer angle drive, instrument end 88.30.15
3. Remove the speedometer 88.30.01
4. From inside the instrument housing, depress the two small plastic prongs which clip the trip reset operating tube into the hole in the instrument housing.
5. Push the reset operating tube out of the mounting hole.

Refitting

6. Reverse the removal procedure.

SPEEDOMETER CABLE ASSEMBLY

Remove and refit   88.30.06

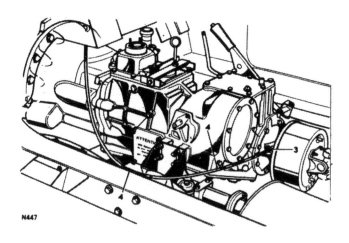

Removing

1. Remove rear cover of instrument housing.
2. Disconnect speedometer cable from instrument.
3. Disconnect gearbox end of speedometer cable by removing nut and retainer fixing cable to gearbox.
4. Remove the two clips securing cable to gearbox casing.
5. Withdraw cable and grommet through the bulkhead.

N447

Refitiing

6. Reverse the removal procedure.

SPEEDOMETER ANGLE DRIVE–INSTRUMENT END

Remove and refit   88.30.15

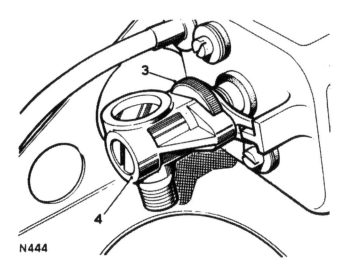

Removing

1. Remove the instrument housing rear cover.
2. Disconnect the speedometer cable from the instrument.
3. Unscrew the captive knurled nut which secures the angle drive to the speedometer.
4. Withdraw the angle drive unit.

Refitting

N444

5. Reverse the removal procedure, ensuring that the slotted base plate of the angle drive unit locates on the peg of the speedometer housing.

CONTENTS

The equipment detailed in this section is mainly of proprietary manufacture and is selected from the list of approved optional equipment for fitting to the Range Rover. Note that later vehicles may be fitted with Unipart marketed optional equipment which may differ in detail to that described.

CENTRE POWER TAKE OFF

– Description

A centre power take off, marketed by Messrs. Fairy Winches Ltd, South Station Yard, Whitchurch Road, Tavistock, Devon, fits to the rear of the transfer gearbox, replacing the existing mainshaft rear bearing housing. It is secured by the bearing housing bolts and operated by a slider control situated above the gearbox cover by the drivers seat.

Remove and Refit

Removing.

1. Remove the exhaust heat shield and silencer.
2. Disconnect any belt drive to the power take off pulley.
3. Remove the cotter pin to disconnect the control rod.
4. Remove the bolts securing the power take off unit to the transfer gearbox.
5. Lift off the power take off unit.

Refitting.

6. Renew the gasket if damaged.
7. Reverse instructions 1 to 5 ensuring that the power take off is in the disengaged position before refitting.

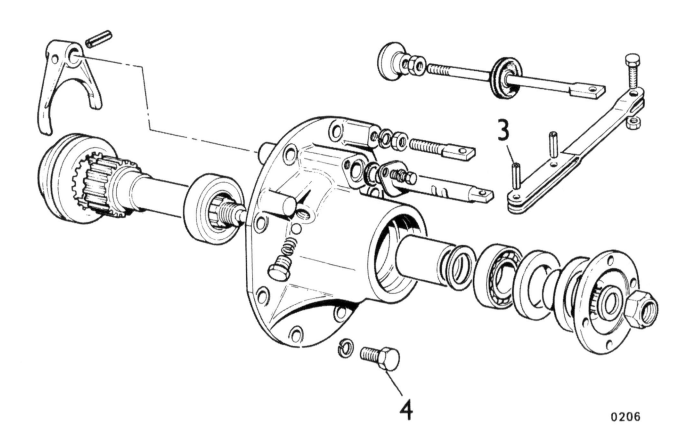

0206

SPLIT CHARGING FACILITY

– Description

The purpose of this equipment is to provide a separate source of 12 volt current for auxiliary use without discharging the main battery.

To achieve this a diode is located between the vehicle alternator and the auxiliary terminal bracket. A second battery may be connected to the terminals and charged by the alternator output. When this second battery is in use, without the engine running, it is effectively isolated from the main vehicle electrical system.

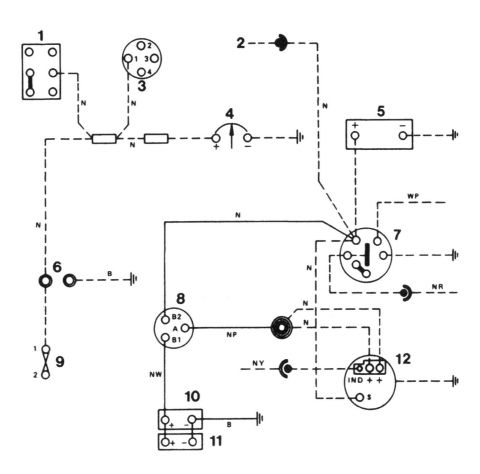

Circuits shows dotted – — — — Are existing and contained in the basic vehicle.
Circuits shown in full ——————— Are additional

1.	Lighting switch.	7.	3M 100 pre-engaged starter.
2.	Connection to starter relay.	8.	Split charge diode.
3.	Steering lock and ignition switch.	9.	Fuse unit.
4.	Voltmeter.	10.	Terminal bracket.
5.	Vehicle battery.	11.	Second bracket.
6.	Inspection sockets.	12.	18ACR or 20ACR Alternator.

Key to cable colour code:

B–Black, R–Red, G–Green, GP–Green/purple, BR–Black/red, LGP–Light green/purple, LGN–Light green/brown, GY–Green/yellow, GR–Green/red, GW–Green/white

THIRD AXLE

− Description

A six wheeled vehicle variant is available by the fitting of a third axle rearwards of the existing back axle. The chassis is extended to accommodate the third axle unit.

The third axle is a standard Range Rover axle case with the differential unit removed and a blanking plate fitted. No drive shafts are used, the hub units being sealed by a cap.

Suspension is similar to the standard Range Rover but additional springs are fitted to both the standard and third rear axle units.

Hydraulic braking facilities are fitted to the third axle, being taken from the existing rear axle braking circuit. To accommodate the change in braking load, the brake master cylinder is fitted with a special piston, obtainable only from the conversion specialists, Messrs. Carmichael and Son Ltd., Gregory's Mill Street, Worcester WR3 8DE.

AXLE UNIT − Remove and refit   *Refer to Operation* 51.25.01

TRAILER FITTINGS

– Description

Two basic fittings are available for this vehicle, a towing plate to which a towing ball may be fitted and a seven pin trailer lighting socket.

The towing plate is a readily understood bolt attachment unit conveying the trailer pulling load to the vehicle chassis by means of tie bars.

The trailer lighting socket is connected into the vehicle lighting and direction indicator circuits to duplicate these on the trailer. In addition a permanent, fused supply is provided for trailer illumination if this is required. A heavy duty flasher unit is fitted to accommodate the additional direction indicator lighting load from the trailer.

The trailer lighting socket is mounted on a bracket attached to the vehicle by a bolt, nut and washers. When refitting this socket ensure that the spring loaded hinge is uppermost.

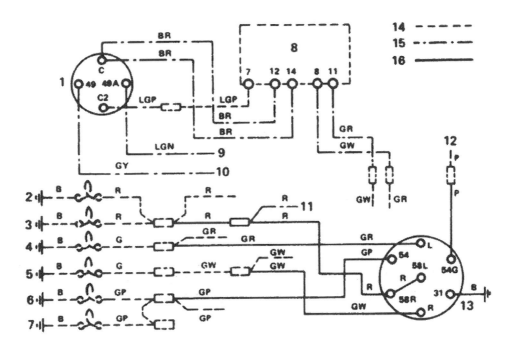

Circuit diagram for trailer lighting

1	Replacement flasher unit	8	Instrument binnacle
2	Side marker lamp, LH	9	From indicator switch
3	Tail lamp, LH	10	From hazard warning switch
4	Indicator lamp, LH	11	To number plate illumination
5	Indicator lamp, RH	12	Feed from fuse A2
6	Stop lamp, LH	13	Seven-pin vehicle socket
7	Stop lamp, RH		

Key to circuit wiring symbols:

14 Dotted lines indicate vehicle wiring

15 Chain dotted lines indicate vehicle wiring repositioned

16 Unbroken lines indicate the conversion harness

Key to cable colour code:

B–Black, R–Red, G–Green, GP–Green/purple, BR–Black/red, LGP–Light green/purple, LGN–Light green/brown, GY–Green/yellow, GR–Green/red, GW–Green/white

WINCH – FRONT – CAPSTAN

– Description

A capstan winch, powered from the engine crankshaft is manufactured by Messrs. Fairy Winches Ltd., South Station Yard, Whitchurch Road, Tavistock, Devon.

A lever alongside the capstan engages a dog clutch drive to turn the capstan. Overload protection is given by a shear pin should the maximum recommended pull of 3,100 lb. (1450 kg) be exceeded at fast idle speed (approximately 1000 rev/min).

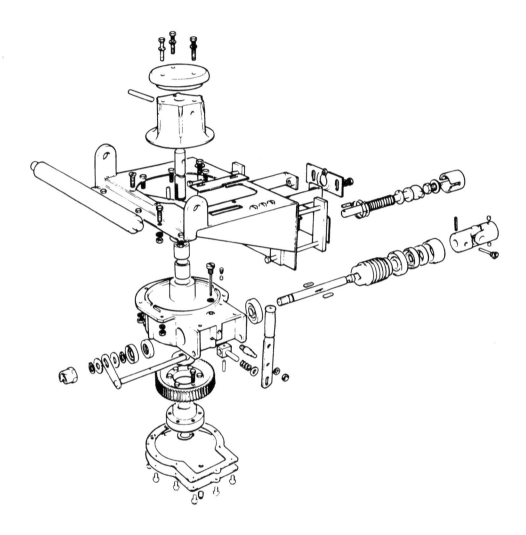

-Remove and refit

Removing

1. Remove the front bumper and grille.
2. Disconnect the drive shaft from the crankshaft pulley.
3. Take the weight of the winch on a suitable hoist.
4. Remove four bolts nuts and washers securing the winch.
5. Lift the winch from the vehicle.

Refitting

6. Reverse instructions 1 to 5.

Servicing

1. At monthly intervals check and top up the oil level. Use SAE 90 oil.
2. Monthly, or as required, grease the drive shaft assembly and other accessible moving parts.

90–6

WINCH – FRONT – ELECTRIC

– Description

Warn electric winches are marketed by Ryders Auto Services (GB) Ltd., Winch Division, 215/217 Knowsley Road, Bootle, Liverpool, Lancs. L20 4NW.
Four types have been available for fitment to the Range Rover, Models 8000 and 8200 to earlier vehicles and Models 8074 and 8274 to later vehicles. In all cases a maximum recommended pull of 5000 lb. (2268 kg) should not be exceeded and overload protection is provided by means of a line fuse in the winch electric feed cables.

-Remove and refit

Removing

1. Disconnect the battery.
2. Remove the radiator grille.
3. Support the winch using a suitable hoist.
4. Disconnect the electrical leads at the winch.
5. Remove eight bolts securing the winch to the vehicle noting the position of the spacers for refitting.
6. Lift the winch clear of the vehicle.

Refitting

7. Reverse instructions 1 to 6.

Servicing

At regular intervals according to use:-

1. Clean and grease the wire rope.
2. Inspect the wire rope for damage and renew as necessary.

This page intentionally left blank

SERVICE TOOLS LIST

Service Tools mentioned in this Manual must be obtained direct from the tool manufacturers:—

Messrs. V.L. Churchill & Co. Ltd.,
P.O. Box 3,
London Road,
DAVENTRY,
Northants.,
England.
Telephone: 03-272 4461
Telex: 31326
Telegrams: Garaquip Daventry Northants Telex

Zenith Carburetter Co. Ltd.,
Honeypot Lane,
STANMORE,
Middlesex.
England.
Telephone: 01-204 3388
Telex: 23571
Telegrams: Zenicarbur, Norphone, London

Tool Number	Description	Tool Number	Description
RO 212	Gearbox bracket	18G 672	Brake piston clamp
RO 1001	Gearbox lifting bracket	18G 1063	Ball joint extractor
RO 1002A	Steering wheel extractor	18G 1150	Gudgeon pin remover/replacer adaptor
RO 1003	Dummy shaft, intermediate gears	18G 1335	Extractor — reverse shaft
RO 1004	Mainshaft spacer extractor	18G 1349	Hub oil seal replacer
RO 1005	Front cover centralising tool, gearbox	47	Multi-purpose press
RO 1006	Ball joint extractor	262757A	Bearing extractor, axle differential
RO 1009	Upper rear main bearing oil seal remove/replacer	262758	Pinion bearing press block
RO 1011	Wrench for carburetter flange nut	274401	Valve guide remover
RO 1012	Hydraulic adaptor -- Power steering	276102	Valve spring compressor
RO 1014	Rear crankshaft oil seal protection sleeve	530105	Differential nut spanner, axle
		530106	Dial indicator gauge bracket
RO 1015	Seal saver, power steering	600959	Valve guide drift
RO 1016	Torque setting tool, power steering	600964 (18G 1001)	Lock-Ring tool, fuel tank gauge
JD 10	Pressure gauge for testing power steering	605004	Differential pinion gauge
		605330A	Carburetter balancer
		605351	Connecting rod guide bolts
JD 10-2	Adaptor and tap for use with JD 10	605774	Distance piece for 600959
18G 2 (6312A)	Remover — Crankshaft gear pulley	606435	Hub nut spanner
18G 47BA	Layshaft bearing extractor	606600	Spanner, power steering adjustment
18G 47BB	Differential bearing extractor	606601	Peg spanner, power steering adjustment
252	Drop arm extractor	606602 (JD 32)	Expander, power steering box seals
18G 79	Clutch centralising tool	606603 (JD 33)	Compressor, power steering box seals
18G 106	Valve spring compressor	606604 (JD 34)	Seal saver, power steering
18G 134 (550)	Assembly tool for 18G 134DG	B 20379 (S 353)	Carburetter adjusting tool (allen key type)
18G 284	Impulse extractor	B 25243 (MS 86)	Carburetter idle speed tool
18G 284AR	Layshaft bearing extractor	B 24667	Carburetter adjusting tool (bottom)
18G 284AT	Intermediate gear shaft extractor	B 25860 (MS 80)	Carburetter adjusting tool (slotted socket type)

NOTE: Zenith tool part numbers commence with suffix 'B'.

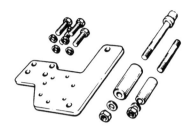

RO 212

RO 1006

RO 1001

RO 1009

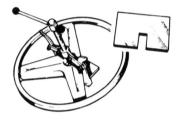

RO 1002A

RO 1011

RO 1003

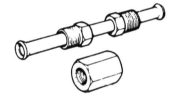

RO 1012

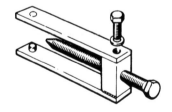

RO 1004

RO 1014

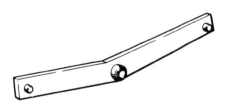

RO 1005

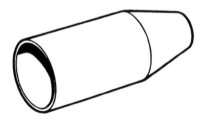

RO 1015

RO 1016

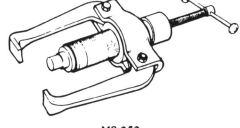

MS 252

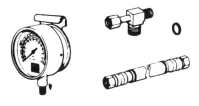

JD 10

18G 79

JD 10-2

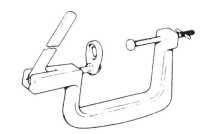

18G 106

18G 2

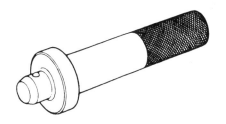

18G 134

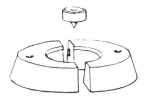

18G 47BA

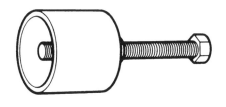

18G 1335

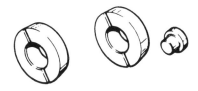

18G 47BB

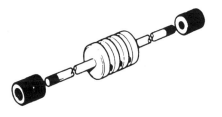

18G 284

99–3

18G 284AR

B 20379 (S 353)

18G 284AT

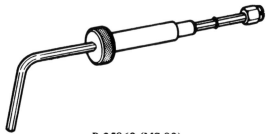

B 25860 (MS 80)

18G 672

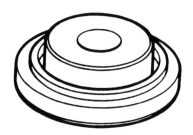

18G 1349

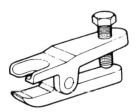

18G 1063

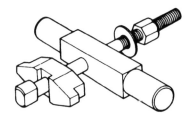

262757A

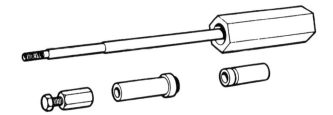

18G 1150

262758

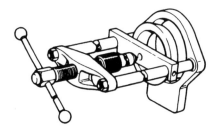

47

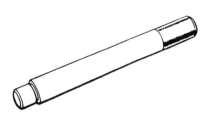

274401

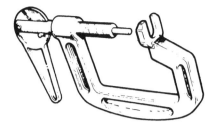

276102

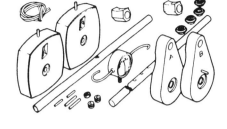

605330A

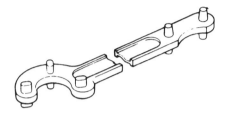

530105

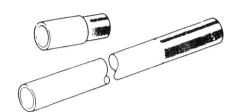

605351

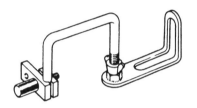

530106

605774

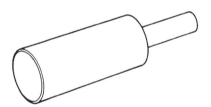

600959

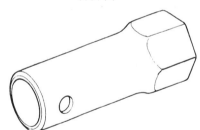

606435

600964

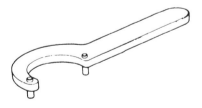

606600

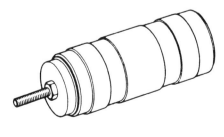

605004

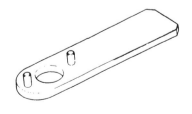

606601

99–5

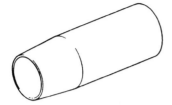

606602

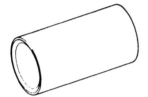

606603

606604

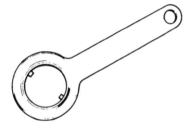

B 24667

B 25243 (MS 86)

SUPPLEMENT TO REPAIR OPERATION MANUAL

INTRODUCTION

This supplement contains information on the Range Rover Five Speed manual gearbox, Transfer gearbox, Automatic gearbox and changes and additions to the Electrical equipment.

When seeking information, reference should first be made to this supplement, all other details are contained in the main Repair Operation Manual.

CONTENTS

FIVE SPEED GEARBOX

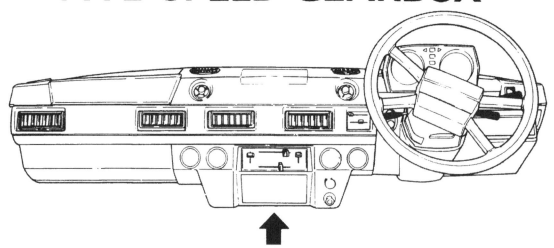

WARNING LABEL FITTED TO VEHICLE

THE DIFFERENTIAL LOCK SHOULD ONLY BE ENGAGED WHEN TRACTION IS LIKELY TO BE LOST. WIDE THROTTLE OPENING SHOULD BE AVOIDED WHEN USED IN CONJUNCTION WITH 1st AND 2nd GEAR LOW RANGE. AS SOON AS THE DIFFICULT SURFACE HAS BEEN CROSSED THE DIFFERENTIAL LOCK MUST BE RELEASED. A SINGLE AXLE ROLLER RIG MAY BE USED FOR SPEEDS UP TO 5 km/h (3 m.p.h.). THE CENTRE DIFFERENTIAL LOCK MUST BE DISENGAGED. FOR ROLLER TESTS OVER 5 km/h (3 m.p.h.). EITHER ALL FOUR WHEELS MUST BE ROTATED AT THE SAME SPEED OR IF ONLY A SINGLE AXLE ROLLER RIG IS AVAILABLE, THE CENTRE DIFFERENTIAL MUST BE LOCKED AND THE PROPELLER SHAFT TO STATIONARY AXLE MUST BE REMOVED.

Main gearbox

Model . . . . LT 77

Type . . . . Five speed, single helical constant mesh with synchromesh on all forward gears

Transfer box

Model . . . . LT 230R or LT 230T

Type . . . . Two-speed reduction on main gearbox output. Front and rear drive permanently engaged via a lockable differential

Gear ratios

Main gearbox . . . .
5th	0.770:1
4th	1.00:1
3rd	1.397:1
2nd	2.132:1
1st	3.321:1
Reverse	3.429:1

Transfer gearbox . . . .
High	1.192:1
Low	3.320:1

Overall ratio (final drive):	In high transfer	In low transfer
5th	3.25:1	9.05:1
4th	4.22:1	11.75:1
3rd	5.89:1	16.41:1
2nd	8.99:1	25.04:1
1st	14.01:1	39.02:1
Reverse	14.46:1	40.27:1

	Nm	lbf ft
CLUTCH		
Clutch cover bolts	27,5	20
Slave cylinder bolts	27,5	20
MAIN GEARBOX (FIVE-SPEED) –	Nm	lbf ft
Oil pump body to extension case	7-10	5-7
Clip to clutch release lever	7-10	5-7
Attachment plate to gearcase	7-10	5-7
Attachment plate to remote housing	7-10	5-7
Extension case to gearcase	22-28	16-21
Pivot plate	22-28	16-21
Remote selector housing to extension case	22-28	16-21
Gear lever housing to remote housing	22-28	16-21
Guide clutch release sleeve	22-28	16-21
Slave cylinder to clutch housing	22-28	16-21
Front cover to gearcase	22-28	16-21
5th support bracket	22-28	16-21
Plunger housing to remote housing	22-28	16-21
Blanking plug extension case	7-10	5-7
Gear lever retainer	7-10	5-7
Yoke to selector shaft	22-28	16-21
Fixing gear lever assembly nut	47-54	35-40
Reverse pin to centre plate nut	47-54	35-40
Clutch housing to gearbox bolt	65-80	48-59
Plug – detent spring	22-28	16-21
Oil drain plug	25-35	19-26
Oil filler plug	25-35	19-26
Plug oil filler – remote housing	25-35	19-26
Breather	7-11	5-8
Oil level plug	25-35	19-26
Blanking plug – reverse switch hole	20-27	15-20
TRANSFER BOX		
Pinch bolt, operating arm	7-10	5-7
Gate plate to grommet plate	7-10	5-7
End cover	7-10	5-7
Speedometer cable retainer	7-10	5-7
Speedometer housing/rear output	See note	
Locating plate to gear change housing	5-7	4-5
Bottom cover to transfer case	22-28	16-21
Front output housing to transfer case	22-28	16-21
Cross shaft housing to front output housing	22-28	16-21
Gear change housing	22-28	16-21
Pivot shaft	22-28	16-21
Connecting rod	22-28	16-21
Retaining plate intermediate shaft	22-28	16-21
Front output housing cover	22-28	16-21
Gear change housing	22-28	16-21
Bracket to extension housing	22-28	16-21
Finger housing to front output housing	22-28	16-21
Mainshaft bearing housing	22-28	16-21

continued

	Nm	lbf ft
Brake drum	22-28	16-21
Gearbox to transfer box	40-50	29-37
Bearing housing to transfer gearbox	40-50	29-37
Speedometer housing to transfer gearbox	40-50	29-37
Selector fork to cross shaft	40-50	29-37
Yoke to selector shaft high/low	22-28	16-21
Selector fork high/low to shaft	22-28	16-21
Operating arm high/low	22-28	16-21
Transmission brake	65-80	48-59
Gearbox to transfer case	40-50	29-37
Gearbox to transfer case	See note	
Oil drain plug	25-35	19-26
Differential case	55-64	40-47
Output flange	146-179	108-132
Link arm and cross shaft lever to ball joint	8-12	6-9
Oil filler/level plug	25-35	19-26
Transfer breather	7-11	5-8

NOTE:– Studs to be assembled into casings with sufficient torque to wind them fully home, but this torque must not exceed the maximum figure quoted for the associated nut on final assembly.

GEARBOX AND TRANSFER BOX

Bell housing to cylinder block bolts	36,6-44,8	27-33
Gearbox casing to bell housing 2 off	146,5-179	108-132
Gearbox casing to bell housing 2 off	85,4-104,4	63-77
Gearbox casing to bell housing nuts	85,4-104,4	63-77
Gearbox casing to bell housing stud and nuts	146,5-179	108-132
Output flange – rear – nut and bolts	43,4-51,5	32-38
Output shaft – rear – nut	146,5-179	108-132
Output shaft – front – nut	146,5-179	108-132
Gear selector spherical seat bolts	13,6-16,3	10-12
Propeller shaft to flange bolts	43,4-51,5	32-38
All other nuts and bolts:		
M6	9,9-11,9	7.3-8.7
M8	23,7-29,7	17.5-21.9
M10	48-58	35.4-42.8

Recommended Lubricants and fluids

These recommendations apply to temperate climates where operational temperatures may vary between
−10°C (14°F) and 32°C (90°F)

COMPONENT	BP	CASTROL	DUCKHAMS	ESSO	MOBIL	PETROFINA	SHELL	TEXACO
5 speed Manual gearbox	BP Autran DX2D	Castrol TQ Dexron IID	Duckhams Fleetmatic CD or Duckhams D-Matic	Esso ATF Dexron IID	Mobil AFT 220 D	Fina Dexron IID	Shell AFT Dexron IID	Texamatic Fluid 9226
*Transfer box	BP Gear Oil SAE 90EP	Castrol Hypoy SAE 90EP	Duckhams Hypoid 90	Esso Gear Oil GX 85W/90	Mobil Mobilube HD 90	Fina Pontonic MP SAE 80W/90	Shell Spirax 90EP	Texaco Multigear Lubricant SAE 85W/90

* Either engine or gearbox oil may be used as an alternative to the specified gear oil for the transfer gearbox and can be mixed together.

All other climates and conditions

			−30	−20	−10	0	10	20	30
5 SPEED MANUAL GEARBOX	Dexron IID		←						→
*TRANSFER GEARBOX	API GL4 or MIL-L-2105A	90 EP			←				→
		80W EP	←						→

Capacities

Component	Litres	Imperial Unit
Main gearbox oil	2.2	3.9 pints
Transfer gearbox oil	2.80	5 pints

MAINTENANCE SCHEDULES

	Every 10 000 km (6000 miles) or 6 months	Every 20 000 km (12 000 miles) or 12 months
UNDERBODY		
Check/top up gearbox and transfer box oil levels.	X	X
Renew gearbox and transfer box oils.	40 000 km (24 000 miles)	

UNDERBODY

Check/top up gearbox oil

Main gearbox oil level
Check oil level daily or weekly when operating under severe wading conditions.

1. From beneath the vehicle remove the oil level/filler plug and top up, if necessary, to the bottom of the hole.
2. Replace the plug. If significant topping up is required check for oil leaks at the drain plug, all joint faces and through the drain hole in the bell housing.

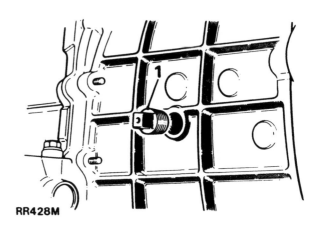

RR428M

Renew gearbox oil

Main gearbox oil
Drain and refill monthly when operating under severe wading conditions.

To change the gearbox oil proceed as follows:

1. Immediately after a run when the oil is warm, drain off the oil into a container by removing the drain plug and washer from the bottom of the gearbox casing.
2. Remove the oil filter.
3. Wash the filter in clean fuel; allow to dry and replace.
4. Refit drain plug and washer and refill gearbox through the oil level/filler plug, with the correct grade of oil, to the bottom of the oil level/filler hole. For capacity see Data, Section

Important: Do not overfill, otherwise leakage may occur.

NOTE:– For details of the transfer gearbox maintenance, refer to the Automatic gearbox section of this supplement.

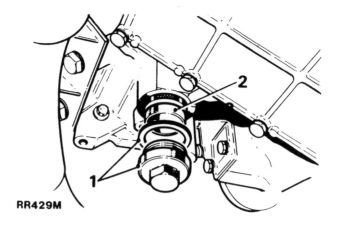

RR429M

GEARBOX/TRANSFER BOX ASSEMBLY

– Five speed manual   37.20.01

Remove and refit

Removing
1.   Drive the vehicle on a ramp.
2.   Open the bonnet and disconnect the battery.
3.   **From inside the vehicle** select low range transfer gear.
4.   Pull off both main and transfer gear/differential lock knobs.
5.   Unclip both ash trays from the centre console.
6.   Remove the four self-tapping screws inside the ash tray bodies to release them and the rear of the console.
7.   Slacken the two shouldered screws retaining the front of the console. If necessary press in the plastic diff lock instruction plate to gain access to the screw heads.
8.   Lift out the complete console assembly.
9.   Remove the four screws retaining the transfer gear/ diff lock lever gaiter clamping plate to the selector housing underneath. Leave the clamp plate in position on the gaiter.
10.   Remove the self-tapping screws to release the frame retaining the combined gaiter assembly which can then be removed.
11.   Using a screwdriver displace the bias spring legs outwards from the main gear lever crosspin.
12.   Remove the bolt and special lock washer retaining the gear lever assembly and lift out of the gearbox.
13.   **From under the bonnet** remove the two bolts (plain washers, spring washer and nut each) securing the fan cowl to the brackets on the radiator top.
14.   Disengage the fan cowl from the slots in the radiator bottom bracket and leave loosely in position.
15.   Slacken the 'Jubilee' clips retaining the two elbows to the air cleaner.
16.   Pull the other ends of the elbows off the carburettor flanges.
17.   Remove the clip screw on top of the air temperature intake control and release the hose.
18.   Release the hose from open clip, on the top of the right hand carburettor.
19.   Pull off the short hose from the one-way breather valve. The other end is attached to the air cleaner.
20.   Pull off the float chamber breather vent pipe and rubber right angle from the top of the right hand carburettor.
21.   Raise the air cleaner from its mounting posts.
22.   Pull off the engine breather hose from the base of the air cleaner.
23.   Remove the air temperature intake sensor unit by lifting the front from its mounting peg and pulling the convoluted hose from the air cleaner.
24.   Slacken the 'Jubilee' clips to release the 'Pulsair' hoses from both ends of the air rails.
25.   Lift off the air cleaner.

continued

26. Release both main and transfer gearbox breather pipes. These are secured to the engine lifting eye on the rear of the right hand cylinder head by a bolt-on clip. Remove the bolt (spring and plain washers) and detach the clip from the breather pipes to prevent it sliding down the pipes.

27. Release the 'Fir tree' rubber clip which retains the harness and breather pipes to the lifting eye.

28. **Raise the vehicle on the ramp.**

29. Place a suitable container under the transmission, remove the three drain plugs, allow the oil to drain and refit the plugs. Clean filter on the extension housing plug before refitting.

30. Remove the chassis cross-member. This is secured to the chassis by eight nuts and bolts (two plain and one spring washer each).
    WARNING:– Leave the two lower bolts loosely in position on each side as a safety measure before proceeding.

31. Spread the chassis with a suitable tool.

32. Remove the four loose bolts and lower the cross-member.

33. Remove the heat shield from the face of the right hand engine mounting bracket. This is retained by two bolts and nuts (with spring and plain washers).

34. Release both front exhaust pipes from the exhaust manifold flanges by removing the six nuts (with spring washers).

35. Release the branch pipe from the front silencer flange by removing the three bolts and nuts (with spring washers).

36. Release the 'U' bolt holding the front of the branch pipe to the gearbox mounting by removing the two nuts (with spring washers) and remove the front pipes and branch pipe assembly from the vehicle.

37. Remove the two nuts and bolts from the support bracket securing the twin rear pipes to the rear of the silencer.

38. Remove the flywheel housing bottom cover. This is secured by nine fixing bolts. The five lower bolts are fitted with nuts.

39. Remove the two bolts (with spring washers) to release the clutch slave cylinder.

40. Remove the single bolt (with spring washer) to release the slave cylinder pipe bracket from the left side of the bell housing. This is the lower bolt.

41. Move the slave cylinder (with its sealing shim) to one side.

42. Mark both pairs of front and rear prop shaft flanges with a punch to facilitate re-assembly.

43. Disconnect both prop shafts, by removing all eight Nyloc nuts and tie the shafts to the chassis.

44. Release the speedometer cable by slackening the Nyloc nut retaining the fork clip.

45. Remove the bolt (and spring washer) securing the cable clip to the transfer box. Remove the clip and replace the bolt and spring washer loosely.

46. Disconnect the speedometer cable and move aside.

continued

47. Manufacture a cradle to the dimensions given in the drawing and attach it to a transmission hoist. To achieve balance of the gearbox and transfer box assembly when mounted on the transmission hoist, it is essential that point A is situated over the centre of the lifting hoist ram. Drill fixing holes B to suit hoist table. Secure the assembly to the lifting bracket at point C, by means of the lower bolts retaining the transfer gearbox rear cover.

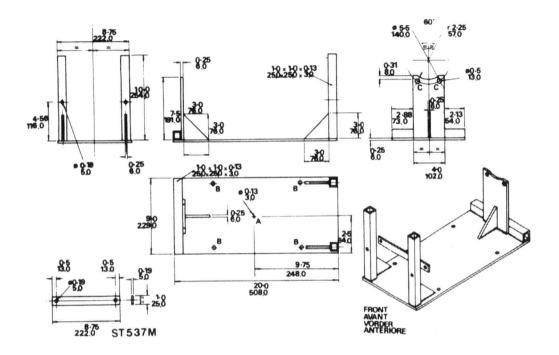

48. Remove the bottom two bolts from the transfer box rear cover and use them to attach the rear end of the cradle to the transfer box.

49. Raise the hoist just enough to take the weight of the gearbox and transfer box assembly.

50. Disconnect both gearbox mounting brackets from the chassis by removing the six nuts and bolts (spring and plain washers).
    NOTE:– The left side upper rear bolt also retains a speedometer cable support clip.

51. Remove the nut (with spring and plain washers) and release the gearbox mounting bracket from the right side rubber mounting only.

52. Lower the hoist sufficiently to gain access to the handbrake fixings on the right side of the transfer box.
    CAUTION:– Do not lower the hoist too far or the handbrake gaiter will be damaged by the handbrake grip.

53. Disconnect the handbrake connecting link from the linkage bracket by removing the split pin, plain and 'Thackery' washers and withdrawing the clevis pin.

54. Remove the three bolts (with spring washers) retaining the handbrake lever mounting bracket and tie it to the chassis.

continued

55. Disconnect the 'Lucar' electrical leads from the differential lock (one green, one black/blue tracer) and the reverse light (one green, one green/brown tracer).

56. Withdraw the reverse light cable from the clip attachment on the selector housing.

57. Release the 'Fir tree' rubber clips securing the cable to the breather pipes.

58. Support the engine with a jack across the ramp.

59. Unscrew the seven remaining bell housing bolts (with spring washers) noting that the third bolt up on the left side also holds a harness clip.

60. Draw the gearbox rearwards on the hoist to disengage it from the engine. At the same time ensure that all connections to the engine and vehicle are disconnected and clear of the gearbox.

61. Lower the hoist as necessary and remove from under the vehicle.

Refitting

62. Fit the cradle to the transmission hoist and the gearbox and transfer box assembly to the cradle, as described in instruction 47.

63. Grease the splines on the gearbox input shaft with Rocol MTS1000.

64. Smear Hylomar PL32 (or Loctite 290) on the bell housing mating face with the engine.

65. Engage third gear.

66. Engage the differential lock, to facilitate lining-up the gearbox input shaft.

67. Raise the gearbox assembly on the hoist whilst guiding the two breather pipes upwards.

68. Rotate the transmission brake drum as required to engage the gearbox input shaft splines and move the gearbox forward on the hoist.

69. Secure the gearbox assembly to the engine with seven of the eight bell housing retaining bolts (with spring washers). The third bolt up on the left side is also fitted with a harness clip. Leave out the lower bolt on the left side at this stage.

70. Locate the reverse light electrical leads (one green, one green/brown tracer) and feed through the harness clip on the left side of the selector housing, connecting the Lucars to the switch.

71. Locate the differential lock leads (one green, one black/blue tracer) and connect the Lucars to the switch on top of the selector housing.

72. Fit the 'Fir tree' rubber clips to secure the harness to the breather pipes in two places.

73. Release the handbrake mounting assembly (tied to chassis) and set in the off position. Apply grease as necessary.

74. Locate the handbrake lever mounting bracket on the side of the gearbox and secure with the three bolts (and spring washers).
    NOTE:– The lower bolt is shorter.

75. Reconnect the handbrake connecting link to the linkage bracket by refitting the clevis pin, Thackery and plain washers and a new split pin.

continued

76. Loosely fit the right side gearbox mounting bracket to the rubber mounting (plain and spring washers and nut).

77. Check that there is no obstruction then raise the hoist until the two gear change levers pass upwards through their respective gaiters and the gearbox mounting brackets are in line with the chassis fixing points.

78. Loosely fit both gearbox mounting brackets to the chassis. Six bolts are inserted from the outer side of the chassis members (plain washers under heads) and retained by nuts (with spring washers).
NOTE:– The upper bolt on the left hand bracket also holds a speedometer cable support clip.

79. Tighten all six mounting bracket nuts.

80. Remove jack from under engine and lower hoist to allow the complete assembly to settle on the rubber mountings.

81. Tighten nut on right side bracket to secure the rubber mounting.

82. Remove the two bolts holding the hoist cradle to the rear of the transfer box and remove the hoist from under the vehicle.

83. Locate the speedometer cable in the transfer box. Fit the cable fork clip and tighten the Nyloc nut.

84. Apply Loctite 290 to the threads of the two bolts and refit to the transfer box rear cover, ensuring that the left bolt also carries the speedometer cable retaining clip.

85. Set gearbox in neutral gear.

86. Release ties from the rear prop shaft.

87. Line up the punch marks and fit rear prop shaft flange to the brake drum flange with four Nyloc nuts.

88. Release ties from the front prop shaft.

89. Line up the punch marks and fit the prop shaft flange to the front output shaft flange on the transfer box with four Nyloc nuts.
WARNING:– Nyloc nuts should only be used once.

90. Apply Hylosil, Hylomar PL32 or Loctite 290 to the spacing plate on the clutch slave cylinder.

91. Locate the slave cylinder in position and loosely retain with two bolts (with spring washers).

92. Align the slave cylinder pipe bracket and fit the eighth bolt through it into the bell housing.

93. Finally, tighten the two bolts to secure the slave cylinder in the casing.

94. Hylomar PL32 or Loctite 290 to the flywheel housing bottom cover.

95. Locate the bottom cover on the front of the gearbox casing and loosely fit the two large dowel bolts (with spring washers) to position it correctly and compress the rubber seal on the top of the cover.

96. Loosely fit the five lower bolts and nuts (with spring washers) from the **rear side** of the gearbox housing.

97. Loosely fit the two top bolts (with spring washers) from the **front side** of the gearbox housing.

98. Tighten all nine nuts and bolts.

99. Release ties from the front silencer.

continued

100. Align the rear of the silencer with the support bracket and from the top fit the two bolts and nuts (with spring washers) and tighten to secure the twin rear pipes to the silencer.

101. Locate the front exhaust pipes and branch assembly and loosely retain the rear end with the 'U' bolt on the gearbox mounting (with two nuts and spring washers).

102. Locate both front pipe flanges over the studs in the manifolds and loosely fit the six nuts (with spring washers).

103. Finally, tighten all exhaust pipe fixings.

104. Fit the heat shield to the face of the right hand engine mounting bracket. This is secured by two bolts and nuts (with spring washers).

105. To fit the chassis cross member, which is secured by eight nuts and bolts (with two plain and one spring washer each), first spread the chassis using a suitable tool.

106. Locate the cross member in the chassis and loosely fit two of the (shorter) bolts each side with their respective washers and nuts. Fit the remaining four bolts (with washers and nuts).
     NOTE:– The two upper rear bolts must be fitted from the inside of the chassis.

107. Check that the three drain plugs are tight and remove the main gearbox and transfer box filler level plugs. Fill the main gearbox with approximately 1,76 litres (3 pints) of a recommended oil or until it begins to run out of the filler level hole. Fit and tighten the filler plug. Similarly remove the transfer box filler level plug and inject approximately 2,6 litres (4.5 pints) of recommended oil or until it runs out of the filler hole. Apply Hylosil, Hymolar PL32 or Loctite 290, to the threads and fit the plug and wipe away any surplus oil.

108. Lower the ramp.

109. **From under the bonnet** locate the fan cowl so that it engages in the two slots in the radiator bottom bracket.

110. Locate both bolts, with the (smaller) plain washers, under the heads, through the front of the two radiator top brackets. Fit the (larger) plain washers, spring washers and nuts to secure the fan cowl.

111. Fit the retaining clip over the top ends of the two gearbox breather pipes and secure it from the back with the bolt (spring and plain washers) to the engine lifting eye on the rear of the right hand cylinder head.

112. Retain the electrical harness to the breather pipes with a 'Fir tree' rubber clip.

113. Locate the air cleaner on the two pegs on the top of the inlet manifold.

114. Connect the engine breather pipe to the base of the air cleaner (push fit).

115. Connect the two hoses (one on each side of the air cleaner) to the Pulsair rails and secure with Jubilee clips.

116. Connect the float chamber breather vent pipe to the right hand carburettor (push fit).

continued

117. Locate the air hose attached to the air cleaner through the open clip on the right hand carburettor and the clip on top of the air temperature control connecting the hose to the air temperature control (push fit).

118. Connect the short tube from the air cleaner to the one-way breather valve on the inlet manifold.

119. Connect the air temperature intake sensor unit to the air cleaner with the large convoluted hose (push fit) and simultaneously engage the front in the mounting peg.

120. Connect the air cleaner to the carburettors with the two aluminium elbows. These are a push fit over the ends of the carburettor flanges and are retained in the air cleaner hoses by 'Jubilee' clips.

121. **From inside the vehicle** apply grease (Duckhams Q5848 or Shell Alvania R1) to the main gear lever ball end and locate the gear lever assembly in the gearbox and retain with the bolt and special lock washer.

122. Using a screwdriver lift the bias spring legs inwards over the gear lever crosspin.

123. To adjust the bias springs first engage third or fourth gear.

124. Slacken the locknuts on the two adjusting screws and screw them **upwards** until both spring legs are lifted clear of the gear lever cross pin. This should allow some slack radial movement of the gear lever in the engaged gear position.

125. Move the gear lever to the **left** until the slack is **completely taken up** and, retaining it in position, screw the **right** hand screw **downwards** sufficiently for the right hand spring leg to just make contact with the right hand end of the gear lever crosspin.

126. Repeat method 125 for the **opposite** gear lever position by adjusting the **left hand** spring leg.

127. At this stage no spring tension is involved and slack radial movement of the gear lever will be restricted at either extreme by contact of the crosspin with the **opposite** spring leg.

128. Screw both adjusting screws **downwards** by equal amounts until slack radial movement is **just** eliminated by spring contact (movement is more easily checked by holding the bottom of the gear lever).

129. Screw both adjusting screws **downwards** a further **two flats** to provide the correct spring tension.

130. Return the gear lever to the neutral position and move it across the gate several times. Upon release the gear lever should return to the third/fourth gear gate position.

131. Finally, tighten the respective lock nuts.

132. Fit the gaiter assembly over the main and transfer gearbox gear levers and locate in the rectangular aperture on the sound insulated tunnel cover.

133. Fit four screws (with plain washers) to secure the transfer box gear lever metal base plate to the selector housing underneath.

134. Fit the retaining frame over the combined gaiter assembly and secure with self-tapping screws.

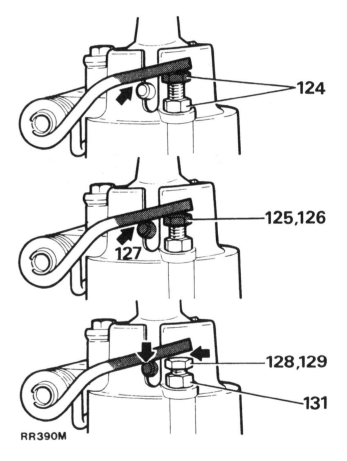

RR390M

continued

135. Locate the complete console over the main and transfer box gear levers and pass the front slotted part under the two shouldered screws.
136. Fit both ashtray bodies in the slots provided (lids open outwards position) and retain with two self-tapping screws, through the base of each body into 'Spire' clips on the gearbox tunnel.
137. Clip the ashtrays into position (cigarette stubbers to the front).
138. Press on both main and transfer box gear lever knobs.
139. Reconnect the battery and **drive the vehicle off the ramp**.

GEARBOX ASSEMBLY

– Five speed manual

Overhaul   37.20.04

Service Tools
18G705; 18G705-1A; 18G705-5; 18G1400; 18G1400-1;
MS47; 18G47BA; 18G47BA-X; 18G284; 18G284AAH;
18G1422 and 18G1431.

Dismantling

1. Place gearbox on a bench with the transfer gearbox removed, ensuring the oil is first drained.
2. Remove the clutch release bearing carrier clip.
3. Remove the clutch release bearing and carrier.
4. Remove the bolt and spring washer securing the clutch release lever clip and remove the clip.
5. Remove the clutch release lever and the slotted washer.
6. Remove the bolts and washers securing the bell housing and remove the bell housing.
7. Remove the three bolts and washers retaining the gear selector housing to the fifth gear extension case. Lift the housing from the case and discard the gasket.
8. Using a suitable pin punch, remove the roll pin retaining the selector yoke.
9. Push the selector shaft forward to engage a gear, and manoeuvre the selector yoke from the shaft. Return the selector shaft to neutral.
10. Remove the circlip which retains the mainshaft oil seal collar located at the rear of the gearbox.
11. Using tools 18G705 and 18G705-1A remove the oil seal collar.
12. Remove the ten bolts and spring washers securing the rear cover to the gearcase; withdraw the rear cover and discard the gasket.
13. Fit two dummy bolts (8 x 35mm) to the casing to retain the centre plate to the main case.
14. Remove the oil seal collar 'O' ring from the mainshaft.
15. Withdraw the oil pump drive shaft.
16. Remove the two bolts and spring washers securing the fifth gear selector fork and bracket.
17. Withdraw the fifth gear selector spool.
18. Withdraw the fifth gear selector fork and bracket.
19. Release the circlip retaining the fifth gear synchromesh assembly to the mainshaft.
20. Using tools 18G1400-1 and 18G1400 withdraw the selective washer, fifth gear synchromesh hub and cone, fifth gear (driven) and spacer from the mainshaft.
21. Remove the split roller bearing assembly from the mainshaft.
22. On early models, remove circlip retaining fifth gear (driving) from the layshaft. On later models, engage reverse gear by turning selector rail anti-clockwise and pulling rearwards. Move the fifth speed synchro hub into mesh with the fifth gear. De-stake the retaining nut securing the fifth gear layshaft and

continued

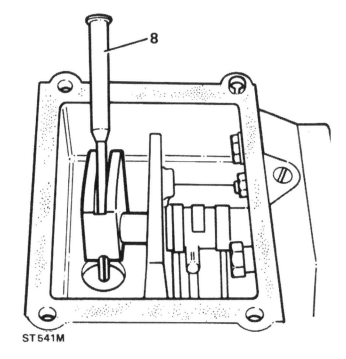

ST541M

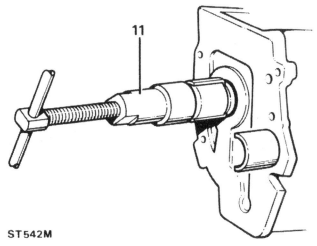

ST542M

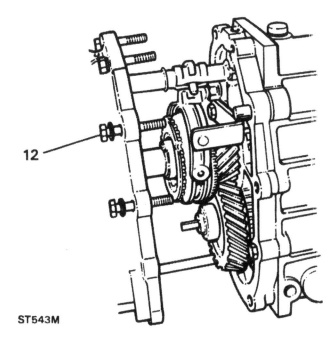

ST543M

remove nut. Select neutral by pushing selector rail inwards and turning clockwise; and return fifth speed synchro hub to its out of mesh position.

23. Using tools 18G705 and 18G705-1A remove the layshaft spacer and layshaft fifth gear.

24. Remove the selector shaft circlip.

25. Fit suitable guide studs (measuring 8 x 60mm) to the main gearbox case.

26. Locate the gearbox to a suitable stand.

27. Remove the six bolts and spring washers from the front cover, withdraw the cover and discard the gasket.

28. Remove the input shaft and layshaft selective washers from the gearcase.

29. Remove the two bolts and washers securing the locating boss for the selector shaft front spool, withdraw the locating boss.

30. Withdraw the selector plug, spring and ball from the centre plate.

31. Remove the dummy bolts and carefully lift the gearcase, leaving the centre plate and gear assemblies in position. Discard the gasket.

32. Insert two slave bolts and nuts to retain the centre plate to the stand; and remove the circlip, pivot pin, reverse lever and slipper pad.

33. Slide the reverse shaft rearwards and lift off the thrust washer, reverse gear and reverse gear spacer.

34. Lift off the layshaft cluster.

35. Remove the input shaft and fourth gear synchromesh cone.

36. Rotate the fifth gear selector shaft clockwise (viewed from above) to align the fifth gear selector pin with the slot in the centre plate.

37. Remove the mainshaft and selector fork assemblies from the centre plate.

38. Detach the selector fork assembly from the mainshaft gear cluster.

39. Remove the slave bolts from the centre plate and lift the centre plate clear of the stand.

Front cover

40. Remove and discard the oil seal from the front cover. Do not fit a new oil seal at this stage.

Layshaft

41. Using press 18G705 and tool 18G705-5 remove the layshaft bearings.

Mainshaft

42. Remove the centre bearing circlip.

43. Using press MS47 and any suitable metal bar, remove the centre bearing, first gear bush, first gear and needle bearings and first gear synchromesh cone.

44. Lift off the first and second gear synchromesh hub assembly, second gear and second gear needle bearing. If a difficulty is experienced in removing the first and second gear synchromesh hub, locate underneath the second gear with a suitable tool; and extract the complete synchromesh hub and second gear assemblies using a suitable press.

continued

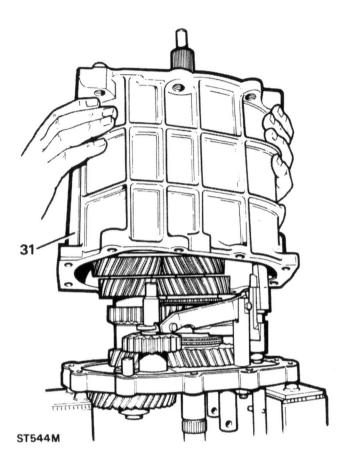

31

ST544M

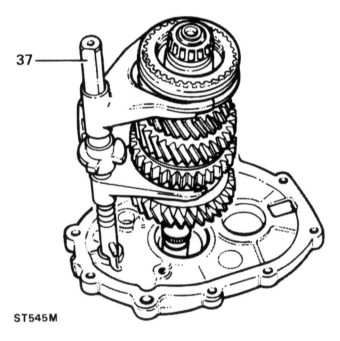

37

ST545M

45. Using press MS47 and extension, remove the pilot bearing, spacer, third and fourth synchromesh hub, third gear synchromesh cone, third gear and third gear needle roller bearing by pressing underneath the third gear.

First and second gear synchromesh assemblies
46. Remove the slipper rings from the front and rear of the first and second gear synchromesh assemblies.
47. Withdraw the slippers and hub from the sleeve.

Third and fourth gear synchromesh assemblies
48. Remove the slipper rings from the front and rear of the assembly.
49. Withdraw the slippers and hub from the sleeve.

Extension case
50. Remove the three oil pump housing bolts, spring washers and oil pump gears.
51. Withdraw the oil pick-up pipe.
52. Remove the plug, washer and filter.
53. Invert casing and extract the oil seal.
54. Press out the ferrobestos bush from the casing.

Input shaft
55. Using tools MS47 and 18G47BA, remove the input shaft bearing.
56. With the aid of tools 18G284AAH and 18G284, extract the pilot bearing track.

Reverse idler gear
57. Remove the circlip from the reverse idler gear.
58. Having noted their positions, remove both needle roller bearings and remaining circlip from the gear.

Fifth gear synchromesh assembly
59. Lever the backing plate off the fifth gear synchromesh assembly.
60. Remove the slipper rings from the front and rear of the assembly.
61. Release the slippers and slide the hub from the sleeve.

Centre plate
62. Remove the layshaft and mainshaft bearing tracks from the centre plate.

Main gearbox casing
63. Remove the mainshaft and layshaft bearing tracks from the main casing.
64. Remove the plastic oil trough from the front face of the casing.

Selector rail
65. The selector rail is supplied complete with first and second selector fork, pin and fifth speed selector pin. The selector rail pins must NOT be removed.

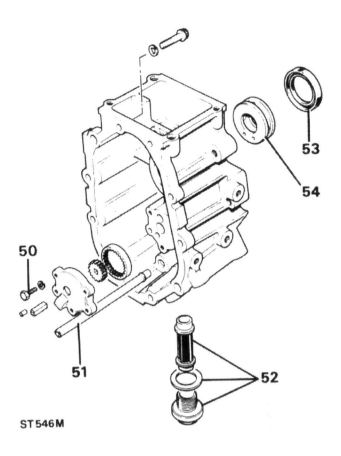

ST546M

continued

Gear selector housing

66. Remove the roll pin and release the bias springs.
67. Remove the two adjusting screws and locknuts.
68. Remove the gear lever extension, secured by a nut (with plain washer).
69. Remove the bolt and special lock washer to release the gear lever shaft from the trunnion housing.
70. Remove the four bolts and spring washers retaining the gear lever housing to the selector housing. Lift off the housing and discard the gasket.
71. Remove the bolts and washers retaining the reverse gear plunger assembly. Care must be taken not to lose the shims underneath the assembly casting. Detach from the selector housing and label components for identification.
72. Remove the locating bolt from the nylon trunnion housing. Pull the selector shaft rearwards and remove the trunnion housing.
73. Release the circlip and detach the nylon insert from the trunnion housing.
74. Invert the gear selector housing and remove the fifth gear spool retainer bolts and spring washers. Lift off the fifth gear spool retainer.
75. Remove the large blanking plug at the rear of the housing.
76. Remove the reverse switch blanking plug.
77. Place the gear selector housing into protected vice jaws, using a suitable pin punch, drift out the selector yoke roll pin. Push the selector shaft forwards and remove the selector yoke. Remove housing from vice.
78. Remove the selector yoke roller circlip and withdraw the pin and rollers.
79. Withdraw the gear selector housing shaft out through the large blanking plug orifice.
80. Remove and discard the gear selector shaft 'O' ring.

Reverse gear plunger assembly

81. Remove the plug, long spring and detent ball from the reverse gear plunger assembly.
82. Detach the circlip which retains the reverse gear plunger, pull out the plunger followed by the short spring.

Cleaning and inspection

83. Clean gearcase thoroughly using a suitable solvent. Inspect case for cracks, stripped threads in the various bolt holes, and machined mating surfaces for burrs, nicks or any condition that would render the gearcase unfit for further service. If threads are stripped, install Helicoil, or equivalent inserts.
84. Inspect all gear teeth for chipped or broken teeth, or showing signs of excessive wear. Inspect all spline teeth on the synchromesh assemblies. If there is evidence of chipping or excessive wear, install new parts on reassembly. Check all slippers and slipper rings for wear or breakage. Replace with new parts if necessary.

ST548M

RR430M

continued

526

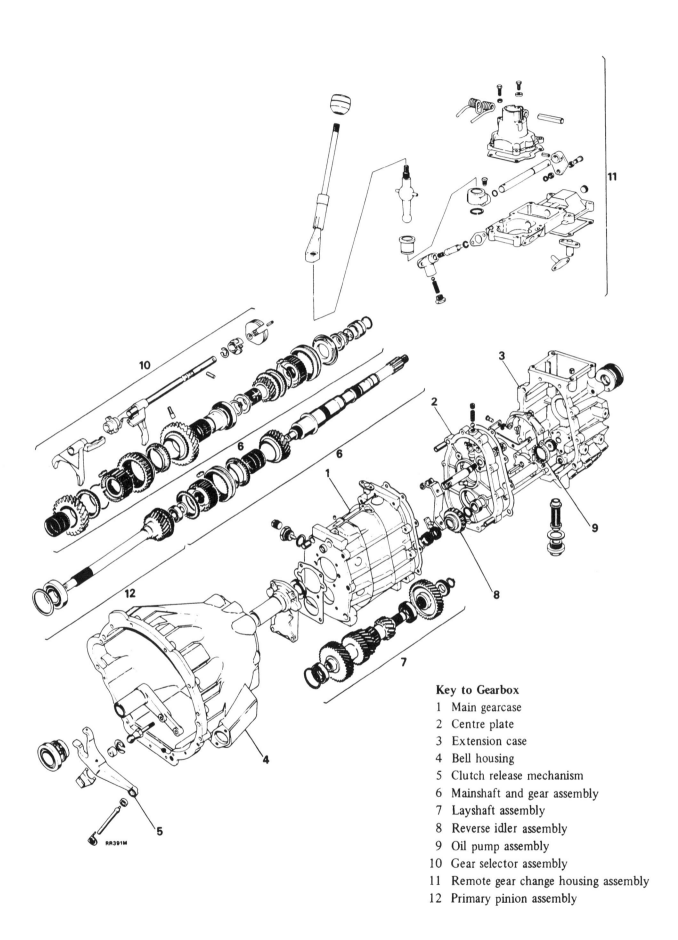

Key to Gearbox
1 Main gearcase
2 Centre plate
3 Extension case
4 Bell housing
5 Clutch release mechanism
6 Mainshaft and gear assembly
7 Layshaft assembly
8 Reverse idler assembly
9 Oil pump assembly
10 Gear selector assembly
11 Remote gear change housing assembly
12 Primary pinion assembly

RR391M

85. Inspect all circlip grooves for burred edges. If rough or burred, remove condition carefully using a fine file.
86. Ensure all oil outlets are clear of sludge or contamination especially the mainshaft oil ways. Clean with compressed air observing the necessary safety requirements.
87. During the rebuild operation, it is recommended that new roller and needle bearings are fitted.

ASSEMBLY

Layshaft
88. Using tools MS47 and a suitable tube, fit new bearing to the layshaft.

Synchromesh assemblies

NOTE: In later gearboxes, the baulk ring fitted towards the main shaft third gear on the third to fourth gear synchromesh assembly, has a molybdenum coated contact face and no internal horizontal grooves. Since this also applies to a replacement synchromesh assembly it is important to ensure that the molybdenum faced ring is indeed fitted towards the third gear. Also when fitting a new synchromesh assembly, with a molybdenum coated baulk ring, it must be matched with a new mainshaft third gear.

89. With the outer sleeve held, a push-through load applied to the outer face of the synchromesh hub should register 8,2-10 kgf/m (18-22 lbf/ft) to overcome the spring detent in either direction.
90. Assemble the first and second synchromesh assembly by locating the shorter splined face towards the second gear.
91. Refit the slippers and locate the slipper rings to each side of the assembly, ensuring that the hooked ends of both slipper rings are located in the same slipper; but running in opposite directions and finishing against the other two slippers.
92. Assemble the third and fourth synchromesh assembly and ensure the hooked ends are located in the same slipper; and run in opposing directions and finally locate against the other two slippers.
93. Refit the fifth synchromesh hub assembly again ensuring the hooked ends of the rings are located in the same slipper, but running in opposite directions. Fit the backplate on to the rear of the synchromesh hub assembly.
94. Check the wear between all the synchromesh cones and gears by pushing the cone against the gear and measuring the gap between the gear and cone. The minimum clearance is 0,64 mm (0.025 in). If this clearance is not met, fit new synchromesh cones.

First gear bush end-float
95. Manufacture a spacer to the dimensions provided in the illustration, this will represent a slave bearing.
96. Lubricate the second gear needle bearing with a light oil and fit the bearing, second gear and synchromesh

cone to the mainshaft. It should be noted that the second gear synchromesh cone has larger slipper slots than the other synchromesh cones.
97. Fit the first and second synchromesh hub assembly with the selector fork annulus to the rear of the mainshaft.
98. Fit the first gear bush and slave bearing spacer and a new circlip to the mainshaft. When fitting the circlip, care must be taken to ensure it is not opened (stretched) beyond the minimum necessary to pass over the shaft.

continued

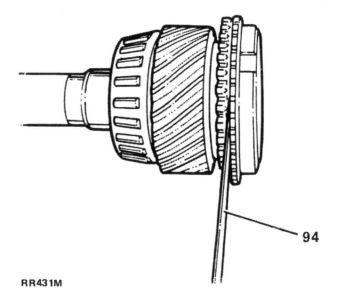

RR431M

94

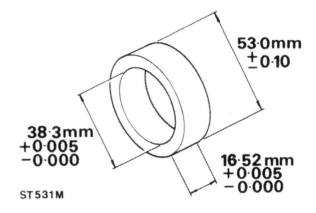

ST531M

53·0mm
± 0·10

38·3mm
+0·005
−0·000

16·52 mm
+0·005
−0·000

99. Press the slave bearing spacer back against the circlip to allow the bush maximum end-float. Measure the clearance between the rear of the first gear bush and front face of the slave bearing spacer with a feeler gauge. The clearance should be within 0,005-0,055 mm. The first gear bush is available with collars of different thickness. Select a bush with a collar to give the required end-float. The bush must be free to rotate easily with the required end-float.

100. Remove the circlip, slave bearing spacer and first gear bush from the mainshaft.

101. First gear bushes are available in the following sizes:

Part No.	Thickness (mm)
FRC5242	40,11-40,16
FRC5243	40,16-40,21
FRC5244	40,21-40,26
FRC5245	40,26-40,31
FRC5246	40,31-40,36

102. Having selected a suitable first gear bush, lubricate the needle bearing and fit to the first gear.

103. Fit the selected bush to the first gear and place first gear synchromesh cone, followed by the first gear assembly to the mainshaft.

104. Using tools MS47, 18G47BA and 18G47BA-X refit the centre bearing and circlip to the mainshaft.

105. Invert the mainshaft, lubricate the third gear needle roller bearing with light oil, fit to the front end of the mainshaft.

106. Fit the third gear to the mainshaft; and locate the third gear synchromesh cone to the third gear.

107. Fit the third/fourth synchromesh assembly (with the longer box of the synchromesh hub to the front of the gearbox) to the mainshaft.

108. Fit the spacer and bearing to the front of the mainshaft.

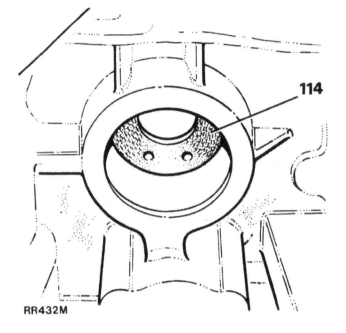

RR432M

Input shaft

109. Using tool MS47 and any suitable tube, refit a new pilot bearing track to the input shaft.

110. Fit the input shaft bearing using tools MS47, 18G47BA and 18G47BA-X.

Reverse gear and shaft

111. Fit a new circlip to the rear of the reverse idler gear, ensuring that the circlip is not stretched beyond the minimum necessary to pass over the shaft.

112. Lubricate with light oil and fit both needle roller bearings. Fit the shorter needle bearing to the rear of the reverse idler gear.

113. Fit a new circlip to the front of the reverse idler gear.

Extension case

114. Using a suitable press, fit a new ferrobestos bush to the case, ensuring the two drain holes are towards the bottom of the case.

115. With the aid of tool 18G1422, fit a new oil seal to the rear of the extension case. Ensure the seal lips are towards the ferrobestos bush. Lubricate the seal lips with a suitable SAE 140 oil.

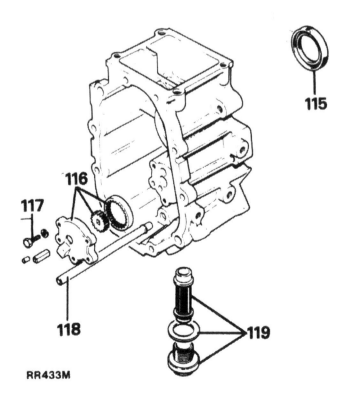

RR433M

continued

116. Assemble and fit the fibre oil pump gears to the oil pump cover, whilst ensuring the centre rotor squared drive faces the layshaft.
117. Fit the three bolts and spring washers to secure the oil pump cover; and tighten to 7 Nm (5 lbf. ft).
118. Refit the oil pick-up pipe to the extension casing, having ascertained it is free of blockages or contamination. Seal with Loctite 290.
119. Fit a new oil filter, fibre gasket and tighten plug to 25 Nm (19 lbf. ft.).

Centre plate

120. Fit the centre plate to a suitable stand and secure with two slave bolts.
121. Place the new mainshaft and layshaft bearing tracks to the centre plate.
122. Lightly lubricate the selector shaft with a light oil.
123. Take the selector shaft complete with the first and second selector fork, front spool and third and fourth selector fork; engage both selector forks in their respective synchromesh sleeves on the mainshaft, simultaneously engaging the selector shaft and mainshaft assemblies in the centre plate, whilst rotating the fifth gear selector pin to align with the slot in the centre plate.
124. Fit the layshaft to the centre plate.
125. Rotate the selector shaft and spool to enable the reverse crossover lever forks to correctly align to the reverse pivot shaft. Reposition the selector shaft and locate the lever between the fork on the reverse gear pivot shaft. Insert pivot pin and fit a new circlip, ensuring that it is not opened beyond the minimum necessary to pass over the shaft.
126. Fit the slipper pad to the reverse lever. If a new reverse lever pivot shaft has been fitted, it will be necessary to ascertain that its radial location is consistent with the reverse pad slipper engagement/clearance. The radial location is determined during initial assembly.
127. Fit the reverse gear spacer and reverse gear assembly, locating the slipper pad lip to the reverse gear groove. Engage the reverse gear shaft from the underside of the centre plate, ensuring the roll pin is aligned with the slot in the centre plate casing.
128. Prior to assembly lubricate the detent ball and spring with light oil, and fit to the top of centre plate. Smear Hylomar PL32 or Loctite 290 to the plug threads and screw the plug flush with the case. Stake the plug to prevent rotation using a suitable centre punch.
129. Release the slave bolts and remove the centre plate and gear assemblies from the stand.
130. Early models – Using a suitable press fit the fifth gear and collar to the layshaft. Fit a new circlip ensuring it is not expanded beyond the minimum necessary to obtain entry over the shaft.
    Later models – Fit the fifth gear to the layshaft using a suitable press and loosely fit a NEW special nut. To tighten the nut, hold the gearbox firmly in a vice and if necessary use a flange holding wrench to restrain the gearbox. Tighten the nut to 204 to 231

continued

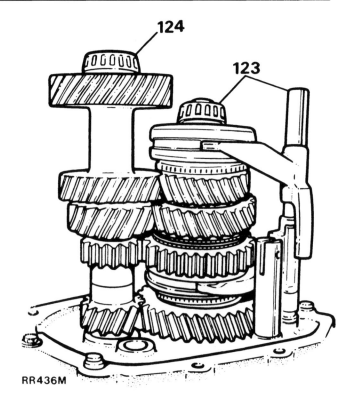

124 123

RR436M

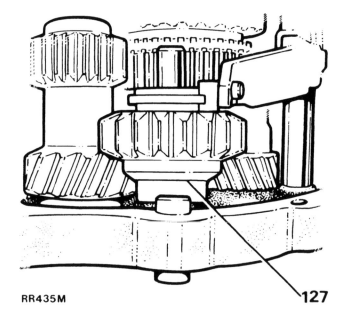

RR435M 127

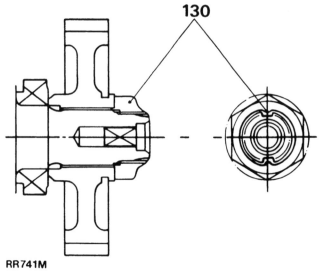

130

RR741M

530

Nm (150 to 170 lb ft). To prevent damage to the adjacent bearings when deforming the nut locking collar, support the fifth gear with a block of timber. Using a round nose punch carefully form the collar into the layshaft grooves, as illustrated.

131. Locate the fourth gear synchromesh cone to the third/fourth synchromesh assembly.
132. Fit the input shaft to the mainshaft.
133. Fit the reverse gear spacer to the reverse gear shaft.
134. Fit a new gasket to the centre plate.

Main gearbox casing

135. Insert a new plastic oil trough to the back of the main gearbox casing, ensuring the open trough faces the top of the case.
136. Carefully lower the gearcase into position over the gear assemblies. DO NOT USE FORCE. Ensure the centre plate dowels and selector shaft are engaged in their respective locations.
137. Fit the layshaft and input shaft bearing outer tracks.
138. Using 8 x 35 mm slave bolts and plain washers to prevent damaging the rear face of the centre plate, evenly draw the gearcase into position on the plate.
139. Fit the locating shaft front spool to the top of the gearcase using Hylomar PL32 to seal between the spool and gearcase. Smear Loctite 290 or Hylomar PL32 to the bolt threads, tighten bolts and spring washers to 7 Nm (5 lbf. ft).

Mainshaft end-float

140. When ascertaining the mainshaft end-float care must be taken when checking the dial gauge readings to ensure that the end-float only, as distinct from side movement, is recorded. To overcome the difficulty in differentiating between end-float and side movement, wrap approximately ten turns of masking tape around the plain portion of the input shaft below the splines. Fit a new gasket and refit the front cover. Ascertain that the rise and fall of the input shaft is not restricted by the tape. Fit, secure and tighten the six bolts and spring washers to 22 Nm (16 lbf. ft).
141. Place a suitable ball bearing in the centre of the input shaft. This facilitates accurate checking of the mainshaft end-float.
142. Mount the dial gauge on the gearcase with the stylus resting on the ball bearing centre. Zero the gauge.
143. Check the end-float by a 'push-pull' action to the input shaft. The required end-float measurement should be between 0,06-0,01 mm. Having determined the end-float, select the required spacer as follows: End-float obtained, minus, End-float required, equals, Spacer thickness required.
144. Fit the calculated spacer required and again check the end-float which must be between 0,06-0,01 mm.
145. Detach the dial gauge equipment and ball bearing, remove the front cover and all tape.

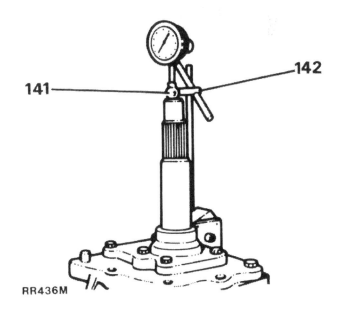

141

142

RR436M

continued

531

146. Fit the required thickness spacer to the mainshaft bearing track. Selective spacers are available in a range of sizes to meet the previously mentioned clearance limits:

Part No.	Thickness (mm)	Part No.	Thickness (mm)
FRC4326	1,48	FRC4349	2,17
FRC4327	1,51	FRC4350	2,20
FRC4328	1,54	FRC4351	2,23
FRC4329	1,57	FRC4352	2,26
FRC4330	1,60	FRC4353	2,29
FRC4331	1,63	FRC4354	2,32
FRC4332	1,66	FRC4355	2,35
FRC4333	1,69	FRC4356	2,38
FRC4334	1,72	FRC4357	2,41
FRC4335	1,75	FRC4358	2,44
FRC4336	1,78	FRC4359	2,47
FRC4337	1,81	FRC4360	2,50
FRC4338	1,84	FRC4361	2,53
FRC4339	1,87	FRC4362	2,56
FRC4340	1,90	FRC4363	2,59
FRC4341	1,93	FRC4364	2,62
FRC4342	1,96	FRC4365	2,65
FRC4343	1,99	FRC4366	2,68
FRC4344	2,02	FRC4367	2,71
FRC4345	2,05	FRC4368	2,74
FRC4346	2,08	FRC4369	2,77
FRC4347	2,11	FRC4370	2,80
FRC4348	2,14		

Layshaft end-float

147. Place a layshaft spacer of nominal thickness 1,02 mm on the layshaft bearing track, fit the front cover and tighten the bolts and spring washers to 22 Nm (16 lbf. ft).

148. Invert the gearbox on the stand, place a suitable ball bearing in the layshaft centre and mount the dial gauge on the gearcase with the stylus resting on the ball bearing centre. Zero the gauge.

149. Check the end-float by a 'push-pull' action to the layshaft. The required layshaft setting is:

0,025 mm end-float,

0,025 mm preload.

Spacer thickness required equals:

Nominal thickness of spacer, plus, end-float obtained.

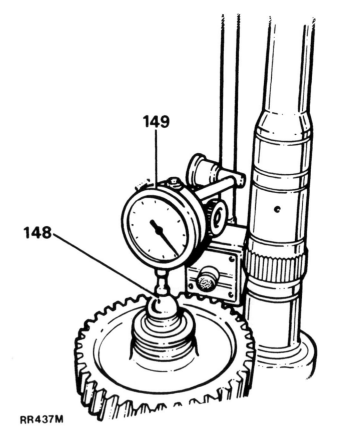

RR437M

continued

150. Having ascertained the end-float, select and fit the layshaft spacer of the appropriate thickness to the layshaft bearing track. Selective spacers are available in a range of sizes to meet the aforementioned clearance limits.

Part No.	Thickness (mm)	Part No.	Thickness (mm)
TKC4632	1,66	TKC4649	2,17
TKC4633	1,69	TKC4650	2,20
TKC4634	1,72	TKC4651	2,23
TKC4635	1,75	TKC4652	2,26
TKC4636	1,78	TKC4653	2,29
TKC4637	1,81	TKC4654	2,32
TKC4638	1,84	TKC4655	2,35
TKC4639	1,87	TKC4656	2,38
TKC4640	1,90	TKC4657	2,41
TKC4641	1,93	TKC4658	2,44
TKC4642	1,96	TKC4659	2,47
TKC4643	1,99	TKC4660	2,50
TKC4644	2,02	TKC4661	2,53
TKC4645	2,05	TKC4662	2,56
TKC4646	2,08	TKC4663	2,59
TKC4647	2,11	TKC4664	2,62
TKC4648	2,14		

151. Fit a new oil seal to the front cover, ensuring the seal lips face towards the gearbox. Lubricate the seal lips with SAE 140 gear oil.
152. Mask the splines with masking tape to protect the oil seal, refit the front cover and remove the spline masking tape.
153. Refit the bolts and spring washers having used Hylomar PL32 or Loctite 290 on the bolt threads. Torque tighten bolts to 22 Nm (16 lbf. ft).
154. Remove gearbox from the stand and place suitably supported on the bench.
155. Remove the guide studs fitted to the centre plate.

Fifth gear to mainshaft
156. Lubricate with light oil and fit a new needle roller bearing to the mainshaft.
157. Fit the front spacer, fifth gear (driven) and fifth gear synchromesh cone to the mainshaft.
158. Using tool 18G1431 fit the fifth gear synchromesh hub assembly, selective spacer and new circlip to the mainshaft. When fitting, care must be taken to ensure the hub assembly and selective spacer are NOT pushed too far on the mainshaft. Only fit with sufficient clearance to allow the circlip to engage in its groove.
159. Using a feeler gauge, measure the clearance between the front spacer and fifth gear (driven), which should be between 0,005-0,055 mm. Use a selective spacer which will provide the required clearance.

continued

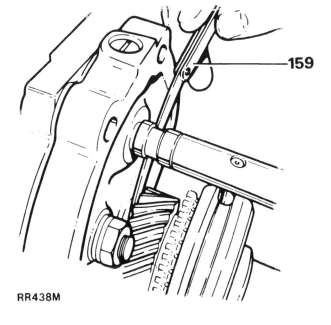

RR438M

160. Selective spacers are available in a range of sizes to meet the aforementioned clearance limits:

Part No.	Thickness (mm)	Part No.	Thickness (mm)
FRC5294	5,40	FRC5284	5,10
FRC5295	5,43	FRC5285	5,13
FRC5296	5,46	FRC5286	5,16
FRC5297	5,49	FRC5287	5,19
FRC5298	5,52	FRC5288	5,22
FRC5299	5,55	FRC5289	5,25
FRC5300	5,58	FRC5290	5,28
FRC5301	5,61	FRC5291	5,31
FRC5302	5,64	FRC5292	5,34
FRC5303	5,67	FRC5293	5,37

Fifth gear selector fork assembly

161. Fit the fifth speed selector fork and bracket to the fifth gear synchromesh hub assembly, ensuring that the largest groove lip is facing the rear of the gearbox.

162. Fit the fifth gear spool to the selector shaft, rotate and engage the selector fork into the groove. It should be noted that the longer shoulder of the spool is fitted towards the front of the gearbox.

163. Fit the fifth speed selector fork bracket bolts and spring washers. Tighten bolts to 22 Nm (16 lbf. ft).

164. Fit a new circlip to the selector shaft ensuring that it is not expanded beyond the minimum necessary to obtain entry.

165. Remove the six dummy bolts securing the centre plate to the main casing.

166. Insert the squared oil pump shaft into the centre of the layshaft.

Extension case

167. Fit a new gasket to the centre plate.

168. Rotate the oil pump to align with the oil pump drive shaft.

169. Carefully fit the extension case ensuring that the oil pump shaft engages the oil pump.

170. Fit the extension case bolts and spring washers; tighten to 22 Nm (16 lbf. ft).

171. Using a large screwdriver, ease the selector shaft forwards to select a gear. It may be found necessary to rotate the mainshaft to ease gear selection.

172. Fit the selector yoke to the selector shaft and secure with a new roll pin. Pull selector shaft rearwards to select a neutral position.

173. Cover the mainshaft splines with masking tape and fit a new oil seal collar 'O' ring. Remove the masking tape.

174. Using tool 18G1431 fit the oil seal collar to the mainshaft, ensuring the collar is NOT pushed too far on the shaft, fit only with sufficient clearance to allow the circlip to engage in its groove.

175. Fit a new gasket to the fifth gear extension case and engage the selection yoke rollers in the selector yoke. Fit the three bolts and spring washers; and tighten to 22 Nm (16 lbf. ft).

continued

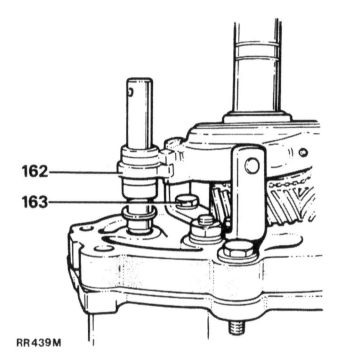

162—

163—

RR439M

Bell housing

176. Locate the bell housing to the dowels and fit the two long bolts (12 x 45 mm) with spring and plain washers to the dowel positions. The remaining four bolts (12 x 30 mm) are fitted with spring washers only. Tighten all bolts to 65 Nm (48 lbf. ft).

177. Prior to reassembly, lubricate the following with Molykote FB180 grease:

    (a) Clutch release lever fulcrum pivot socket.
    (b) The clutch release lever and the faces.
    (c) Ball end of the clutch operating push rod.

178. Locate the slotted washer on the pivot and fit the clutch release lever.

179. Fit the spring clip on the release lever and secure with the bolt and spring washer.

180. Lubricate the inner face of the clutch release bearing carrier with Molykote FB180 grease and fit to the front cover spigot, locating the clutch release lever to the carrier recesses.

181. Fit a new nylon clutch release carrier clip.

182. Refit the gearbox oil level plug, and tighten to 25 Nm (19 lbf. ft).

183. Refit the gearbox oil drain plug and fit new fibre washer. Tighten plug to 25 Nm (19 lbf. ft).

Reverse gear plunger assembly

184. Lubricate the short spring and plunger with BP Energrease L2 or similar prior to assembly.

185. Fit the spring into the plunger base and slide the assembly into the reverse gear plunger housing. Fit a new circlip to retain the plunger.

186. Lubricate the detent ball with light oil and fit into the recess.

187. Refit the short spring and plug, coat the plug threads with Loctite 290 or Hylomar PL32, and tighten to 22 Nm (16 lbf. ft).

Gear selector housing

188. Lubricate the gear selector housing shaft with light oil and fit a new 'O' ring.

189. Refit the gear selector rollers, pin and new circlip ensuring the circlip is not expanded beyond minimum necessary to obtain entry.

190. Insert shaft through the large blanking plug orifice, ensuring the shaft indent is uppermost.

191. Place the gear selector housing into protected vice jaws and fit the selector yoke to the shaft, using a suitable pin punch and new roll pin. Remove the housing assembly from the vice on completion.

192. Fit the reverse switch and large blanking plugs. Coat plug threads with Loctite 290 and tighten to 22 Nm (16 lbf. ft).

193. Refit the fifth gear spool retainer and tighten the bolts and washers to 7 Nm (5 lbf. ft).

194. Fit a new nylon insert into the trunnion housing and secure with a new circlip.

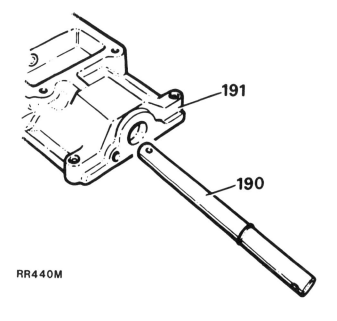

RR440M

continued

195. Invert the gear selector housing and fit the trunnion housing to the selector shaft, ensuring the locating bolt aligns with the shaft indent. Coat the bolt threads with Loctite 290. Tighten bolt to 22 Nm (16 lbf. ft).

196. Fit a new gear lever gasket and locate the gear lever housing, spring washers and bolts. Tighten bolts to 22 Nm (16 lbf. ft).

197. Coat the upper and lower balls of the gear lever shaft with BP Energrease L2 or similar. Push lever into the trunnion nylon bush and retain with the special lock-washer and bolt.

198. Fit the two bias spring adjusting screws and locknuts loosely.

199. Locate the bias springs and retain with the roll pin.

200. Using a screwdriver raise the spring legs on to the top of the respective spring adjusting screws.

201. Refit the gear lever extension and secure the nut and plain washer.

202. Select first or second gear. It may be necessary to rotate the mainshaft whilst manipulating the gear lever.

203. Locate the reverse gear plunger assembly until light contact with the gear lever yoke is felt. Whilst maintaining a light finger pressure, measure the clearance between the plunger assembly casting and gear selector casting. Select suitable thickness shim(s) to equal the gap.

204. Remove the reverse plunger assembly, fit the required thickness shim(s) refit the plunger assembly, spring washers and bolts. Apply Hylomar PL32 or Loctite 290 to the bolt threads and tighten to 22 Nm (16 lbf. ft).

205. Select third or fourth gear and adjust the gear lever bias springs as detailed in 'Remove and Refit of five speed manual gearbox with transfer box' from paragraph 123.

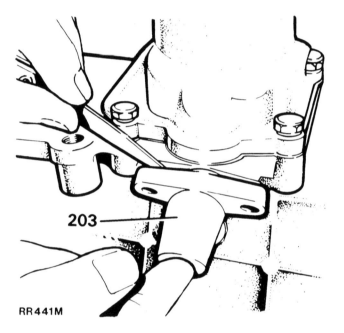

RR441M

TRANSFER GEARBOX
FOR
FIVE SPEED MANUAL
GEARBOX
AND
AUTOMATIC GEARBOX

Two types of Transfer gearbox are in use. The LT230R is fitted to early models and is described in the following pages.

NOTE: Automatic gearbox version is illustrated, manual gearbox version is similar.

A modified design, LT230T type, is fitted to later models and is described from page 607 onward.

All Transfer gearboxes have a serial number stamped on the left hand side. For identification, a suffix 'B' has been added to the LT230T serial number.

TRANSFER BOX

– Five Speed Manual and Automatic models

Overhaul   37.29.28

Dismantling

1. Remove transfer box from vehicle, operation 37.29.25 (Automatic models).
2. Alternatively, remove complete gearbox and transfer box assembly, operation 37.20.01 (Five speed manual) and 44.20.04 (Automatic). In this event the transfer box is separated from the gearbox on the work bench. The following items should be removed with the transfer box still attached to the hydraulic hoist.

Hand brake linkage removal

3. Slacken off the inner locknut on the hand brake adjuster link a few threads.
4. Remove the four bolts securing the hand brake lever mounting bracket.
5. Remove the remaining lower bolt securing the hand brake mounting. (The other mounting bolts were removed in situ when detaching the bell housing tie rod.)
6. Supporting the hand brake and linkage assembly allow it to pivot downwards on the clevis pin connected to the brake operating lever.
7. Finally remove the clevis pin to release the hand brake and linkage assembly.

Transfer box mounting removal

8. Remove the bolts retaining the right hand rubber mounting plate.

Gear change housing removal

*9. Disconnect the bottom of the differential lock cross shaft lever with its short connecting link from the differential lock lever by removing the Nyloc nut. The latest cranked link is retained by a split pin at each end.
10. Remove the two bolts retaining the differential lock cross shaft lever pivot bracket.
11. Disconnect the bottom of the high/low connecting rod from the high/low operating arm by removing the clevis pin.
12. Remove the plastic bushes from the operating arm.
13. Remove the four bolts securing the gearchange housing and remove it complete with linkage.

* The latest cranked short connecting link is retained by a split pin at each end.

Continued

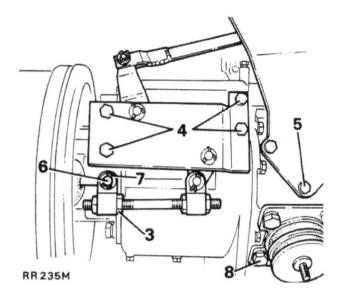

RR235M

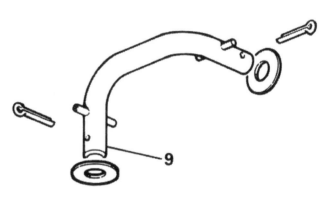

RR392M

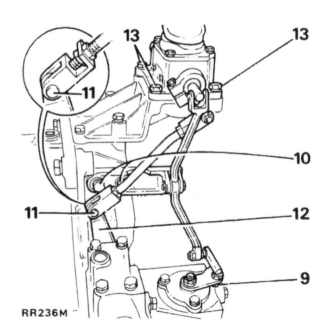

RR236M

Extension housing removal

14. Remove the four bolts and two nuts securing the extension housing.

NOTE:– Having removed the extension housing remove the upper of the two locating dowels in the transfer box case, which is a loose fit.

Transmission brake removal

15. Remove the two countersunk brake drum retaining screws and pull off the drum.
16. Remove the four bolts securing the brake backplate, two of these also retain the oil drip plate.

Continued

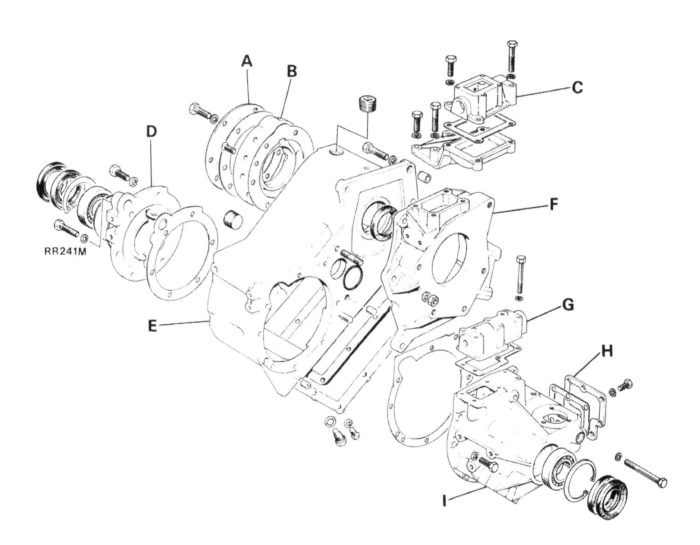

RR241M

A	Power take-off cover	G	High/low selector housing
B	Mainshaft rear bearing housing	H	Differential lock selector side cover
C	Transfer gear change housing	I	Front output shaft housing

D Speedo/rear output shaft housing
E Transfer box
F Extension housing

Moving transfer box to work bench

17. Pass chain around the transfer box and using suitable lifting equipment support its weight.
18. Remove the four bolts retaining the transfer box to the hydraulic hoist (locally made) adaptor plate.
19. Lift the transfer box off the hydraulic hoist on to the work bench.

NOTE:— To facilitate removal of various items on the work bench, obtain suitable wooden blocks to enable the transfer box to be turned and propped up as required.

Bottom cover removal

20. Remove the six remaining bolts retaining the bottom cover, the outer four were removed with the adaptor plate (see item 18).

Intermediate shaft removal

21. Remove the shaft lock plate retained by a single bolt at the front face of the transfer box.
22. Withdraw the intermediate shaft, using a screw driver in the slotted end. Where the shaft cannot be easily withdrawn use extractor RO605862.
23. Lift out the intermediate gear train.
24. Remove the thrust washers.
25. Remove the 'O' ring from the intermediate shaft and the transfer box case.

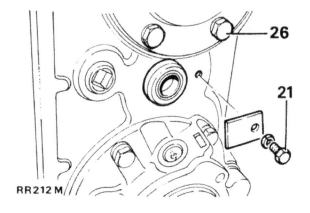

RR212M

Power take off cover removal

26. Remove the six bolts retaining the circular P.T.O. cover.
27. Remove the gasket.

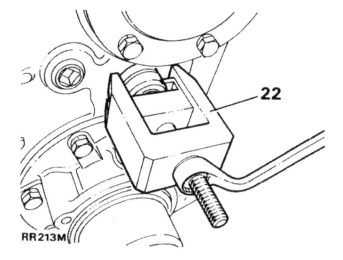

RR213M

Input gear removal

28. Remove the two countersunk screws and release the mainshaft bearing housing.
29. Remove the gasket.
30. Remove the input gear assembly.
31. Prise out and discard the oil seal at the front of the transfer case using service tool 18G1271.
32. Drift out the input gear front bearing track.

High/low selector housing removal

33. Remove the six bolts to release the selector housing.
34. Remove the gasket.

Continued

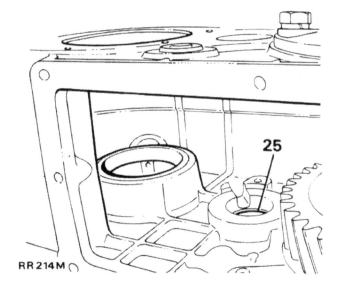

RR214M

Front output shaft housing removal

35. Slacken the set screw securing the yoke to the high/low selector shaft inside the high/low selector housing aperture.
36. If necessary use a screw driver to move the selector shaft rearwards and allow the yoke to be lifted out.
37. Remove the eight bolts to release the front output shaft housing assembly. The upper middle bolt is longer.

NOTE:– The 'radial' dowel in the transfer box face should not be disturbed.

Centre differential removal

38. Remove the high/low selector shaft detent plug, spring and ball.

NOTE:– The ball may be more easily retrieved from inside the transfer case after the selector shaft is taken out.

39. Remove the centre differential unit with the selector shaft/fork assembly.

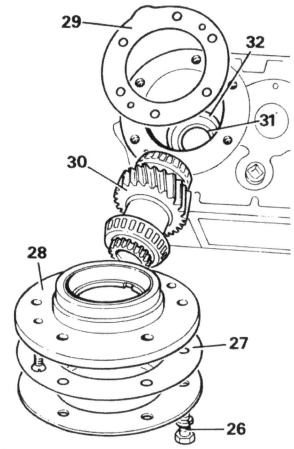

RR 242M

Rear output shaft housing assembly removal

40. Remove the six screws to release the housing. The upper screw is longer.
41. Remove the gasket.

NOTE:– Removal of the above housing will reveal the centre differential rear bearing track in the transfer box casing. Before drifting out, either unscrew the two studs and radial dowel projecting from the transfer box front face or use suitable wooden blocks to support the box to avoid damage to these items.

42. Drift out the differential rear bearing track.

NOTE:– If it is required to completely strip down the transfer box to the basic casting, remove the level, filler and drain plugs.

IMPORTANT:– Clean all parts ensuring any traces of Loctite are removed from faces and threads. Renew oil seals and examine all other parts for wear or damage, renew as necessary.

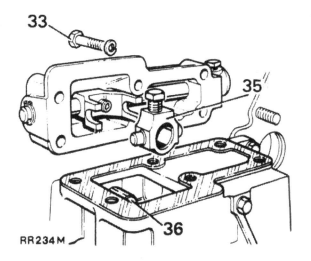

RR234M

Reassembling

43. Ensure that all faces of the transfer box are clean.
44. Check that level filler and drain plugs are in position.
45. Fit the two studs which are used for part retention of the extension housing.
46. Screw in the 'radial' dowel. It is important that its projecting blade is set radially in line with the tapped fixing hole centres in the transfer box casing.

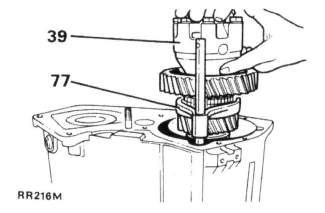

RR216M

Continued

REAR OUTPUT SHAFT HOUSING – OVERHAUL
(Items 47 to 73)

Dismantling

47. Using flange wrench 18G1205 remove the flange nut, steel and felt washers.

NOTE:– Ensure flange bolts are fully engaged in the wrench.

48. Remove the output shaft flange with circlip attached.

NOTE:– The circlip need only be released if the flange bolts are to be renewed.

49. Remove the speedometer spindle housing. This can be prised out with a screw driver.
50. Drift the rear output shaft out of the housing from the flange end.
51. Carefully prise off the oil catch ring using a screw driver in the slot provided.
52. Prise out and discard the two oil seals using tool 18G1271.
53. Using circlip pliers 18G257 remove the circlip retaining the bearing.
54. Drift out the bearing from the back of the housing.
55. Remove the speedometer driven gear and spindle from the spindle housing.
56. Remove the 'O' ring and oil seal.
57. Slide off the spacer and speedometer drive gear from the output shaft.
58. Clean all parts, renew oil seals and Nyloc flange nut and examine all other parts for wear or damage, renew as necessary.

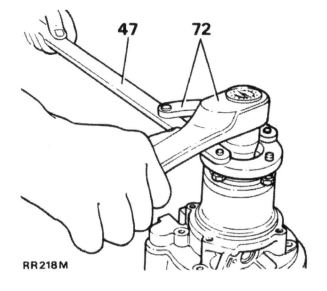

RR218M

Reassembling

59. Fit the output bearing by drifting it in with a suitable tube.
60. Fit the bearing retaining circlip using circlip pliers 18G257.
61. Fit the two new oil seals simultaneously using replacer tool 18G1422. Fit the seals with both sealing lips inwards as illustrated.

NOTE:– On later production a single dual-lipped oil seal is fitted.

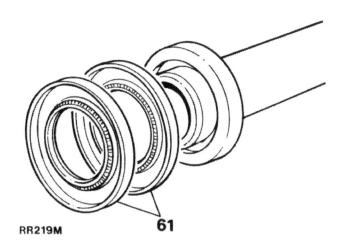

RR219M

62. Charge the lips of both seals with grease.
63. Fit oil catcher ring on to housing.
64. Fit oil seal into speedometer spindle housing with a suitable tube.
65. Fit 'O' ring to speedometer spindle housing.
66. Lubricate seal and 'O' ring with oil.
67. Locate speedometer driven gear and spindle in spindle housing and push into position.
68. Slide speedometer drive gear and spacer on to output shaft.
69. Locate output shaft through back of housing.

Continued

70. Fit flange on to output shaft (with bolts in position).
71. Fit flange felt washer, plain washer and a new Nyloc nut.
72. Using flange wrench 18G1205 and a torque wrench pull up output shaft to correct position.

NOTE:– Ensure flange bolts are fully engaged in the wrench.

73. Locate speedometer spindle housing assembly in the output shaft housing and push in flush with housing face.

NOTE:– Before fitting the rear output shaft housing to the transfer box casing the centre differential rear bearing track must be fitted.

74. Drift the centre differential rear bearing track into the transfer box casing 3 mm (1/8 in) below the outer face of the casing. Check the depth before proceeding.

Fitting rear output shaft housing to transfer box

75. Grease and fit housing gasket and locate the housing in position on the transfer box.

NOTE:– If the differential rear main bearing track has been correctly fitted there will be a gap between the housing face and the gasket at this stage.

76. Apply Loctite Dri-loc 290 to the threads of the six housing securing screws, noting that the upper screw is longer. Fit the screws (with spring washers) evenly tightening them to the specified torque. This will press in the rear main bearing track to the correct position and seat the housing.

CENTRE DIFFERENTIAL UNIT – OVERHAUL
(Items 77 to 130)

Dismantling

77. Detach the high/low selector assembly.
78. **Secure the differential unit in a vice** with the 'stake' nut uppermost.
79. Drill the 'stake' nut flange to facilitate removal of the nut.
80. Remove the 'stake' nut using tool 18G1423.
81. **Remove the differential unit from the vice.**
82. **Secure hand press 47 in vice** with collars 18G47BB/1 and using button 18G47BB/3 remove rear taper roller bearing.
83. Substituting collars 18G47BB/2 remove front taper roller bearing.
84. **Remove the hand press from the vice.**
85. Remove the high range (smallest) differential gear and its bush.
86. Remove the high/low selector sleeve.

Continued

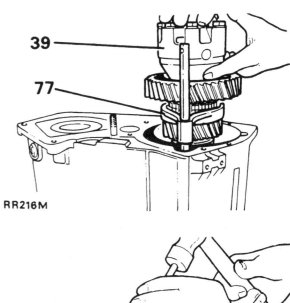

RR216M

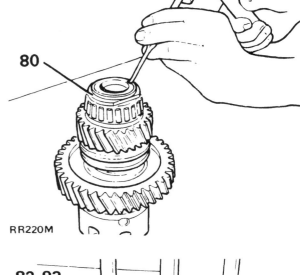

RR220M

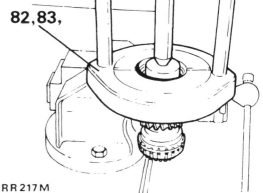

RR217M

87. Using a suitable press behind the low range (largest) gear carefully remove both high/low hub and low range gear together.

88. **Secure the differential unit in the vice.**

89. Remove the eight retaining bolts and lift off the front differential case.

90. Lift off the front (upper) bevel gear and thrust washer.

91. Remove both pairs of side gears with their respective shafts and dished washers together.

92. Lift out the remaining rear (lower) bevel gear and thrust washer.

93. **Remove the rear differential case from the vice.**

94. Clean all parts, examine for wear or damage, renew as necessary.

Obtaining differential backlash by checking bevel gear end float

95. **Secure the rear differential case in the vice.**

96. Ensure that all differential components are dry to assist in checking end float.

97. Using a micrometer measure one of the bevel gear thrust washers and note thickness.

98. Fit the thrust washer and bevel gear to the rear (lower) differential case.

99. Assemble the side gears and dished washers on their respective shafts and fit to the rear case.

100. Measure the remaining bevel gear thrust washer, noting its thickness.

101. Fit the thrust washer and bevel gear to the rear case.

102. Fit and align the front differential case tightening the eight securing bolts to the specified torque.

103. Ensure that the front bevel gear is fully in mesh by tapping it down, using a punch through the front differential case.

104. Measure the front bevel gear end float with feeler gauges through the slots provided in the front differential case. This must be between zero and −0,07 mm maximum (zero and 0.003 in).

105. **Invert differential unit in vice** and repeat the above procedure (items 103 and 104) for the rear bevel gear in the rear differential case.

106. **Return the differential unit to its former position in the vice** i.e. with the front differential case uppermost.

107. Remove the eight securing bolts and lift off the front differential case.

108. Remove the bevel gears and thrust washers, and side gear assemblies.

109. Select correct thrust washers required for final reassembly.

Continued

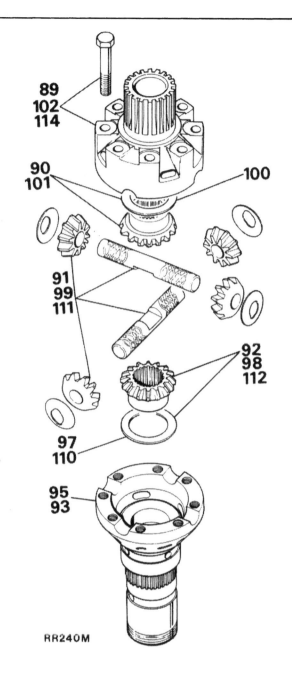

RR240M

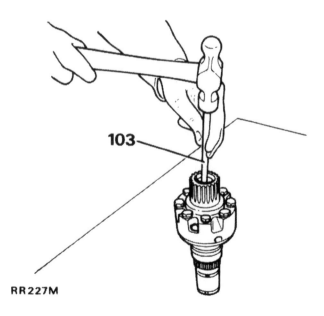

RR227M

Reassembling

110. Fit the selected thrust washer and bevel gear to the rear (lower) differential case.
111. Assemble the side gears and dished washers on their respective shafts and fit to the rear case.
112. Fit the other selected thrust washer and bevel gear to the rear case.
113. Lubricate all parts with oil.
114. Fit and align the front differential case, locate the eight securing bolts and tighten to the specified torque.
115. Finally check that the differential gears revolve freely.
116. Place the front (outer) differential bearing on the front differential case and drift into position using larger end of tool 18G1424.
117. **Invert the differential unit in the vice.**
118. Fit the low range gear (largest) to the rear differential case (with its 'dog' teeth uppermost).
119. Drift the high/low hub on to the splined area of the case
    *Check end float of low range gear.
120. Slide the high/low selector sleeve on to the hub outer splines.
121. Fit the bush into the high range (smallest) gear and slide the bushed gear on to the rear differential case.
    *Check end float of high range gear and running clearance of gear on bush.
122. Place the rear (inner) differential bearing on the rear differential case and drift into position using the smaller end of tool 18G1424.
123. Fit the 'stake' nut using tool 18G1423 and tighten to the specified torque.
124. Peen the nut flange into the slot provided.
125. **Remove the differential unit from the vice.**
126. Lubricate gears, bearings, sleeve and bush with oil.
127. Clean and check the high/low selector fork assembly for wear and renew if necessary.
128. To renew the selector fork remove the set screw retaining it to the selector shaft and ensure any traces of Loctite are removed from the threads.
129. Refit the selector fork to the lower of the two blind holes in the selector shaft, i.e. nearest to the selector shaft detent grooves.
130. Apply Loctite Driloc 290 to the set screw threads and fit the set screw.
131. **Prop up the transfer box so that its front side is uppermost.**
132. Fit the selector fork to the high/low selector sleeve in the differential assembly.
133. Fit the high/low selector shaft ball from inside the transfer case.
134. Locate the differential assembly and high/low selector fork assembly into the transfer case.
135. Fit the selector shaft spring, apply Loctite Driloc 290 to the detent plug threads and fit the detent plug.

* See Technical Data, page 561.

Continued

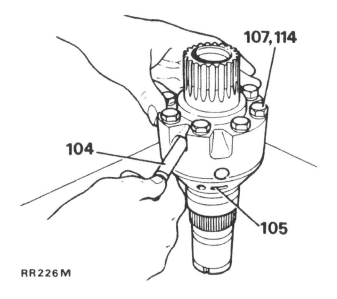

RR226M

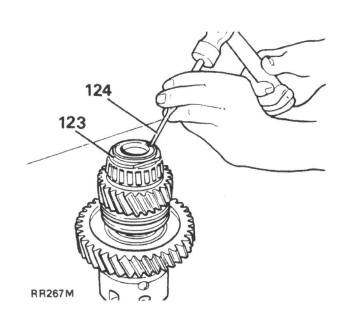

RR267M

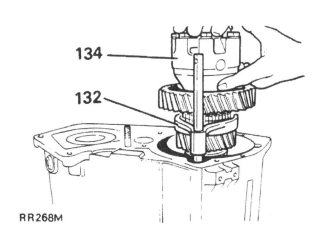

RR268M

FRONT OUTPUT SHAFT HOUSING – OVERHAUL
(Items 136 to 190)

Dismantling

136. Remove the seven screws securing the differential lock selector side cover and release the side cover and gasket.
137. Remove the three screws securing the differential lock finger housing and lift out the complete assembly.
138. Remove the 'O' ring from the assembly.
139. Slacken the lock nut retaining the differential lock switch and unscrew the switch.
140. Remove the detent plug from the top of the housing and lift out the spring and ball.
141. Compress the selector fork spring inside the housing and slide out the spring locating 'C' caps.
142. Slide the selector shaft out of the rear of the housing.
143. Remove the selector fork and spring through the side cover aperture.
144. Lift out dog sleeve from the back of the output shaft housing.
145. Using the flange wrench 18G1205 remove the flange nut, steel and felt washers.

NOTE:– Ensure that flange bolts are fully engaged in the wrench.

146. Remove the output shaft flange with oil seal shield.

NOTE:– These parts need not be separated unless the flange bolts are to be renewed.

147. Drift the front output shaft rearwards out of the housing.
148. Slide off the collar from the output shaft.
149. **Secure the front output shaft housing in the vice.**
150. Prise out and discard the two oil seals, using service tool 18G1271.
151. Using circlip pliers 18G257 release the circlip and remove the bearing.
152. **Remove the housing from the vice** and drift out the bearing from inside.

Continued

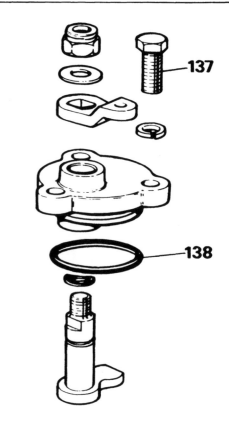

RR286M

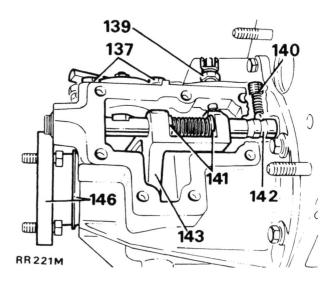

RR221M

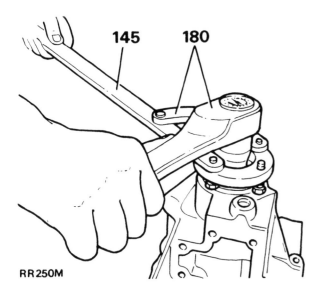

RR250M

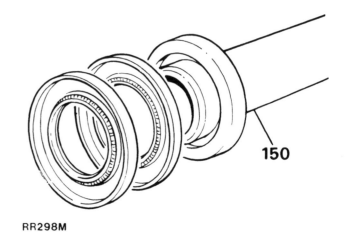

RR298M

153. Drift out the differential front bearing track and remove the shim behind it.

154. Clean all parts ensuring that any traces of Loctite are removed from faces and threads. Renew oil seals and examine all parts for wear or damage, renew as necessary.

Obtaining bearing pre-load

155. Measure the original differential front bearing track shim, noting its thickness.

156. Select a slightly thinner trial shim than the original in order to obtain an end float condition and fit to the housing.

157. Drift the differential front bearing track into the housing.

158. Grease and fit gasket and locate front output shaft housing on the transfer box.

159. Fit four of the seven securing bolts only and tighten to the specified torque.

160. Fit a dial gauge mounting bracket on to the housing, bolting it to one of the side cover fixing holes.

161. Fit a dial gauge with bracket RO530106 to the mounting bracket.

162. Align the gauge pointer on the end of the output shaft, setting the gauge to '0'.

163. Using a screwdriver via the bottom cover aperture, lift the gear assembly to record the end float.

164. Remove the dial gauge assembly and mounting bracket.

165. Remove the four bolts retaining the housing.

166. Drift out the differential front bearing track from inside the housing and remove and discard the trial shim.

167. Select and fit a shim of the required thickness to obtain the correct pre-load of 0,02–0,07 mm (0.001 –0.003 in) on reassembly. This is achieved by adding the thickness of the trial shim and the end float obtained to the pre-load specified.

168. Finally drift the differential front bearing track into position.

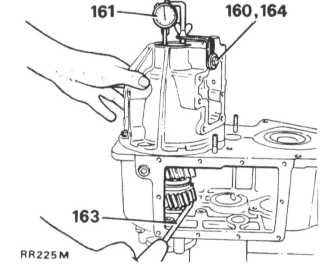

RR225M

Reassembling

169. **Secure the front output shaft housing in the vice.**

170. Drift the front bearing into the housing.

171. Fit the bearing retaining circlip, using circlip pliers 18G257.

172. Fit the two new oil seals simultaneously using replacer tool 18G1422. Fit the seals with both sealing lips inwards as illustrated.

NOTE:– On later production a single dual-lipped seal is fitted.

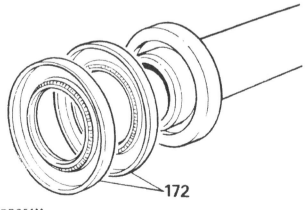

RR251M

Continued

173. Charge the lips of both seals with grease.
174. **Remove the housing from the vice.**
175. Slide the collar on to the front of the output shaft with its chamfered edge to the front.
176. Fit the output shaft through the back of the housing.
177. Fit the flange/oil seal shield assembly on to the output shaft.
178. Fit the flange felt washer, plain washer and a new Nyloc nut.
179. **Refit the housing in the vice.**
180. Using flange wrench 18G1205 and a torque wrench pull up output shaft to correct position. Check that the oil seal shield does not foul the housing.

NOTE:— Ensure that flange bolts are fully engaged in the wrench.

181. **Remove the housing from the vice.**
182. Slide the dog sleeve on to the rear of the output shaft ensuring that the groove in the dog sleeve is to the front.
183. Compress the differential lock selector shaft spring, and fit it between the selector fork lugs.
184. Locate the selector fork inside the side cover aperture in the housing engaging the groove in the dog sleeve on the output shaft.
185. Fit the differential lock selector shaft into the housing from the back, grooved (detents) end last, and pass it through the selector fork lugs and spring and into the front of the housing.
186. Rotate the selector shaft until the two flats are uppermost.
187. Compress the spring slightly between the fork lugs and fit the two locating 'C' caps.
188. Fit the detent ball and spring via the tapped hole in the top of the housing fully home.
189. Apply Loctite Driloc 290 to the detent plug threads and fit the detent plug.
190. Loosely fit the differential lock switch in the tapped hole on top of the housing, ready for adjustment.

DIFFERENTIAL LOCK FINGER HOUSING – OVERHAUL (Items 191 to 197)

Dismantling

191. Remove and discard the Nyloc nut securing the housing assembly and release the lock lever and selector 'finger' from the 'finger' housing.
192. Remove and discard the 'O' ring from the selector finger.
193. Clean remaining parts, examine for wear or damage, renew as necessary.

Continued

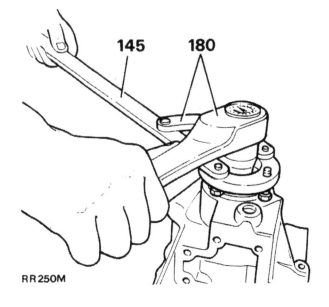

RR250M

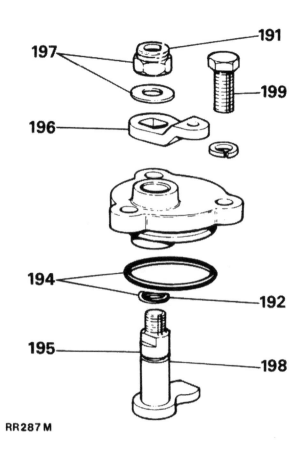

RR287M

Reassembly

194. Fit new 'O' rings to the finger housing and selector finger and lubricate with oil.
195. Locate the selector finger in the finger housing.
196. Fit the differential lock lever over the flats on the selector finger so that it will face forward in the operating position.
197. Fit the plain washer and a new Nyloc nut.
198. Fit the differential lock 'finger' housing assembly into the round aperture in the front output shaft housing locating the selector 'finger' on the flat on the selector shaft inside the housing.
199. Apply Loctite Driloc 290 to the 'finger' housing screw threads and fit the three securing screws (with spring washers).
200. Grease and fit the differential lock selector side cover gasket and fit the side cover, securing it with the seven bolts (with spring washers).
201. **Prop up the transfer box on the bench with the front side uppermost.**
202. Grease and fit the front output shaft housing gasket and locate the housing on the transfer box.
203. Apply Loctite Driloc 290 to the threads of the housing securing bolts and fit the eight securing bolts (with spring washers). Note that the upper middle bolt is longer.
204. **Turn the transfer box into its normal operating position.**
205. Using a screwdriver inside the housing move the high/low selector shaft rearwards (i.e. into high range position) to provide access for fitting the yoke over the end of the selector shaft.
206. Locate the yoke on the selector shaft, apply Loctite Driloc 290 to the yoke set screw and fit the set screw to the specified torque.

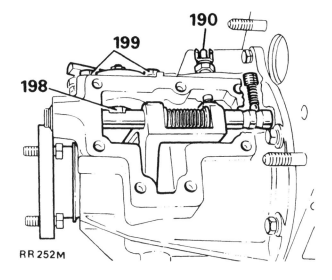

RR 252M

HIGH/LOW SELECTOR HOUSING – OVERHAUL
(Items 207 to 223)

Dismantling

207. Remove the selector fork grub screw **completely**.
208. Remove the cross shaft retaining circlip.
209. Withdraw the cross shaft from the selector housing with the operating arm attached.
210. Lift out the selector fork from the housing.
211. Remove the two 'O' rings from the cross shaft.
212. Remove the operating arm from the cross shaft by removing the retaining set screw.
213. Clean parts ensuring that all traces of Loctite are removed, examine for wear or damage, renew as necessary.

Continued

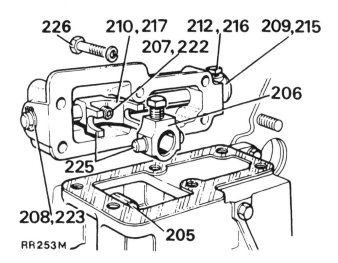

RR253M

Reassembling

214. Fit the 'O' ring to the operating arm end of the cross shaft.
215. Locate operating arm on the shaft blind hole.
216. Apply Loctite Driloc 290 to the operating arm set screw threads and fit the set screw.
217. Locate the selector fork inside the housing.
218. Slide the cross shaft into the housing passing it through the selector fork.
219. Fit the 'O' ring to the fork end of the cross shaft (inside the housing) and lubricate both 'O' rings.
220. Position the cross shaft fully home.
221. Locate the selector fork on the shaft blind hole.
222. Apply Loctite Driloc 290 to the fork grub screw threads and fit the grub screw.
223. Fit the circlip on the end of the cross shaft.
224. Grease and fit the gasket to the high/low selector housing aperture on the front output shaft housing.
225. Position the high/low selector housing so that the projecting selector fork engages the yoke side pins inside the housing.
226. Fit the six selector housing retaining bolts (with plain washers).
227. **Prop up the transfer box on the bench with front side uppermost.**
228. Fit the oil seal into the front of the transfer box (seal lip to rear of case) using replacer tool 18G1422.
229. **Prop up the transfer box on the bench with rear side uppermost.**
230. Drift in the input gear front bearing track from inside the back of the transfer box, using a suitable punch.

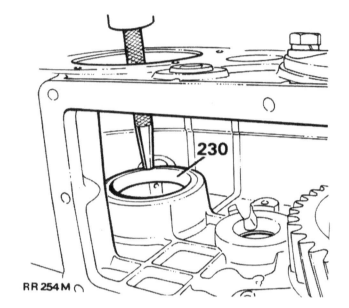

RR 254 M

INPUT GEAR – OVERHAUL
(Items 231 to 237)

Dismantling

231. **Secure hand press 47 in vice** and using collars and buttons 18G47-7 remove the **front** taper roller bearing from the input gear assembly.
232. Reverse input gear assembly in hand press and remove the **rear** taper roller bearing.
233. **Remove the hand press from the vice.**
234. Clean all parts, examine for wear and damage, renew as necessary.

Continued

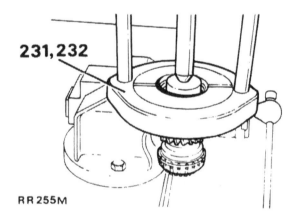

231,232

RR 255M

Reassembly

235. Locate the **front** taper roller bearing on the input gear assembly and drift the bearing fully home.
236. Repeat above procedure and fit the **rear** taper roller bearing.
237. Lubricate both bearings with oil.
238. Fit input gear assembly into the transfer box. From the rear (larger gear to the front).

Obtaining bearing pre-load

239. **Secure the mainshaft bearing housing in the vice.**
240. Drift out the **rear** input gear bearing track and remove the shim behind it.
241. Clean the main bearing housing and measure original shim, noting its thickness.
242. Select a slightly thinner trial shim than the original in order to obtain an end float condition and fit to the main bearing housing.
243. Locate the **rear** bearing track on the main bearing housing and drift it fully home.
244. Apply grease to the gasket and fit on to the transfer box casing.
245. Fit the main bearing housing and tighten the two securing screws to the specified torque.
246. Fit a dial gauge mounting bracket on to the mainshaft bearing housing with a single bolt.
247. Fit a dial gauge with bracket RO530106 to the mounting bracket.
248. Align the gauge pointer on the end of the gear, setting the gauge to '0'.
249. Lift the large gear by hand to record the end float.
250. Remove the dial gauge assembly and mounting bracket.
251. Remove the two screws retaining the mainshaft bearing housing.
252. Drift out the rear bearing track from the bearing housing and remove and discard the trial shim.
253. Select a shim to the required thickness to obtain the correct pre-load of 0,02–0,07 mm (0.001–0.003 in) on reassembly. This is achieved by adding the thickness of the trial shim and the end float obtained to the pre-load specified.
254. Fit the shim to the main bearing housing and then drift the rear bearing track into position.
255. Fit the main bearing housing and tighten the two securing screws to the specified torque.
256. Grease and fit P.T.O. cover gasket and finally fit the P.T.O. cover securing it with six bolts (with spring washers) to the specified torque.

Continued

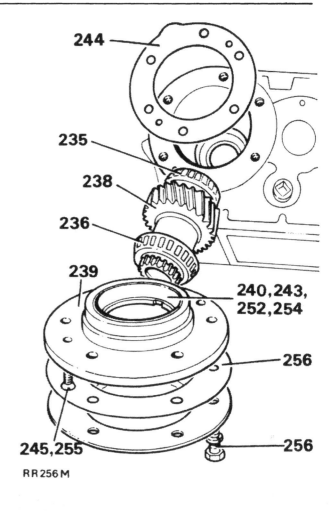

RR256M

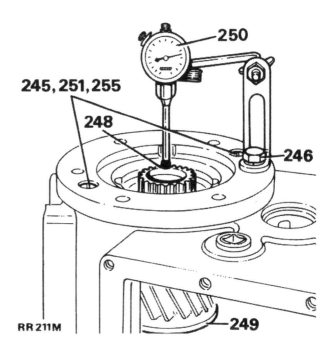

RR211M

Intermediate gear assembly – reassembly

257. First remove the needle roller bearings and spacer from the gear assembly.
258. Clean the parts, including the thrust washers and lock plate and examine for wear or damage, renew as necessary.
259. Fit the 'O' ring to the intermediate shaft.
260. Fit the 'O' ring into the front of the transfer case.
261. Lubricate thrust washers, bearings, shaft and spacer.
262. Fit needle bearings with plain edge diameter outwards and spacer interposed.
263. Fit front thrust washer to slot in transfer case (plain side to case).
264. Locate gear assembly partially into the transfer case so that it rests on the front thrust washer.
265. Locate rear thrust washer (plain side uppermost) into slot in transfer case.
266. Gently push gear assembly into mesh.
267. Using a screwdriver through the intermediate shaft hole guide the locating tab on the rear thrust washer into the slot provided in the transfer case.
268. Align gear and thrust assembly and slide the intermediate shaft into the transfer box from the rear.
269. Align the shaft so that the lock plate slot in the end is on top.
270. Apply Loctite Driloc 290 to the lock plate bolt threads. Locate lock plate into position and fit securing bolt (with spring washer).
271. Using a screwdriver via the bottom of the transfer case lift up the gear assembly and measure the end float with feeler gauges. This should be between 0,08 – 0,35 mm (0.003 – 0.014 in).
272. Grease and fit the bottom cover gasket.
273. Apply Loctite Driloc 290 to six of the ten bottom cover fixing bolts only.
274. Clean and fit the bottom cover, using the six bolts (with spring washers) leaving the four centre holes free.

NOTE:– The remaining four bolts are used later to secure the transfer box assembly to the locally made adaptor plate on the hydraulic hoist.

Transmission brake assembly

275. Clean brake backplate and oil drip plate and locate the backplate on the rear output shaft/speedometer housing so that the brake operating lever is on the offside rear.
276. Fit the four Brako Durlok bolts, the lower two also retain the oil drip plate. Tighten to specified torque.
277. Clean and fit brake drum and fit the two countersunk retaining screws.

Continued

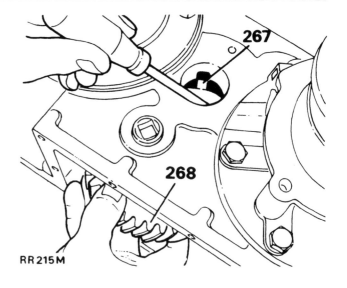

RR 215 M

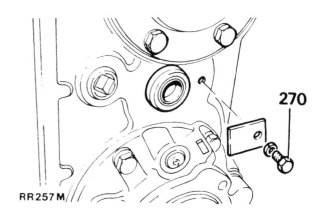

RR 257 M

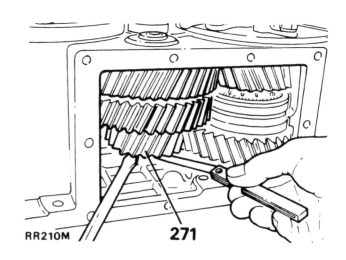

RR 210 M

Differential Lock Switch Adjustment

278. Select differential locked position by moving the differential lock lever towards the right side of the transfer box.

279. Obtain a battery and connect a test lamp circuit to the differential lock switch.

280. Slacken the lock nut off and screw in the lock switch until the bulb is illuminated.

281. Turn the lock switch another half turn and tighten the lock nut against the housing.

282. Disconnect the battery and move the differential lock lever towards the left side of the transfer box to disengage the differential lock.

Extension Housing – refitting

283. First ensure that the lower fixed dowel and the upper loose dowel are in position.

284. Locate the extension housing over the two studs in the transfer box front casing.

285. Secure the extension housing with two nuts (with spring washers) and four bolts (with spring washers).

NOTE:– The upper bolt on the left side of the transfer box is longer.

HIGH/LOW GEAR CHANGE HOUSING – OVERHAUL
(Items 286 to 327)

Dismantling

286. Remove the split pin from the clevis pin at the top of the differential lock cross shaft lever which secures it to the gear change cross shaft.

287. Remove the washer and clevis pin and the anti-rattle nylon strip.

288. Mark the position of the high/low gear change operating arm on the splined shaft of the gear change crank arm.

289. Slacken the clamp bolt and remove the operating arm.

†290. Remove the four bolts from the top of the gear change housing and lift off the grommet plate, grommet, gate plate and gasket.

291. Remove the four **lower** gear change housing retaining bolts and the harness clip from the longer front bolt.

292. Remove the housing and gasket.

293. Remove the split pin from the gear change crank arm clevis pin and remove the clevis pin.

294. Remove the circlip from the high/low gear change lever bush.

295. Withdraw the gear change lever from the housing, with ball and socket bush.

296. Remove the two countersunk screws from the housing end cover.

297. Remove the housing end cover.

†On manual gearbox models the grommet and grommet plate illustrated are not used. The gate plate is retained to the housing by two screws. The four main bolts are fitted in-situ through a floor mounted gaiter assembly.

Continued

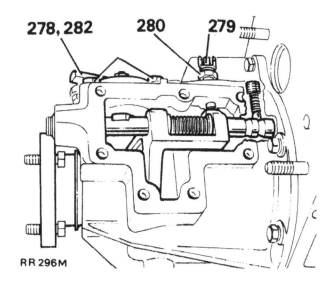

RR 296M

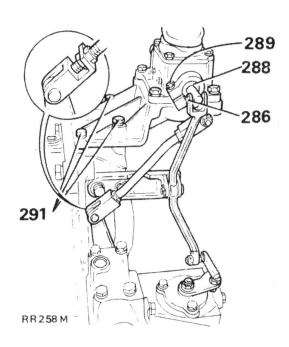

RR 258 M

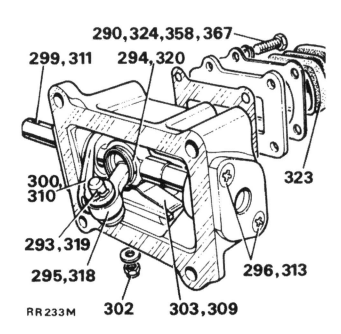

RR 233M

298. Remove the two 'O' rings from the end cover.
299. Remove the cross shaft from the housing.
300. Compress the detent spring and remove the gear change arm from inside the housing.
301. Remove the two 'O' rings from the crank arm.
302. Remove and discard the two Nyloc nuts retaining the detent plate.
303. Remove the detent plate and spring from the housing.
304. Clean all parts, examine for wear or damage, renew as necessary.

Reassembly

305. Fit the two 'O' rings to the housing end cover.
306. Fit the two 'O' rings to the gear change crank arm.
307. Lubricate 'O' rings with oil.
308. Clip the detent spring on to the detent plate.
309. Fit detent plate assembly into housing and retain from outside with two Nyloc nuts (with plain washers).
310. Compress the detent spring and fit the gear change crank arm in the housing.
311. Fit cross shaft into position locating one end in the crank arm (bush).
312. Fit the housing end cover (bush) to support the other end of the cross shaft.
313. Finally secure the housing end cover with the two countersunk screws.
314. Before refitting the gear change lever remove the clevis pin bushes and the Nylon socket bush and ball.
315. Clean all parts, examine for wear and damage, renew as necessary.
316. Fit and grease gear lever ball and Nylon socket bush to gear lever.
317. Fit and grease clevis pin bush.
318. Locate gear change lever assembly in cross shaft (do not fit socket bush retaining circlip at this stage).
319. Align gear change lever end with crank arm fork and fit clevis pin and split pin.
320. Finally secure Nylon socket bush with circlip.
321. Grease and fit gasket to gear change housing face.
†322. Fit the gate plate.
†323. Fit the grommet.
†324. Fit the grommet plate and retain with the four securing bolts (with spring washers).

NOTE:— On earlier production where a different grommet and grommet plate are fitted, adhesive is used to keep the grommet positively seated.

325. Clean the **lower** gear change housing.
326. Grease and fit the gasket to the housing.
327. Fit the gear change housing to the **lower** gear change housing and secure with the four bolts (with spring washers). The longer front bolt also retains the harness clip.

NOTE:— Do not fit the lower gear change housing to the extension housing at this stage.

†On manual gearbox models the grommet and grommet plate (illustrated on the previous page) are not used. The gate plate is retained to the housing by two screws. The four main bolts are fitted in-situ through a floor mounted gaiter assembly. *Continued*

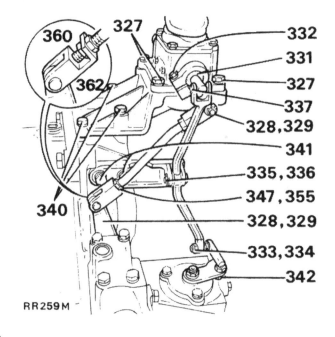

RR259M

328. Before refitting the high/low connecting rod and gear change operating arm remove the respective clevis pins and Nylon bushes. Clean and examine for wear or damage, renew as necessary.

329. Grease and fit the Nylon bushes to the high/low connecting rod and the gear change operating arm.

330. Assemble the operating arm to the connecting rod with clevis pin, plain washer and split pin.

331. Slacken the operating arm clamp bolt and fit the operating arm on to the splined shaft projecting from the high/low gear change housing, carefully aligning it to the marks on both components.

332. Tighten the clamp bolt to the specified torque.

*333. Before refitting the differential lock cross shaft lever and pivot bracket remove the respective clevis pins and Nylon bushes. Also remove and discard the Nyloc nut retaining the short connecting link and pull off the link. Clean and examine for wear or damage, renew as necessary.

*334. Fit the short connecting link to the **bottom** of the cross shaft lever and secure with a new NYLOC nut.

335. Grease and fit the Nylon bushes to the **middle** pivot of the cross shaft lever.

336. Fit the cross shaft lever to the (loose) pivot bracket with the clevis pin, washer and split pin.

337. Fit the cross shaft lever fork (top) to the gear change cross shaft.

338. Locate the anti-rattle Nylon strip and fit the clevis pin, plain washer and split pin.

339. Locate the complete gear change housing assembly in position on top of the extension housing with the high/low connecting rod and the differential lock cross shaft lever attached.

340. Fit the four bolts (with spring washers) to secure the lower gear change housing to the extension housing.

341. Fit the pivot bracket to the extension housing using the two securing bolts (with spring washers).

*342. Fit the short connecting link at the bottom of the differential lock cross shaft lever to the lock lever and retain it with a new Nyloc nut.

343. Grease and fit the Nylon bushes to the operating arm on the selector housing cross shaft.

Adjustment of original high/low connecting rod (by fork rotation)

344. Align the high/low connecting rod to the operating arm and temporarily fit the clevis pin.

345. The gear lever grommet is retained by adhesive and can be immediately removed for checking that the gear lever does not foul the gate plate when high or low range is selected.

346. If adjustment is required the following procedure should be carried out.

347. Slacken the locknut on the operating arm fork.

348. Remove the loose clevis pin temporarily connecting the operating arm to the high/low connecting rod.

*The latest cranked short connecting link is retained by a split pin at each end.

Continued

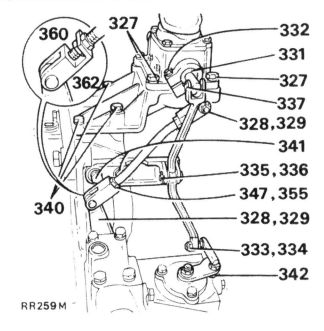

RR259M

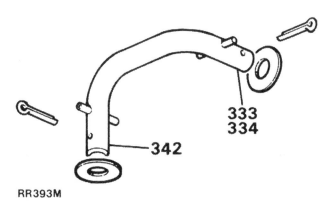

RR393M

**349. Lift up the operating arm and rotate the fork to shorten or lengthen the connecting rod as required.

350. Refit the clevis pin temporarily.

351. Move gear change lever into high range (rearwards) and move the operating arm on the selector housing cross shaft into high range (forwards).

352. Check that the gear change lever does not foul the gate plate in this position.

353. Engage and check low range in the same way.

354. After adjustment return the gear change lever to the high range position.

355. Tighten the locknut on the connecting rod fork.

356. Finally fit the plain washer and split pin to the clevis pin to secure the connecting rod to the operating arm.

**Alternative method of adjusting original high/low connecting rod length

Where difficulty is experienced in releasing or replacing the clevis pin and split pin securing the connecting rod fork end to the operating arm, due to its proximity to the transfer box casing, the following method may be substituted.

(a) Disconnect the top of the differential lock cross shaft lever from the gear change cross shaft.

(b) Remove the four bolts retaining the gear change housing.

(c) Lift up the housing assembly (with the connecting rod attached) and rotate it as required to vary the length of the connecting rod.

Adjustment of latest type high/low connecting rod (in situ by locknuts)

357. Align the high/low connecting rod to the operating arm and fit the clevis pin, plain washer and split pin.

†358. Remove the four bolts from the top of the gear change housing and lift off the gear change lever grommet plate and the gear change lever grommet. Replace the four bolts temporarily to retain the gate plate in position.

359. Check that the gear lever does not foul the gate plate when high or low range is selected. If adjustment is required carry out the following procedure.

360. Slacken off the connecting rod locknuts.

361. Move gear change lever into high range (rearwards) and move the operating arm on the selector housing cross shaft into high range (forwards).

362. Tighten both locknuts.

363. Check that the gear change lever does not foul the gate plate in this position.

364. Engage and check low range in the same way.

365. After adjustment return the gear change lever to the high range position.

†366. Remove the four bolts retaining the gate plate and refit the grommet and grommet plate.

†367. Refit the four bolts (with spring washers).

†On manual gearbox models the grommet and grommet plate illustrated are not used. The gate plate is retained to the housing by two screws. The four main bolts are fitted in-situ through a floor mounted gaiter assembly.

Continued

Transfer box mounting – refitting

368. Fit the rubber mounting plate to the right side of the front output housing by fitting the four securing bolts (with spring washers).

Handbrake linkage – refitting

369. Locate the hand brake linkage bracket in position on the right hand side of the transfer box casing and secure with the two long and two short bolts (with spring washers).

370. Fit the hand brake lever mounting bracket on the right side of the front output shaft housing. It is secured by three bolts (with spring washers) but only the bottom bolt is fitted on the bench.

NOTE:– The remaining bolts are fitted on the vehicle when securing the bell housing tie-rod.

371. Align the handbrake adjustment linkage to the mounting position. If necessary the length of the operating link may be varied by slackening the appropriate locknuts.

372. Insert the clevis pin through the brake operating lever, fit the Thackeray (double spring) washer, handbrake link, plain washer and split pin.

373. Ensure that adjusting link locknuts are tight.

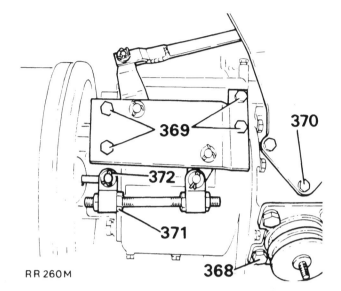

RR 260M

557

TRANSFER GEARBOX LT230R

SPECIAL TOOLS

TRANSFER GEARBOX REMOVE AND REFIT	Mounting plate for transmission hoist Local manufacture dimensions in workshop manual
INTERMEDIATE SHAFT REMOVE	Extractor for transfer box intermediate shaft RO605862
STAKE NUT REMOVE AND REFIT	Spanner centre differential stake nut 18G1423
DRIVE FLANGES REMOVE AND REFIT	Adjustable flange holding wrench 18G1205
OIL SEALS REMOVE	Input and output seals 18G1271
OIL SEAL REFIT	Drift Input and output seals 18G1422
CIRCLIP	Circlip pliers 18G257
PRE-LOAD BEARING CHECK	Centre differential and input bearings RO530106 bracket plus suitable dial gauge
LOCATE TRANSFER GEARBOX TO AUTO-TRANSMISSION	Guide studs to eliminate damage to input oil seal 18G1425
CENTRE DIFFERENTIAL BEARING) CONE REAR REMOVE) CENTRE DIFFERENTIAL BEARING) CONE FRONT REMOVE)	Tool 47 Hand press 18G47BB Bearing remover
CENTRE DIFFERENTIAL BEARINGS CONE REPLACE	Drift on bearings both ends of drift used due to different sized bearings 18G1424
INPUT GEAR ASSEMBLY BEARING) CONES REMOVE)) INPUT GEAR ASSEMBLY BEARING) CONES REFIT)	Tool 47 hand press 18G47-7

AUTOMATIC GEARBOX

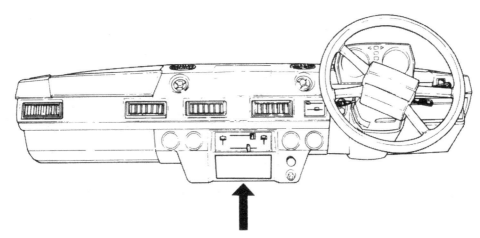

WARNING LABEL FITTED TO VEHICLE

WARNING

IN 'P' DO NOT IDLE FOR MORE THAN 10 MINUTES, OR RUN ENGINE ABOVE IDLE. USE 'N' FOR PROLONGED IDLING.

TO CHANGE TRANSFER RATIO REDUCE SPEED TO BELOW 8KPH (5MPH), SELECT AUTO **'N'**, MOVE HIGH/ LOW LEVER RAPIDLY TO REQUIRED POSITION. SELECT AUTO GEAR. ALTERNATIVELY, STOP VEHICLE, STOP ENGINE, MAKE SELECTION. IN CASE OF DIFFICULTY REFER TO HANDBOOK.

FOR OPTIMUM ENGINE BRAKING, SELECT AUTO '1', KEEP ENGINE RUNNING.

USE DIFF LOCK ONLY WHEN TRACTION IS LIKELY TO BE LOST.

SINGLE AXLE ROLLER RIGS MAY BE USED WITH DIFF LOCK DISENGAGED FOR SPEEDS UP TO 5KPH (3MPH). FOR OTHER CONDITIONS SEE HANDBOOK.

REPAIR OPERATION

Operation No.

Technical data
Torque wrench settings
Recommended service lubricants
Maintenance schedules

AUTOMATIC GEARBOX

Automatic and transfer gearbox assembly – remove and refit	44.20.04
Automatic gearbox – overhaul	44.02.06
Brake band, front – adjust	44.30.07
Brake band, rear – adjust	44.30.10
Gearbox fluid – drain and refill	44.24.02
Gear selector assembly – remove and refit	44.15.04
Gear selector rod and cable – adjust	44.30.04
Gearbox tunnel side plate – remove and refit	76.10.93
Gearbox tunnel top cover – remove and refit	76.25.07
Hydraulic pressure – check and adjust	44.30.15
Line pressure – adjust	44.30.26
Reverse starter switch – test	44.15.17
Shift speeds and line pressure charts	
Sump – remove and refit	44.24.04
Throttle linkage – adjust	44.15.03
Throttle pressure – check and adjust	44.30.25

ENGINE

Engine mountings – remove and refit	12.45.04
Left hand mounting – remove and refit	12.45.07
Right hand mounting – remove and refit	12.45.09

TRANSFER GEARBOX

Transfer gearbox – remove and refit	37.29.25

SPECIAL TOOLS

Automatic gearbox

TECHNICAL DATA

AUTOMATIC GEARBOX

Type	A727 Automatic three speed & reverse epicyclic gearbox with fluid torque converter.
Lubrication	Pump (rotor type)
Pump clearances:–	
Rotor side clearance	0,090 to 0,190mm (0.0035 to 0.0075 inch)
Rotor tip clearance	0,120 to 0,240mm (0.0045 to 0.0095 inch)
Rotor face clearance	0,030 to 0,063mm (0.0010 to 0.0025 inch)
Gear train end play	0,300 to 0,630mm (0.0100 to 0.0250 inch)
Input shaft end play	0,920 to 2,130mm (0.0360 to 0.0840 inch)
Thrust washers, input shaft	
(Selective) Natural	1,550 to 1,600mm (0.0610 to 0.0630 inch)
Red	2,140 to 2,180mm (0.0840 to 0.0860 inch)
Yellow	2,600 to 2,640mm (0.1020 to 0.1040 inch)
Snap rings:–	
Rear clutch snap ring	
(Selective)	1,530 to 1,570mm (0.0600 to 0.0620 inch)
	1,880 to 1,930mm (0.0740 to 0.0760 inch)
	2,240 to 2,280mm (0.0880 to 0.0900 inch)
	2,700 to 2,740mm (0.1060 to 0.1080 inch)
Output shaft (forward end)	
(Selective) Natural	1,220 to 1,320mm (0.0480 to 0.0520 inch)
Blue	1,400 to 1,500mm (0.0550 to 0.0590 inch)
Green	1,570 to 1,670mm (0.0620 to 0.0660 inch)
Clutch plate clearance	
Front clutch	1,940 to 3,120mm (0.0760 to 0.1230 inch)
Rear clutch	0,640 to 1,140mm (0.0250 to 0.0450 inch)
Number of front clutch plates	3 (Plain)
Number of front clutch discs	3 (With grooved friction faces)
Number of front clutch springs	9
Number of rear clutch plates	3 (Plain)
Number of rear clutch discs	4 (With friction faces)
Band adjustments	
Number of turns backed off from 8,1Nm (72 lbfin)	2.50

TRANSFER GEARBOX

Type	LT230R. Two speed reduction on main gearbox output. Front & rear drive permanently engaged via a lockable differential.
Input gear	26 Teeth
Intermediate gear	19 x 41 x 44 Teeth
Output gear	40 x 28 Teeth
Input shaft bearing pre-load	0,02 to 0,07mm (0.001 to 0.003 inch)
Intermediate gear end float	0,08 to 0,35mm (0.003 to 0.014 inch)
Differential pinions backlash	Zero to 0,07mm (Zero to 0.003 inch)
Output shaft bearing pre-load	0,02 to 0,07mm (0.001 to 0.003 inch)
Low range gear end-float)	
High range gear end-float)	0,05 to 0,15mm (0.002 to 0.006 inch)
High range gear/bush running clearance diameter	0,03 to 0,09mm (0.0012 to 0.0035 inch)

GEAR RATIOS

Automatic gearbox	Top	1.00:1
	Second	1.45:1
	First	2.45:1
	Reverse	2.20:1
Transfer gearbox	High	1.003:1
	Low	3.320:1

Overall ratio	In high transfer	In low transfer
Top	3.55	11.75
Second	5.15	17.04
First	8.70	28.79
Reverse	7.81	25.86

AUTOMATIC SHIFT SPEEDS AND LINE PRESSURE CHARTS

Automatic shift speeds in high range transfer			
Automatic selector position	**Throttle position and shift response**	**Approximate road speed**	
		km/h	m.p.h.
	Kickdown range		
D	1 − 2 Upshift	55 to 74	34 to 46
D	2 − 3 Upshift	90 to 117	56 to 73
D	3 − 2 Downshift	82 to 106	51 to 66
D	3 − 1 Downshift	42 to 58	26 to 36
2	2 − 1 Downshift	40 to 51	25 to 32
	Part throttle		
D	3 − 2 Downshift	45 to 61	28 to 38
	Light throttle		
D	1 − 2 Upshift	10 to 18	6 to 11
D	2 − 3 Upshift	21 to 29	13 to 18
	Throttle closed		
D	3 − 1 Roll out	8	5
1	2 − 1 Roll out	37 to 53	23 to 33

Stall speed ..... 2000 rev/min.

HYDRAULIC PRESSURES (Line pressure)		
Check with:− Transfer gearbox in neutral Automatic gearbox in 'D' Throttle linkage disconnected at gearbox Engine speed 1000 rev/min		
Throttle valve position (move manually)	lbf/in^2	kgf/cm^2
Throttle valve fully forward	54 to 60	3,8 to 4,2
Throttle valve fully rearward	90 to 96	6,3 to 6,7

TORQUE WRENCH SETTINGS – AUTOMATIC GEARBOX

COMPONENT	DESCRIPTION	QUANTITY	Nm	lbf ft
Adaptor ring to engine	3/8 UNC x 1.625 inch bolt		36 to 48	26 to 34
Gearbox to adaptor ring	1/2 inch UNF nut	2	56 to 70	40 to 50
Gearbox to adaptor ring	1/2 UNC x 1.250 inch screw	4	56 to 70	40 to 50
Cover plate to adaptor ring	6 x 25,0 mm screw	7	7 to 10	5 to 7
Cover plate to adaptor ring	screw	2	7 to 10	5 to 7
Tie plate to sump (engine sump)	8 x 20,0 mm screw	3	22 to 28	16 to 21
Tie plate to sump (engine sump)	8 x 16,0 mm screw	3	22 to 28	16 to 21
Kick down pivot bracket adaptor ring	8 x 16,0 mm screw	1	22 to 28	16 to 21
Lower gear change housing – adaptor housing	8 mm screw and bolts	2 of each	22 to 28	16 to 21
Spacer to crankshaft	7/16 UNF x 0.75 inch screw	6	77 to 90	55 to 65
Spigot aligner to spacer	10 x 25 mm screw	4	40 to 50	29 to 37
Starter ring to drive place	screw	10	30 to 37	22 to 27
Drive plate to torque converter	5/16 UNF x 1.125 inch screw	4	25 to 35	18 to 25
Oil filter extension	stud	3	3,5 to 4,5	2.5 to 3.2
Oil filter to spacer	10 UNF self lock nut	3	3,5 to 4,5	2.5 to 3.2
Breather connector to gearbox	6 x 16 mm screw	1	7 to 10	5 to 7
Breather pipe (metal) to connector block	sleeve nut and olive	1	8 to 10	5.7 to 7.2
Clamping plate to oil pump housing	5/16 inch UNC x 1.50 inch screw	2	21 to 28	15 to 20
Breather pipe to gearbox (external)	banjo bolt	1	7 to 11	5 to 8
Packer to torque converter housing	1/4 inch UNC x 0.875 inch screw	3	5,6 to 8,4	4 to 6
Sump to gearbox	5/16 inch UNC x 0.750 inch screw	14	21 to 28	15 to 20
Drain plug	16 mm	1	25 to 35	19 to 26
Mainshaft nut	special nut	1	126 to 154	90 to 110
Coupling shaft to mainshaft nut	12 x 120 mm bolt	1	65 to 80	48 to 59
Adaptor housing to gearbox	10 x 30 mm screw	1	40 to 50	29 to 37
Adaptor housing to gearbox	10 x 40 mm screw	3	40 to 50	29 to 37
Adaptor housing to gearbox	bolt (fitted)	1	40 to 50	29 to 37
Lower bracket to gearbox	1/2 inch UNF nut and screw	1	35 to 42	25 to 30
Bridge pipe for oil cooler	sleeve nuts	2	7 to 9	5 to 6.5
Tie plate to gearbox sump	8 x 20,0 mm screw	2	22 to 28	16 to 21
Torque connector housing sealing	grub screw	2	Top of thread to be flush with outer face of torque converter housing recess.	
Throttle lever clamping bolt	5 x 25 mm bolt	1	5 to 7	4 to 5
Gear change pivot retention	8 mm self lock nut	1	22 to 28	16 to 21
Oil cooler adaptors	1/8 inch taper thread	2	10 to 12	7.2 to 8.5
Oil cooler elbow to cooler adaptors	sleeve nut	2	10 to 12	7.2 to 8.5

Continued

TORQUE WRENCH SETTINGS – TRANSFER GEARBOX

COMPONENT	DESCRIPTION	QUANTITY	Nm	lbf ft
Pinch bolt, operating arm	6 x 25,0 mm bolt	1	7 to 10	5 to 7
Gate plate to grommet plate	6 x 20,0 mm screw	4	7 to 10	5 to 7
End cover	6 x 20,0 mm screw	2	7 to 10	5 to 7
Speedometer cable retainer	6 mm nut	1	7 to 10	5 to 7
Rear output/speedometer housing	6 x 30,0 mm stud	1	See note	
Locating plate to gear change housing	5 mm self lock nut	2	5 to 7	4 to 5
Bottom cover to transfer case	8 x 30,0 mm bolt	10	22 to 28	16 to 21
Front output housing to transfer case	8 x 30,0 mm bolt	7	22 to 28	16 to 21
Front output housing to transfer case	8 x 90,0 mm bolt	1	22 to 28	16 to 21
Cross shaft housing to front output housing	8 x 55,0 mm bolt	6	22 to 28	16 to 21
Gear change housing	8 x 55,0 mm	2	22 to 28	16 to 21
Pivot shaft	8 mm nut	1	22 to 28	16 to 21
Connecting rod	8 mm nut	1	22 to 28	16 to 21
Retaining plate intermediate shaft	8 x 20,0 mm screw	1	22 to 28	16 to 21
Front output housing cover	8 x 25,0 mm screw	7	22 to 28	16 to 21
Gear change housing	8 x 25,0 mm screw	2	22 to 28	16 to 21
Bracket to extension housing	8 x 25,0 mm screw	2	22 to 28	16 to 21
Finger housing to front output housing	8 x 25,0 mm screw	3	22 to 28	16 to 21
Mainshaft bearing housing	8 x 25,0 mm screw	2	22 to 28	16 to 21
Brake drum	8 x 20,0 mm screw	2	22 to 28	16 to 21
Gearbox to transfer box	10 x 40,0 mm bolt	3	40 to 50	29 to 37
Gearbox to transfer box	10 x 45,0 mm bolt	1	40 to 50	29 to 37
Bearing housing to transfer gearbox	10 x 35,0 mm bolt	6	40 to 50	29 to 37
Speedometer housing to transfer gearbox	10 x 30,0 mm screw	5	40 to 50	29 to 37
Speedometer housing to transfer gearbox	10 x 45 mm screw	1	40 to 50	29 to 37
Selector fork to cross shaft	10 x 16 mm grub screw	1	40 to 50	29 to 37
Yoke to selector shaft high/low	10 mm tapered screw	1	22 to 28	16 to 21
Selector fork, high/low to shaft	10 mm tapered screw	1	22 to 28	16 to 21
Operating arm high/low	10 mm tapered set screw	1	22 to 28	16 to 21
Transmission brake	10 x 25,0 mm bolt	4	65 to 80	48 to 59
Gearbox to transfer case	10 mm nut	2	40 to 50	29 to 37
Gearbox to transfer case	10 mm studs	2	See note	
Oil drain plug	12 x 14 mm hexagon head	1	25 to 35	19 to 26
Differential case	10 x 60 mm bolt	8	55 to 64	40 to 47
Output flange	20 mm self locking nut	2	146 to 179	108 to 132
Differential case rear and shaft main drive 2/4 wheel drive	50 mm nut	1	66 to 80	50 to 60
Link arm and cross shaft lever to ball joint	1/4 inch UNF self locking nut	2	8 to 12	6 to 9
Oil filler/level plug	3/4 inch taper thread	2	25 to 35	19 to 26
Transfer breather	1/8 inch B.S.P.		7 to 11	5 to 8

NOTE:– Studs to be assembled into casings with sufficient torque to wind them fully home, but this torque must not exceed the maximum figure quoted for the associated nut on final assembly.

RECOMMENDED LUBRICANTS AND FLUIDS

U.K. and EUROPE

COMPONENT	BP	CASTROL	DUCKHAMS	ESSO	MOBIL	SHELL	TEXACO	PETROFINA
Automatic Gearbox	BP Autran DX2D	Castrol TQ DEXRON 11D	Duckhams Fleetmatic C/D or D-Matic	Esso ATF DEXRON 11D	Mobil ATF 22 OD	Shell ATF DEXRON 11D	Texamatic 9226 DEXRON 11D	DEXRON 11D
Transfer Gearbox	BP Gear Oil SAE 90 EP	Castrol Hypoy 90 EP	Duckhams Hypoid 90	Esso Gear Oil GX85W/90	Mobil Mobilube HD 90	Shell Spirax 90 EP	Texaco Multigear Lubricant 85W/90	FINA Pontonic MP SAE80W/90

ALL OTHER TERRITORIES

COMPONENT	Service Classification		Ambient Temperature Centigrade
	Performance Level	SAE Viscosity	
Automatic Gearbox	DEXRON D 11		
*Transfer Gearbox	MIL-L-2105A	90 EP 80 WEP	

*NOTE: – Either engine or automatic gearbox oil may be used as an alternative to the specified gear oil for the transfer gearbox and can be mixed together.

CAPACITIES

	Litres	U.K. Pints
Automatic Gearbox	8.52	15
Transfer Gearbox	2.80	5

*IMPORTANT:– If gearbox is drained approx. 7 pints (up to 4 litres) of oil will remain in the torque convertor. Initially refill gearbox with 8 pints (4,5 litres) only and check with dipstick to ensure correct level.

MAINTENANCE SCHEDULES

	Every 10 000 km (6000 miles) or 6 months	Every 20 000 km (12 000 miles) or 12 months
ENGINE COMPARTMENT		
Check/top up automatic gearbox oil level	X	X
UNDERBODY		
Drain converter housing if drain plug is fitted for wading	X	X
Renew automatic gearbox oil and change filter	Every 40 000 km	
Normal use	(24 000 miles)	
Arduous use		X
Check/adjust automatic gearbox brake bands	Every 40 000 km	
Normal use	(24 000 miles)	
Arduous use		X
Check/top up transfer gearbox oil level	X	X
Renew transfer gearbox oil	Every 40 000 km (24 000 miles)	
Check handbrake for security and operation: adjust if necessary	X	X
PASSENGER COMPARTMENT		
Check automatic gearbox parking pawl engagement	X	X

ENGINE COMPARTMENT

Automatic gearbox oil level

Check oil level daily or weekly when operating under severe wading conditions.

Always check the oil level with the vehicle standing on level ground and use the correct type of automatic gearbox fluid.

1. Apply the handbrake and select N (Neutral).
2. Start and run the engine until normal operating temperature is reached.
3. Still with the handbrake and footbrake applied, and with the engine idling, move the selector through all the gear positions, pausing momentarily in each position.
4. Select N (Neutral) and with the engine still running at idling speed withdraw the dipstick from the filler tube and wipe the blade with a piece of clean paper or non-fluffy cloth.
5. Re-insert the dipstick fully and withdraw it immediately and check the fluid level indication. This must be between the 'FULL' mark and the 'ADD ONE PINT' mark.
6. Top up the gearbox with fluid through the filler tube, then repeat the procedure above until the fluid level is correct. **Do not overfill.**
7. After checking fluid level be sure that the dipstick is reseated correctly to prevent dirt or water from entering the gearbox.

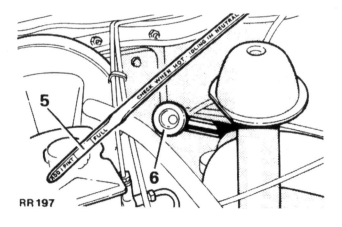

RR 197

566

UNDERBODY

Drain converter housing if drain plug is fitted for wading

1. The converter housing can be completely sealed to exclude mud and water under severe wading conditions by means of a plug fitted in the bottom of the housing.
2. When not in use, the plug is screwed into the tie-plate between the gearbox and engine sumps, and should only be fitted when the vehicle is expected to do wading or very muddy work.
   When the plug is in use it must be removed periodically and all oil allowed to drain off before the plug is replaced.

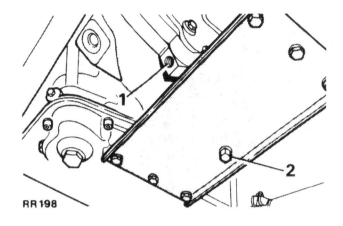

RR 198

Renew automatic gearbox oil, change filter and adjust brake bands

Drain, renew filter and refill monthly when operating under severe wading conditions.

1. Immediately after a run when the fluid is warm, drain off the fluid into a container by removing the drain plug and washer from the transmission sump.
2. Slacken the sump bolts and tap the sump at one corner to break it loose, then remove the sump and gasket.
3. Remove and discard the filter from the bottom of the valve body.

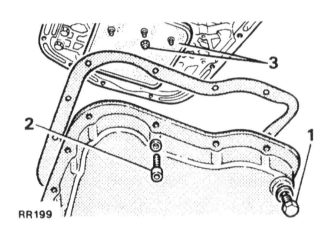

RR199

Adjusting kickdown band

The kickdown band adjusting screw is located on the left side of the gearbox case.

4. Slacken the adjusting screw locknut and back off the nut approximately five turns. Check that the adjusting screw turns freely in the case.
5. Using a torque wrench, tighten the adjusting screw to 8 Nm (72 lbf in).
6. Back off the adjusting screw 2½ turns. Hold the adjusting screw in this position and tighten the locknut to 41 Nm (30 lbf ft).

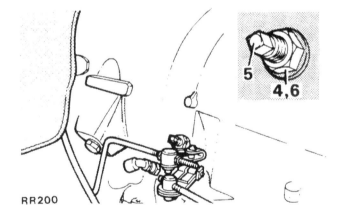

RR200

Adjusting low and reverse band

The low and reverse band adjusting screw is located inside the gearbox case at the bottom.

7. Slacken the adjusting screw locknut and back off the nut approximately five turns. Check that the adjusting screw turns freely in the lever.
8. Using a torque wrench, tighten the adjusting screw to 8 Nm (72 lbf in).
9. Back off the adjusting screw 2½ turns. Hold the adjusting screw in this position and tighten the locknut to 41 Nm (30 lbf ft).

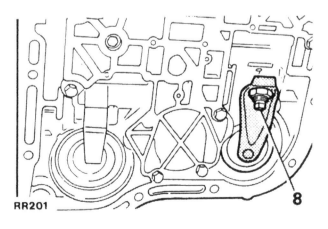

RR201

continued

Reassembling and filling

10. Install a new filter on the bottom of the valve body and tighten the retaining nuts to 11 Nm (100 lbf in).
11. Clean the gearbox sump and refit using a new gasket. Tighten the sump bolts to 18 Nm (160 lbf in).
12. Clean and refit the sump drain plug and washer.
13. Pour five litres (one gallon) of the correct type of automatic transmission fluid through the filler tube.
14. Start the engine and allow to idle for at least two minutes. Then, with the handbrake and footbrake firmly applied, move the selector through all the gear positions, pausing momentarily in each position, ending in the N (Neutral) position.
15. With the engine running at idle speed and Neutral selected, add sufficient fluid to bring the level to the 'ADD ONE PINT' mark on the dipstick.
16. With the engine idling, neutral still selected, check the fluid level after the transmission is at normal operating temperature. The level should be between the 'FULL' mark and 'ADD ONE PINT' mark. DO NOT OVERFILL.
17. After checking fluid level be sure that the dipstick is reseated correctly to prevent dirt or water from entering the gearbox.

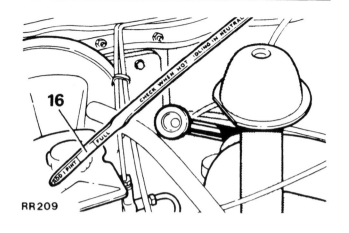

RR209

Check/top up transfer gearbox oil level

Check oil level daily or weekly when operating under severe wading conditions.

1. To check oil level: remove the oil level plug, located on the rear of the transfer box casing; oil should be level with the bottom of the hole.
2. If necessary, top up through the oil level plug hole using a pump type oil can. If significant topping up is required, check for oil leaks at drain and filler plugs.

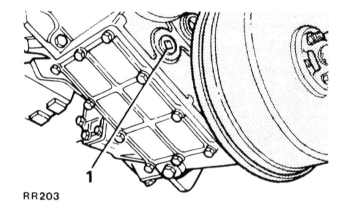

RR203

Renew transfer gearbox oil

Drain and refill monthly when operating under severe wading conditions.

1. Immediately after a run when the oil is warm, drain off the oil into a container by removing the drain plug and washer from the bottom of the transfer box.
2. Replace the drain plug and washer and refill the transfer box through the oil level plug hole with the correct grade of oil, to the bottom of the oil level plug hole. For capacity see Data, Section

IMPORTANT: Do not overfill, otherwise leakage may occur.

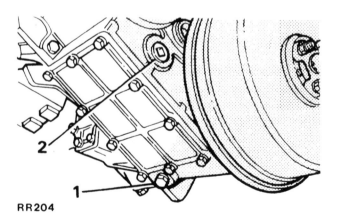

RR204

Check/adjust transmission handbrake

If handbrake movement is excessive, adjust as follows:—

1. Release the handbrake. The adjuster protrudes from the front of the brake backplate.
2. During rotation of the adjuster a click will be felt and heard at each quarter revolution. Rotate adjuster in a clockwise direction until the brake shoes contact the drum. Then unscrew the adjuster two clicks and give the handbrake a firm application to centralise the shoes.

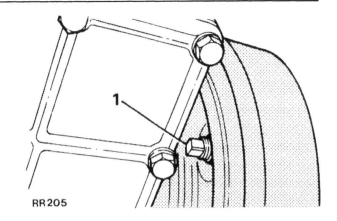

RR205

PASSENGER COMPARTMENT

Check automatic gearbox parking pawl engagement

1. Stand the vehicle on a level surface. Switch off the engine and release the handbrake.
2. Move the gear lever to P (Park).
3. Attempt to push the vehicle backwards and forwards; the parking pawl must hold. Consult your Distributor or Dealer should the pawl not hold.

ENGINE MOUNTING FRONT

– Remove and refit   12.45.04

Removing

1. Remove the air cleaner to prevent it fouling the bulkhead when the engine is raised.
2. Raise the vehicle on a ramp or axle stands and support the engine under the sump with a jack.
3. Remove the rubber mounting nuts, top and bottom, on both sides.
4. Remove, from both sides, the three bolts securing the mounting brackets to the engine.
5. Raise the engine sufficiently to enable the rubber mountings to be removed.

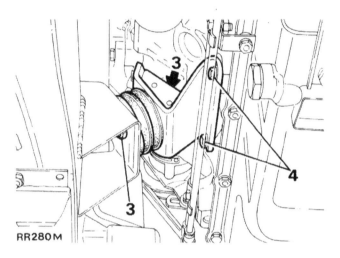

RR280M

Refitting

6. Loosely attach the rubber mountings to the engine brackets.
7. Secure the brackets to the engine.
8. Locate the rubber mounting studs into the chassis brackets and lower the engine.
9. Fit and tighten the rubber mounting lower nuts and tighten the top nuts.
10. Lower the vehicle and fit the air cleaner.

TRANSFER GEARBOX LH MOUNTING

– Remove and refit   12.45.07

Removing

1. Raise vehicle on ramp or axle stands.
2. Support the gearbox.
3. Remove the three bolts retaining the mounting bracket to the chassis.
4. Remove the single nut retaining the rubber mounting to the chassis bracket.
5. Remove the four bolts securing the bracket to the gearbox and remove the bracket and rubber.
6. Remove the single nut retaining the rubber mounting to the gearbox bracket.

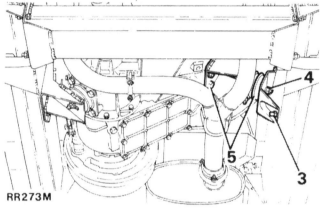

RR273M

Refitting

7. Loosely attach the rubber mounting to the gearbox bracket.
8. Loosely secure the rubber mountings, with the gearbox bracket attached, to the chassis bracket.
9. Secure the bracket to the chassis with the three nuts and bolts.
10. Secure the bracket to the gearbox with the four bolts.

NOTE:– To facilitate instructions 9 and 10 it may be an advantage to slacken the exhaust pipe attachment to the bracket adjacent to the gearbox mounting.

11. Remove the support from the gearbox and tighten the two rubber mounting retaining nuts.
12. Tighten the exhaust pipe attachment, if slackened.

TRANSFER BOX RH MOUNTING

– Remove and refit   12.45.09

Removing

1. Raise the vehicle on a ramp or axle stands.
2. Support the gearbox.
3. Remove the single nut securing the mounting rubber to the chassis bracket.
4. Remove the three nuts and bolts securing the chassis bracket and remove the bracket.
5. Remove the four bolts holding the mounting plate to the gearbox and remove the plate complete with rubber mounting.
6. Remove the rubber mounting from the plate.

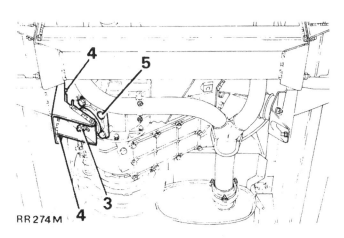

TRANSFER GEARBOX

– Remove and refit   37.29.25

Special tools: Guide studs 18G1425

Removing

1. Drive vehicle on to a ramp and disconnect the battery.
2. Remove the ash trays and carpet covering the transmission tunnel.
3. Remove ten screws and withdraw the hand brake gaiter.
4. Remove high-low transfer knob and four screws and withdraw the gaiter.
5. Remove the air cleaner.

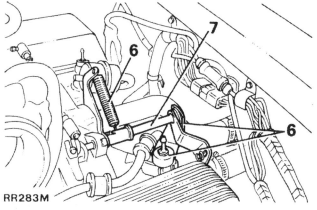

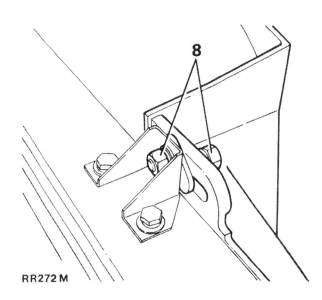

8. Slacken the two outboard nuts retaining the fan cowl and turn the two inboard nuts anticlockwise to raise the cowling sufficiently to prevent fouling when the engine is lowered.

Continued

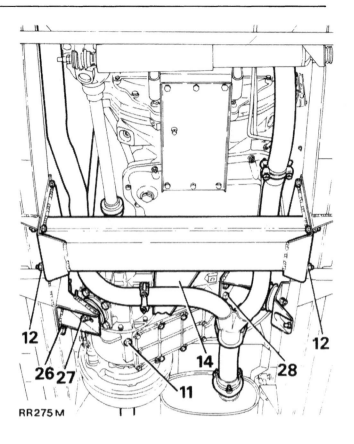

RR275M

9. Release the induction manifold to radiator hose clipped to the alternator bracket.
10. Raise the vehicle on the ramp.
11. Drain the transfer box and replace the plug.
12. Remove the eight nuts and bolts securing the chassis crossmember and using a suitable means of parting the chassis side members, remove the crossmember.
13. Remove the complete front exhaust pipe system.
14. Remove the steady plate between the gearbox and transfer box.

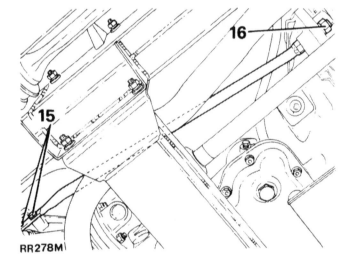

RR278M

15. Remove the two bolts retaining the rear end of the tie bar to the transfer housing.
16. Remove the nut and washer securing the tie bar to the bell housing.

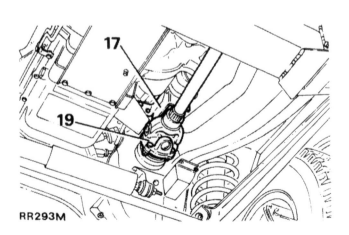

RR293M

17. Remove the starter motor solenoid heat shield.
18. Release the speedometer cable from clamps and disconnect it from the transfer box.
19. Mark for reassembly and disconnect the front propeller shaft from the transfer box flange.

Continued

20. Mark for reassembly and disconnect the rear propeller shaft from the transmission brake drum flange.

21. Remove the nut retaining the gear selector cable to the fulcrum arm and release the cable. Also release outer cable from L.H. mounting bracket.

22. Manufacture an adaptor plate in accordance with the drawing to attach to the gearbox hoist and transfer box to facilitate removal.

23. Attach the adaptor plate to the gearbox hoist.

24. Remove four transfer box bottom cover bolts and move the hoist into position and attach the adaptor plate to the transfer box.

25. Adjust the hoist to take the weight of the transfer box.

26. Remove the six nuts and bolts (three each side) securing the transfer box mounting brackets to the chassis.

27. Remove the nuts retaining the R.H. mounting rubber to the bracket and remove the bracket leaving the rubber attached to the transfer box.

28. Remove the four bolts securing the L.H. mounting bracket to the transfer box and remove the bracket with rubber mounting attached.

continued

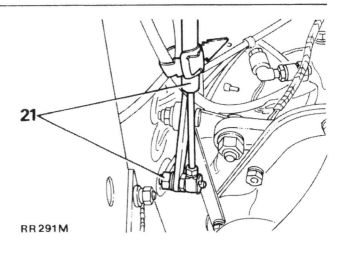

RR291M

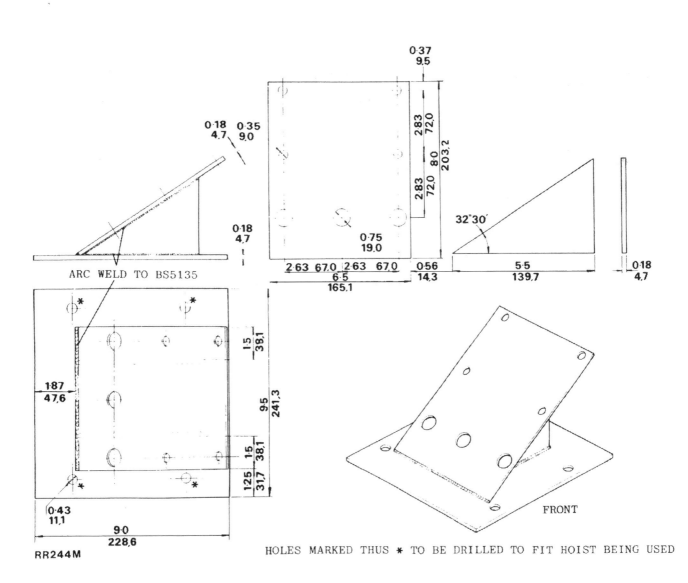

ARC WELD TO BS5135

RR244M

HOLES MARKED THUS * TO BE DRILLED TO FIT HOIST BEING USED

FRONT

29. Lower the complete transmission assembly so that high-low transfer lever is approximately 12 mm (0.5 ins) below the transmission tunnel.
30. Place a jack below the automatic gearbox with a piece of timber between the jack pad and under side of gearbox to prevent damage.
31. Separate the two snap connectors for the gear selector illumination.
32. Release the differential lock harness from the clips and disconnect the leads from the switch on the transfer box.
33. Disconnect the transfer box breather banjo union.

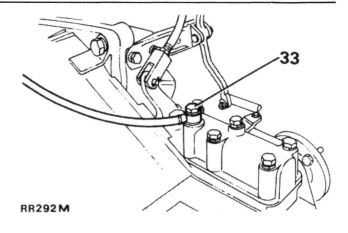

RR292M

34. Support the front silencer and release the rear exhaust system from the hanger bracket below the rear wheel arch. Also release the attachment at the rear of the front silencer and move the assembly aside.

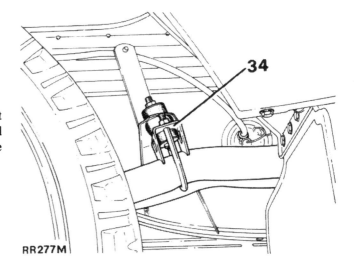

RR277M

35. Move the selector lever to PARK on the automatic gearbox.
36. Remove the five bolts securing the transfer box to the automatic gearbox.
37. Withdraw the transfer box rearwards from the vehicle.

NOTE:— If the transfer box will not release, proceed as follows.

38. Remove the six bolts retaining the rear cover plate and remove the cover to expose the coupling shaft bolt.
39. Remove the coupling shaft retaining bolt and 'O' ring and withdraw the transfer box.

NOTE:— If the coupling shaft remains attached to the automatic gearbox it must be extracted ready for fitment to the transfer box input gear. Damage to the oil seals could occur if it is attempted to fit the transfer box with the coupling shaft fitted to the automatic gearbox mainshaft.

Continued

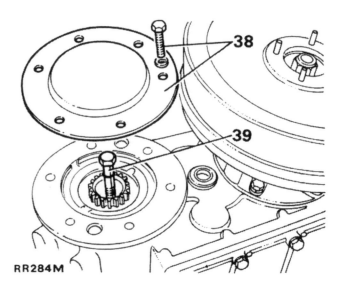

RR284M

Refitting

40. Fit the transfer box to the adaptor plate on the transmission hoist.
41. Clean the transfer box and automatic gearbox mating faces and fit the guide rods to the transfer casing, as illustrated, to facilitate the fitting of the transfer box and to prevent damage to the oil seal.
42. Slide the transfer box into position on the automatic gearbox and secure with the five bolts.
43. Move the gear selector lever into the neutral position.
44. Raise the rear exhaust system into position and secure the assembly to the hanger bracket at the rear of the front silencer. Secure the assembly to the rear hanger bracket under the rear wheel arch.
45. Fit breather pipe banjo union to transfer box.
46. Secure wiring harness with the clips.
47. Connect differential lock leads to the switch on the transfer box.
48. Connect the two snap connectors for the gear selector illumination.
49. Raise the transfer box into position and remove the jack in support of the automatic gearbox.
50. Fit the L.H. mounting bracket to the transfer gearbox and secure with the four bolts.
51. Fit the R.H. mounting bracket to the rubber mounting with the single nut.
52. Fit the L.H. and R.H. mounting brackets to the chassis and secure with the six nuts and bolts (three each side).
53. Remove the bolts securing the adaptor plate to the transfer box and remove the transmission hoist. Refit the lower cover bolts.
54. Fit the selector cable to the fulcrum arm and secure the outer cable to the clamp on the L.H. mounting bracket.
55. Fit the rear propeller shaft to the transmission brake drum flange ensuring that the reassembly marks coincide.
56. Line up the reassembly marks and fit the front propeller shaft.
57. Fit the speedometer cable to transfer box and fasten the outer cable securing clips.
58. Fit the tie bar to the bell housing but leave nut slack.
59. Fit the rear end of the tie bar to the transfer box and secure with the two bolts and tighten the bell housing nut.
60. Fit the steady plate between the automatic gearbox and transfer box.
61. Refit the front exhaust pipe system.
62. Fit the starter motor solenoid heat shield.
63. Expand the chassis side members and fit the cross-member and secure with the eight nuts and bolts.
64. Remove the transfer box filler level plug and fill the gearbox with oil of the correct grade as specified in the lubrication chart and refit the level plug.
65. Lower the vehicle.
66. Reposition the radiator cowl, reversing the procedure in instruction 8.
67. Secure the induction manifold to radiator hose in the clip attached to the alternator bracket.

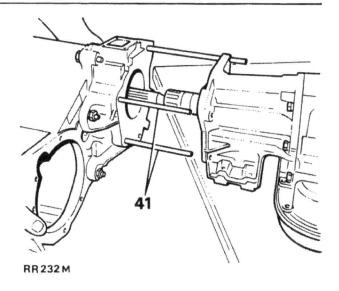

RR 232 M

68. Fit the throttle valve coupling shaft support bracket and shaft.
69. Connect the coupling shaft to the counter shaft.
70. Secure the assembly with the air cleaner L.H. mounting peg.
71. Connect the return spring to the coupling shaft lever.
72. Fit the air cleaner assembly.
73. Fit the hand brake lever gaiter.
74. Fit the high-low transfer knob and gaiter.
75. Fit the ash trays and transmission tunnel carpet. Reconnect the battery.

AUTOMATIC GEARBOX

General   44.00.00

The automatic gearbox combines a torque convertor and a fully automatic three speed system. The gearbox comprises two multiple disc clutches, an overrunning clutch, two servos and bands; and two planetary gear sets to provide three forward ratios and a reverse ratio. A common sun gear of the planetary gear set is connected to the front clutch by a driving shell which is splined to the sun gear and to the front clutch retainer. The hydraulic system consists of an oil pump and a valve body. Gearbox venting is accomplished by a passage through the upper part of the oil pump housing.

The torque convertor is a non-serviceable sealed unit, the cooling of which is accomplished by circulating the transmission fluid (DEXRON D 11) through an oil cooler, located on the front of the vehicle. Transmission fluid is filtered by an internal "Dacron type" filter attached to the lower side of the valve body assembly. Engine torque is transmitted through the input shaft to the multiple disc clutches in the gearbox. The power flow is dependant upon the application of the clutches and bands.

Service requirements

(a) For all operations, whether dismantling, inspecting or reassembling operations, high standards of cleanliness are essential.

(b) Cloths used in wiping and cleaning components must be lint free, nylon cloths are preferable.

(c) Prior to dismantling the gearbox, thoroughly wash the exterior of the casing with a suitable solvent.

(d) During dismantling and inspection operations, all components must be cleaned thoroughly with a suitable industrial solvent.

(e) All defective items must be replaced.

(f) New joint washers should be used during reassembly operations.

(g) All screws, washers and nuts must be tightened to the recommended torque figure.

(h) Thrust washers and bearings should only be coated with petroleum jelly to facilitate retaining them in position during assembly operations. Unless specified grease should not be used as it may be insoluble in the transmission fluid, causing blockage of the fluid passages and contamination of the brake bands and clutch facings.

(i) Use only the recommended transmission fluid of the correct grade and type. Other grades of fluid although the same proprietary brand, could cause damage to the gearbox.

TORQUE CONVERTOR

NOTE:— Following an automatic gearbox failure, debris from a burnt out clutch or brake band material will have probably been deposited in the transmission fluid. It is virtually impossible to ensure that some particles are not left inside the torque convertor, causing contamination of the gearbox, even after flushing. It is strongly recommended that a new torque convertor is fitted and under no circumstances should old transmission fluid be used.

GEARBOX ASSEMBLY OVERHAUL 44.02.06

Dismantling

1. Remove transmission unit and transfer gearbox assembly, see operation 44.20.04.
2. Remove transfer gearbox, see operation 44.20.04.
3. Release the bolts and remove the torque convertor retaining strap.
4. Remove torque convertor assembly.
5. Release and remove the bolt securing the angle stop bracket fitted to the torque convertor retaining strap.
6. Refit the torque convertor retaining strap to the gearbox casing.
7. Measure the input shaft end play before proceeding further, this will indicate when the thrust washer is worn and requires changing. The thrust washer is located between the reaction shaft support and front clutch retainer.
8. Attach the dial gauge brackets (and dial gauge) to the torque convertor retaining strap with the dial plunger seated against the end of the input shaft.
9. Move the shaft in and out to obtain the end float readings. The end float should be between 0,914 mm to 2,133 mm (0.036 to 0.084 in). Any variation outside these measurements should be rectified by changing the thrust washer. Three thrust washer thicknesses are available to bring the end float to specification.
   1,549 mm – 1,8 mm (0.061 in – 0.063 in) – natural.
   2,133 mm – 2,184 mm (0.084 in – 0.063 in) – red.
   2,59 mm – 2,641 mm (0.102 in – 0.104 in) – yellow.
   Note the end float measurement for future reference when reassembling the transmission.

14. Remove the main nut and bolt which secures the selector pivot bracket.
15. Remove all the linkages.
16. Detach the breather junction box bolt and washer fitted to the exterior of the gearbox casing.
17. Detach the breather block and pipe away from the inside of the gearbox bell housing.
18. Unscrew the two bolts securing the breather pipe clamp and remove the breather assembly from the pump housing.
19. Turn the gearbox around, suitably supported, with the fluid pan uppermost.
20. Select park and remove the coupling shaft bolt and 'O' ring.
21. Withdraw the coupling shafts.

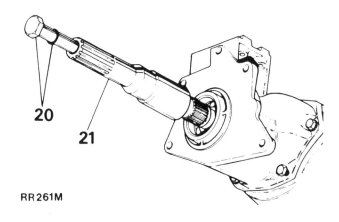

20

21

RR261M

22. Release and remove the extension case bolts.
23. Carefully detach the extension case housing and gasket.

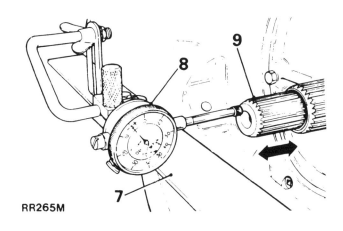

8

9

7

RR265M

10. Remove the dial gauge assembly, brackets and torque convertor retaining strap.
11. With suitable spanner slacken the throttle valve lever pinch bolts.
12. Slacken the selector lever clamp bolt.
13. Detach the selector pivot bracket pinch bolt.

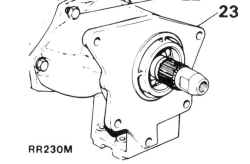

22

23

RR230M

continued

24. Remove small circlip from the weight end of the governor valve shaft and detach.
25. Remove the large circlip which retains the governor and parking gear to the output shaft.
26. Slide the governor and parking gear off the output shaft.

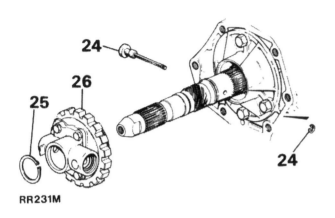

RR231M

27. Release and remove the sump pan bolts and lift off the sump cover gasket.
28. Remove the three 'Nyloc' nuts which hold the filter pad in position.
29. Remove the filter pad and spacer block.
30. Unscrew and remove the inhibitor switch and seal. When refitting use a new seal.
31. Release and remove the ten bolts which hold the valve block to the gearbox casing. Hold the valve body in position whilst removing the bolts.
32. Whilst removing the valve block from the casing, it may be found necessary to manoeuvre the valve block and parking brake rod forward out of the casing.
33. Remove the spring and accumulator piston.
34. Slacken the front band adjuster lock nut.
35. Lightly tighten the front band nuts using adaptor number CBW 547A-50-2A. This prevents the clutch retainer from coming out with the pump, and possibly damaging the clutches.
36. Remove the five remaining oil pump housing bolts.
37. Locate the impulse hammers tool number 18G1387 to the threaded holes in the pump housing flange.
38. Bump outward evenly with the two weights to withdraw the pump and reaction shaft support assembly from the case.
39. Lift the pump and reaction shaft support clear, remove gasket and detach impulse hammers.
40. Using tool CBW 547A-50-2A, slacken the front band adjuster.
41. Remove the front band anchor and strut.
42. Slide the band strut out of the case.
43. Slide the front clutch assembly out of the case.
44. Grasp the input shaft and slide the input shaft and rear clutch assembly out of the casing. Care should be taken not to lose the thrust washer located between the rear end of the input shaft and the forward end of the output shaft.

45. Whilst supporting the output shaft, carefully remove the output shaft assembly comprising the planetary gear and driving shaft.
46. Remove the low and reverse drum.
47. Push the low and reverse band linkage shaft out towards the rear of the gearbox.
48. Remove the 'O' rings from the shaft.
49. Lift out the rear servo lever and strut.
50. Remove the rear band, link and anchor.
51. At the opposite end of the gearbox unscrew the four bolts and remove the output shaft support.
52. Prior to removal, note the position of the overrunning clutch rollers and springs. Carefully slide out the clutch hub.
53. Remove the 6,3 mm (1/4 in) square plug.
54. Slide out the kickdown lever shaft towards the front of the casing and withdraw shaft.
55. Lift out the kickdown servo lever.
56. Compress the kickdown servo spring and remove circlip, allow piston to lift under controlled spring pressure. Take care not to damage the piston rod or guide.
57. Lift out the kickdown servo piston taking care not to break the rings.
58. Compress the low and reverse servo spring and remove circlip, allow piston to rise under controlled spring pressure.
59. Remove the spring retainer, spring, servo piston and plug assembly from the casing.

Inspection

60. Clean casing thoroughly using a suitable solvent, dry with compressed air. Inspect case for cracks, stripped threads in various bolt holes, and machined mating surfaces for burrs, nicks, or any condition that would render the case unfit for further service. The front mating surface should be smooth, any burrs present should be removed using a fine mill file. If threads are stripped, install Helicoil, or equivalent inserts.

continued

EXTENSION HOUSING OVERHAUL

Dismantling

61. Fit the extension housing in a vice.
62. Remove the parking pawl shaft.
63. Detach the parking pawl and spring.
64. Release and remove reaction plug circlip.
65. Lift out reaction plug and pin from the casing.
66. Remove the bearing retaining circlip.
67. Release housing from vice and using a suitable press, extract the bearing.
68. Remove the oil seal.

Inspection

69. Inspect the parking pawl shaft for scores and free movement in the housing.
70. Check the spring for distortion.
71. Inspect the square lug on the parking pawl for broken edges.
72. Replace parts that are damaged.

Reassembly

73. Replace oil seal using tool 18G1421.
74. Using a press, refit new bearing.
75. Fit housing in vice and replace the bearing retaining circlip. Ensure that the circlip ends do not obstruct the oil return passage.
76. Refit reaction plug, pin and circlip. Locate the parking pawl and spring ensuring the spring is positioned so that it moves the pawl away from the gear.
77. Remove extension housing from vice.

GOVERNOR AND PARKING GEAR OVERHAUL

Dismantling

NOTE:— Although the governor and parking gear had been partially dismantled during the gearbox dismantling procedures, an asterisk denotes where the information is repeated for sequential clarity.

*78. Remove the small circlip from the weight end of the governor valve shaft.
*79. Slide the valve and shaft assembly out of the governor body.
*80. Remove the large circlip from the weight end of the governor body, lift out the governor weight assembly.
81. Carefully prise the circlip from inside the governor weight.
82. Remove the inner weight and spring from the outer weight.
83. Detach and remove the circlip from behind the governor body, then slide the governor and support assembly off the output shaft.
84. Remove the four bolts and separate the governor body and screen from the parking gear.

Inspection

85. Thoroughly clean all parts in a suitable solvent.
86. Inspect all parts for damage and wear.
87. Check the inner weight for free movement in the outer weight.
88. Verify that the weights and valve fall freely in the bores when clean and dry.
89. Wash the governor screen in a suitable solvent.
90. Inspect the governor seal rings for wear on their sides and outer diameter.
91. Check the governor weight spring for distortion.
92. Inspect the lugs on the support gear for broken edges or other damage.

Reassembly

93. Reverse procedures 84–81 noting that Loctite 290 is applied to the four bolt threads prior to torque tightening the bolts to 11,29 Nm (100 lbf. in). Procedures 80–78 are not implemented until the gearbox is reassembled.

Continued

VALVE BODY OVERHAUL

Dismantling

94. Do not clamp any portion of the valve body or transfer plate in a vice. Any slight distortion of the aluminium body or transfer plate will result in sticking valves and/or excessive leakage. When removing or installing valves or plugs, slide them in or out carefully. Do not use force.

95. Remove the top and bottom screws from the spring retainer and adjustment screw bracket. Hold the spring retainer firmly against the spring pressure whilst removing the screw from the side of the valve body.

96. Without disturbing the setting, remove the spring retainer complete with the line and the throttle pressure adjusting screws and the line pressure and torque convertor regulator springs.

97. Slide the torque convertor and line pressure valves out of their bores.

98. Remove the transfer plate retaining screws and lift off the transfer plate and separator plate assemblies.

99. Undo the two stiffener plate retaining screws and the three screws holding the separator plate. Remove plate for cleaning.

100. Remove the rear clutch ball check valve from the transfer plate.

101. Release the valve screen filter from the separator plate for cleaning.

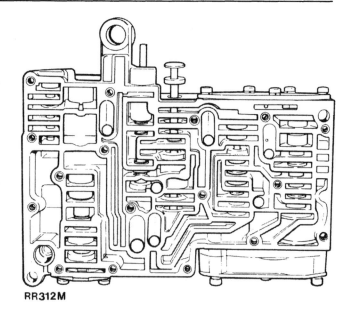

RR312M

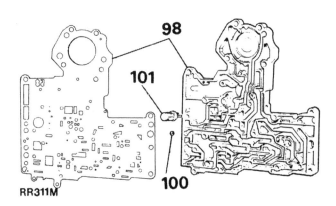

RR311M

102. Remove the seven balls from the valve body.
103. Turn over the valve body and undo the six screws from the shuttle valve cover plate.
104. Remove the governor plug end plate and slide out the shuttle valve throttle plug and spring, the 1–2 shift valve governor plug and 2–3 shift valve governor plug.
105. Detach the circlip and slide the shuttle valve out of its bore, in addition to the secondary springs and guides which were retained by the circlip.
106. Remove the circlip from the park control rod.
107. Remove the circlip and washer from the throttle lever shaft.

108. Whilst holding the manual lever detent ball and spring in their bore, slide the manual lever off the throttle shaft and remove the seal.
109. Remove the detent ball and spring.
110. Slide the manual lever out of its bore.
111. Slide out the kickdown detent, kickdown valve, throttle valve spring and throttle valve.
112. Remove the line pressure regulator valve end plate and slide out the regulator valve sleeve, line pressure and throttle pressure plugs and spring.
113. Remove the end plate and downshift housing assembly and detach the throttle plug from the housing.
114. Slide the retainer from the housing and remove the limit valve and spring.
115. Remove the springs and 1–2 shift control valve, 1–2 shift valve and 2–3 shift valve.

Inspection

116. Allow all parts to soak in suitable clean solvent. Wash thoroughly and blow dry using compressed air. Ensure all passages are clean and free from obstructions.

117. Inspect all mating surfaces for burrs, distortion and warping. Slight distortion may be corrected using a surface plate.

118. Check all valve springs for distortion and collapsed coils.

119. When the bores, valves and plugs are clean and dry, the valves and plugs should fall freely in the bores. The valve bores do not change dimensionally with use. Therefore, a valve body that was functioning properly when the vehicle was new, will operate correctly if it is properly and thoroughly cleaned. There is no need to replace the valve body unless it is damaged.

Continued

Reassembly

120. Slide the shift valves and springs into their appropriate bores.

121. Sub-assemble the down shift housing as follows:—
    (a) Install the limit valve and spring into the housing.
    (b) Slide the spring retainer into the groove in the housing.
    (c) Insert the throttle plug in the housing bore and position the whole assembly against the shift valve springs.

122. Install the end plate and tighten the screws to 3,95 Nm (35 lbf. in).

123. Fit the throttle pressure plug, line pressure plug and sleeve; and refit end plate to valve body. Tighten screws to 3,95 Nm (35· lbf. in).

124. Install the throttle valve and spring, kickdown valve and kickdown detent.

125. Slide the manual valve into its bore.

126. Install the throttle lever assembly on the valve body.

127. Insert the detent spring and ball in its bore on the valve body. Depress the ball and spring and slide the manual lever over the throttle shaft, so that it engages the manual valve and detent bore. Fit a new seal, retaining washer and circlip on the throttle shaft.

128. Place the 1—2 and 2—3 shift valve governor plugs in their respective bores.

129. Fit the shuttle valve and hold it in the bore whilst installing on the opposite end of the secondary spring with guides and retaining circlip.

130. Install the primary shuttle valve spring and throttle plug and refit the governor plug end plate. Tighten the five remaining screws to 3,95 Nm (35 lbf. in).

131. Replace the shuttle valve cover plate and tighten the six retaining screws to 3,95 Nm (35 lbf. in).

132. Fit the seven balls in the valve body in the positions noted during dismantling.

133. Replace the rear clutch ball check valve in the transfer plate and regulator valve screen in the separator plate.

134. Refit and tighten the screws in the stiffener and separator plate to 3,95 Nm (35 lbf. in).

135. Place the transfer plate assembly on the valve body. Care should be exercised to align the filter screen. Fit the smaller screws to the transfer plate, tighten to 3,95 Nm (35 lbf. in).

136. Fit and tighten the longer screws starting at the centre and working outwards, to 3,95 Nm (35 lbf. in).

137. Slide the torque convertor and line pressure valves and springs into their bores.

138. Install the pressure adjusting screw and bracket assembly on the springs and tighten the screws to 3,95 Nm (35 lbf. in).

139. Using the gauge block tool 18G1385, measure the throttle pressure.
    (a) Insert gauge pin of tool between the throttle lever cam and kickdown valve.
    (b) By pushing in on tool compress the kickdown valve against its spring so that the throttle valve is completely bottomed inside the valve body.
    (c) As force is being exerted to compress the spring, turn the throttle lever stop screw with a suitable allen key, until the head of the screw touches the throttle lever tag with the throttle lever cam touching the tool and the throttle valve bottomed.

NOTE:— The throttle pressure setting must be accurate otherwise a wrong setting will cause an incorrect line pressure reading, despite the line pressure being correct.

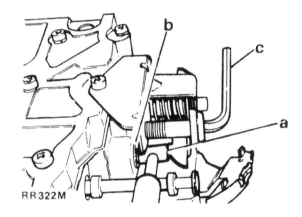

RR322M

140. To obtain the correct line pressure adjustments measure from the valve body to the inner edge of the adjusting plate. The correct measurement is 33,3 mm (1.5/16 in) however this measurement may be varied to obtain the specified line pressure.
    (a) Turn the adjusting screw with a suitable allen key, one complete turn of the screw will alter the closed throttle line pressure by approximately 0,1167 Kgf/cm^2 (1.66 lbf. in^2). Turning the adjusting screw clockwise decreases the pressure and anti-clockwise increases the pressure.

141. Refit the parking lock rod and circlip to the manual lever.

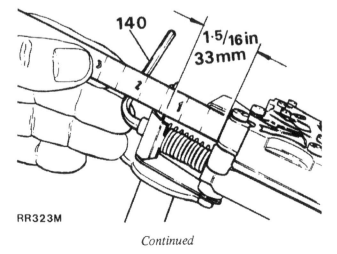

RR323M

Continued

OIL PUMP AND REACTION SHAFT SUPPORT OVERHAUL

Dismantling

142. Remove the six reaction shaft support retaining bolts and lift the support off the pump.
143. Slide the inner and outer rotors from the pump body.
144. Remove the rubber seal ring from pump body flange.
145. Drive out the oil seal with a blunt punch.
146. Carefully remove the two interlocking seal rings fitted to the shaft support.
147. Remove the selective thrust washer.

Inspection

148. Inspect the interlocking seal rings for wear or broken locks. When reassembled, ascertain they rotate freely in their grooves.
149. Check the machined surfaces on the pump body and reaction shaft support for damage.
150. Inspect the pump rotors for scoring or pitting.

Reassembly

151. Install the outer and inner rotors in the oil pump body. Place a straight edge across the face of the rotors and pump body. Use a feeler gauge to measure the clearance between the straight edge and the face of the rotors. The clearance limits are 0,025 mm to 0,05 mm (0.001 in to 0.002 in).

152. Measure the rotor tip clearance between the inner and outer teeth. The clearance limits are between 0,127 mm to 0,254 mm (0.005 in to 0.01 in).
153. The clearance permitted between the rotor and its bore in the oil pump body should be 0,101 mm to 0,203 mm (0.004 in to 0.008 in).
154. Fit a new rubber sealing ring to the pump body.
155. Refit the reaction shaft support. Install the six retaining bolts and tighten to 18,07 Nm (160 lbf. in).
156. Fit a new oil seal into the opening of the pump body, ensuring the oil seal lips face inwards, using tools 18G134-3 and 18G134.
157. Replace the thrust washer or one of the appropriate thickness depending upon the initial end float measurement.
158. Refit the interlocking seal rings.

FRONT CLUTCH OVERHAUL

Dismantling

159. Remove the large waved circlip that secures the pressure plate in the clutch retainer.
160. Lift the pressure plate and clutch plates out of the retainer.

Continued

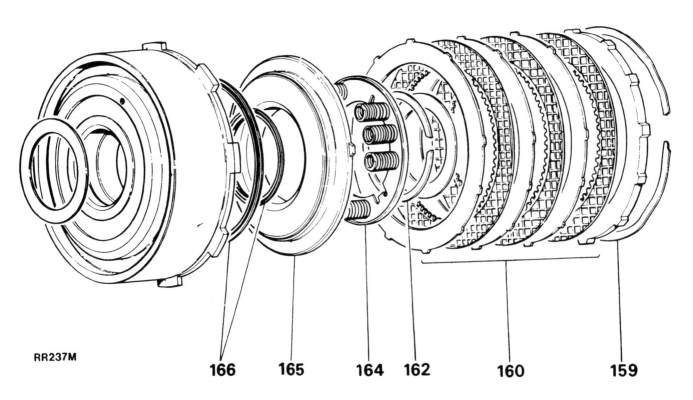

RR237M

| 166 | 165 | 164 | 162 | 160 | 159 |

161. Fit the spring compressor tool 18G1386 over the piston ring retainer.
162. Compress the ring and remove the circlip.
163. Slowly release the spring tool until the spring retainer is free of the hub.
164. Remove the tool, retainer and springs.
165. Invert the piston retainer assembly and bump it on a wood block to dislodge the piston.
166. Remove the seals from the piston retainer hub.

Inspection

167. Check the clutch plate and disc for flatness and facing material on all driving discs. Change all discs, plates that are charred, glazed or heavily pitted.
168. Inspect the disc splines for wear or damage and check the steel plate lug grooves for smoothness. The plates must be able to travel freely in the grooves.
169. Examine the inside bore of the piston for score marks, if light scoring is apparent, it can be removed by the use of extra fine wet and dry paper.

Reassembly

170. Lubricate and install new inner seal on the hub of the clutch retainer, using tool 18G1421. Ensure the seal lip faces down and is properly seated.

171. Fit new outer seal on the clutch piston, ensuring the seal lip is towards the bottom of the clutch retainer.
172. Apply a coating of transmission fluid to the outer edge of seal and press seal to the bottom of its groove around the piston diameter. Place piston assembly in retainer and carefully seat the piston into the bottom of retainer.
173. Fit springs on to piston.
174. Position spring retainer and circlip over springs, compress springs using tool 18G1386 and seat circlip in hub groove. Remove tool 18G1386.
175. Lubricate all clutch plates with transmission fluid and locate one steel plate followed by a lined disc until all plates are correctly located.
176. Install pressure plate and circlip, ensuring the circlip is correctly seated.
177. Insert a feeler gauge between the pressure plate and waved circlip to measure the maximum clearance where the circlip is waved away from the pressure plate. The clearance should be between 1,77 mm to 3,27 mm (0.070 in to 0.129 in).

Continued

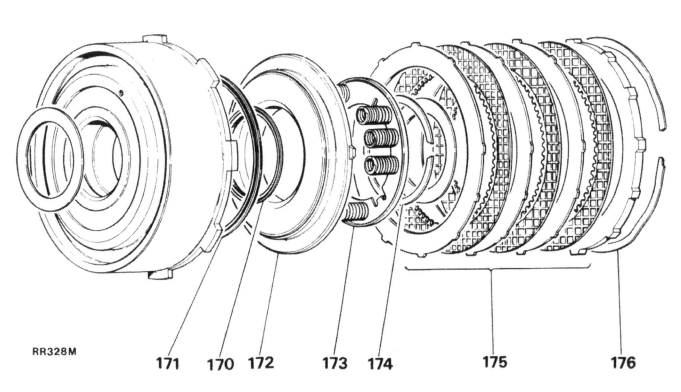

RR328M

171　　**170**　**172**　　　**173**　**174**　　　　**175**　　　　　**176**

REAR CLUTCH OVERHAUL

Dismantling

178. Remove the large selective circlip that secures the pressure plate in the clutch retainer.
179. Lift off the pressure plate, four clutch plates and inner pressure plate out of the clutch retainer.
180. Carefully prise one end of the wave spring out of the groove and remove the spring.
181. Remove the nylon spacer ring and clutch piston spring out of the retainer.
182. Slide the clutch retainer off the piston retainer.
183. Invert the clutch piston retainer assembly and bump it on a suitable wood block to dislodge the piston. Remove piston.
184. Remove the inner and outer seals from the piston.
185. Remove the thrust washer.
186. Detach the three interlocking sealing rings and tabbed thrust washer from the input shaft.
187. If necessary, remove the circlip and press input shaft from the clutch piston retainer.

Inspection

188. Examine the clutch plates and driving discs for flatness. They must not be warped or cone shaped.
189. Inspect the facing material on all driving discs. Replace if charred, glazed or heavily pitted. Discs should also be replaced if they show evidence of material flaking off or if facing material can be scraped off easily.
190. Verify the steel plates and pressure plate surfaces are not scored, burnt or have damaged driving lugs, replace if necessary.
191. Inspect steel plate lug grooves in the clutch retainer for smooth surfaces, the plates must travel freely in the grooves.
192. Verify the ball check in the piston moves freely.
193. Inspect the seal surfaces in the clutch retainer for nicks or deep scratches. Light scratches will not interfere with the sealing of the seals.
194. Examine the piston spring, wave spring and spacer for distortion or breakage.
195. Inspect the interlocking seal rings for wear and broken locks, and turn freely in their grooves when refitted.
196. The rear clutch to front clutch thrust washer should be measured for wear. Washer thickness should be 1,54 mm to 1,6 mm (0.061 in to 0.063 in) replace if necessary.

Continued

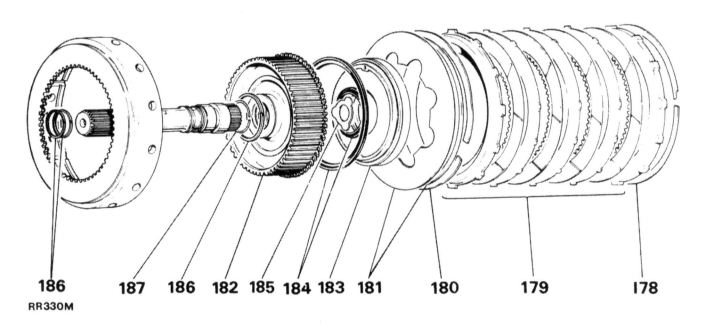

186   187   186   182   185   184   183   181   180   179   178

RR330M

Reassembly

197. If removed, press input shaft into the clutch piston retainer.

198. Refit tabbed washer, using petroleum jelly to hold in place.

199. Replace the interlocking seal rings and ensure they turn freely in their grooves.

200. Fit new inner and outer seals on the clutch piston and install seals on clutch piston. Make sure the seal lips face towards the head of the clutch retainer, and are properly seated in the piston grooves.

201. Place piston assembly in the retainer and with twisting motion seat the piston in the bottom of the retainer.

202. Position the clutch retainer over the piston retainer splines and support the assembly so that the clutch retainer remains in place.

203. Place the clutch piston spring and spacer ring on top of the piston in the clutch retainer, ensuring the spring and spacer ring are positioned in the retainer recess. Start one end of the wave spring in the retainer groove and progressively tap or push spring into place, making sure it is fully seated in the groove.

204. Install inner pressure plate in the clutch retainer with the raised portion of the plate resting on the spring.

205. Lubricate all clutch plates in transmission fluid, install one lined disc followed by a steel plate until all plates are installed. Install outer pressure plate and selective circlip.

206. Measure the rear clutch plate clearance by pressing down firmly on the outer pressure plate to seat the plates, insert a feeler gauge between the plate and the circlip. The permitted clearance is 0,63 mm – 1,14 mm (0.025 in – 0.045 in). If necessary, install a new selective circlip of the proper thickness to obtain the specified thickness. Low limit clearance is desirable.

207. The rear clutch plate clearance is very important in obtaining the proper clutch operation. The clearance can be adjusted by the use of various thickness outer circlips, which are available in the following thicknesses

> 1,52 mm (0.060 in)
> 1,87 mm (0.074 in)
> 2,23 mm (0.088 in)
> 2,69 mm (0.106 in)

Continued

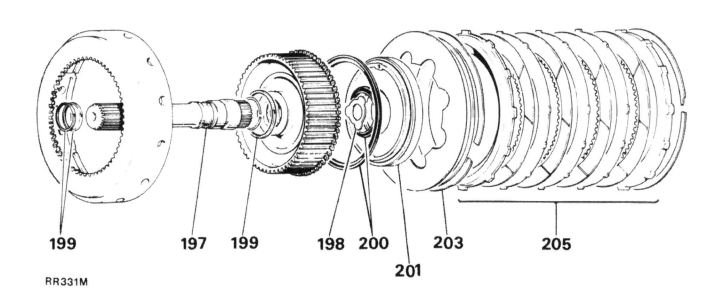

199   197 199   198 200   203   205

201

RR331M

PLANETARY GEAR TRAIN OVERHAUL

Dismantling

208. Measure the end play of the planetary gear assemblies, sun gear and driving shell prior to removing these parts from the output shaft.
209. Support assembly upright with the forward end of output shaft on a wood block so that all parts will move forward against the selective circlip at the front of the shaft.
210. Insert a feeler gauge between the rear annulus gear hub support and shoulder on the output shaft. The clearance should be 0,3 mm – 0,63 mm (0.01 – 0.025 in). If the clearance exceeds the specification replace the thrust washers.
211. Remove the thrust washer from the forward end of the output shaft.
212. Detach the selective circlip from the shaft, and slide the front planetary assembly off the shaft.
213. Slide front annulus gear off the planetary gear set. Remove the thrust washer from the rear side of the planetary gear set.
214. Lift sun gear, driving shaft and rear planetary assembly off the output shaft.
215. Lift the sun gear and driving shell off the rear planetary gear assembly. Remove the thrust washer from inside the driving shell.
216. Remove the circlip and thrust washer from the sun gear (rear side of the driving shell) and slide the sun gear out of the shell.
217. Remove the front snap ring from the sun gear. It should be noted that the front end of the sun gear is longer than the rear.
218. Remove the thrust washer from the forward side of the rear planetary gear assembly.
219. Remove the planetary gear set and thrust plate from the annulus gear.

Inspection

220. Inspect the bearing surfaces on the output shaft for nicks, burrs, scores or other damage. Light scratches nicks or burrs can be removed with extra fine wet and dry paper or a fine stone.
221. Ensure all oil passages in the shaft are open and clean.
222. Measure the thrust washers and replace if worn below specification. Thrust washer numbers refer to the paragraphs mentioned.
    No 211. 1,57–1,62mm (0.062–0.064in)
    No 213. 1,49–1,57mm (0.059–0.062in)
    No 215. 1,52–1,57mm (0.060–0.062in)
    No 216. 0,86–0,91mm (0.034–0.036in)
    No 218. 1,49–1,57mm (0.059–0.062in)
    No 219. 0,86–0,91mm (0.034–0.036in)
223. Inspect the thrust faces of the planetary gear carriers for wear, scoring or other damage.
224. Check the planetary gear carrier for cracks and pinions for broken pinion shaft lock pins.
225. Inspect annulus gear and driving gear teeth for damage, replace unit if damaged.

Reassembly

226. Install the rear annulus gear on the output shaft.
227. Apply a thin coat of petroleum jelly to the thrust plate, place plate on the shaft and in the annulus gear, ensuring the plate teeth are correctly located over the shaft splines.
228. Position the rear planetary gear assembly in the rear annulus gear.
229. Place the thrust washer on the front side of the planetary gear assembly.
230. Install the circlip in the front groove (long end) of the sun gear. Fit the sun gear through the front side of the driving shell.
231. Fit the rear thrust plate and circlip.
232. Carefully slide the driving shell and sun gear assembly on to the output shaft, engaging the sun gear teeth with the rear planetary pinion teeth.
233. Position the thrust washer inside the front driving shell.
234. Place the thrust washer on the rear hub of the front planetary gear set, then slide the assembly into the front annulus gear.
235. Carefully install the front planetary and annulus gear assembly to the output shaft, meshing the planetary pinions with the sun gear teeth.
236. With all the components properly positioned, install the selective snap ring on the front end of the output shaft. Remeasure the end play of the assembly.
237. If the end play tolerance of 0,22 mm – 1,11 mm (0.009 in – 0.044 in) is not met, and the thrust washers have been measured and found to be within the required specification, then the end float clearance can be adjusted by the use of various thickness circlips. Circlips are in the following thicknesses:–
    1,21 mm (0.048 in)
    1,39 mm (0.055 in)
    1,57 mm (0.062 in)

Continued

212 **216 218 219**

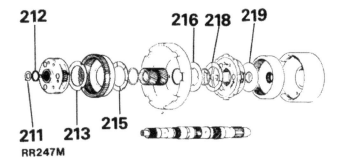

211 213 215
RR247M

OVERRUNNING CLUTCH OVERHAUL

Inspection

238. Inspect the clutch rollers for smooth round surfaces, they must be free of flat spots and chipped edges. Replace if damaged.
239. Check roller contacting surfaces in the cam and race for brinelling.
240. Inspect the roller springs for distortion, wear or other damage. Replace if damaged.
241. Inspect the cam set screw for tightness.

KICKDOWN SERVO AND BAND OVERHAUL

Dismantling

242. Remove the sealing ring from the piston rod guide.
243. Remove the small circlip which retains the piston rod to the servo piston.
244. Remove the washer, springs, piston rod.
245. Withdraw the piston rod from the servo piston.
246. Remove the 'O' ring from the piston rod and piston rings from the servo piston.

Inspection

247. Examine all piston and guide seal rings for wear, and during reassembly ensure they turn freely in the grooves.
248. Inspect the piston and piston bore for cracks, scores or wear.
249. Examine the fit of guide on the piston rod and inspect the piston spring for distortion.
250. Inspect the band lining for wear and bond of lining to band.
251. Examine the lining for black burn marks, glazing, a non uniform wear pattern and flaking. If the band lining is worn so that the grooves are not visible at the ends, or any portion of the bands, replace the band.
252. Inspect band for distortion or cracked ends.

Reassembly

253. Grease the new 'O' ring and install on piston rod, use petroleum jelly as a greasing agent.
254. Install piston rod into the servo piston.
255. Locate inner spring, flat washer, outer spring, piston rod guide and secure assembly with the circlip.
256. Refit the piston rings to the servo piston and piston rod guide.

LOW/REVERSE SERVO AND BAND OVERHAUL

Dismantling

NOTE:– Although the low/reverse servo had been partially dismantled during the gearbox dismantling procedures, an asterisk denotes where the information is repeated for sequential clarity.

*257. Remove the circlip, retainer and spring.
*258. Lift out the piston plug and spring assembly.
259. Compress the spring and remove the circlip fitted to the rear of the piston.
260. Release the spring slowly and dismantle the piston assembly, comprising piston plug, spring and piston.

Inspection

261. Inspect the seal for deterioration, wear and hardness.
262. Check piston plug and piston for cracks and burrs.
263. The piston plug must operate freely in the piston.
264. Inspect the piston bore for damage and springs for distortion.
265. Inspect the band lining for wear and the condition of the bond of lining to the band. If lining is worn so that the grooves are not visible at the ends of any portion of the band, replace the band.

Reassembly

266. Lubricate and insert piston plug and spring into the piston, compress spring and secure with small circlip.
267. Renew the seal around the piston.

ACCUMULATOR PISTON OVERHAUL

Inspection

268. Remove the sealing rings from the piston and inspect for damage.
269. Inspect spring for distortion.

Continued

PLANETARY GEAR TRAIN OVERHAUL

Dismantling

208. Measure the end play of the planetary gear assemblies, sun gear and driving shell prior to removing these parts from the output shaft.
209. Support assembly upright with the forward end of output shaft on a wood block so that all parts will move forward against the selective circlip at the front of the shaft.
210. Insert a feeler gauge between the rear annulus gear hub support and shoulder on the output shaft. The clearance should be 0,3 mm – 0,63 mm (0.01 – 0.025 in). If the clearance exceeds the specification replace the thrust washers.
211. Remove the thrust washer from the forward end of the output shaft.
212. Detach the selective circlip from the shaft, and slide the front planetary assembly off the shaft.
213. Slide front annulus gear off the planetary gear set. Remove the thrust washer from the rear side of the planetary gear set.
214. Lift sun gear, driving shaft and rear planetary assembly off the output shaft.
215. Lift the sun gear and driving shell off the rear planetary gear assembly. Remove the thrust washer from inside the driving shell.
216. Remove the circlip and thrust washer from the sun gear (rear side of the driving shell) and slide the sun gear out of the shell.
217. Remove the front snap ring from the sun gear. It should be noted that the front end of the sun gear is longer than the rear.
218. Remove the thrust washer from the forward side of the rear planetary gear assembly.
219. Remove the planetary gear set and thrust plate from the annulus gear.

Inspection

220. Inspect the bearing surfaces on the output shaft for nicks, burrs, scores or other damage. Light scratches nicks or burrs can be removed with extra fine wet and dry paper or a fine stone.
221. Ensure all oil passages in the shaft are open and clean.
222. Measure the thrust washers and replace if worn below specification. Thrust washer numbers refer to the paragraphs mentioned.
    No 211. 1,57–1,62mm (0.062–0.064in)
    No 213. 1,49–1,57mm (0.059–0.062in)
    No 215. 1,52–1,57mm (0.060–0.062in)
    No 216. 0,86–0,91mm (0.034–0.036in)
    No 218. 1,49–1,57mm (0.059–0.062in)
    No 219. 0,86–0,91mm (0.034–0.036in)
223. Inspect the thrust faces of the planetary gear carriers for wear, scoring or other damage.
224. Check the planetary gear carrier for cracks and pinions for broken pinion shaft lock pins.
225. Inspect annulus gear and driving gear teeth for damage, replace unit if damaged.

Reassembly

226. Install the rear annulus gear on the output shaft.
227. Apply a thin coat of petroleum jelly to the thrust plate, place plate on the shaft and in the annulus gear, ensuring the plate teeth are correctly located over the shaft splines.
228. Position the rear planetary gear assembly in the rear annulus gear.
229. Place the thrust washer on the front side of the planetary gear assembly.
230. Install the circlip in the front groove (long end) of the sun gear. Fit the sun gear through the front side of the driving shell.
231. Fit the rear thrust plate and circlip.
232. Carefully slide the driving shell and sun gear assembly on to the output shaft, engaging the sun gear teeth with the rear planetary pinion teeth.
233. Position the thrust washer inside the front driving shell.
234. Place the thrust washer on the rear hub of the front planetary gear set, then slide the assembly into the front annulus gear.
235. Carefully install the front planetary and annulus gear assembly to the output shaft, meshing the planetary pinions with the sun gear teeth.
236. With all the components properly positioned, install the selective snap ring on the front end of the output shaft. Remeasure the end play of the assembly.
237. If the end play tolerance of 0,22 mm – 1,11 mm (0.009 in – 0.044 in) is not met, and the thrust washers have been measured and found to be within the required specification, then the end float clearance can be adjusted by the use of various thickness circlips. Circlips are in the following thicknesses:–
    1,21 mm (0.048 in)
    1,39 mm (0.055 in)
    1,57 mm (0.062 in)

Continued

212   **216 218 219**

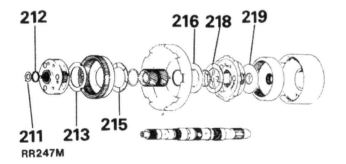

211   213   215
RR247M

OVERRUNNING CLUTCH OVERHAUL

Inspection

238. Inspect the clutch rollers for smooth round surfaces, they must be free of flat spots and chipped edges. Replace if damaged.
239. Check roller contacting surfaces in the cam and race for brinelling.
240. Inspect the roller springs for distortion, wear or other damage. Replace if damaged.
241. Inspect the cam set screw for tightness.

KICKDOWN SERVO AND BAND OVERHAUL

Dismantling

242. Remove the sealing ring from the piston rod guide.
243. Remove the small circlip which retains the piston rod to the servo piston.
244. Remove the washer, springs, piston rod.
245. Withdraw the piston rod from the servo piston.
246. Remove the 'O' ring from the piston rod and piston rings from the servo piston.

Inspection

247. Examine all piston and guide seal rings for wear, and during reassembly ensure they turn freely in the grooves.
248. Inspect the piston and piston bore for cracks, scores or wear.
249. Examine the fit of guide on the piston rod and inspect the piston spring for distortion.
250. Inspect the band lining for wear and bond of lining to band.
251. Examine the lining for black burn marks, glazing, a non uniform wear pattern and flaking. If the band lining is worn so that the grooves are not visible at the ends, or any portion of the bands, replace the band.
252. Inspect band for distortion or cracked ends.

Reassembly

253. Grease the new 'O' ring and install on piston rod, use petroleum jelly as a greasing agent.
254. Install piston rod into the servo piston.
255. Locate inner spring, flat washer, outer spring, piston rod guide and secure assembly with the circlip.
256. Refit the piston rings to the servo piston and piston rod guide.

LOW/REVERSE SERVO AND BAND OVERHAUL

Dismantling

NOTE:– Although the low/reverse servo had been partially dismantled during the gearbox dismantling procedures, an asterisk denotes where the information is repeated for sequential clarity.

*257. Remove the circlip, retainer and spring.
*258. Lift out the piston plug and spring assembly.
259. Compress the spring and remove the circlip fitted to the rear of the piston.
260. Release the spring slowly and dismantle the piston assembly, comprising piston plug, spring and piston.

Inspection

261. Inspect the seal for deterioration, wear and hardness.
262. Check piston plug and piston for cracks and burrs.
263. The piston plug must operate freely in the piston.
264. Inspect the piston bore for damage and springs for distortion.
265. Inspect the band lining for wear and the condition of the bond of lining to the band. If lining is worn so that the grooves are not visible at the ends of any portion of the band, replace the band.

Reassembly

266. Lubricate and insert piston plug and spring into the piston, compress spring and secure with small circlip.
267. Renew the seal around the piston.

ACCUMULATOR PISTON OVERHAUL

Inspection

268. Remove the sealing rings from the piston and inspect for damage.
269. Inspect spring for distortion.

Continued

589

325. Locate the linkage to the manual lever.
326. Fit the pivot and bracket bolt to the casing, but do not tighten.
327. Remove the sump bolt adjacent to the linkage, fit linkage plate and tighten sump bolt to within the specified tolerances of 21 to 28 Nm (15 to 25 lb.ft).
328. Tighten the pivot bracket nut to 34 to 41 Nm (25 to 30 lb.ft).
329. Tighten the pinch bolt fitted to the selector linkage to 7 to 10 Nm (5 to 7 lb.ft).
330. Replace the throttle valve lever to the throttle valve shaft and tighten bolts to 5 to 7 Nm (4 to 5 lb.ft).

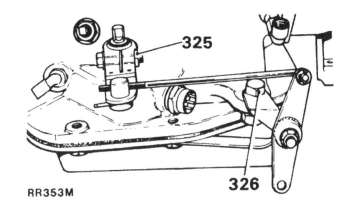

RR353M

OIL COOLER AND TUBES – FLUSHING

331. When an automatic gearbox failure has contaminated the fluid, the oil cooler should be flushed to ensure that metal particles or sludged oil are not later transferred back into the reconditioned gearbox.
332. Place a length of hose over the end of the oil cooler input, insert hose securely into a waste oil container.
333. Apply very short sharp blasts of clean dry compressed air to the output connector tube.
334. Pump approximately 0,6 litre (1 pint) of DEXRON D11 transmission fluid into the oil cooler and apply compressed air in short blasts.
335. Remove tube and reconnect oil cooler to gearbox.

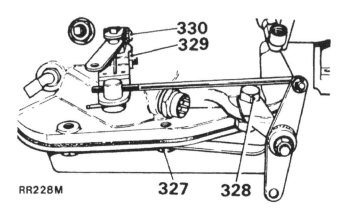

RR228M

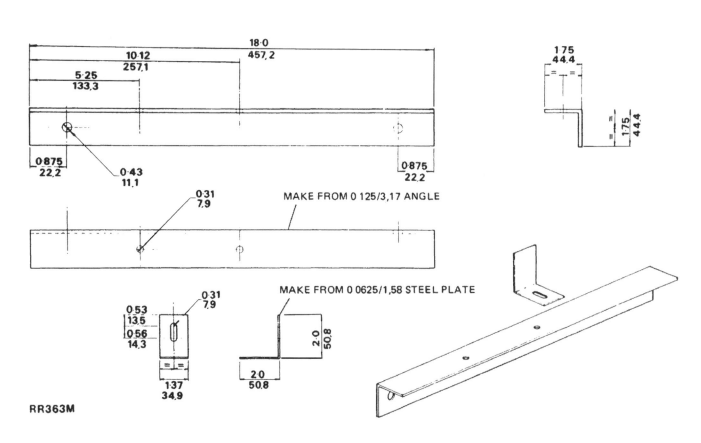

RR363M

590

THROTTLE VALVE LINKAGE

– Adjust   44.15.03

1. Remove the air cleaner to gain access to the throttle valve upper linkage.
2. Remove the twelve screws and withdraw the gear box tunnel side plate.
3. Remove the split pin and washers securing the trunnion to the throttle valve lever.
4. Remove the trunnion from the throttle valve lever.
5. With assistance, depress the accelerator pedal to give full throttle position and check that the down link is at the bottom of the slot in the coupling shaft lever.
6. Whilst holding full throttle, move the throttle valve lever on the gearbox fully rearwards and adjust the position of the trunnion on the rod so that it drops into the throttle valve lever hole.
7. Temporarily fit the trunnion to the throttle valve lever. Release the accelerator pedal then depress it again to full throttle. If the adjustment is correct, the throttle lever should still be in the fully rearward position such that it cannot be moved back any further. If further rearward movement is possible, repeat the adjustment procedure.
8. Release the accelerator and secure the trunnion with the washer and a new split pin.
9. Fit the side cover ensuring that the sealant is evenly distributed round the aperture to prevent ingress of water.
10. Fit the air cleaner.

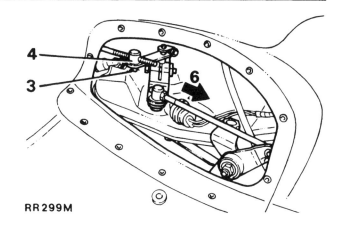

RR299M

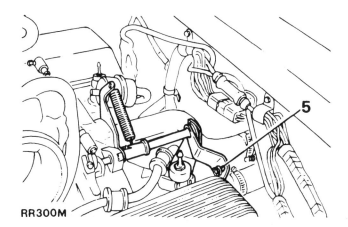

RR300M

GEAR SELECTOR ASSEMBLY

– Remove and refit   44.15.04

Removing

1. Remove the gear box tunnel top cover assembly.
2. Remove the two socket headed nuts and bolts securing the cable clamp to the selector assembly quadrant.
3. Remove the split pin and washer retaining the inner cable to the cross shaft lever.
4. Disconnect at the bullet connection, the feed wire to the illumination bulb.
5. Withdraw the selector assembly.

Refitting

6. Fit the outer cable to the selector assembly quadrant with the clamp and two bolts.
7. Connect the inner cable to the cross shaft lever with washer and split pin.
8. Connect the illumination feed wire.
9. Fit the top cover assembly.

Continued

REVERSE STARTER SWITCH

– Test   **44.15.17**

1. Remove the gearbox tunnel side plate.
2. Disconnect the multi plug from the switch.
3. Obtain a slave battery and test lamp.
4. Connect the negative terminal of the battery to the transmission case (earth).
5. Connect the test lamp circuit to the battery positive terminal and the other end to the switch centre pin.
6. Move the gear selector to 'P' (Park) to check continuity, which is evident if the lamp illuminates.
7. Select 'N' (Neutral) and check continuity.
8. Connect the earth lead from the battery to one of the outer pins on the switch and the test lamp to the other outer pin.
9. Select Reverse to establish continuity.
10. Remove test equipment, connect the multi plug to switch and refit gear box tunnel side plate.

AUTOMATIC GEARBOX AND TRANSFER BOX

– Remove and refit   **44.20.04**

Removing

1. Drive vehicle on to a ramp and disconnect the battery.
2. Remove the ashtrays and carpet covering the transmission tunnel.
3. Remove ten screws and withdraw handbrake gaiter.
4. Remove high/low transfer knob and four screws and withdraw the gaiter.
5. Remove air cleaner.
6. Unscrew the air cleaner L.H. mounting peg and remove the coupling shaft support bracket. Release the throttle valve linkage return spring.
7. Disconnect the throttle valve coupling shaft.
8. Slacken the two outboard nuts retaining the fan cowl and turn the two inboard nuts anti-clockwise to raise the cowling sufficiently to prevent fouling when the engine is lowered.
9. Release the hose from the clip attached to the alternator bracket.

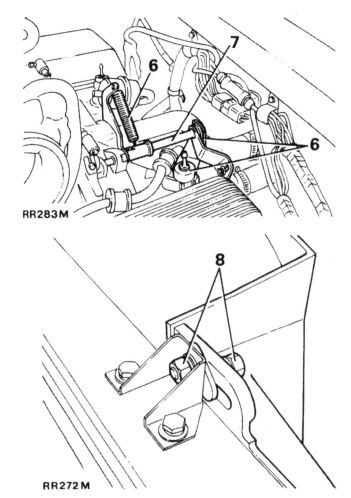

RR283M

RR272M

Continued

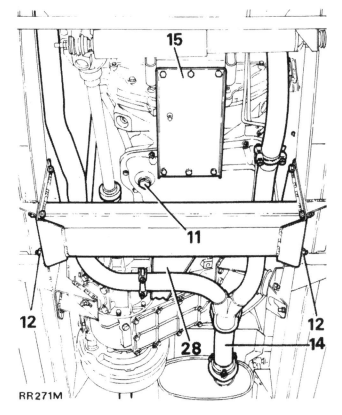

RR271M

10. Raise the vehicle on the ramp.
11. Place a suitable container beneath the gear box sump, remove the drain plug and allow the fluid to drain. Refit the drain plug.
12. Remove the eight nuts and bolts securing the chassis cross member and using a suitable means of parting the chassis side member, remove the cross member.

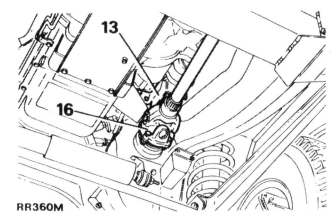

RR360M

13. Remove the starter motor solenoid heat shield.
14. Remove the front exhaust pipe system complete.
15. Remove the tie-plate between engine and gear box.
16. Mark for reassembly and disconnect the front propellor shaft from the transfer box flange.

17. Mark for reassembly and disconnect the rear propellor shaft from the transmission brake drum.
18. Remove the nut and washer securing the tie bar to the bell housing.
19. Remove the two bolts retaining the rear end of the tie bar to the transfer housing.
20. Disconnect the speedometer cable from the transfer box and release the outer cable from its retaining clip.

Continued

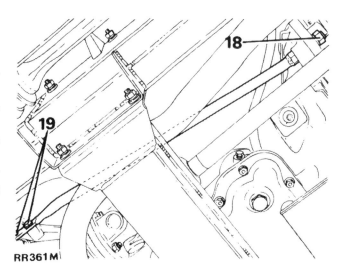

RR361M

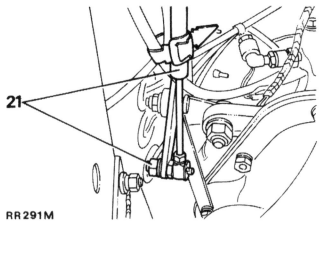

RR291M

21. Remove the nut retaining the gear selector cable from the fulcrum arm and release the outer cable from the securing clamp and move the cable aside.
22. Remove the split pin and disconnect the vertical rod linking the upper and lower throttle valve linkage.
23. Remove the bell housing nut and bolt that retains the throttle valve linkage bracket and move the assembly aside.

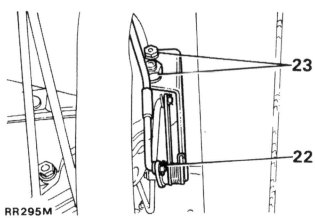

RR295M

24. Disconnect the oil cooler pipes from unions at L.H. side of engine sump and release the steel pipes from the clamps.
25. Disconnect the oil pipes from the gear box unions and remove the complete pipe assembly from the vehicle. Cover all exposed pipe ends to prevent entry of dirt.
26. Withdraw the nine bolts and remove the bell housing cover plate.
27. Working through the bell housing aperture, remove the four bolts securing the torque converter to the drive plate. After removing the last bolt mark the hole in the drive plate and torque converter relative to the drive plate housing and bell housing to facilitate reassembly.
28. Remove the tie-plate between the gear box and transfer box.

Continued

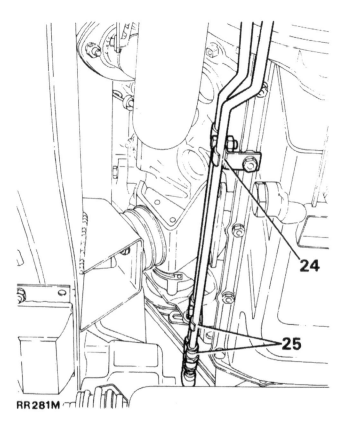

RR281M

29. Manufacture an adaptor plate in accordance with the drawing to attach to a transmission lift, in the interest of safety.
30. Bolt the adaptor plate to the under side of the gear box.
31. Move the transmission lift into position and attach the adaptor plate.
32. Raise the lift to just take the weight of the gear box.
33. Remove the transmission mounting bracket bolts both sides.
34. Lower the engine and transmission sufficiently to enable the following instructions to be carried out.
35. Separate the two snap connectors for the gear selector illumination.
36. Remove the inhibitor switch and reverse light switch multi connector.
37. Release the differential lock harness from the two clips on the transfer box and disconnect the warning light connectors.
38. Remove the transfer box breather tube banjo connection.
39. Remove the gear box breather tube banjo connection.
40. Remove the bell housing bolt securing the gear box filler tube and release the tube from the gear box and cover the hole.

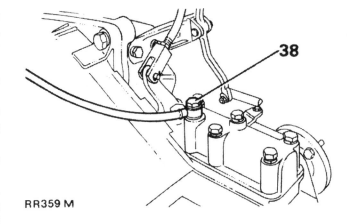

RR359 M

Continued

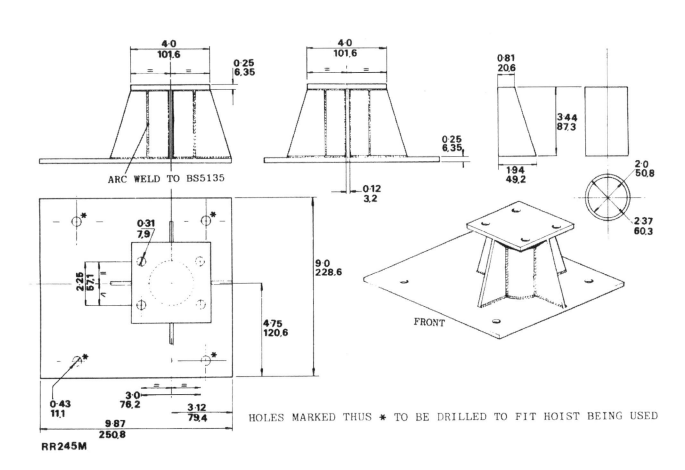

HOLES MARKED THUS * TO BE DRILLED TO FIT HOIST BEING USED

RR245M

595

41. Support the engine with a suitable jack.
42. Remove the three remaining bell housing bolts.
43. Lower the gear box and transfer box from the vehicle, whilst checking that all connections to the chassis and engine are clear.
44. To separate the transfer box from the gear box proceed as follows:—
45. Remove the five bolts retaining the transfer box to the gear box and withdraw the transfer box.
46. Should difficulty be experienced in separating the two nuts at this stage, continue as follows:—
47. Remove the six bolts retaining the cover plate and remove the cover to expose the coupling shaft bolt.
48. Remove the coupling shaft bolt.
49. Withdraw the transfer box from the gear box leaving the extension casing attached to the transfer box.

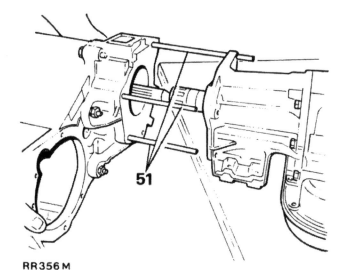

RR358M

Refitting

50. Clean the transfer box and automatic gear box mating faces.
51. To prevent damage to the oil seal and to facilitate the fitting of the transfer box, fit the three guide studs to the transfer gear box extension casing, as illustrated, noting that one stud is longer and must be fitted on the left hand side of the casing. See operation 37.29.25.
52. Place the automatic gear box into PARK and slide the transfer box into position passing the guide studs through the corresponding holes in the automatic gearbox.
53. Secure the transfer box with two bolts, remove the guide studs and fit the remaining bolts and tighten.
54. Fit the adaptor plate to the transmission and mount the assembly on the transmission lift.
55. If the same torque converter is being used line up the marked hole with the mark made on the bell housing and check that the marked hole in the drive plate still coincides with the one made on the drive plate housing.
56. Offer up the transmission to the engine and locate the dowels in the bell housing to the corresponding holes in the drive plate housing.
57. Place in position, with the head towards the rear, the bolt also used to secure the throttle valve linkage lower bracket.
58. Fit the gear box fluid filler tube.
59. Fit and tighten the remaining bell housing bolts.
60. Remove the support from the engine.
61. Connect the multi plug connector to the combined inhibitor and reverse light switch.
62. Connect the two differential lock warning light leads to the transfer box switch and secure with cable clips.
63. Fit the banjo breather connection for the gear box.
64. Fit the banjo breather connection for the transfer box. Tighten both banjo bolts to 7—11 Nm (5—8 lb.ft).

RR356M

Continued

65. Connect the gear selector illumination leads.

66. Raise the transmission to enable the mounting bracket bolts to be fitted. Note that the speedometer cable clip is attached to the top L.H. forward bolt.

67. Remove the bolts securing the adaptor to the transmission and remove the lift.

68. Fit the four bolts securing the torque convertor to the drive plate and tighten to 30.5 Nm (22.5 lb.ft).

69. Fit the transmission fluid cooler pipes to the flexible connections at the front of the engine and to the unions on the gearbox. Secure the pipes with the clamps to the engine.

70. Fit the throttle valve linkage lower bracket to the bell housing bolt and tighten the nut, see torque settings.

71. Using a new split pin connect the throttle valve vertical control rod to the lower linkage.

72. Secure the gear selector outer cable with the clip to the L.H. rear mounting brackets.

73. Connect gear selector inner cable to the fulcrum lever.

74. Fit the speedometer cable to the transfer box.

75. Fit the tie-rod to the bell housing but do not tighten the nut.

76. Secure the rear end of the tie-rod to the transfer box with the two bolts.

77. Tighten the nut at the bell housing end of the tie rod.

78. Line up the marks and connect the front propellor shaft to the transfer box flange.

79. Connect the rear propellor shaft to the transmission brake drum flange.

80. Fit the steady plate between the gear box and transfer box.

81. Fit the steady plate between the engine and gear box.

82. Fit the front exhaust pipe system to the engine manifold flanges and front silencer.

83. Fit the starter motor solenoid heat shield.

84. Using a suitable means to expand the chassis, fit the cross member and secure with the eight nuts and bolts.

85. Lower the ramp.

86. Fit the throttle valve coupling shaft support bracket to the shaft.

87. Connect the coupling shaft to the outer shaft.

88. Secure the assembly with the air cleaner L.H. mounting peg.

89. Connect the return spring to the coupling shaft lever.

90. Fit the air cleaner.

91. Reposition the fan cowl, reversing the procedure in instruction 8.

92. Secure the water hose to clip on alternator bracket.

93. Fit the handbrake gaiter.

94. Fit the high/low transfer gaiter and knob.

95. Fit the ash trays and the carpet to the transmission tunnel.

96. Refill the automatic gearbox with new clean fluid of the recommended type and specification as described in operation 44.24.02.

AUTOMATIC GEARBOX FLUID

– Drain and refill   44.24.02

Drain

1. Raise vehicle and place a suitable container beneath the gearbox sump.
2. Remove the drain plug and allow time for the fluid to completely drain.
3. Refit drain plug and tighten to 25 – 35 Nm (19 – 26 lb.ft).

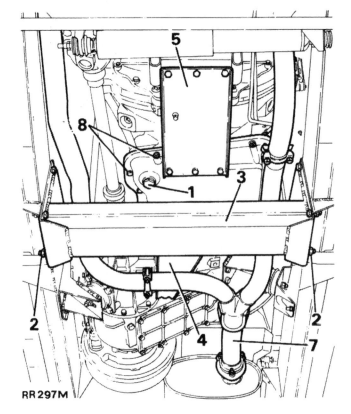

RR 297M

Refill

4. Remove the dipstick and fit a filling funnel.
5. Fill with 5 litres (9 pints) of transmission fluid of the correct make and grade, see data section.
6. Remove the funnel and fit the dipstick.
7. Select 'N' and start the engine and run for two minutes at 600 r.p.m.
8. Apply the handbrake and move the transmission lever to each gear in turn, including 'P' (Park), pausing momentarily in each position and finishing in 'N'.
9. With the engine still running remove the dipstick and clean. Replace the dipstick and check the level.
10. Using the funnel, carefully add more fluid while continually checking the level as in instruction 9 until it reaches the "add one more pint" mark. DO NOT OVERFILL.
11. Drive the vehicle on the road and when the normal operating temperature of the transmission is reached recheck the fluid level and top up if necessary.

AUTOMATIC GEARBOX FLUID SUMP

– Remove and refit   44.24.04

Removing

1. Raise the vehicle on a ramp or axle stands and drain the gearbox fluid, and refit the plug.
2. Remove the eight bolts securing the chassis cross-member.
3. Using a suitable spreader, prise apart the chassis side members sufficiently to enable the crossmember to be removed.
4. Remove the steady plate between the gearbox and transfer box.
5. Remove the steady plate between the engine and gearbox.
6. Disconnect the fluid temperature sensor lead.
7. Release the front exhaust pipe from the two manifold connections and from the flange connection forward of the front silencer. Also remove the 'U' bolt and disconnect the system from the exhaust hanger bracket adjacent to the gearbox L.H. mounting. Allow the system to drop sufficiently to permit access to the sump bolts.
8. Remove the twelve bolts and remove the sump. Take precautions against the entry of foreign matter into the gearbox whilst the sump is removed.

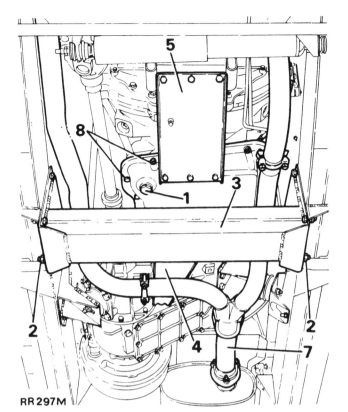

RR297M

Refitting

9. Ensure that the sump is clean and that the gearbox and sump mating faces are free from old joint washer material.
10. Secure the sump to gearbox, evenly tightening the bolts, noting that the two hexagon headed bolts are fitted on the L.H. side of the sump. Torque $20 - 28$ Nm ($15 - 20$ lb.ft).
11. Fit the exhaust pipe assembly.
12. Fit the front and rear stiffener plates.
13. Fit the electrical connection to the fluid temperature sensor.
14. Spread the chassis members apart to enable the crossmember to be fitted. Note that the hole in the centre of the crossmember should be to the rear and welded seam to the front. Secure with the four nuts and bolts.
15. Refit the transmission with new clean fluid of the recommended type and specification as described in operation 44.24.02.

GEARBOX SELECTOR ROD AND CABLE

– Adjust   44.30.04

Selector rod

1. Remove the transfer lever knob and gaiter.
2. Move the gear selector to 'N' (Neutral) and maintain it in this position throughout the following instructions.
3. Remove the screws retaining the gearbox tunnel cover and lift the cover from the front sufficiently to expose the cable attachment to the gear selector quadrant.
4. Remove the spring clip retaining the cable trunnion to the quadrant and remove the trunnion from the quadrant.
5. Remove the gearbox tunnel side plate.
6. Ensure that the gear selector lever on the side of the gearbox has not moved from its previously set neutral position.
7. Check that the dimension 'A' between the centre of the trunnion and the centre of the gear selector rod is 159 to 161 mm.

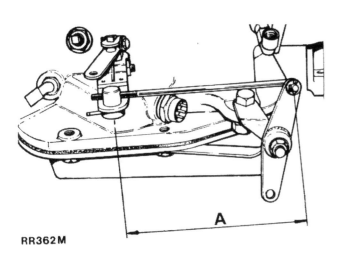

RR362M

8. If adjustment is required remove the split pin and washer from the selector rod and remove the rod from the arm. Adjust the rod in the trunnion until the correct dimension is achieved.
9. Secure the selector rod to the arm with the washer and a new split pin.

Selector cable

10. Check that the gear selector is still in 'N' (Neutral).
11. If necessary adjust the position of the trunnion along the thread of the cable until the trunnion fits cleanly into the corresponding hole in the gear selector quadrant.
12. Fit the trunnion retaining spring clip and temporarily fit the gearbox top cover with a few screws.

13. Move the gear selector lever through the full range of gear positions whilst checking that the cable attachment to the link arm does not foul the support bracket.
14. Apply the handbrake and switch on the ignition and check the engine can be started in both 'N' (Neutral) and 'P' (Park) positions.
15. If the engine cannot be started in either or both positions repeat the selector rod and cable adjustment instructions and set the rod dimension to another figure within the tolerance.
16. Finally fit the gearbox top cover.
17. Fit the transfer lever and gaiter knob.
18. Fit the gearbox side cover.

FRONT BRAKE BAND

– Adjust   44.30.07

1. Remove the gearbox tunnel side plate.
2. Slacken the brake band adjuster locknut.
3. Using a square socket tighten the brake band adjuster clockwise to 13 Kg/cm (72 lb.in) torque and back off the adjuster two and a half turns.
4. Tighten the locknut.
5. Fit the tunnel side plate.

REAR BRAKE BAND

– Adjust   44.30.10

1. Raise the vehicle on a ramp.
2. Remove the automatic gearbox sump, see operation 44.24.04.
3. Slacken the band adjuster and tighten the adjuster to 13 Kg/cm (72 lb.in) and back off two turns and tighten the locknut.
4. Using a new joint washer fit the gearbox sump and evenly tighten the retaining bolts to 21 – 28 Nm (15 – 20 lb.ft).
5. Fit the engine to gearbox steady plate.
6. Fit the automatic gearbox to transfer box steady plate.
7. Fit the front exhaust pipe system.
8. Fit the chassis crossmember.
9. Fill the transmission with new fluid of the correct make and grade as described in operation 44.24.04.

HYDRAULIC PRESSURES

— Test   44.30.15

Special equipment:
  Tachometer
  Pressure gauge and adaptors plus long hose CBWIC
  (18G502) 0 to 300 psi (0 − 21 kgf/cm²)
  Gauge and adaptors plus long hose (18G1427)
  0 − 100 psi (0 − 7 kgf/cm²)
  Adaptor hose (18G677-2)
  Hose (long) (18G502K)

WARNING: − The 0 to 100 psi (0 − 7 kgf/cm²) pressure gauge must only be used for the governor test. It is dangerous to use it for other tests or at speeds in excess of 70 m.p.h. (110 km/h).

Pressure testing is an important part of determining the condition of the gearbox and diagnosing of faults. Before commencing the following tests, check the condition and fluid level in the gearbox. If the fluid condition is unknown or suspect, take a sample for examination. If necessary drain and refill with new fluid of the correct make and grade described in operation 44.24.02.

All the following tests must be carried out with the fluid at normal operating temperature of 80 degrees C (175 degrees F).

1. Drive vehicle on to a ramp.
2. Place automatic gearbox and transfer box in 'N' (Neutral).
3. Block front and rear wheels for safety.
4. Remove the transmission tunnel side plate.
5. Remove the split pin and washer from the throttle valve trunnion and disconnect the trunnion from the lever.
6. Connect the tachometer to the engine.
7. Raise the ramp.

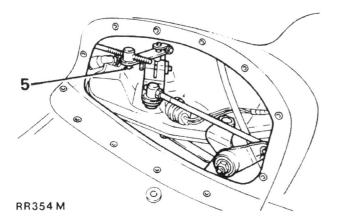

5

RR354 M

Line pressure

Test 1

8. Fit the 0 − 300 psi (0 − 21 kgf/cm²) gauge to the line pressure port and fasten the hose clear of exhaust.
9. Start the engine and run at 1000 r.p.m.
10. Position the throttle valve lever on the side of the gearbox, fully forward.
11. Move the gear lever to '1' and note the pressure which should be 54−60 psi (3,73 − 4,22 kgf/cm²)
12. Move the throttle valve lever, by hand, to fully rearward position and the pressure should gradually increase to 90−96 psi (6,33 − 6,75 kgf/cm²)
13. Stop engine, select 'N' and move throttle lever to forward position.

Test 2

14. Start the engine and run at 1000 r.p.m.
15. Select '2'.
16. With the throttle lever fully forward the pressure should be 54−60 psi (3,73 − 4,22 kgf/cm²)
17. Move the throttle lever fully rearward and a gradual increase in pressure to 90−96 psi (6,33 − 6,75 kgf/cm²) should be observed.
18. Stop engine, select 'N', and move throttle lever fully forward.

Test 3

19. Start engine and run at 1000 r.p.m.
20. Move gear lever to 'D'.
21. The pressure should be 54−60 psi (3,73−4,22 kgf/cm²) with the throttle lever fully forward.
22. Move the throttle lever fully rearward and the pressure should gradually increase to 90-96 psi (6,33-6,75 kgf/cm²)
23. Select 'N', stop engine and move throttle lever forward.

Front servo

24. Remove the gauge connection from the line pressure port to the front servo port.

Test 4

25. Start engine and run at 1000 r.p.m.
26. Select 'D'.
27. With the throttle lever forward the pressure should read 54−60 psi (3,73 − 4,22 kgf/cm²)
28. Move throttle lever fully rearward and up to the downshift point the pressure should gradually increase to 90−96 psi (6,33 − 6,75 kgf/cm²)
29. Select 'N' move throttle lever forward and stop engine.

Continued

Rear servo

30. Remove the steady plate between automatic gearbox and transfer box.
31. Remove the gauge connection from the front servo port.
32. Fit the gauge connection to the rear servo port and fasten clear of the exhaust system.

Test 5

33. With 'N' selected start the engine and run at 1600 r.p.m.
34. Select 'R'.
35. Move the throttle lever fully rearward and the pressure should gradually increase to 235–270 psi (16,52-18,98 kgf/cm²)
36. Select 'D' and with the engine at 1600 r.p.m. the pressure should remain at zero with the throttle lever forward and rearward.
37. Select 'N' stop engine and move throttle forward.

Test 6

38. With 'N' selected start the engine and run at 1000 r.p.m.
39. Select '1'.
40. With the throttle lever fully forward the pressure should read 54–60 psi. (3,73 – 4,22 kgf/cm²)
41. Move the throttle lever fully rearward and the pressure should increase to 90–96 psi. (6,33 – 6,75 kgf/cm²).

NOTE:– The figures obtained in test 6 should be the same as those in test 1 within 3 psi (0,21 kgf/cm²)

42. Select 'N' and stop engine.

Governor pressure

43. Disconnect tachometer.
44. Fit trunnion to throttle valve lever and secure with the washer and new split pin.
45. Fit the tunnel side plate.
46. Disconnect the gauge connection from the rear servo port.
47. Connect a 0 to 100 psi (0 – 7 kgf/cm²) pressure gauge to the governor port and fasten the hose clear of the exhaust system.
48. Refit the steady plate.
49. Position the pressure gauge in the driving compartment so that it can be read whilst driving.
50. Remove blocks from wheels and drive vehicle from ramp.

Test 7

51. Select high range in the transfer box.
52. Check the pressure figures against the following chart.

Governor pressure related to speeds

km/h	bar
15	0,1
25	0,7
30	1,4
40	1,7
50	2
55	2,4
65	2,75
72	3,1
80	3,5
88	3,7
100	3,8
105	3,9
110	3,9

53. Remove the pressure gauge connector from the governor port.

THROTTLE PRESSURE

– Adjust   44.30.25

Special tools: Throttle pressure gauge

1. Drive the vehicle on to a ramp.
2. Remove the gearbox tunnel side plate.
3. Remove the split pin and washer from the throttle valve trunnion and remove it from the lever.

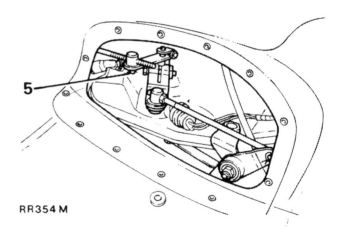

RR354M

continued

4. Raise the vehicle and drain the transmission fluid and refit the drain plug tightening to 25–35 Nm (19–26 lb.ft).
5. Remove the automatic gearbox sump, see operation 44.24.04.
6. Remove the nuts and withdraw the filter and spacer.
7. Insert the throttle pressure gauge tool between the throttle lever cam and kickdown valve and move the tool sideways (in direction of arrow) towards the centre of the valve block thus compressing the kickdown valve rod to its fullest extent.
8. Whilst compressing the valve turn the throttle lever stop screw, using an allen key, until the underside of the screw head touches the tag on the throttle lever and the cam is hard against the gauge tool with zero clearance. Remove the gauge.

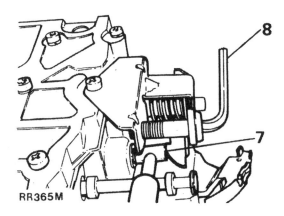

RR365M

9. Fit the spacer and fluid filter using new Nyloc nuts and tighten evenly to 3.5 to 4.5 Nm (2.5 to 3.2 lb.ft).
10. Using a new joint washer fit the gearbox sump and evenly tighten the retaining bolts to 21–28 Nm (15–20 lb.ft).
11. Fit the engine to gearbox steady plate.
12. Fit the automatic gearbox to transfer box steady plate.
13. Fit the front exhaust pipe system.
14. Fit the chassis crossmember.
15. Fill the transmission with new fluid of the correct make and grade as described in operation 44.24.04.

LINE PRESSURE

– Adjust   44.30.26

NOTE:– An incorrect throttle pressure setting and throttle valve linkage adjustment will cause incorrect line pressure readings. Always check these adjustments before adjusting the pressure.

1. Drive the vehicle on to a ramp.
2. Raise the vehicle and drain the transmission fluid and refit the drain plug.
3. Remove the automatic gearbox sump, see operation 44.24.04.
4. Remove three nuts and withdraw the filter and spacer.
5. To adjust the line pressure, turn the line pressure regulation valve adjusting screw, with an allen key, clockwise to decrease the pressure and anti-clockwise to increase the pressure. One complete turn of the screw alters the closed throttle line pressure approximately 1.2/3 psi (0,075 kgf/cm^2). Therefore the direction and amount of turn of the screw depends on the pressure gauge reading.
6. As a guide, the approximate adjustment of the regulation valve is 25.5 mm (1.5/16 in) measured from the valve body to the inner edge of the adjusting plate, as illustrated.

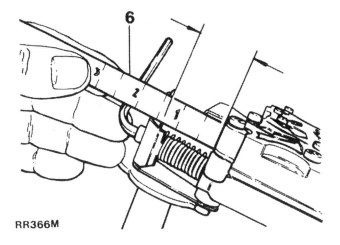

RR366M

7. Fit the spacer and fluid filter using new Nyloc nuts and tighten evenly to 3,5 to 4,5 Nm (2.5 to 3.2 lb.ft).
8. Using a new joint washer fit the gearbox sump and evenly tighten the retaining bolts to 21–28 Nm (15–20 lb.ft).
9. Fit the engine to gearbox steady plate.
10. Fit the automatic gearbox to transfer box steady plate.
11. Fit the front exhaust pipe system.
12. Fit the chassis crossmember.
13. Fill the transmission with new fluid of the correct make and grade as described in operation 44.24.04.

GEARBOX TUNNEL SIDE COVER

– Remove and refit   76.10.93

Removing

1. Fold back L.H. front footwell rubber mat.
2. Fold back gearbox tunnel carpet to expose plate.
3. Remove the twelve cover retaining bolts and remove the cover.

Refitting

4. Scrape off sealant from tunnel and cover plate and re-disperse it round the cover plate.
5. Fit the cover plate and evenly tighten the retaining bolts.
6. Fit the tunnel carpet and footwell mats.

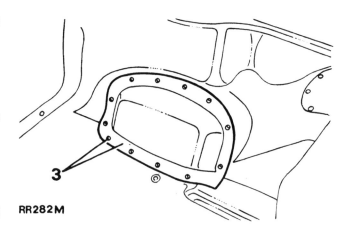

RR282M

GEARBOX TUNNEL TOP COVER

– Remove and refit   76.25.07

Removing

1. Unscrew the high/low transfer differential lock knob.
2. Lift out the ash trays.
3. Remove the four screws retaining the ash tray frame.
4. Remove the three screws beneath the cubby box carpet and remove the cubby box.
5. Lift out the carpet and sound deadening material.
6. Remove the four screws retaining the gear selector bezel and remove plate and trim.
7. Remove the four bolts securing the gear selector assembly to the top cover.
8. Push the selector assembly downwards to release it from the top cover.
9. Remove the fixing bolts retaining the top cover.

Refitting

10. Scrape off the sealant from the tunnel and cover and re-disperse it round the tunnel aperture.
11. Fit and secure the cover with the fixing bolts.
12. Secure the selector assembly to the cover with the four bolts.
13. Fit the plate and trim and secure the selector bezel with the four screws.
14. Fit the high/low differential lock knob.
15. Fit the tunnel carpet and sound deadening material.

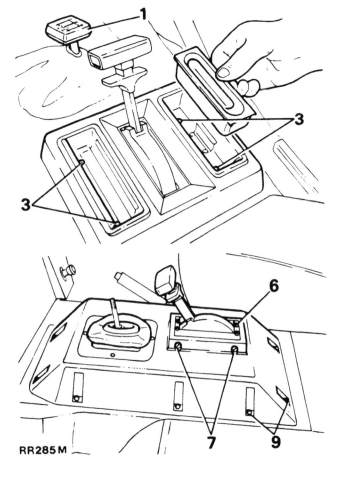

RR285M

AUTOMATIC GEARBOX

SPECIAL TOOLS

AUTOMATIC GEARBOX AND TRANSFER GEARBOX ASSEMBLY – REMOVE AND REFIT	Mounting plate for gearbox hoist Dimensions in workshop manual
TORQUE CONVERTOR	Retaining strap for torque convertor Dimensions in workshop manual
INPUT SHAFT END FLOAT – CHECK	Use torque convertor retaining strap (as above) and mounting bracket RO530106 plus suitable dial gauge
PUMP HOUSING AND REACTION SHAFT SUPPORT – REMOVAL	Use impulse hammers (2 off) 18G1387
PUMP HOUSING AND REACTION SHAFT SUPPORT – REFIT	Use guide studs (2 off) 18G1384
FRONT CLUTCH SPRINGS REMOVE/REFIT	Use spring compressor 18G1386
THROTTLE PRESSURE SETTING	Use gauge block 18G1385
PUMP HOUSING OIL SEAL REFITTING	Use drift 18G134-3 plus 18G134 (MS 550)
REAR EXTENSION HOUSING OIL SEAL – REFITTING	Use drift 18G1421
BAND ADJUSTMENT – FRONT AND REAR	Use socket CBW 547 A50-2A plus 3/8 x 6 in. extension bar
GOVERNOR PRESSURE TEST EQUIPMENT 0–100 psi GAUGE (0 – 7 kgf/cm²)	18G1427
LINE PRESSURE TEST EQUIPMENT 0–300 psi GAUGE (0 – 21 kgf/cm²)	18G502A
HOSE AND ADAPTORS	18G502K
TACHOMETER	Any suitable type of Tachometer

This page intentionally left blank

LT230T TRANSFER GEARBOX

FOR

FIVE SPEED MANUAL GEARBOX

AND

AUTOMATIC GEARBOX

NOTE: TWO TYPES OF TRANFER GEARBOX ARE IN USE.
SEE PAGE 537 OF THIS SUPPEMENT FOR AN EXPLANATION.

TORQUE WRENCH SETTINGS

DESCRIPTION	TYPE OF FIXING AND SIZE (METRIC)	SPECIFIED TORQUE		QUANTITY
LT230T Transfer Gearbox				
Fixings securing mounting brackets to gearbox		80-100	59-73	
Pinch bolt operating arm	M6 x 25.0 Bolt	7-10	5-7	1
Gate plate to grommet plate	M6 x 20.0 Screw	7-10	5-7	4
Bearing housing to transfer case	M6 x 20 C/sunk head screw	7-10	5-7	2
Speedometer cable retainer	M6 nut	7-10	5-7	1
Speedometer housing	M6 x 30.0 Stud	■ See note		1
Locating plate to gear change	M5 Nut Nyloc	5-7	4-5	2
Bottom cover to transfer	M8 x 30.0 bolt	22-28	16-21	10
Front output housing to transfer	M8 x 30.0 bolt	22-28	16-21	7
Front output housing to transfer	M8 x 90.0 bolt	22-28	16-21	1
Cross shaft housing to front output housing	M8 x 55.0 bolt	22-28	16-21	6
Gear change	M8 x 55.0 bolt	22-28	16-21	1
Gear change	M8 x 25.0 screw	22-28	16-21	3
Cross shaft to high/low lever	M8 nut	22-28	16-21	1
Pivot shaft to link arm	M8 nut	22-28	16-21	1
Connecting rod	M8 nut (locknut)	22-28	16-21	2
Anti-rotation plate intermediate shaft	M8 x 20.0 screw	22-28	16-21	1
Front output housing cover	M8 x 25.0 screw	22-28	16-21	7
Pivot bracket to extension housing	M8 x 25.0 screw	22-28	16-21	2
Finger housing to front output housing	M8 x 25.0	22-28	16-21	3
Mainshaft bearing housing to transfer case	M8 x 25.0 C/sunk head screw	22-28	16-21	2
Brake drum to coupling flange	M8 x 20.0 C/sunk head screw	22-28	16-21	2
Gearbox to transfer case	M10 x 40.0 bolt	40-50	29-37	3
Gearbox to transfer case	M10 x 45.0 bolt	40-50	29-37	1
End cover bearing housing to transfer case	M10 x 35.0 bolt	40-50	29-37	6
Speedometer housing to transfer	M10 x 30.0 screw	40-50	29-37	5
Speedometer housing to transfer	M10 x 45.0 bolt	40-50	29-37	1
Selector finger to cross shaft (high/low)	M10 grub screw	22-28	16-21	1
Selector fork high/low to shaft	M10 grub screw	22-28	16-21	1
Transmission brake to speedometer housing	M10 x 25.0 bolt	65-80	48-59	4
Gearbox to transfer case	M10 nut	40-50	29-37	2
Transfer case assembly	M10 stud	■ See note		2
Oil drain plug	M12 x 14−1.5 Pitch hex head A/F 17.0	25-35	19-26	1
Detent plug	M12 x 10.0 plug	Plus to be coated with Hylomar and peened. Screw plug fully in (spring solid) then turn two complete turns back		2
Differential casings	M10 x 60.0 bolt	55-64	40-47	8
Front and rear out flange	Nyloc nut M20	146-179	108-132	2
Differential case rear	M50 nut − 1.5 pitch	66-80	50-59	1
Oil filler and level plug transfer	Plug-R ¾in taper thread	25-35	19-26	1
Transfer breather	1/8 B.S.P. bolt	7-11	5-8	
Inter shaft stake nut	M20 nut A/F 30.0	130-140	96-104	1

■ Studs to be assembled into casing with sufficient torque to wind them fully home, but this torque must not exceed the maximum figure quoted for the associated nut on final assembly.

RECOMMENDED LUBRICANTS AND FLUIDS

Service instructions for temperate climates – ambient temperature range –10°C to 35°C

COMPONENTS	BP	CASTROL	DUCKHAMS	ESSO	MOBIL	PETROFINA	SHELL	TEXACO
LT230/T transfer box (used with five-speed gearbox)	BP Autran DX2D or BP Visco 2000 15W/40 or BP Visco Nova 10W/30 or BP Vanellus C3 Multigrade 15W/40 or BP Gear Oil SAE 90EP	Castrol TQ Dexron IID or Castrol GTX 15W/50 or Castrol Hypoy SAE 90EP	Duckhams Fleetmatic CD or Duckhams D-Matic or Duckhams 15W/50 Hypergrade Motor Oil or Duckhams Hypoid 90	Esso ATF Dexron IID or Esso Superlube 15W/40 or Esso Gear Oil GX 85W/90	Mobil ATF220 D or Mobil Super 15W/40 or Mobil Mobilube HD 90	Fina Dexron IID or Fina Supergrade Motor Oil 15W/40 or 20W/50 Fina Pontonic MP SAE 80W/90	Shell ATF Dexron IID or Shell Super Motor Oil 15W/40 or Shell Spirax 90 EP	Texamatic Fluid 9226 or Havoline Motor Oil 15W/40 or Texaco Multigear Lubricant EP 85W/90

Capacities	Litres	Imperial Unit
LT230T Transfer gearbox	2,80	4.9 pints

Check/top up oil level (LT230T Transfer gearbox)

Check oil level daily or weekly when operating under severe wading conditions.

1. To check oil level: remove the oil level plug, located on the rear of the transfer box casing; oil should be level with the bottom of the hole.

2. If necessary, top up through the oil level plug hole using a pump type oil can. If significant topping up is required, check for oil leaks at drain and filler plugs.

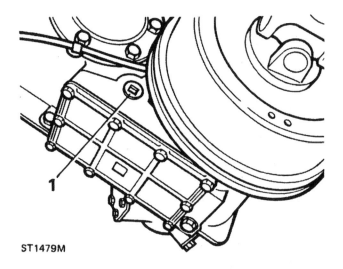

ST1479M

Renew oil (LT230T Transfer gearbox)

Drain and refill monthly when operating under severe wading conditions.

1. Immediately after a run when the oil is warm, drain off the oil into a container by removing the drain plug and washer from the bottom of the transfer box.

2. Replace the drain plug and washer and refill the transfer box through the oil level plug hole with the correct grade of oil, to the bottom of the oil level plug hole. For capacity see Data Section.

IMPORTANT: Do not overfill, otherwise leakage may occur.

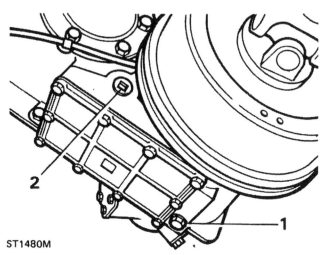

ST1480M

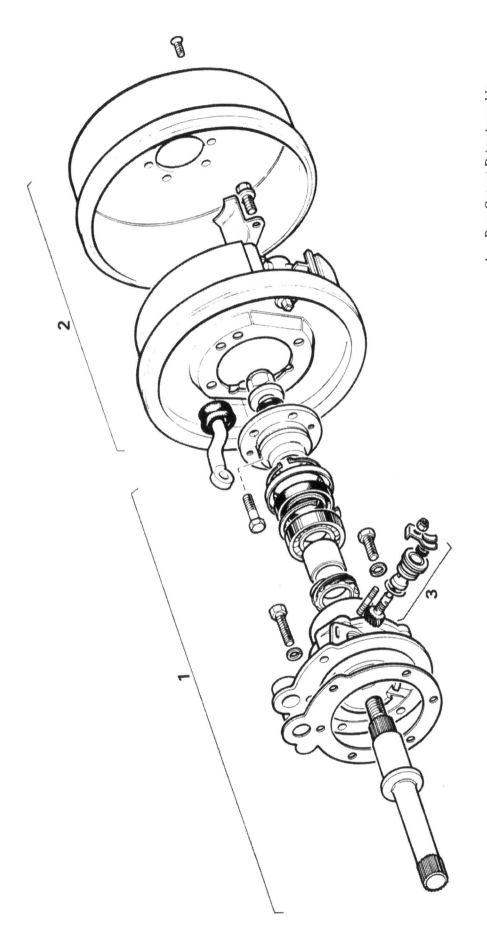

1. Rear Output Drive Assembly
2. Transmission Brake Drum Assembly
3. Speedo Housing Assembly

ST1650M

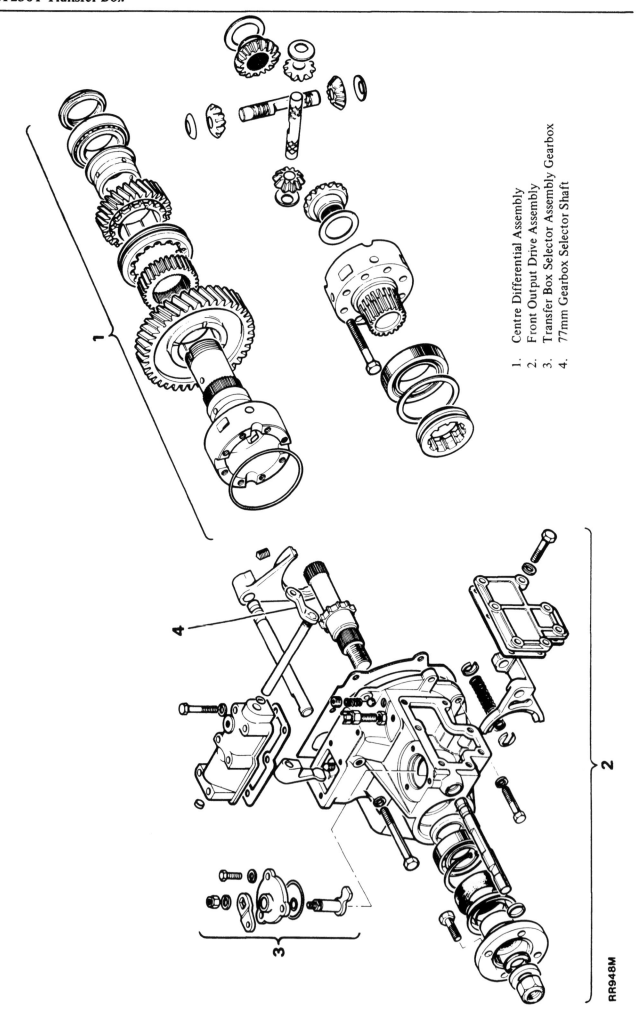

1. Centre Differential Assembly
2. Front Output Drive Assembly
3. Transfer Box Selector Assembly Gearbox
4. 77mm Gearbox Selector Shaft

RR948M

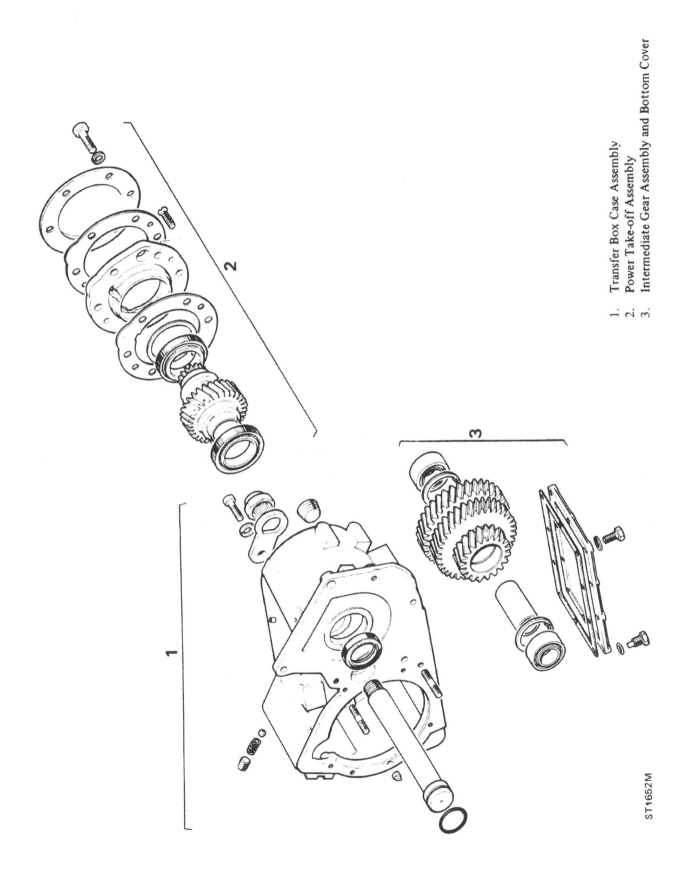

1. Transfer Box Case Assembly.
2. Power Take-off Assembly.
3. Intermediate Gear Assembly and Bottom Cover

ST1652M

REMOVE LT230T TRANSFER GEARBOX

Special tool: 18G 1425 – Guide studs (3)
Also, locally manufactured adaptor plate, see below.

Adaptor for removing transfer gearbox
The transfer gearbox should be removed from underneath the vehicle, using a hydraulic hoist. An adaptor plate for locating the transfer gearbox on to the hoist can be manufactured locally to the drawing below. If a similar adaptor plate was made for the LT230R transfer gearbox, it can be modified to suit both the LT230R and LT230T gearboxes by making the modifications shown by the large arrows.

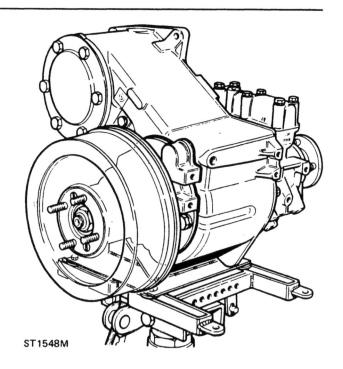

ST1548M

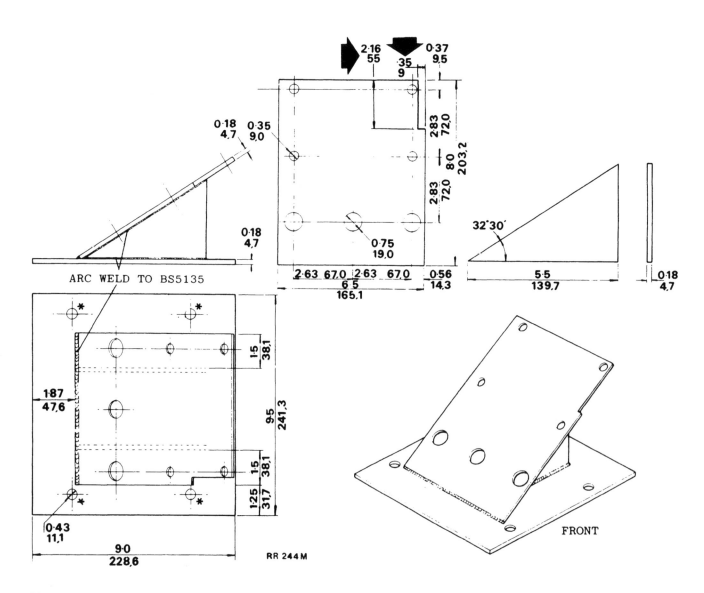

ARC WELD TO BS5135

RR 244 M

Material. Steel plate BS 1449 Grade 4 or 14
Holes marked thus * to be drilled to fit hoist being used.

◆ = MODIFICATION

LT230T TRANSFER GEARBOX OVERHAUL

Service Tools:

18G47-7 – Input gear cluster bearing cones remover/replacer
18G47BB-1 – Adaptor centre differential bearing remover
18G47BB-3 – Adaptor centre differential bearing remover button
18G257 – Circlip pliers
18G1205 – Prop flange wrench
18G1271 – Oil seal remover

18G1422 – Mainshaft rear oil seal replacer
18G1423 – Adaptor/socket centre differential stake nut remover/replacer
18G1424 – Centre differential bearing replacer
MS47 – Hand press
MS550 – Bearing and oil seal replacer handle
LST47-1 – Adaptor centre differential bearing remover
LST104 – Intermediate gear dummy shaft
LST105 – Input gear mandrel
LST550-4 – Intermediate gear bearing replacer

TRANSFER BOX DATA

Front bevel gear end-float	0,025 to 0,075mm (0.001 to 0.003in)
Rear bevel gear end-float	0,025 to 0,075mm (0.001 to 0.003in)
Rear output housing clearance	1,00mm (0.039in)
High range gear end-float	0,05 to 0,15mm (0.002 to 0.006in)
Front differential bearing pre-load	1,36 to 4,53 kg (3 to 10 lbs)
Input gear bearing pre-load	2,26 to 6,80 kg (5 to 15 lbs)
Intermediate shaft bearing pre-load	1,81 to 4,53 kg (4 to 10 lbs)

Transmission brake removal

1. Remove two countersunk screws and withdraw brake drum.
2. Remove four bolts securing the brake back-plate; the two bottom fixings retain the oil catcher.

   NOTE: An hexagonal type socket should be used for these bolts.

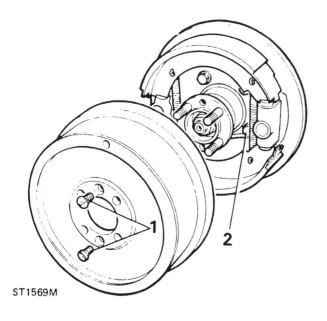

ST1569M

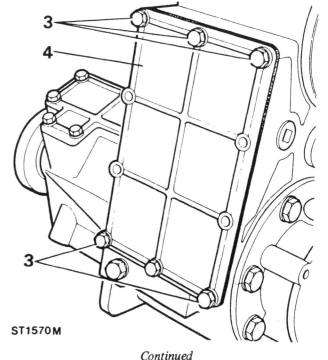

ST1570M

Bottom cover removal

3. Remove the six bolts and washers retaining the bottom cover.
4. Remove the gasket and bottom cover.

Continued

615

Intermediate shaft and gear cluster removal

5. Release stake nut from recess in intermediate shaft and remove stake nut and discard.
6. Unscrew the single bolt and remove anti-rotation plate at the rear face of the transfer box.

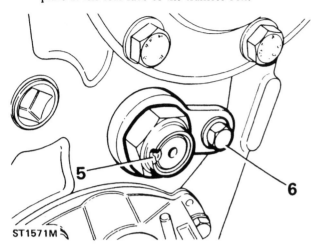

ST1571M

7. Tap the intermediate gear shaft from the transfer box.

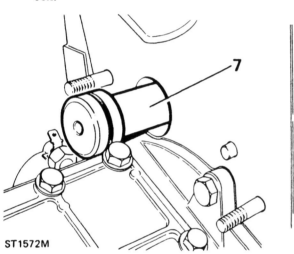

ST1572M

8. Lift out the intermediate gear cluster and bearing assembly.
9. Remove the 'O' rings from the intermediate gear shaft and from inside the transfer box.

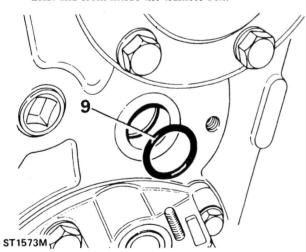

ST1573M

10. Remove taper roller bearings and bearing spacer from the intermediate gear cluster assembly.

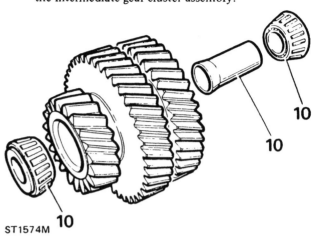

ST1574M

Power take-off cover removal

11. Remove six bolts and washers retaining the take-off cover and speedo cable clips.
12. Remove the gasket and cover.

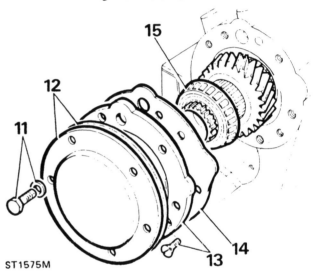

ST1575M

Input gear removal

13. Remove the two countersunk screws and detach the main shaft bearing housing.
14. Remove the gasket.
15. Withdraw the input gear assembly.
16. Prise out and discard the oil seal at the front of the transfer box casing using service tool 18G1271.
17. Drift out the input gear front bearing track.

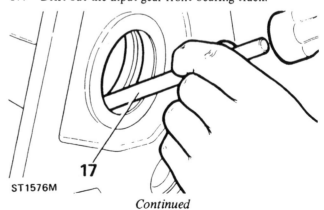

ST1576M

Continued

616

High/low cross-shaft housing removal

18. Remove the six bolts and washers retaining the cross-shaft housing and earth lead.
19. Remove the gasket and cross-shaft housing.

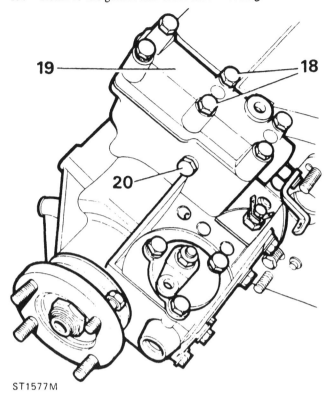

ST1577M

Front output housing removal

20. Remove the eight bolts and washers and detach the output housing from the transfer box casing, taking care not to mislay the dowel.

Centre differential removal

21. Remove high/low selector shaft detent plug, spring and retrieve the ball with a suitable magnet.

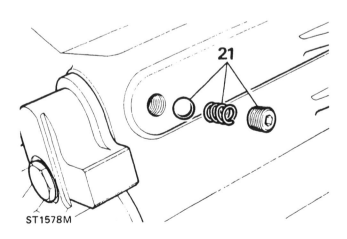

ST1578M

22. Withdraw the centre differential and selector shaft/fork assembly.

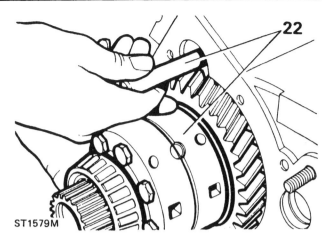

ST1579M

Rear output housing removal

23. Remove six bolts and washers and detach the rear output housing and shaft assembly from the transfer casing.
24. Remove the gasket.

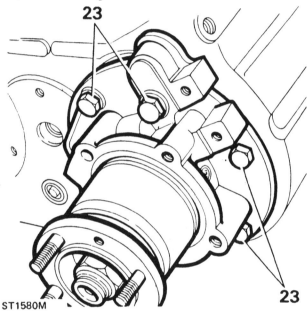

ST1580M

Transfer case overhaul – dismantling

25. Remove the studs and dowels.
26. Remove the magnetic drain plug and filler/level plug.

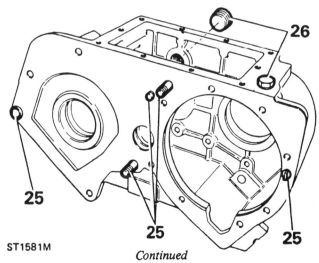

ST1581M

Continued

617

27. Drift out differential rear bearing track.
28. Clean all areas of the transfer casing ensuring all traces of "Loctite" are removed from faces and threads.

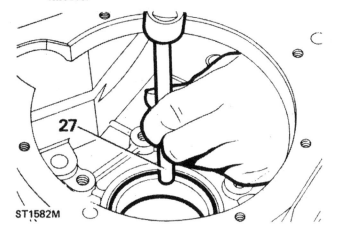

ST1582M

Transfer case overhaul – re-assembling

29. Fit studs and dowels to front face of the transfer casing.

    NOTE: The position of the radial dowel blade is set in line with the circle which is formed by the front output housing fixing holes.

30. Refit magnetic drain plug with new copper washer and tighten to the specified torque, loosely fit the filler/level plug.

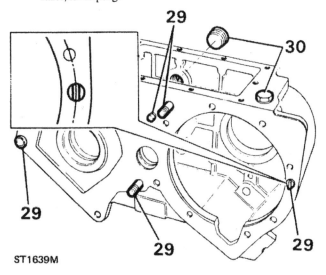

ST1639M

Rear output housing overhaul – dismantling

1. Using flange wrench 18G1205 and socket spanner, remove the flange nut, steel and felt washers. Ensure flange bolts are fully engaged in the wrench.
2. Remove output flange with circlips attached. If necessary, use a two-legged puller.

    NOTE: The circlip need only be released if the flange bolts are to be renewed.

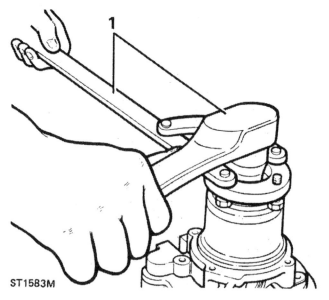

ST1583M

3. Remove speedo-drive housing. This can be prised out with a screwdriver.
4. Remove housing from the vice and drift out the output shaft, by striking the flange end of the shaft.
5. Carefully prise off the oil catch ring using a screwdriver in the slot provided.

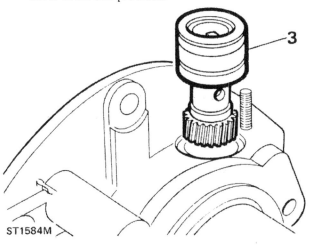

ST1584M

6. Prise out and discard the seal from the output housing using tool 18G1271.
7. Using circlip pliers 18G257, remove the circlip retaining the bearing.

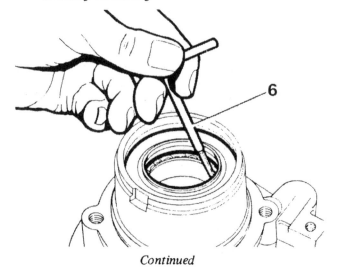

Continued

8. Drift out the bearing from the rear of the housing.

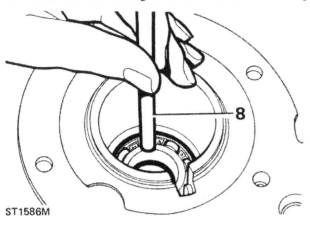

ST1586M

9. Remove speedometer gear (driven) from its housing.
10. Remove the 'O' ring and oil seal and discard.

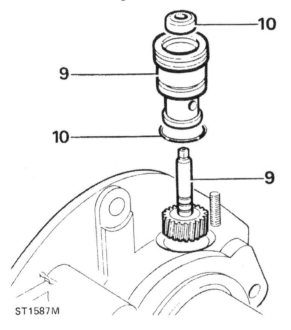

ST1587M

11. Slide off spacer and speedometer drive gear from output shaft.
12. Clean all parts, renew the 'O' ring, oil seals, felt seal and flange nut. Examine all other parts for wear or damage and renew, if necessary.

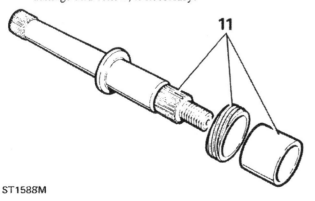

ST1588M

Reassembling

13. Press output bearing into the housing. Do not use excessive force. To facilitate fitting the bearing, heat the output housing case. (This is not to exceed 100°C).
14. Retain bearing with circlip, using circlip pliers 18G257.
15. Fit new seal (open side inwards) using tool 18G1422. The seal should just make contact with the bearing circlip.
16. Carefully charge the lips of the seal with clean grease and refit oil catch ring on to output housing.

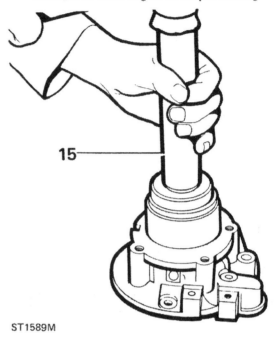

ST1589M

17. Fit the 'O' ring and oil seal (open side inwards) to speedometer housing.
18. Lubricate the 'O' ring and seal with oil.
19. Locate speedometer gear (driven) in housing and press into position.

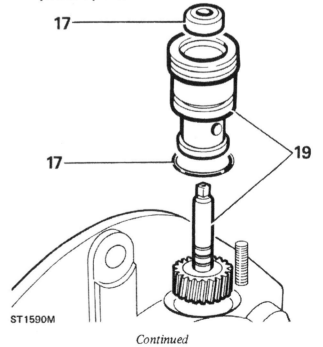

ST1590M

Continued

20. Slide drive gear and spacer on to the output shaft.
21. Locate output shaft into the bearing in the housing and drift into position.
22. Locate speedometer gear (driven) housing assembly into the output housing and press in until flush with the housing face.

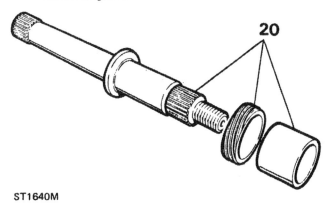

ST1640M

Centre differential unit overhaul – dismantling

1. Secure centre differential unit to a vice fitted with soft jaws, and release stake nut from recess.
2. Remove stake nut using tool 18G1423 and suitable socket wrench.
3. Remove the differential unit from the vice.

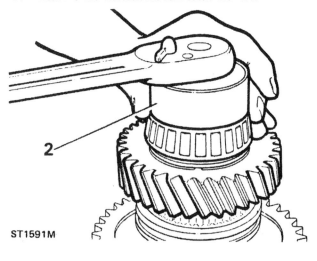

ST1591M

4. Secure hand press MS47 in vice with collars 18G47BB/1 and using button 1847BB/3 remove the rear taper bearing and collars.

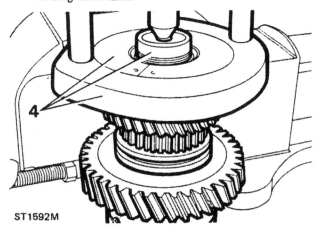

ST1592M

5. Remove the high range gear and bush, taking care not to disturb the high/low sleeve.
6. Mark. the relationship of the high/low sleeve to the hub and then remove the sleeve.
7. Using a suitable press behind the low range gear carefully remove the high/low hub and low range gear.

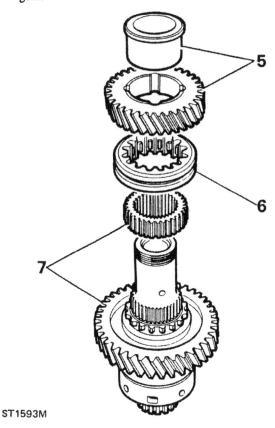

ST1593M

8. Substituting collar LST47-1 remove front taper roller bearing.
9. Remove hand press from the vice.
10. Using soft jaws secure the differential unit in the vice by gripping the hub splines.

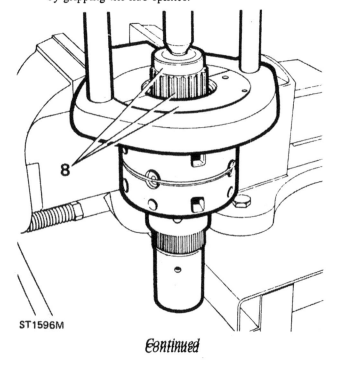

ST1596M

Continued

11. Remove the eight retaining bolts and lift off the front part of the differential unit.
12. Release the retaining ring and remove front upper bevel gear and thrust washer.
13. Remove the pinion gears and dished washers along with the cross shafts.
14. Remove the rear lower bevel gear and thrust washer from the rear part of the differential unit.

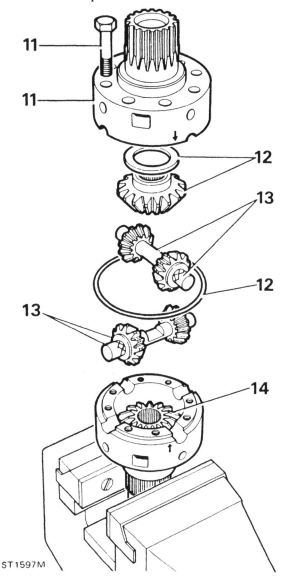

ST1597M

15. Remove the rear differential unit from the vice and clean all components; examine for wear or damage and renew if necessary.
16. Clean all components; examine for wear or damage and renew if necessary.
17. Using soft jaws secure the rear differential unit in the vice by gripping the hub splines.
18. Ensure that all differential components are dry to assist in checking end-float.
19. Using a micrometer, measure one of the bevel gear thrust washers and note the thickness.
20. Fit the thrust washer and bevel gear to the rear lower differential unit.
21. Assemble both pinion assemblies and dished washers on to their respective shafts and fit to the rear differential unit.

22. Measure the front upper bevel gear thrust washer and note the thickness.
23. Fit the thrust washer and bevel gear to the front unit.
24. Refit the retaining ring and front differential unit, aligning the two engraved arrows marked on both halves of the unit.
25. Fit four bolts equi-spaced and torque to the correct figure.

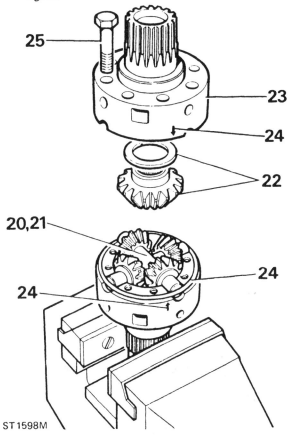

ST1598M

26. Measure the front bevel gear end-float with feeler gauges through the slots provided in the front differential unit. The end-float must be 0,025 to 0,075mm (0.001 to 0.003in) maximum. When measuring use two sets of feeler gauges, one on each side of the front differential unit. This will give a true reading of the end-float.

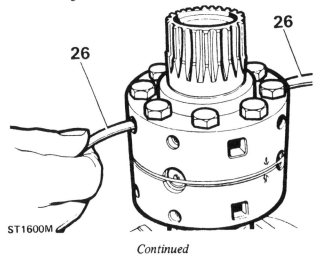

ST1600M

Continued

621

27. Invert the differential unit and repeat operation 26 for the rear bevel gear end-float.
28. Invert the differential unit and secure in vice and remove the four bolts and lift off the front differential unit.
29. Remove the retaining ring, bevel gear and the washer and both pinion assemblies.
30. Select the correct thrust washers required for final assembly.

Reassembling

31. Fit the selected thrust washer and bevel gear into the rear lower differential unit.
32. Assemble both pinion assemblies and dished washers on to their respective shafts and fit the rear differential unit. Secure the assemblies with the retaining ring.
33. Lubricate all the components.
34. Fit the selected thrust washer and bevel gear into the front upper differential unit.

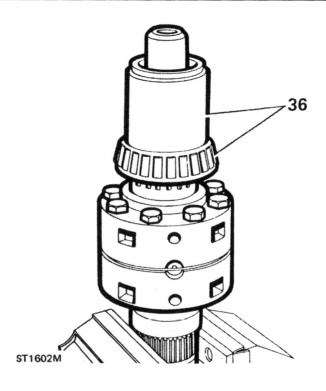

ST1602M

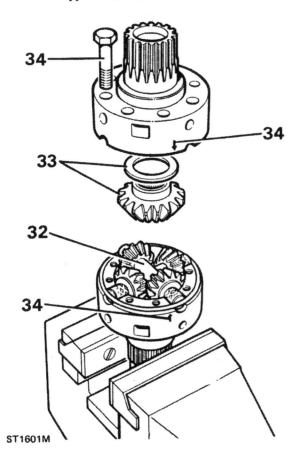

ST1601M

38. Fit the low range gear, with its dog teeth uppermost to the differential assembly.
39. Press the high/low hub on to the differential splines.
40. Slide the high/low selector sleeve on to the high/low hub ensuring that the alignment marks are opposite each other.
41. Fit the bush into the high range gear so that the flange is fitted on the opposite side of the gear to the dog teeth. Slide the bushed gear on to the differential assembly with the dog teeth down.

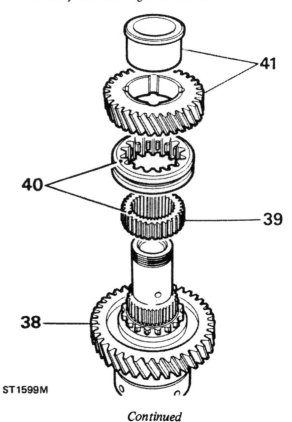

ST1599M

35. Align both units as previously described and secure with the eight bolts to the specified torque.
36. Finally check that the differential gears rotate freely. Locate the front differential bearing on to the front, upper differential shaft and press into position using larger end of tool 18G1424 as shown.
37. Invert the differential unit and secure in the vice.

NOTE: During the following sequences all parts should be lubricated as they are fitted.

Continued

622

42. Locate the rear differential bearing on to the hub and press it into position using the smaller end of tool 18G1424.

43. Fit the stake nut and tighten to the specified torque using tool 18G1423.

44. Check the end float of the high and low range gears 0,05 to 0,15mm (0.002 to 0.005in).

NOTE: If the clearances vary from those specified in the data, the assembly must be rebuilt using the relevant new parts.

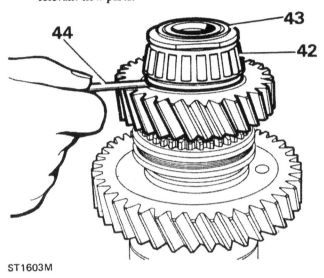

ST1603M

45. Peen the stake nut collar by carefully forming the collar of the nut into the slot as illustrated.

CAUTION: A round nose tool must be used for this operation to avoid splitting the collar of the nut.

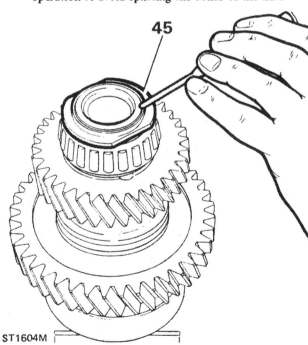

ST1604M

46. Clean and check high/low selector fork assembly for wear and renew if necessary.

47. To renew the selector fork remove the square set screw and slide the fork from the shaft.

48. Fit the new selector fork with its boss towards the three detent grooves. Align the tapped hole in the fork boss with the indent in the shaft nearest to the detent grooves.

49. Apply Loctite 290 to the set screw threads and fit the set screw and tighten to the specified torque.

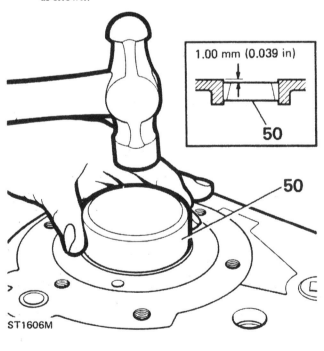

ST1605M

Centre differential rear bearing track

50. Fit the differential rear bearing track 1,00mm (0.039 in) below the outer face of casing using a suitable tool as shown.

1.00 mm (0.039 in)

50

50

ST1606M

Rear output housing – refit

1. Grease output housing gasket and position on to the rear face of the transfer box casing.

Continued

2. Fit output housing and ensure clearance of 1,00mm (0.039in) between housing face and gasket.
3. Fit the six output housing bolts with Loctite 290 on the threads, with washers and tighten evenly to the correct torque, which will pull the rear bearing into position.

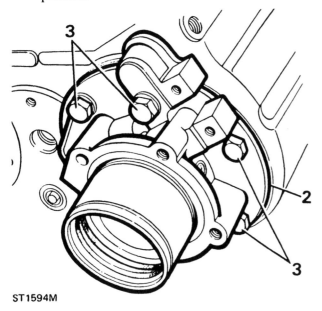

ST1594M

Centre differential unit refit

1. Fit the selector fork/shaft assembly to the high/low selector sleeve on the differential assembly, with detent groove to the rear of the differential assembly.
2. Locate the differential assembly complete with selector fork into the transfer box casing. It may be necessary to rotate the output shaft to ease fitment, and engage selector shaft into its hole.

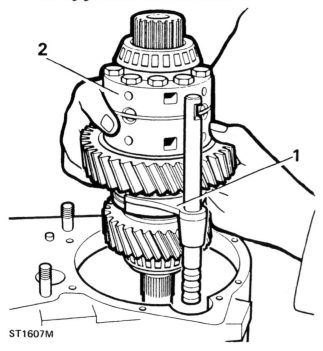

ST1607M

3. Fit selector shaft ball and spring through the side of the transfer box casing.

4. Apply Loctite 290 to detent plug; fit and locate, by screwing gently fully home and then unscrewing two turns.

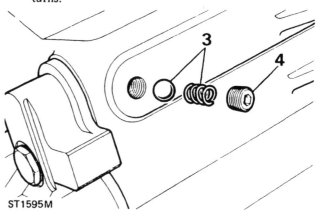

ST1595M

Front output housing overhaul − dismantling

1. Unscrew seven retaining bolts and washers and remove the differential lock selector side cover and gasket.
2. Unscrew three retaining bolts and washers and lift the differential lock finger housing and actuator assembly from the front output housing.
3. Slacken the locknut and unscrew the differential lock warning light switch.
4. Remove selector shaft detent plug, spring and ball using a suitable magnet.

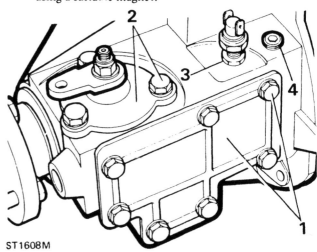

ST1608M

5. Compress the selector fork spring and remove the two spring retaining caps.

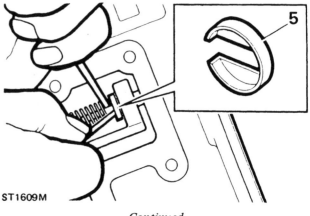

ST1609M

Continued

6. Withdraw the selector shaft from the rear of the output housing.
7. Remove the selector fork and spring through the side cover aperture.
8. Remove lock-up sleeve from the rear of the output housing.

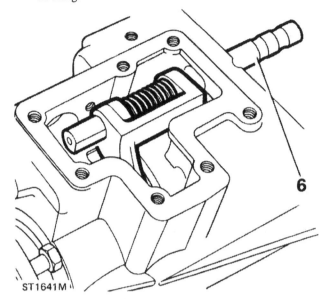

ST1641M

9. Using flange wrench 18G1205 and socket wrench, remove the flange nut, steel and felt washers.

   NOTE: Ensure that flange bolts are fully engaged in the wrench.

10. Remove the output flange with oil seal shield.

    NOTE: These parts need not be separated unless the flange bolts are to be renewed.

11. Drift output shaft rearwards from housing using a soft headed mallet.
12. Slide off the collar from the output shaft.

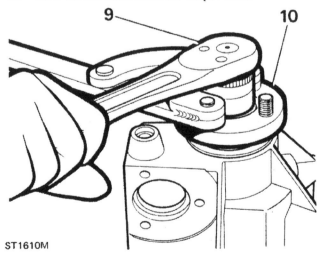

ST1610M

13. Prise out and discard oil seal from output housing using service tool 18G1271.
14. Remove circlip with circlip pliers 18G257.

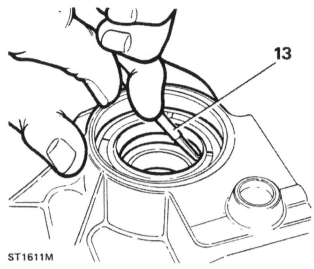

ST1611M

15. Invert housing and drift out bearing from inside the case as shown.

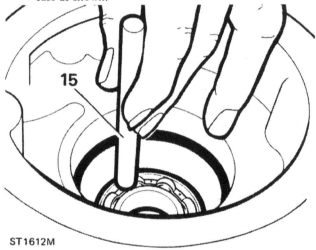

ST1612M

16. Drift out centre differential front taper roller bearing track and shim.
17. Drift out selector shaft cup plug from housing.
18. Clean all components ensuring all traces of "Loctite" are removed from faces and threads.
19. Examine components for wear or damage and renew if necessary.

    NOTE: Renew oil seal and felt seal and flange nut.

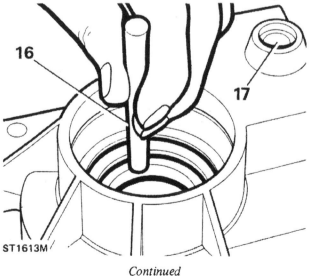

ST1613M

Continued

Reassembling

20. Press the bearing into the housing; do not use excessive force. To facilitate fitting the bearing, heat the front output housing. (This is not to exceed 100°C).
21. Using circlip pliers 18G257, fit the bearing retaining clips.
22. Fit a new oil seal (open side inwards) using replacer tool 18G1422, until the seal just makes contact with the circlip.
23. Carefully charge the lips of the seal with clean grease.
24. Slide collar on to the output shaft, with its chamfered edge towards the dog teeth.
25. Fit the output shaft through the bearing and drift home.

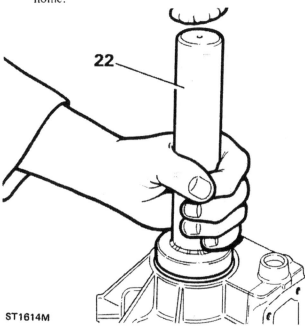

ST1614M

Adjusting front differential bearing pre-load
26. Measure original differential front bearing track shim.
27. Refit original shim into input housing.
28. Drift differential front bearing track into the housing.

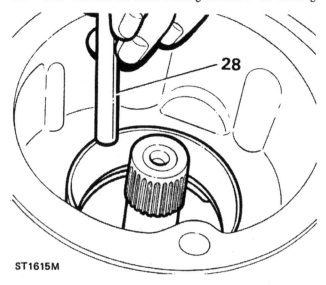

ST1615M

29. Grease and fit new gasket and locate the front output housing on the transfer box casing.
30. Secure housing with the eight retaining bolts and washers, the upper middle bolt being longer than the rest. Do not tighten the bolts at this stage.
31. Engage high or low gear.
32. Check the rolling resistance of the differential using a spring balance and a length of string wound around the exposed splines of the high/low hub.

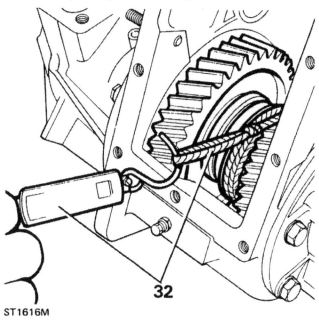

ST1616M

33. With the correct shim fitted the load to turn should be 1,36 kg to 4,53 kg (3 lbs to 10 lbs). This applies to new or used bearings. (New bearings will register at the top end and used bearings will register at the low end.)
34. If the reading is in excess of the above measurements, remove the front output housing assembly from the transfer box casing.
35. Using a suitable extractor, withdraw the centre differential bearing track and change the shim for one of a suitable thickness. (A thinner shim will reduce the rolling resistance.)
36. Fit the new shim and drift the differential bearing track back into its housing until fully home.
37. Having obtained the load to turn prop-up the transfer box casing on the bench with the front face uppermost.
38. Apply Loctite 290 to the threads of the housing retaining bolts and fit the eight bolts and washers into the front output housing and secure to transfer box casing.
39. Fit front output flange, felt washers, steel washers and flange nut.
40. Using flange wrench 18G1205 and torque wrench pull the output shaft up to the correct position. Check that the oil seal shield does not foul the housing.

NOTE: Ensure that the flange bolts are fully engaged in the wrench.

Continued

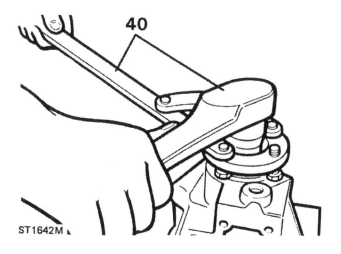

ST1642M

41. Repeat the above operation for the rear output flange.

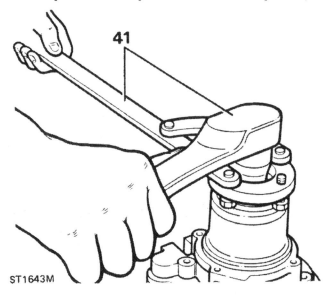

ST1643M

42. Compress the selector shaft spring and fit to the selector fork.
43. Locate selector fork through front output housing side cover aperture, ensuring that the fork engages in the groove of the lock-up sleeve.
44. Fit selector shaft through the aperture in the front of the output housing and pass it through the selector fork lugs and spring into the rear part of the housing.
45. Rotate the selector shaft until the two flats for the spring retaining caps are at right angles to the side cover plate face.

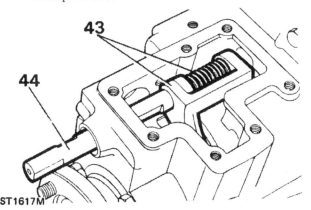

ST1617M

46. Compress the spring between the fork lugs and slide the retaining caps on to the shaft ensuring the spring is captured with the "cupped" side of the caps.
47. Drift selector shaft seal cup into position.

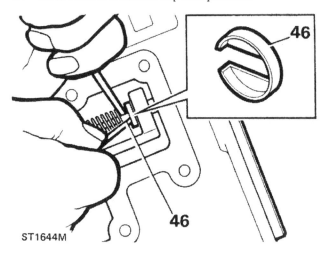

ST1644M

48. Fit selector shaft detent ball and spring in the tapped hole on top of the output housing.
49. Apply Loctite 290 to detent plug threads. Screw detent plug gently home and then unscrew two turns.

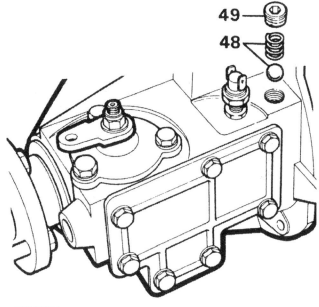

ST1645M

Differential lock finger housing overhaul — dismantling

1. Unscrew and discard the "nyloc" nut and remove the operating lever and washer.
2. Remove the pivot shaft from lock finger housing.
3. Remove the 'O' rings from the pivot shaft and housing and discard.
4. Clean all components; examine for wear or damage and renew if necessary.

Reassembling

5. Fit new 'O' rings on to pivot shaft and lock finger housing and lubricate with oil.
6. Locate the pivot shaft in the housing.

Continued

7. Fit the differential lock lever over the pivot shaft so that the lever will face forward to the bend upwards. This lever is then in the correct operating position.
8. Retain the lever with a plain washer and new nyloc nut.

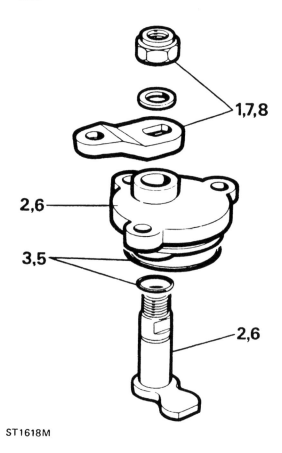

ST1618M

9. Fit the differential lock finger housing into its seating on the front output housing, ensuring that the selector finger is located in the flat of the selector shaft.
10. Apply Loctite 290 to the bolt threads and retain the lock finger housing with the three bolts and washers to the specified torque.

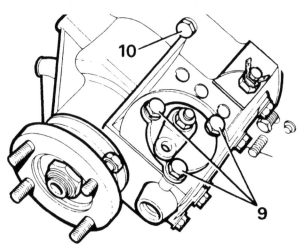

ST1619M

High/low cross-shaft housing overhaul

1. Remove the selector finger grub screw and withdraw the cross-shaft from the cross-shaft housing and remove the selector finger.
2. Remove the 'O' ring from the cross-shaft.
3. Drift out selector housing cup plug if necessary.
4. Clean all the components and check for damage or wear, replace if necessary.
5. Apply sealant to a new cup plug and fit so that the cup is just below the chamfer for the cross-shaft bore.
6. Fit new 'O' ring to cross-shaft.
7. Lubricate the shaft and insert into the cross-shaft housing.
8. Fit selector finger ensuring that it aligns with the recess in the cross-shaft.
9. Apply Loctite 290 to the grub screw and secure the selector finger to the cross-shaft and fully tighten to the specified torque.

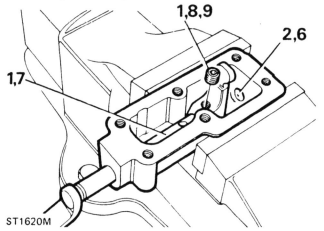

ST1620M

10. Grease and fit the high/low selector housing gasket on the front output housing.
11. Fit high/low cross-shaft housing, ensuring that the selector finger locates in the slot of the selector shaft, and secure with six bolts and washers to the specified torque.

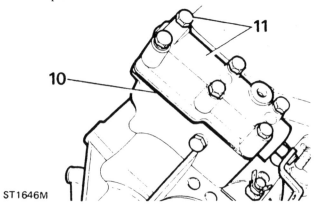

ST1646M

Input gear overhaul – dismantling

1. Clean the input gear assembly and examine for wear or damage. Remove the bearings only if they are to be renewed.

Continued

2. Secure hand press MS47 in the vice and using collars 18G47-7 and button 18G47-BB/3, remove rear taper roller bearing from input gear assembly.
3. Invert input gear assembly in hand press and remove front taper roller bearing.
4. Clean input gear.

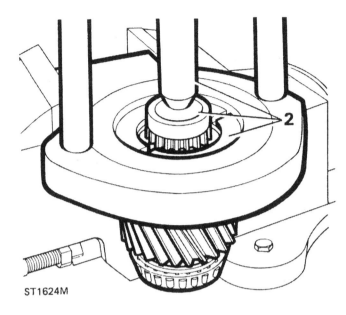

ST1624M

Reassembling

5. Position rear taper roller bearing on input gear and using hand press MS47 and collars 18G47-7 press the bearing fully home.
6. Invert input gear and fit the front taper roller bearing using the press and collars.

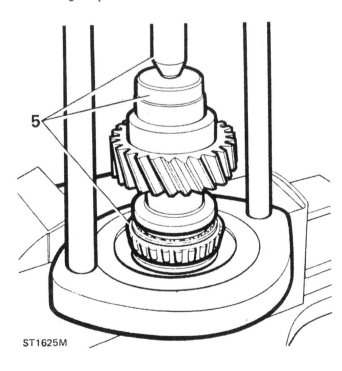

ST1625M

7. Prop up the transfer box casing on the bench with the rear face uppermost.
8. Drift in the front taper bearing track.

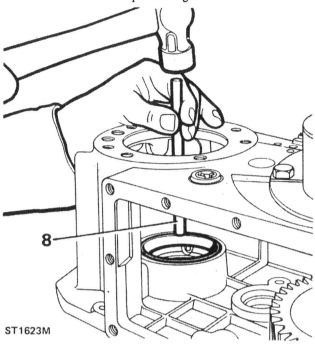

ST1623M

9. Reposition transfer box casing so the front face is uppermost and fit oil seal (open side inwards) using replacer tool 18G1422.

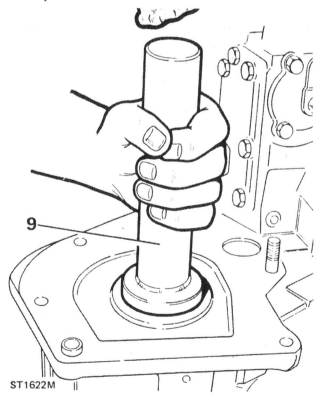

ST1622M

10. Lubricate both bearings with clean oil.
11. Fit the input gear assembly into the transfer box casing with the dog teeth uppermost.

Continued

Checking input gear bearing pre-load

12. Secure bearing support plate in the vice. Drift out input gear bearing track, and remove shim.
13. Clean bearing support plate and shim. Measure original shim and note its thickness.
14. Fit the original shim to the support plate.
15. Locate the bearing track in the support plate and press fully home.

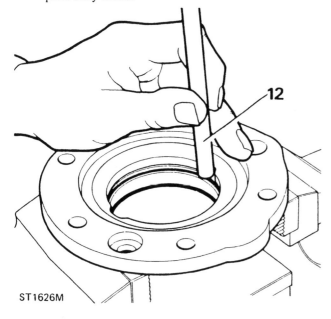

ST1626M

16. Apply grease to the gasket and fit on to the transfer box casing.
17. Fit the bearing support plate on to the transfer box casing and secure with the six bolts, but do not tighten.
18. Fit the service tool LST105 to input gear and engage the spline.
19. Tie a length of string to the split pin and fit it to the service tool as shown.
20. Attach a spring balance to the string and carefully tension the spring until a load to turn the input gear is obtained. A pull of 2,26 kg to 6,80 kg (5 lbs to 15 lbs) is required.
21. If the reading obtained is outside the above limits, the original shim must be changed.

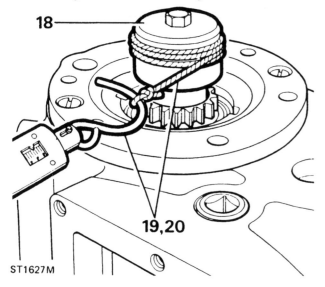

ST1627M

22. Remove the spring balance, string and service tool.
23. Remove the six bolts and the bearing support plate.
24. Drift out the input gear bearing track from the support plate and discard original shim.
25. Select the correct size shim to obtain a load to turn of 2,26 kg to 6,80 kg (5 lbs to 15 lbs).
26. Fit shim to support plate, locate bearing track and press home.
27. Fit bearing support plate and secure to transfer box casing with the six bolts (do not tighten).
28. Repeat the rolling resistance check as previously described, and note the value obtained.

Intermediate gear assembly overhaul

1. Drift out intermediate gear bearing tracks.
2. Remove circlips.

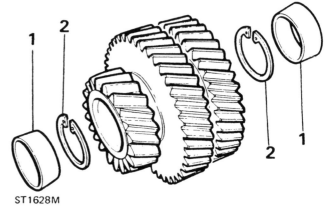

ST1628M

3. Clean all intermediate gear components and lock plate. Check for damage or wear and replace as necessary.
4. Fit new circlips into the intermediate gear cluster.
5. Using tools LST550-4 and MS550 fit bearing tracks into the intermediate gear cluster.

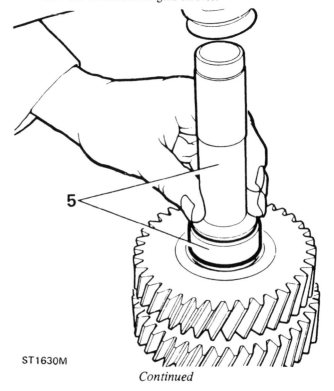

ST1630M

Continued

630

6. Fit the 'O' rings to the intermediate shaft and into the intermediate shaft bore at the front of the transfer box casing.

ST1648M

Intermediate gear reassembly

7. Check for damage to the intermediate shaft thread and if necessary clean up with a fine file or stone.
8. Lubricate the taper roller bearings and intermediate gear shaft.
9. Insert new bearing spacer to gear assembly, followed by the taper roller bearings.

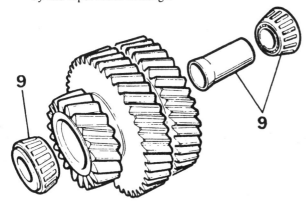

ST1649M

10. Fit dummy shaft LST104 into the intermediate gear cluster.
11. Locate the gear assembly into the transfer box casing from the bottom cover aperture.
12. Insert intermediate shaft from the front of the transfer box casing, pushing the dummy shaft right through as shown and remove. (Making sure that the intermediate gear cluster meshes with the input gear and high range and low range gears).

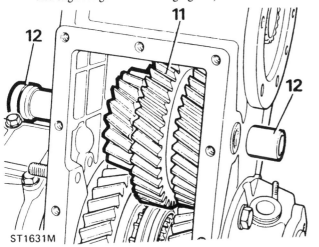

ST1631M

13. Turn the intermediate shaft to allow fitting of retaining plate.
14. Fit retaining plate and secure with retaining bolt and washer.
15. Fit the intermediate gear shaft retaining stake nut.

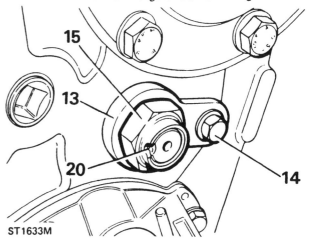

ST1633M

Adjusting intermediate gear torque-to-turn

16. Select neutral.
17. Fit service tool LST105 to input gear and engage spline.
18. Tie a length of string to a split pin and fit to the service tool as shown. Attach the spring balance to the string.
19. To obtain the correct figures and to collapse the spacer within the intermediate gear cluster, tighten the intermediate shaft nut until the load-to-turn has increased by 3,7 kg (7 lbs) ± 1,63 kg (± 3 lbs) on that noted when checking input shaft load-to-turn. The torque to tighten the retaining nut will be approximately 203 Nm (150 lb ft).
20. Peen the stake nut by carefully forming the collar of the nut into the intermediate shaft recess, as illustrated.

CAUTION: A round nose tool must be used for this operation to avoid splitting the collar of the nut.

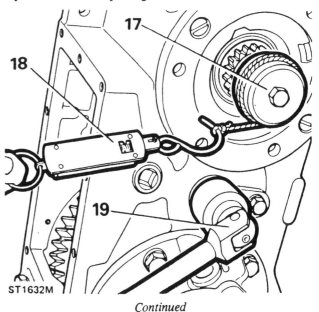

ST1632M

Continued

Power take-off cover – reassemble
21. Clean power take-off cover and gasket face.
22. Fit the two countersunk screws and tighten.
23. Remove the six bolts from the bearing support plate.
24. Apply sealant to the cover plate gasket and fit it to the bearing support plate.
25. Apply Loctite 290 to bolt threads and secure the power take-off cover with the six bolts and washers.

Bottom cover – reassemble
26. Clean bottom cover and gasket face.
27. Apply sealant to cover gasket and fit to transfer box casing.
28. Apply Loctite 290 to bolt threads and secure the bottom cover with six bolts and washers.

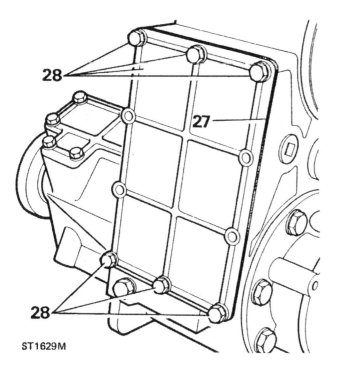

ST1629M

Differential lock switch adjustment
1. Select differential locked position by moving the lock taper towards the right side of the transfer box casing.
2. Apply sealant to the differential lock warning light switch and fit to the top of the front output housing.
3. Connect a test lamp circuit to the differential lock switch.
4. Screw in the lock switch until the bulb is illuminated.
5. Turn in the switch another half a turn and tighten with the locknut against the housing.

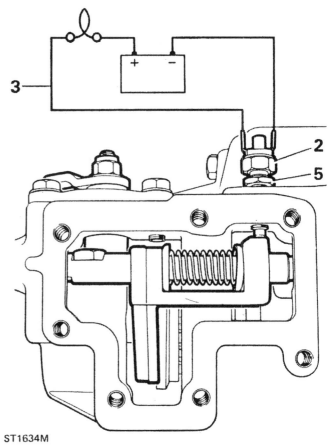

ST1634M

6. Disconnect the battery and move the differential lock lever to the left to disengage differential lock.
7. Clean the front output housing side cover.
8. Grease and fit side cover gasket.
9. Apply Loctite 290 to bolt threads, fit side cover and secure with seven bolts and washers.

Transmission brake – reassemble
1. Clean brake backplate and oil catcher and apply sealant to the catcher joint face.
2. Locate brake backplate on the rear output housing with the brake operating lever on the right side of the transfer box casing.
3. Secure the backplate (including the oil catcher) with the four special bolts and tighten using a hexagonal socket to the specified torque.
4. Clean and fit brake drum and secure with two countersunk screws.

ELECTRICAL

This section of the manual contains information on important changes that have been made to the electrical equipment fitted to Range Rovers. The changes were introduced in two main groups, as follows:—

1982 to mid 1984
New circuit diagram for automatic gearbox models
New ignition distributor and timing procedure for 9.35:1 high compression engine
Central door locking on four-door models

Mid 1984 to 1985
New circuit diagrams for all models
New instrument binnacle, auxiliary switch panel and fuse box
Electrically operated exterior mirrors – Option
Electrically operated side window lifts – Option
New central door locking on four-door models, controlled by driver's door only
Electronic ignition

CONTENTS   Page

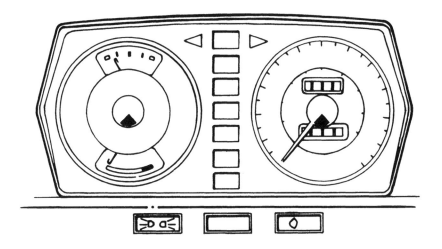

**FOR ALL MODELS WITH THE ABOVE TYPE OF INSTRUMENT BINNACLE
FROM 1982 TO 1984**

KEY TO CIRCUIT DIAGRAM – AUTOMATIC GEARBOX MODELS – 1982 to 1984

1 Oil temperature transmitter
2 LH Front fog lamp
3 RH Front fog lamp
4 Automatic gearbox oil cooler temperature switch
5 Automatic gearbox graphics illumination
6 Battery voltmeter illumination
7 Oil temperature gauge illumination
8 Cigar lighter illumination
9 Clock illumination
10 Oil pressure indicator
11 LH Front side lamp
12 RH Front side lamp
13 Number plate illumination
14 Number plate illumination
15 LH Rear tail lamp
16 RH Rear tail lamp
17 Underbonnet illumination (where fitted)
18 LH Direction indicator side repeater
19 LH Front indicator lamp
20 Horns
21 RH Headlamp main beam
22 LH Headlamp main beam
23 RH Headlamp dipped beam
24 LH Headlamp dipped beam
25 LH Rear indicator lamp
26 RH Direction indicator side repeater
27 RH Front indicator lamp
28 RH Rear indicator lamp
29 LH Reverse lamp
30 RH Reverse lamp
31 Front fog lamps pick-up point
32 Oil cooler temperature warning light
33 Oil temperature gauge
34 Rear fog lamp switch
35 Panel lighting switch
36 Direction indicator switch
37 Underbonnet lamps (one lamp or two, according to specification)
38 Trailer warning light
39 Panel illumination
40 Panel illumination
41 Differential lock warning light
42 Main beam warning light
43 LH indicator warning light
44 RH indicator warning light
45 Neutral, Start & Reverse lighting switch
46 Voltage stabiliser
47 Rear fog warning light
48 Side lamps warning light
49 Lighting switch
50 Indicator, headlamp dip and horn switch
51 Clock
52 Differential lock warning light switch
53 Water indicator
54 Fuel indicator
55 Choke warning light
56 Oil warning light

57 Ignition warning light
58 Brake warning light
59 Fuel warning light
60 Cigar lighter
61 Hazard warning switch
62 Radio fuse (when fitted)
63 Radio (when fitted)
64 Fuse unit
65 Hazard unit
66 Battery voltmeter
67 Ignition switch
68 Oil pressure indicator
69 Trailer socket connection
70 Interior lights
71 Brake fluid check switch
72 Choke switch
73 Heated rear screen switch
74 Heater fuse
75 Stop lamp switch
76 Alternator
77 Starter relay
78 Resistive wire
79 Coil
80 Oil pressure transmitter
81 Courtesy light delay unit (when fitted)
82 Park brake warning light (when fitted)
83 Rear screen fuse
84 Front wash/wiper switch
85 Programmed wash/wipe control unit
86 Rear wash/wiper switch
87 Starter motor
88 Electric fuel pump
89 Distributor
90 Inspection sockets
91 Courtesy light switch (4 door only)
92 Courtesy light switch
93 Interior light switch
94 Courtesy light switch (4 door only)
95 Brake circuit check relay
96 Oil pressure switch
97 Oil pressure switch
98 Park brake pick up point
99 Park brake switch (if fitted)
100 Fuel gauge transmitter
101 Water temperature transmitter
102 Heated rear screen relay
103 Heated rear screen element
104 Front screen washer motor
105 Front screen wiper motor
106 Rear screen washer motor
107 Rear screen wiper motor
108 Heater motor
109 LH Stop lamp
110 RH Stop lamp
111 LH Rear fog lamp
112 RH Rear fog lamp
113 Battery

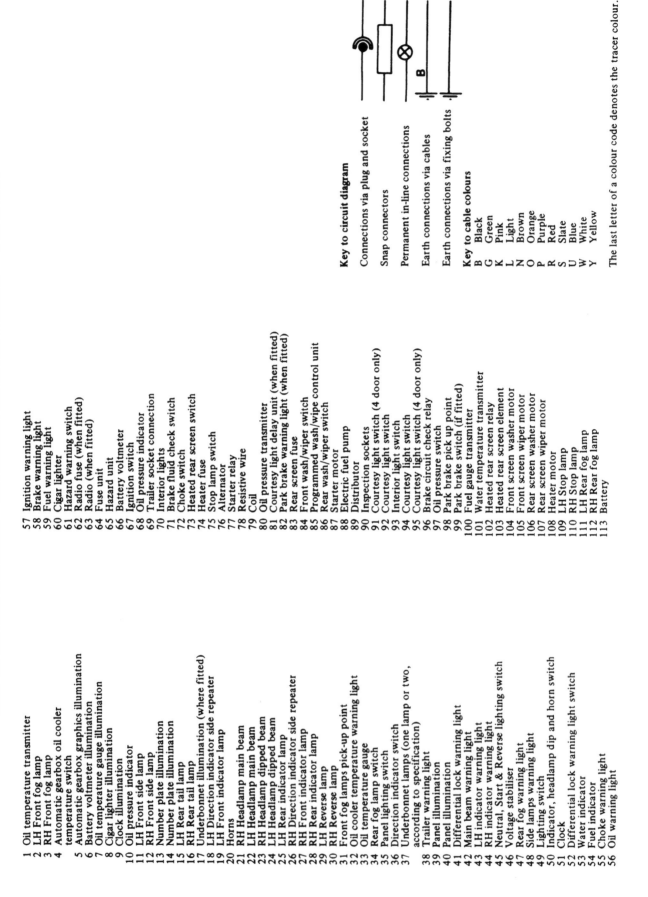

Key to circuit diagram

Connections via plug and socket

Snap connectors

Permanent in-line connections

Earth connections via cables

Earth connections via fixing bolts

Key to cable colours

B   Black
G   Green
K   Pink
L   Light
N   Brown
O   Orange
P   Purple
R   Red
S   Slate
U   Blue
W   White
Y   Yellow

The last letter of a colour code denotes the tracer colour.

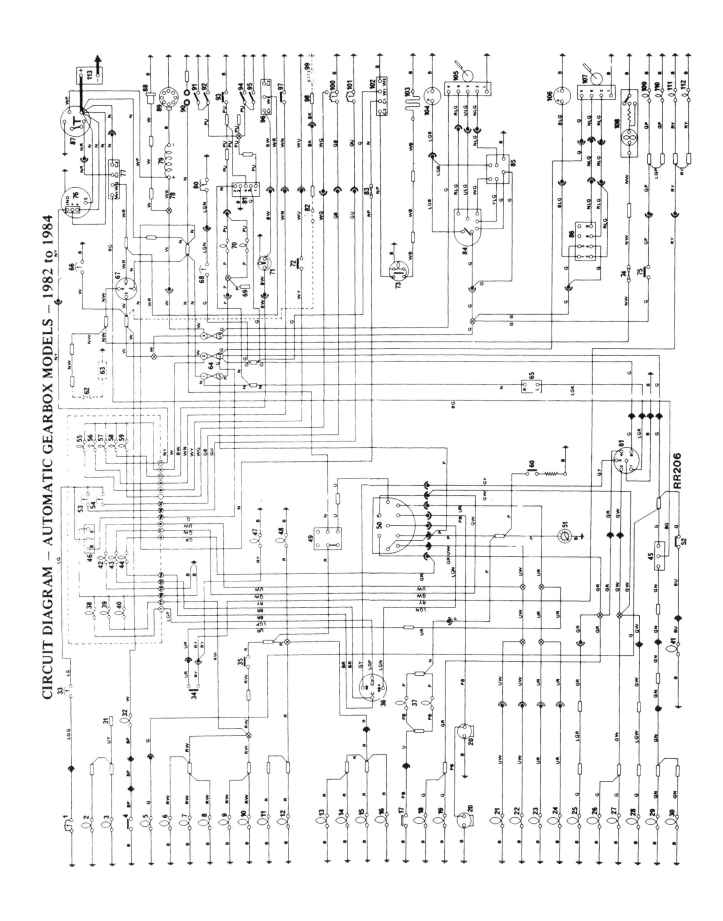

CIRCUIT DIAGRAM – AUTOMATIC GEARBOX MODELS – 1982 to 1984

RR206

DISTRIBUTOR AND IGNITION TIMING
FOR HIGH COMPRESSION ENGINE — 1982 to 1984

Renew distributor contact breaker points

Sliding contact type
To obtain satisfactory engine performance it is most important that the contact points are adjusted to the correct dwell angle, see Data Section using suitable workshop equipment. This work should be carried out by your local Land Rover Distributor or Dealer.

1. Release the clips and remove the distributor cap.
2. Remove the rotor arm from the cam spindle.
3. Remove the retaining screw and washers and lift the complete contact breaker assembly from the moveable plate.
4. Remove the nut and plastic bushes from the terminal post to release the leads and spring.
5. Discard the old contact breaker assembly.
6. Clean the new points with petrol to remove the protective coating.
7. Connect the leads to the terminal post in the following sequence:
   (a) lower plastic bush
   (b) red lead tab
   (c) contact breaker spring eye
   (d) black lead tab
   (e) upper plastic bush
   (f) retaining nut
8. Fit the contact breaker assembly to the moveable plate ensuring that the pegs underneath locate in the holes in the moveable plate.
9. The sliding contact actuating fork must also be located over the fixed peg in the adjustable base plate.
10. Fit the retaining screw plain and spring washer to secure the contact breaker assembly to the moveable plate.
11. When new contact points have been fitted, the dwell angle must be checked after a further 1,500 km (1,000 miles) running.

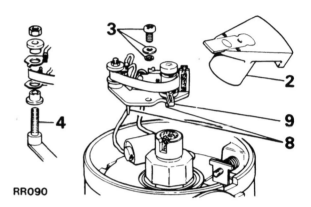

RR090

Check security of distributor vacuum unit line and operation of vacuum unit (9.35:1 compression ratio engines)
To ensure that the vacuum capsule is functioning correctly, it is important that the following procedure is carried out:
Remove the vacuum retard pipe from the distributor (this will increase the idle speed). If the vacuum capsule is functioning correctly, an advance to between 6 to 14 degrees B.T.D.C. should be noted.

continued

Check/adjust dwell angle and ignition timing using electronic equipment — 1982 to 1984 — High and Low compression engines

1. The accurate setting of ignition timing is of extreme importance, and the correct functioning of the emission control system relies to a large extent on its accuracy. It is necessary to set the ignition timing dynamically with the engine at idling speed. This requires the use of a suitable tachometer, for determining the engine speed, and a stroboscopic lamp for determining the points in the engine cycle at which the ignition sparks occur. It is obvious therefore that this work should be carried out by a Land Rover Distributor or Dealer.

2. Engines with emission control have a special ignition distributor included in the specification. The distributor provides a retarded ignition setting at the lower speed range whilst maintaining the normal advance characteristics at higher engine speeds. The distributor, together with the other modifications embodied, reduces exhaust emissions to an acceptable level. Failure to set ignition timing correctly, as subsequently described, will almost certainly result in the vehicle failing to comply with emission control regulations and can also lead to engine damage and increase fuel consumption.

3. Set the contact breaker gap to 0,35 mm (0.014 in). See Data Section engine details for ignition timing for the various engine versions.

    Carry out item 4 only if distributor has been disturbed.

4. Set ignition timing statically as specified in Data Section prior to the engine being run, by the basic timing lamp method. (This sequence is to give only an approximation in order that the engine may be run. The engine must not be started after distributor replacement until this check has been carried out.)

5. Start and run the engine until it is at normal operating temperature, that is, warm water flowing through the radiator top hose.

6. **8.13:1 Compression ratio engines**
    Set the idle speed to 600 to 650 rev/min with distributor vacuum pipes connected.

7. **Emission engines (9.35:1 compression ratio)**
    (a) Set the idle speed to 550 to 650 rev/min. with vacuum pipes connected.
    (b) Disconnect both vacuum pipes from the distributor.
    (c) The idle speed will increase with the vacuum pipes disconnected and must be reduced to below 750 rev/min. as follows.
    (d) Disconnect one of the breather pipes from a carburetter. The idle speed should now be below 750 rev/min., if not, reduce the idle speed by progressively altering **(equally)** both idle adjustment screws.

8. Set dwell angle as follows.

9. Set selector knob to 'calibrate' position on the tach/dwell meter. Adjust calibration knob to give a zero reading on the meter.

continued

639

10. Couple meter to engine following manufacturer's instructions.

11. Set selector knob to 8 cylinder position and tach/dwell selector knob to 'dwell'. The dwell angle should be 26 degrees to 28 degrees. If the dwell angle is outside the limits of 24 degrees to 30 degrees repeat the foregoing procedure from item 3 but with a contact breaker setting of 0,3 mm (0.012 in). If the dwell angle is still outside the limits of 24 degrees to 30 degrees the distributor must be renewed or overhauled.

12. Uncouple tach/dwell meter.

Set ignition timing as follows:

13. Couple a stroboscopic timing lamp to the engine following the manufacturer's instructions, with the high tension lead attached into No. 1 cylinder plug lead.

    (a) Slacken distributor clamping bolt.

    (b) Turn the distributor body until the stroboscopic lamp synchronises the timing pointer and the applicable timing mark on the vibration damper rim.

    (c) Arrow (R) indicates direction to retard ignition. Arrow (A) indicates direction to advance ignition.

    (d) Re-tighten the distributor clamping bolt.

14. **Emission engines (9.35:1 compression ratio)**

    (a) Reconnect the carburetter breather and vacuum retard pipes.

    (b) Check the engine idle speed and reset (equally) as necessary, see Data Section and re-seal the carburetters.

    (c) Check the dynamic ignition timing with the vacuum retard pipe connected, it should be 4 degrees to 8 degrees A.T.D.C., otherwise there is a fault in the vacuum system.

15. **All engines**

    Switch off the engine and disconnect stroboscopic timing lamp and tachometer.

    NOTE:— Engine speed accuracy during ignition timing is of paramount importance. Any variation from the specified idle speed, particularly in an upward direction, will lead to wrongly set ignition timing.

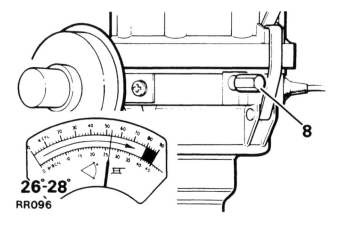

26°-28°
RRO96

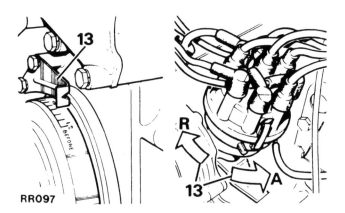

RRO97

CENTRAL DOOR LOCKING (4 door only) – 1982 to 1984

An electrically driven central door locking system is fitted as standard to four door vehicles. The system is activated from either front door by operation of the exterior key lock or the interior locking button and is effective on all four passenger doors; tailgate locking arrangements are unchanged. In the event of electrical failure manual locking remains operative.

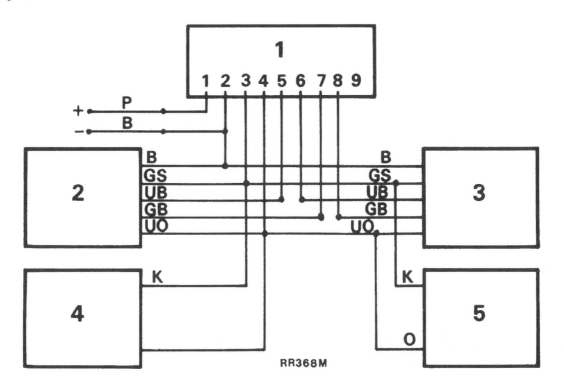

RR368M

Key to cable colours

B – Black   G – Green   K – Pink   O – Orange
P – Purple   S – Slate   U – Blue
The last letter of a colour code denotes the tracer colour

1. Door lock electrical control unit (multi-plug connector)
2. Driver's front door lock actuator
3. Passenger's front door lock actuator   Five pin plug connector
4. Passenger's rear door lock actuator
5. Passenger's rear door lock actuator   Two pin plug connector

The circuit shown is connected into the main vehicle circuit protected by No. 1 fuse (35 amp). The purple (+) wire is connected to an existing connector and the black earth (–) wire to an existing earthing point. Both connections are located behind the lower fascia panel adjacent to the door lock electrical control unit on the outer side of the steering column support.

Both front door actuator motor drive units incorporate an integral master switch. **Either** switch will control all four door locks when operated by the key from outside or by the sill locking control inside the vehicle.

Independent manual operation of rear door sill locking controls will be over-ridden by subsequent electrical switching.

continued

Failure of an actuator will not affect the electric locking of the remaining three doors. The door with the inoperative actuator can still be unlocked or locked manually.

NOTE:– The door lock electrical control unit and door lock actuator units contain non-serviceable parts. If a fault should occur replace the unit concerned with a new one. Before carrying out any maintenance work disconnect the battery.

DOOR LOCK ELECTRICAL CONTROL UNIT

– Remove and refit   86.26.08

Removing
1. Release the lower fascia panel by removing the five self-tapping screws.
2. Disconnect the electrical leads by releasing the multi-plug from the bottom end of the control unit.
3. Remove the two self-tapping screws securing the control unit to the outer side of the steering column support bracket and remove the control unit.

Refitting
4. Reverse instructions 1 to 3.

FRONT DOOR ACTUATOR UNITS

Remove and refit

Removing
1. Remove the two screws holding the armrest to the inner door panel.
2. Wind the glass up into the fully closed position, then remove the single screw retaining the window regulator handle.
3. Carefully prise out the upper and lower halves of the inside door release handle bezel by depressing the trim pad slightly.
4. Unscrew the sill locking knob.
5. Release the door trim pad by inserting a screwdriver between the trim pad and the inner door panel, gently prising out the plastic clips from their respective holes around the edges of the trim pad.
6. Unplug the two speaker connections inside the door and remove the door trim pad complete with speaker.
7. Peel back the top of the plastic weather sheet at the rear of the inner door panel to expose the lock actuator unit.
8. Remove the four screws securing the lock actuator mounting plate to the inner door panel.
9. Release the clip retaining the electrical cable.
10. Manoeuvre the actuator assembly to detach the operating rod 'eye' from the hooked end of the actuator link on the door lock.
11. Withdraw the actuator assembly from the door until the electrical cable is pulled out of its channel sufficiently to expose the connectors which can then be detached.
12. Remove the actuator assembly from the door.
13. The actuator unit may be changed if necessary by

removing the two rubber mounted screws which secure it to the mounting plate.

Refitting
14. Locate the actuator assembly in the inner door panel and fit the electrical cable connectors. The cable, and connectors, are pulled back into the channel from the front end and the cable clip refitted.
15. Manoeuvre the actuator assembly to engage the operating rod 'eye' on the hooked actuator link.
16. Loosely fit the actuator mounting plate to the inner door panel with the four screws, setting the mounting plate in the centre of the slotted holes.
17. Ensure that manual operation of the sill locking control is not restricted by the operation of the actuator operating rod and vice versa, resetting the mounting plate as necessary.
18. Reconnect the vehicle battery.
19. Check that electrical operation of the door lock occurs when the sill locking control is moved through half of its total movement. Reset the mounting plate if necessary and tighten the four screws.
    NOTE:– The above adjustment ensures that the full tolerance on the switching operation is utilised.

REAR DOOR ACTUATOR UNITS

Remove and refit
Instructions as for front doors with the following exceptions:
20. No radio speaker is involved.
21. No mounting plate is used, the rear actuator is secured directly to the inner door panel by the two actuator rubber mounted screws.
22. The electrical cable and plug is retained to the inner door panel by two spring clips and is immediately accessible through the large aperture in the door.
23. Instruction 19 does not apply to rear actuator units which are not fitted with switches.

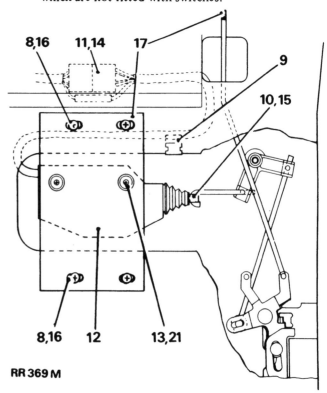

RR 369 M

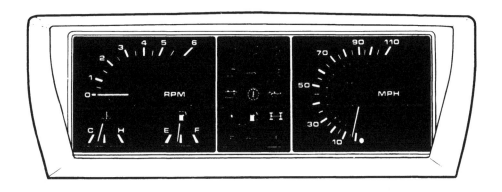

FOR ALL MODELS WITH NEW INSTRUMENT BINNACLE
FROM MID 1984 TO 1985

MAIN CIRCUIT DIAGRAM – ALL MODELS – MID 1984 TO 1985

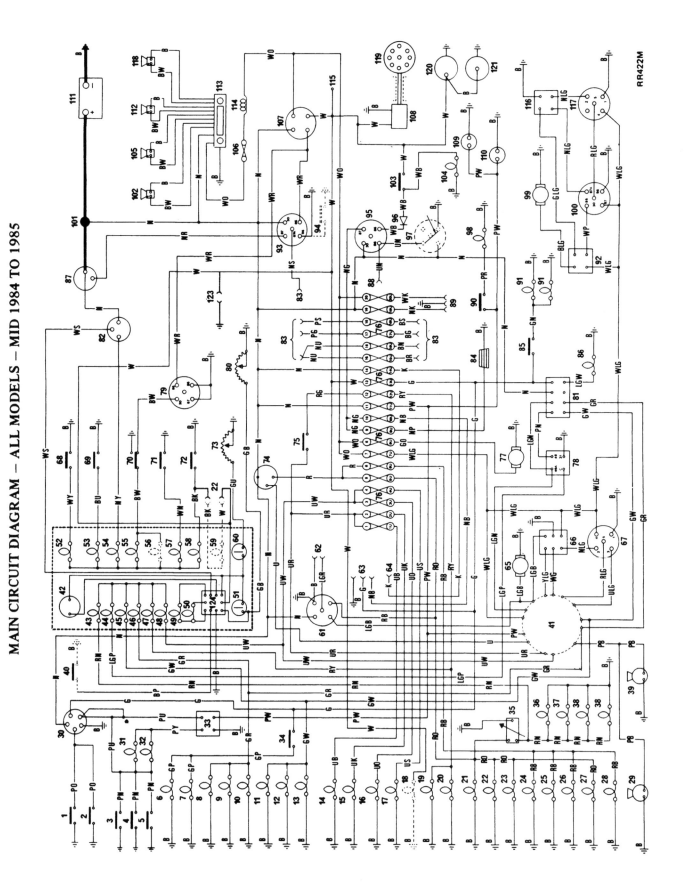

RR422M

MAIN CIRCUIT DIAGRAM – ALL MODELS – MID 1984 TO 1985

1 Left hand front door switch
2 Right hand front door switch
3 Tailgate switch
4 Left hand rear door switch
5 Right hand rear door switch
6 Right hand stop lamp
7 Left hand stop lamp
8 Left hand front indicator lamp
9 Left hand rear indicator lamp
10 Left hand side repeater lamp
11 Right hand front indicator lamp
12 Right hand rear indicator lamp
13 Right hand side repeater lamp
14 Right hand head lamp dip
15 Left hand head lamp dip
16 Right hand head lamp main
17 Left hand head lamp main
18 Automatic gear selector graphics illumination
19 Right hand rear fog lamp
20 Left hand rear fog lamp
21 Right hand number plate lamp
22 Right hand side lamp
23 Right hand tail lamp
24 Left hand number plate lamp
25 Left hand side lamp
26 Left hand tail lamp
27 Radio illumination
28 Switch illumination
29 Right hand horn
30 Interior lamp delay
31 Front interior lamp
32 Rear interior lamp
33 Interior lamp switch
34 Stop lamp switch
35 Rheostat
36 Cigar lighter illumination
37 Clock illumination
38 Heater illumination
39 Left hand horn
40 Transfer oil temperature switch
41 Steering column switches
42 Tachometer
43 Instrument illumination

44 Trailer warning light
45 Right hand indicator
46 Left hand indicator
47 Rear fog warning light
48 Head lamp warning light
49 High transfer oil temperature warning light
50 Low fuel warning light
51 Fuel indicator gauge
52 Cold start warning light
53 Differential lock warning light
54 Ignition warning light
55 Brake failure warning light
56 Brake failure warning light (Australia)
57 Oil pressure warning light
58 Park brake warning light
59 Park brake warning light (Australia)
60 Water temperature gauge
61 Head lamp wash timer (Option)
62 Head lamp wash pump (Option)
63 Heated electric mirrors (Option)
64 Trailer socket (Option)
65 Front screen wash
66 Wiper delay
67 Wiper motor
68 Cold start warning lamp switch
69 Differential lock switch
70 Brake failure switch
71 Oil pressure switch
72 Park brake switch
73 Water temperature transducer
74 Light switch
75 Rear fog lamp switch
76 Fuses
77 Heater motor switch
78 Flasher unit
79 Brake failure warning lamp check relay
80 Fuel tank unit
81 Hazard switch
82 Alternator
83 Air conditioning (Option)
84 Heated rear screen
85 Reverse lamp switch
86 Hazard warning lamp
87 Starter solenoid

88 Split charge relay (Option)
89 Electric windows and central door locking (Options)
90 Hood switch
91 Reverse lamps
92 Rear wipe wash switch
93 Starter solenoid relay
94 Start inhibitor switch (automatic)
95 Heated rear window relay
96 Diode
97 Voltage switch (Option) to fit, remove (arrowed) link first
98 Bonnet lamp
99 Rear screen wash motor
100 Rear wiper relay
101 Terminal post
102 Left hand rear speaker (Option)
103 Heated rear window switch
104 Heated rear window warning lamp
105 Right hand rear speaker (Option)
106 Radio fuse
107 Ignition/heat start switch
108 Constant energy ignition unit
109 Cigar lighter
110 Clock
111 Battery
112 Left hand front speaker
113 Radio (Option)
114 Radio choke
115 Split charge relay (Option)
116 Rear wiper delay
117 Rear wiper motor
118 Right hand front speaker
119 Distributor
120 Fuel pump
121 Fuel pump capacitor
122 Pick-up point (2 pin plug) for Australian park brake warning light lead (right hand drive vehicles only)
123 Pick-up point (2 pin plug) for high speed sensor and front seat belt buzzer circuit (left hand drive vehicles only)
124. Multi-function unit in instrument case (chain dotted area)

Key to cable colours

B Black   G Green   K Pink   L Light   N Brown   O Orange   P Purple   R Red   S Slate   U Blue   W White   Y Yellow

The last letter of a colour code denotes the tracer colour

Connectors via plug and socket

Snap connectors

Permanent in-line connections

Earth connections via cables

Earth connections via fixing bolts

OPTIONAL ELECTRICAL EQUIPMENT – RANGE ROVER 2 AND 4 DOOR MODELS – MID 1984 TO 1985

SPLIT CHARGE CIRCUIT DIAGRAM

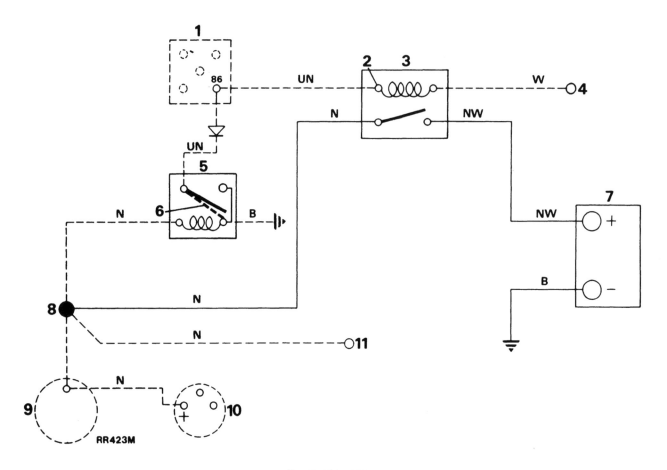

Key to cable colours

B Black G Green K Pink L Light N Brown O Orange P Purple R Red S Slate U Blue W White Y Yellow

The last letter of a colour code denotes the tracer colour

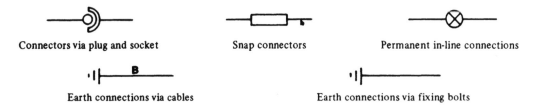

Connectors via plug and socket   Snap connectors   Permanent in-line connections

Earth connections via cables   Earth connections via fixing bolts

1 • Heated rear window relay
2 Pick-up point for split charge relay (Item 88 on main circuit diagram)
3 Split charge relay
4 Fuse box
5 Voltage sensitive switch
6 Link wire (Removed from plug when voltage sensitive switch is fitted)
7 Terminal box auxiliary battery
8 Terminal post
9 Starter motor
10 Alternator
11 Vehicle battery

NOTE:– Chain dotted lines indicate existing parts.

OPTIONAL ELECTRICAL EQUIPMENT — RANGE ROVER 4 DOOR MODELS
— MID 1984 TO 1985

ELECTRIC MIRRORS CIRCUIT DIAGRAM

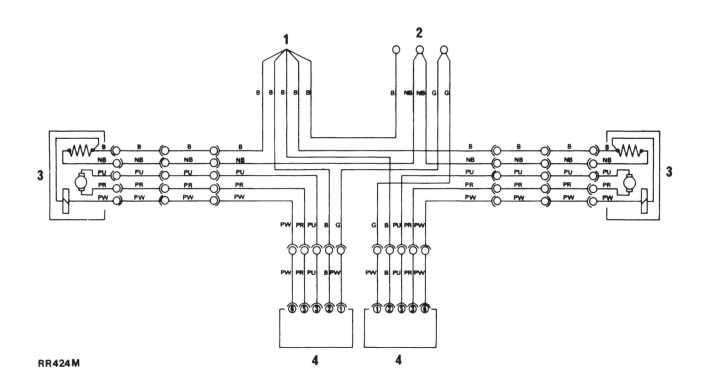

RR424M

Key to cable colours

B Black   G Green   K Pink   L Light   N Brown   O Orange   P Purple   R Red   S Slate   U Blue   W White   Y Yellow

The last letter of a colour code denotes the tracer colour

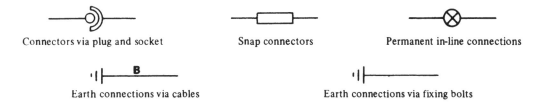

Connectors via plug and socket   Snap connectors   Permanent in-line connections

Earth connections via cables   Earth connections via fixing bolts

1   Clinch
2   Main cable connections (Item 63 on main circuit diagram)
3   Mirrors
4   Mirror switches

OPTIONAL ELECTRICAL EQUIPMENT — RANGE ROVER 2 AND 4 DOOR MODELS
— MID 1984 TO 1985

RECIRCULATORY AIR CONDITIONING CIRCUIT DIAGRAM

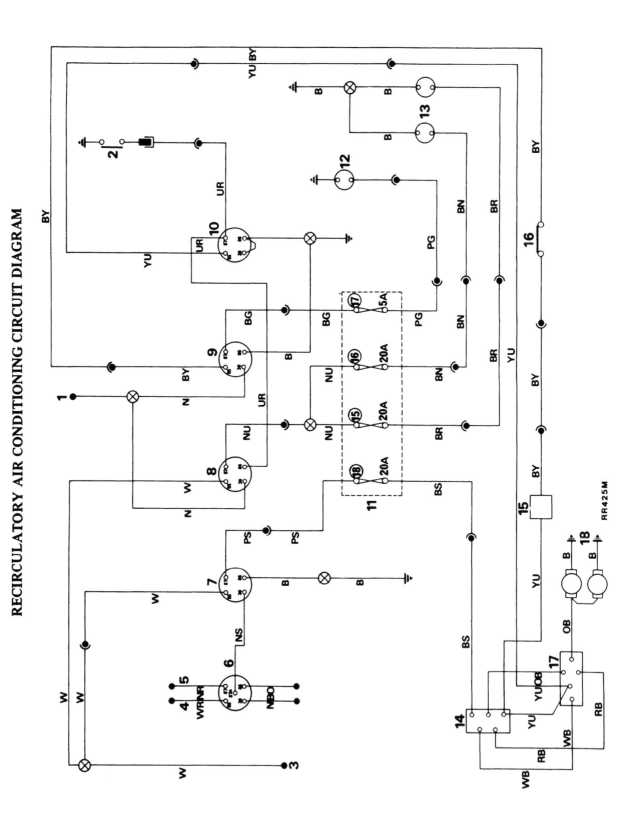

OPTIONAL ELECTRICAL EQUIPMENT – RANGE ROVER 2 AND 4 DOOR MODELS – MID 1984 TO 1985

RECIRCULATORY AIR CONDITIONING CIRCUIT DIAGRAM

1 Terminal post
2 Engine water temperature switch (Automatic only)
3 Ignition feed
4 Crank feed
5 Starter solenoid
6 Starter relay (Item 33, air conditioning inhibited supply, main circuit diagram)
7 Air conditioning relay (ignition controlled)
8 Fan relay
9 Compressor clutch relay
10 Air conditioning controlled fan relay
11 Fuse box
12 Compressor clutch
13 Fans
14 Control switch
15 Thermostat.
16 High pressure switch
17 Resistor
18 Blower motors

Key to cable colours

B Black G Green K Pink L Light N Brown O Orange P Purple R Red S Slate U Blue W White Y Yellow

The last letter of a colour code denotes the tracer colour

Connectors via plug and socket

Snap connectors

Permanent in-line connections

Earth connections via cables

Earth connections via fixing bolts

OPTIONAL ELECTRICAL EQUIPMENT – RANGE ROVER 4 DOOR MODELS
– MID 1984 TO 1985

WINDOW LIFTS AND DOOR LOCKS CIRCUIT DIAGRAM (DOOR LOCKING CONTROLLED BY EITHER FRONT SIDE DOOR SWITCH)

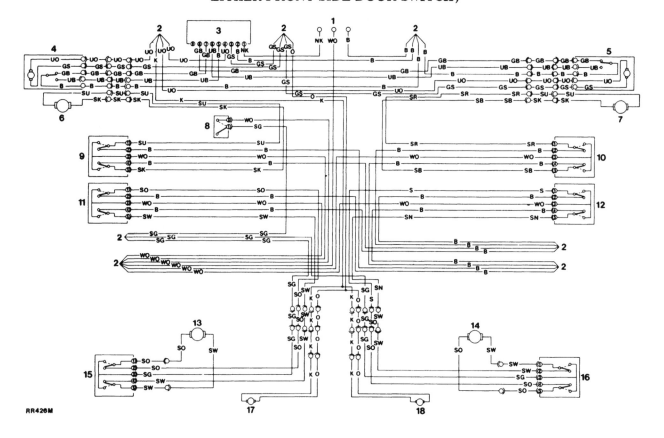

RR426M

1   Main cable connections (Item 89 NK, WK, B on main circuit diagram)
2   Clinches
3   Control box central door locking
4   Switch unit central door locking left hand front
5   Switch unit central door locking right hand front
6   Window lift motor left hand front
7   Window lift motor right hand front
8   Isolator switch
9   Window lift switch left hand front
10 Window lift switch right hand front
11 Window lift switch left hand rear
12 Window lift switch right hand rear
13 Window lift motor left hand rear
14 Window lift motor right hand rear
15 Window lift switch left hand rear door
16 Window lift switch right hand rear door
17 Lock unit central door locking left hand rear
18 Lock unit central door locking right hand rear

Key to cable colours

B Black   G Green   K Pink   L Light   N Brown   O Orange   P Purple   R Red   S Slate   U Blue   W White   Y Yellow

The last letter of a colour code denotes the tracer colour

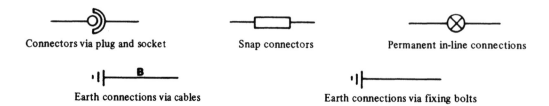

Connectors via plug and socket   Snap connectors   Permanent in-line connections

Earth connections via cables   Earth connections via fixing bolts

OPTIONAL ELECTRICAL EQUIPMENT – RANGE ROVER 4 DOOR MODELS
– MID 1984 TO 1985

WINDOW LIFTS AND DOOR LOCKS CIRCUIT DIAGRAM (DOOR LOCKING CONTROLLED BY FRONT DRIVERS DOOR SWITCH ONLY)

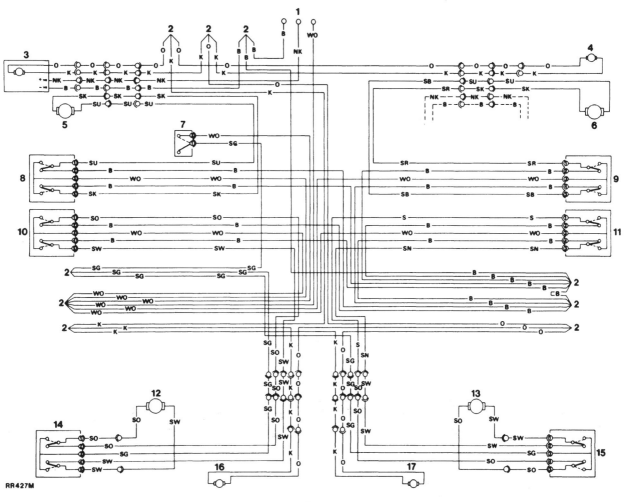

RR427M

1	Main cable connections (Item 89 NK, WK, B on main circuit diagram)
2	Clinches
3	Switch unit central door locking (drivers door)
4	Lock unit central door locking (front passenger door)
5	Window lift motor left hand front
6	Window lift motor right hand front
7	Isolator switch
8	Window lift switch left hand front
9	Window lift switch right hand front
10	Window lift switch left hand rear
11	Window lift switch right hand rear
12	Window lift motor left hand rear
13	Window lift motor right hand rear
14	Window lift switch left hand rear door
15	Window lift switch right hand rear door
16	Lock unit central door locking left hand rear
17	Lock unit central door locking right hand rear

Key to cable colours

B Black G Green K Pink L Light N Brown O Orange P Purple R Red S Slate U Blue W White Y Yellow

The last letter of a colour code denotes the tracer colour

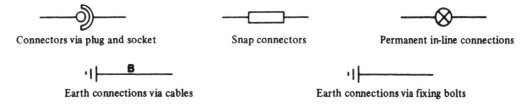

Connectors via plug and socket   Snap connectors   Permanent in-line connections

Earth connections via cables   Earth connections via fixing bolts

FAULT DIAGNOSIS

SYMPTOM	POSSIBLE CAUSE	CURE
A– Battery in low state of charge	1. Broken or loose connection in alternator circuit	1. Examine the charging and field circuit wiring. Tighten any loose connections and renew any broken leads. Examine the battery connection
	2. Current voltage regulator not functioning correctly	2. Adjust or renew
	3. Slip rings greasy or dirty	3. Clean
	4. Brushes worn not fitted correctly or wrong type	4. Renew
B– Battery overcharging, leading to burnt-out bulbs and frequent need for topping-up	1. Current voltage regulator not functioning correctly	1. Renew
C– Lamps giving insufficient illumination	1. Battery discharged	1. Charge the battery from an independent supply or by a long period of daylight running
	2. Bulbs discoloured through prolonged use	2. Renew
D– Lamps light when switched on but gradually fade out	1. Battery discharged	1. Charge the battery from an independent supply or by a long period of daylight running
E– Lights flicker	1. Loose connection	1. Tighten
F– Failure of lights	1. Battery discharged	1. Charge the battery from an independent supply or by a long period of daylight running
	2. Loose or broken connection	2. Locate and rectify
G– Starter motor lacks power or fails to turn engine	1. Stiff engine	1. Locate cause and remedy
	2. Battery discharged	2. Charge the battery either by a long period of daytime running or from an independent electrical supply
	3. Broken or loose connection in starter circuit	3. Check and tighten all battery, starter and starter switch connections and check the cables connecting these units for damage
	4. Greasy or dirty slip rings	4. Clean
	5. Brushes worn not fitted correctly or wrong type	5. Renew
	6. Brushes sticking in holders or incorrectly tensioned	6. Rectify
	7. Starter pinion jammed in mesh with flywheel	7. Remove starter motor and investigate
H– Starter noisy	1. Starter pinion or flywheel teeth chipped or damaged	1. Renew
	2. Starter motor loose on engine	2. Rectify, checking pinion and the flywheel for damage
	3. Armature shaft bearing	3. Renew
J– Starter operates but does not crank the engine	1. Pinion of starter does not engage with the flywheel	1. Check operation of starter solenoid. If correct, remove starter motor and investigate
K– Starter pinion will not disengage from the flywheel when the engine is running	1. Starter pinion jammed in mesh with the flywheel	1. Remove starter motor and investigate

FAULT DIAGNOSIS

SYMPTOM	POSSIBLE CAUSE	CURE
L− Engine will not start	1. The starter will not turn the engine due to a discharged battery	1. The battery should be recharged by running the car for a long period during daylight or from an independent electrical supply
	2. Sparking plugs faulty, dirty or incorrect plug gaps	2. Rectify or renew
	3. Defective coil or distributor	3. Remove the lead from the centre distributor terminal and hold it approximately 6mm (¼in) from some metal part of the engine while the engine is being turned over. If the sparks jump the gap regularly, the coil and distributor are functioning correctly. Renew a defective coil or distributor
	4. Fault in the low tension wiring circuit	4. Examine all the ignition cables and check that the bottom terminals are secure and not corroded
	5. Faulty amplifier	5. Check or renew
	6. Air gap out of adjustment	6. Adjust
	7. Controls not set correctly or trouble other than ignition	7. See Starting Procedure in the Owner's Instruction Manual
M− Engine misfires	1. Distributor incorrectly set	1. Adjust
	2. Faulty coil or reluctor	2. Renew
	3. Faulty sparking plugs	3. Rectify
	4. Faulty carburetter	4. Check and rectify
N− Frequent recharging of the battery necessary	1. Alternator inoperative	1. Check the brushes, cables and connections or renew the alternator
	2. Loose or corroded connections	2. Examine all connections, especially the battery terminals and earthing straps
	3. Slipping fan belt	3. Adjust
	4. Voltage control out of adjustment	4. Renew
	5. Excessive use of the starter motor	5. In the hands of the operator
	6. Vehicle operation confined largely to night driving	6. In the hands of the operator
	7. Abnormal accessory load	7. Superfluous electrical fittings such as extra lamps, etc.
	8. Internal discharge of the battery	8. Renew
P− Alternator not charging correctly	1. Slipping fan belt	1. Adjust
	2. Voltage control not operating correctly	2. Rectify or renew
	3. Greasy, charred or glazed slip rings	3. Clean
	4. Brushes worn, sticking or oily	4. Rectify or renew
	5. Shorted, open or burnt-out field coils	5. Renew
Q− Alternator noisy	1. Worn, damaged or defective bearings	1. Renew
	2. Cracked or damaged pulley	2. Renew
	3. Alternator out of alignment	3. Rectify
	4. Alternator loose in mounting	4. Rectify
	5. Excessive brush noise	5. Check for rough or dirty slip rings, badly seated brushes, incorrect brush tension, loose brushes and loose field magnets. Rectify or renew
R− Defective distributor (refer to distributor overhaul and test procedure)	1. Air gap incorrectly set	1. Adjust
	2. Distributor cap cracked	2. Renew
	3. Faulty pick-up or reluctor	3. Renew
	4. Excessive wear in distributor shaft bushes, etc.	4. Renew
	5. Rotor arm and flash shield cracked or showing signs of tracking	5. Renew

FAULT DIAGNOSIS

	SYMPTOM	POSSIBLE CAUSE	CURE
S–	Mixture control warning light fails to appear when engine reaches running temperature	1. Mixture control already pushed in 2. Broken connection in warning light circuit 3. Blown bulb 4. Faulty thermostat switch (at cylinder head) 5. Faulty manual switch (at mixture control) 6. Broken operating mechanism at manual switch	1. In the hands of the operator 2. Rectify 3. Renew 4. Renew 5. Renew 6. Rectify
T–	Mixture control warning light remains on with engine at running temperature	1. Mixture control out 2. Faulty manual switch 3. Broken operating mechanism at manual switch	1. Push control right in 2. Renew 3. Rectify
U–	Poor performance of horns	1. Low voltage due to discharged battery 2. Bad connections in wiring 3. Loose fixing bolt 4. A faulty horn	1. Recharge 2. Carefully inspect all connections and horn push 3. Rectify 4. Adjust or renew
V–	Central door locking does not operate (on all four doors)	1. Battery discharged 2. Control unit in drivers door lock actuator faulty 3. Loose or broken connection in drivers door 4. Blown fuse	1. Recharge 2. Renew 3. Locate and rectify 4. Rectify
W–	Central door locking does not operate (on one door only)	1. Loose or broken connection 2. Lock actuator failure 3. Faulty lock 4. Mechanical linkages disconnected	1. Locate and rectify 2. Renew 3. Rectify 4. Locate and rectify
X–	Window lift will not operate	1. Motor failure 2. Loose or broken connection 3. Faulty switch 4. Mechanical linkage faulty	1. Renew 2. Locate and rectify 3. Renew 4. Rectify
Y–	Exterior mirrors fail to operate	1. Loose or broken connection 2. Faulty switch 3. Mirror motor failure	1. Locate and rectify 2. Renew 3. Renew

ELECTRONIC IGNITION — MID 1984 TO 1985

A Lucas model 35DM8 distributor is employed. This has a conventional advance/retard vacuum unit and centrifugal automatic advance mechanism.

A pick-up module, in conjunction with a rotating timing reluctor inside the distributor body, generates timing signals. These are applied to an electronic ignition amplifier unit fitted under the ignition coil mounted on top of the left front wing valance.

MAINTENANCE

80,000 km (48,000 miles)
Remove the distributor cap and rotor arm and wipe inside with a nap-free cloth.

DO NOT DISTURB the clear plastic insulating cover which protects the magnetic pick up module.

1. Cap
2. HT Brush and spring
3. Rotor arm
4. Insulation cover (Flash shield)
5. Pick-up and base plate assembly
6. Vacuum unit
7. 'O' ring oil seal

RR458M

LUCAS CONSTANT ENERGY IGNITION SYSTEM 35DM8 PRELIMINARY CHECKS

Inspect battery cables and connections to ensure they are clean and tight. Check battery state of charge if in doubt as to its condition.

Inspect all LT connections to ensure that they are clean and tight. Check the HT leads are correctly positioned and not shorting to earth against any engine components. The wiring harness and individual cables should be firmly fastened to prevent chafing.

PICK-UP MODULE AIR GAP SETTINGS

Air gap settings vary according to vehicle application.

NOTE:– The gap is set initially at the factory and will only require adjusting if tampered with or when the pick-up module is replaced.

Test Notes
(i) The ignition must be switched on for all checks.
(ii) Key to symbols used in the charts for Tests 2.

 Correct Reading   High Reading   Low Reading

(iii) Use feeler gauges manufactured from a non-magnetic material when setting air gaps.

TEST 1:
Check HT Sparking
Remove coil/distributor HT lead from distributor cover and hold approximately 6mm (0.25in) from the engine block. Switch the ignition 'on' and operate the starter. If regular sparking occurs, proceed to Test 6. If no sparking proceed to Test 2.

Test 1

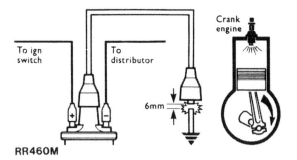

RR460M

TEST 2:
Amplifier Static Checks
Switch the ignition 'on'.
(a) Connect voltmeter to points in the circuit indicated by the arrow heads and make a note of the voltage readings.

NOTE: Only move the voltmeter positive lead during tests 2, 3 and 4.

(b) Compare voltages obtained with the specified values listed below:–

Test 2

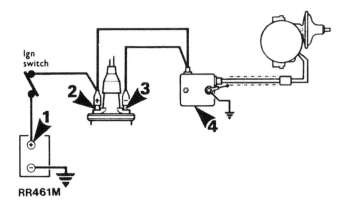

RR461M

EXPECTED READINGS
1 More than 11.5 volts
2 1 volt max below volts at point 1 in test circuit
3 1 volt max below volts at point 1 in test circuit
4 0 volt – 0.1 volt

(c) If all readings are correct proceed to Test 3.
(d) Check incorrect reading(s) with chart to identify area of possible faults, i.e. faults listed under heading "Suspect".

1	2	3	4	SUSPECT
L	✓	✓	✓	Discharged battery
✓	L	L	✓	Ign. switch and/or wiring
✓	✓	L	✓	Coil or amplifier
✓	✓	✓	H	Amplifier earth

TEST 3:
Check Amplifier Switching
Disconnect the High Tension lead between the coil and distributor.

Connect the voltmeter between battery positive (+ve) terminal and HT coil negative (–ve) terminal: the voltmeter should register zero volts. Switch the ignition 'on' and crank the engine: the voltmeter reading should increase just above zero, in which case proceed with Test 5.

If there is no increase in voltage during cranking proceed to Test 4.

continued

Test 3

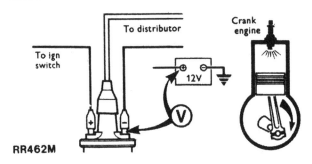

RR462M

TEST 4:
Pick-up Coil Resistance
Applications with Separate Amplifier

Disconnect the pick-up leads at the harness connector. Connect the ohmmeter leads to the two pick-up leads in the plug.

The ohmmeter should register between 2k and 5k ohms if pick-up is satisfactory. Change the amplifier if ohmmeter reading is correct. If the engine still does not start carry out Test 5.

Change the pick-up if ohmmeter reading is incorrect. If the engine still does not start proceed to Test 5.

Test 4

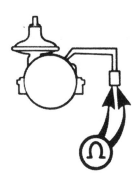

RR463M

TEST 5:
Check HT Sparking

Remove existing coil/distributor HT lead and fit test HT lead to coil chimney. Hold free end about 6mm (0.25in) from the engine block and crank the engine.

HT sparking good, repeat test with original HT lead, if then no sparking, change HT lead. If sparking is good but engine will not start, proceed to Test 6.

If no sparking, replace coil.

If engine will not start carry out Test 6.

continued

Test 5

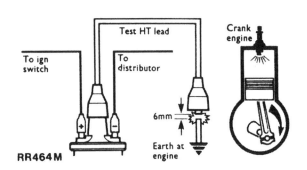

RR464M

TEST 6:
Check Rotor Arm

Remove distributor cover. Disconnect coil HT lead from cover and hold about 3mm (0.13in) above rotor arm electrode and crank the engine. There should be no HT sparking between rotor and HT lead. If satisfactory carry out Test 7.

HT sparking, replace rotor arm.

If engine will not start carry out Test 7.

Test 6

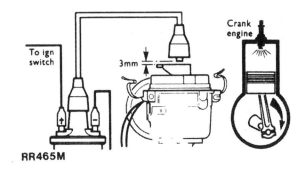

RR465M

TEST 7:
Visual and HT Cable Checks

Examine:	Should be:
1. Distributor Cover	Clean, dry, no tracking marks.
2. Coil Top	Clean, dry, no tracking marks.
3. HT Cable Insulation	Must not be cracked chafed or perished.
4. HT Cable Continuity	Must not be open circuit.
5. Sparking Plugs	Clean, dry, and set to correct gap.

NOTE:

1. Reluctor	Must not foul pick-up or leads.
2. Rotor and Flash Shield	Must not be cracked or show signs of tracking marks.

DISTRIBUTOR

– Remove and refit

Removing
1. Disconnect battery.
2. Disconnect vacuum pipe(s).
3. Remove distributor cap.
4. Disconnect low tension lead from coil.
5. Mark distributor body in relation to centre line of rotor arm.
6. Add alignment marks to distributor and front cover.
   NOTE: Marking distributor enables refitting in exact original position, but if engine is turned while distributor is removed, complete ignition timing procedure must be followed.
7. Release the distributor clamp and remove the distributor.

Refitting
NOTE: If a new distributor is being fitted, mark body in same relative position as distributor removed.
8. Leads for distributor cap should be connected as illustrated.
   Figures 1 to 8 inclusive indicate plug lead numbers.
   RH – Right hand side of engine, when viewed from the rear.
   LH – Left hand side of engine, when viewed from the rear.
9. If engine has not been turned whilst distributor has been removed, proceed as follows (items 10 to 17).
10. Fit new 'O' ring seal to distributor housing.
11. Turn distributor drive until centre line of rotor arm is 30° anti-clockwise from mark made on top edge of distributor body.
12. Fit distributor in accordance with alignment markings.
    NOTE: It may be necessary to align oil pump drive shaft to enable distributor drive shaft to engage in slot.
13. Fit clamp and bolt. Secure distributor in exact original position.
14. Connect vacuum pipe to distributor and low tension lead to coil.
15. Fit distributor cap and check that the plug leads are fitted in the order illustrated, or the engine will misfire.

continued

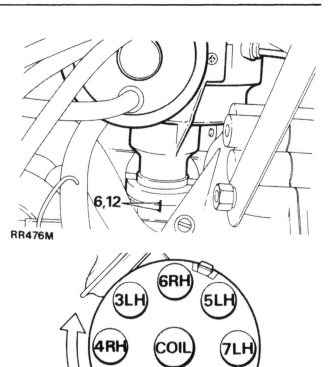

RR476M

RR616M

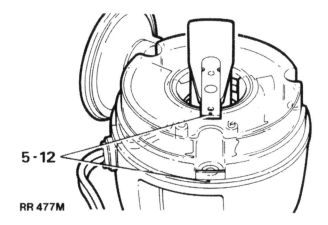

RR 477M

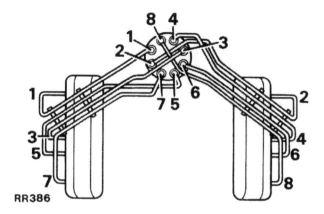

RR386

658

16. Reconnect battery.
17. Using suitable electronic equipment, set ignition timing as follows.
18. If, with distributor removed, engine has been turned it will be necessary to carry out the following procedure.
19. Set engine – No. 1 piston to static ignition timing figure (see section 05) on compression stroke.
20. Turn distributor drive until rotor arm is approximately 30° anti-clockwise from number one sparking plug lead position on cap.
21. Fit distributor to engine.
22. Check that centre line of rotor arm is now in line with number one sparking plug lead on cap. Reposition distributor if necessary.
23. If distributor does not seat correctly in front cover, oil pump drive is not engaged. Engage by lightly pressing down distributor while turning engine.
24. Fit clamp and bolt leaving both loose at this stage.
25. Set the timing statically to within 2–3° of T.D.C.
26. Connect vacuum pipe(s) to distributor.
27. Fit low tension lead to coil.
28. Fit distributor cap.
29. Reconnect battery.
30. Using suitable electronic equipment, set the ignition timing.

IGNITION COIL

– Remove and refit

Removing
1. Disconnect the battery.
2. Disconnect the electrical leads from the coil.
3. Remove the two retaining bolts with washers securing the coil to the amplifier.
4. Lift the coil off the amplifier.

Refitting
5. Reverse the removal procedure.

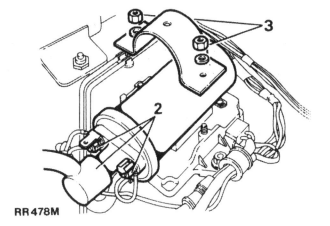

RR 478M

659

DISTRIBUTOR — LUCAS 35DM8

— Overhaul

Distributor Cap
1. Unclip and remove cap.
2. Renew cap if known to be faulty.
3. Clean cap with a nap-free cloth.

Rotor Arm
4. Pull rotor arm from keyed shaft.
5. Renew rotor arm if known to be faulty.

Insulation Cover (Flash Shield)
6. Remove cover, secured by three screws.
7. Renew cover if known to be faulty.

Vacuum Unit
8. Remove two screws from vacuum unit securing bracket, disengage vacuum unit connecting rod from pick-up base plate connecting peg, and withdraw vacuum unit from distributor body.

Pick-up & Base Plate Assembly
9. Use circlip pliers to remove the circlip retaining the reluctor on rotor shaft.
10. Remove the flat washer and then the 'O' ring recessed in the top of the reluctor.
11. Insert the blade of a small screwdriver beneath the reluctor and prise it partially along the shaft, sufficient to enable it to be gripped between the fingers and withdrawn from the shaft.
    NOTE: Coupling ring fitted beneath reluctor.
12. Remove pick-up and base plate assembly, secured by 3 support pillars.
    NOTE: Do not disturb the 2 barrel nuts securing the pick-up module, otherwise the air gap will need re-adjustment.
13. Renew pick-up and base plate assembly if module is known to be faulty, otherwise check pick-up winding resistance (2k—5k ohms).

Re-assembly
14. This is mainly a reversal of the dismantling procedure, noting the following points.

Lubrication
Apply clean engine oil:
(a) 3 drops to felt pad reservoir in rotor shaft.

Apply Chevron SR1 (or equivalent) grease
(b) Auto advance mechanism.
(c) Pick-up plate centre bearing.
(d) Pre-tilt spring and its rubbing area (pick-up and base plate assembly).
(e) Vacuum unit connecting peg (pick-up and base plate assembly) and
(f) The connecting peg hole in vacuum unit connecting rod.

Apply Rocal MHT (or equivalent) grease

(g) Vacuum unit connecting rod seal (located in vacuum unit where connecting rod protrudes).
NOTE: Applicable only to double acting vacuum units.

Fitting Pick-up & Base Plate Assembly

15. Pick-up leads must be prevented from fouling the rotating reluctor. Both leads should be located in plastic carrier as illustrated. Check during re-assembly.

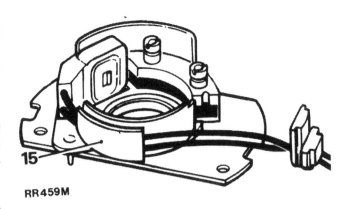

RR459M

Refitting Reluctor

16. Slide reluctor as far as it will go on rotor shaft, then rotate reluctor until it engages with the coupling ring beneath the pick-up base plate. The distributor shaft, coupling ring and reluctor are 'keyed' and rotate together.

Pick-up Air Gap Adjustment

17. The air gap between the pick-up limb and reluctor teeth must be set within the specified limits, using a non-ferrous feeler gauge.
NOTE: When the original pick-up and base plate assembly has been refitted the air gap should not normally require resetting as it is pre-set at the factory. When renewing the assembly the air gap will require adjusting to within the specified limits. See technical data section. Refer to 'Engine Tuning Data' for ignition timing data.

TECHNICAL DATA

Firing angles	$0-45-90°$ etc. $\pm 1°$
Application	12V Negative earth
Pick up air gap adjustment	0.20mm–0.35mm (0.008in–0.014in)
(Pick up limb/reluctor tooth)	
Pick up winding resistance	·2k–5k ohms

TIGHTENING TORQUES

Pick up bearing plate support pillars	1.0–1.2 Nm (9–11 lb in)
Pick up barrel nuts	1.1–1.5 Nm (10–12 lb in)

AMPLIFIER

– Remove and refit

Removing

18. Disconnect the battery.
19. Disconnect the electrical leads from the amplifier and coil.
20. Remove the two retaining bolts with washers securing the coil to the amplifier.
21. Remove the two bolts, nuts, spring washers and plain washers securing the amplifier to the valance.
22. Remove the amplifier from the valance.

Refitting

23. Reverse the removal procedure, ensuring that all electrical leads are correctly reconnected.
NOTE: The amplifier is not serviceable, in the event of a fault a new amplifier must be fitted.

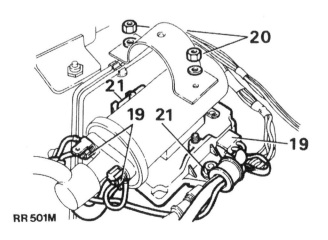

RR501M

Ignition Timing

1. It is essential that the following procedures are adhered to. Inaccurate timing can lead to serious engine damage and additionally create failure to comply with the emission regulations applying to the country of destination. If the engine is being checked in the vehicle and is fitted with an air conditioning unit the compressor must be isolated.

2. On initial engine build, or if the distributor has been disturbed for any reason, the ignition timing must be set statically to within 2–3° of T.D.C. (This sequence is to give only an approximation in order that the engine may be started). **ON NO ACCOUNT MUST THE ENGINE BE STARTED BEFORE THIS OPERATION IS CARRIED OUT.**

Equipment required
Calibrated Tachometer
Stroboscopic lamp

3. Couple stroboscopic timing lamp and tachometer to engine following the manufacturers instructions.

4. Disconnect the vacuum pipes from the distributor.

5. Start engine, with no load and not exceeding 3,000 rpm, run engine until normal operating temperature is reached. (Thermostat open). Check that the normal idling speed falls within the tolerance specified in Engine Tuning Data.

6. Idle speed for timing purposes must not exceed 750 rpm, and this speed should be achieved by removing a breather hose **NOT BY ADJUSTING CARBURETTER IDLE SETTING SCREWS.**

7. With the distributor clamping bolt slackened, turn distributor until the timing flash coincides with the timing pointer and the correct timing mark on the rim of the torsional vibration damper as shown in the Engine Tuning Data.

8. Retighten the distributor clamping bolt securely. Recheck timing in the event that retightening has disturbed the distributor position.

9. Refit vacuum pipes.

10. Disconnect stroboscopic timing lamp and tachometer from engine.

FUSE BOX – MID 1984 TO 1985

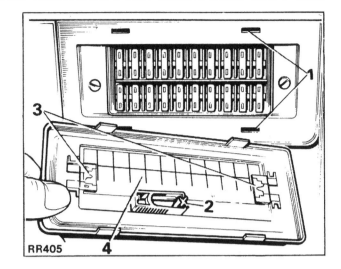

1. The fuse box on the lower fascia contains twenty 'Autofuse' type fuses. To gain access remove the clip-on cover.
2. The cover contains a clip-in fuse extractor to facilitate fuse removal.
3. A spare 10 amp and 20 amp fuse are also clipped into the cover.
4. A label identifying the position and values of the twenty fuses is attached to the cover recess for easy reference. Each fuse is colour coded and continuous current rating is specified on the fuse box.

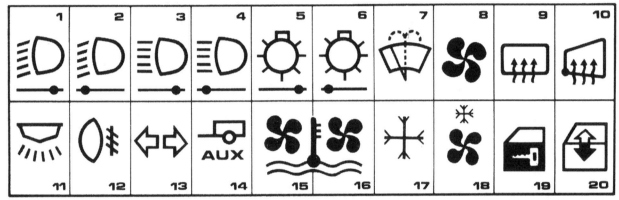

Key to fuse box circuits

FUSE NO.	COLOUR CODE	FUSE VALUE	CIRCUIT SERVED	IGNITION KEY CONTROLLED
1	Brown	7.5 amp	RH headlamp dipped beam	–
2	Brown	7.5 amp	LH headlamp dipped beam	–
3	Brown	7.5 amp	RH headlamp main beam	–
4	Brown	7.5 amp	LH headlamp main beam	–
5	Tan	5 amp	RH side and panel lights	–
6	Tan	5 amp	LH side lights	–
7	Light Blue	15 amp	Front and rear wiper motors	AUX
8	Yellow	20 amp	Heater motor	AUX
9	Light Blue	15 amp	Heated rear screen	IGN
10	Violet	3 amp	Electric mirror heating elements – option	IGN
11	Light Blue	15 amp	Interior lights, under bonnet illumination, clock, cigar lighter, headlamp flash, horns	–
			Cigar lighter (Australia only)	IGN
12	Red	10 amp	Rear fog guard (from dipped headlamps)	–
13	Light Blue	15 amp	Directional indicators, stop lights, reverse lights, and electric mirror motors	IGN
14	Light Blue	15 amp	Auxiliary circuit to trailer	–
15	Yellow	20 amp	Air conditioning fan – option	IGN
16	Yellow	20 amp	Air conditioning fan – option	IGN
17	Tan	5 amp	Air conditioning compressor clutch – option	IGN
18	Yellow	20 amp	Air conditioning blower motor – option	IGN
19	Red	10 amp	Central door locking – option	–
20	White	25 amp	Electric window lifts – option	AUX

NOTE: Radio/cassette combination – option. An in-line type 7 amp fuse is incorporated in the power input lead of the unit.

FUSE BOX – MID 1984 TO 1985

– Remove and refit

Removing

1. Disconnect the battery.
2. Unclip the fuse box cover.
3. Remove the two screws retaining the fuse box body to the lower fascia.
4. Manoeuvre the fuse box body to enable it to be withdrawn through the fuse box aperture. Withdraw only as far as the leads will permit.
5. Remove all the fuses from the fuse box.
6. Remove the leads from the fuse box, by inserting a small screwdriver into each fuse socket to depress the small retaining tab on the back of the lucar connectors, withdraw the leads from the rear of the fuse box.

Refitting

7. This is a reversal of the removal instructions 1 to 6. Ensure that all leads are refitted to the correct fuse socket, (refer to main circuit diagram).
   NOTE: When refitting the leads to the fuse box, the retaining tabs on the back of the lucar connectors must be in their raised position to prevent the leads being pushed out of the rear of the fuse box when the fuse is refitted.

AUXILIARY SWITCH PANEL – MID 1984 TO 1985

The auxiliary switch panel on the centre console contains four 'paddle' type switches which incorporate integral symbols for identification. The symbols are illuminated by a fibre optic light source which becomes operational when the vehicle lights are on.

The hazard warning and heated rear screen switches (1 and 4) are also provided with individual warning lights, illuminated when the switches are operated.

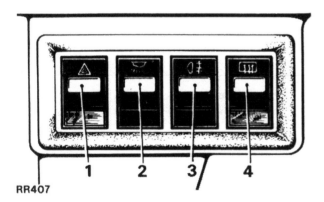

RR407

1. Hazard warning
2. Interior roof and tailgate lights
3. Rear fog guard lamps
4. Heated rear screen

Auxiliary switch panel warning lights – bulb replacement (switches 1 and 4)

To replace either bulb:

1. Disconnect the battery.
2. Carefully prise out the auxiliary switch panel from the fascia panel.
3. Remove the warning light bulb holder located in the bottom of the appropriate light guide clip moulding (see remove and refit Item 4).
4. Pull the bulb from the holder.
5. Renew the bulb and refit the holder.
6. Press the auxiliary switch panel back into the fascia.
7. Reconnect the battery.

The correct bulb type is a 1.2 watt 'wedge' base.

Auxiliary switch panel illumination

A fibre optic light source is employed to illuminate the four switch symbols.

To replace the single bulb:

8. Disconnect the battery.
9. Remove the lower fascia panel by releasing the six screws.
10. Pull the bulb holder from the rear of the optic light source unit. This unit is attached to the bottom of the central console moulding (see Remove and Refit Item 11 for full details).
11. Remove the bayonet type bulb from the holder.
12. Fit new bulb and refit holder.
13. Replace the lower fascia panel and retaining screws.

The correct bulb type is a 5 watt bayonet fitting.

AUXILIARY SWITCH PANEL — MID 1984 TO 1985

— Remove and refit

Removing
1. Disconnect the battery.
2. Carefully prise the switch panel surround from the centre console.
3. Withdraw the auxiliary switch panel assembly from the console to gain access to the connections at the rear of the switch panel.
4. If necessary the two warning light bulbs (switches 1 and 4 illustrated) can be removed at this stage.
5. Disconnect the multi-plugs at the rear of the switches by depressing the retaining lugs at the top and bottom of the plugs to release them from their sockets.
6. Remove the four fibre optic light guide clip mouldings from their intermediate positions on the rear of the switches. The fibre optic tubes and the two warning light leads will still be attached. The auxiliary panel can then be released from the centre console.
7. To remove each individual switch from the auxiliary panel surround, apply a little pressure to the rear of the switch, releasing the four retaining lugs. The switch can then be removed from the front of the surround.

   NOTE: If it is necessary to renew the fibre optic light tubes and the optic light source unit, these may be removed as follows.

8. Pull the light tubes from their respective light guide clip mouldings.
9. Similarly, release the other end of the light tubes from the optic light source unit and withdraw them through the auxiliary switch panel aperture.

   NOTE: To refit the light tubes to the optic light source unit, it is necessary to remove the unit from the back of the centre console in order to refit the light tubes.

10. Remove the lower fascia panel by releasing the six self tapping screws.
11. Pull the bulb from the rear of the optic light source unit. This is located forward of the auxiliary panel aperture attached to the bottom of the central console moulding.
12. Remove the two self tapping screws securing the optic light source unit and withdraw it from the central console.

Refitting
13. Reverse operations 1 to 11. Ensure that when refitting the optic light source unit (items 8 to 11) the optic light tubes are connected prior to locating the unit in the central console.
14. The light guide clips must be refitted in their intermediate position (registered) on the switch body.
15. It is important that the multi-plugs are reconnected to their respective sockets (see main circuit diagram).

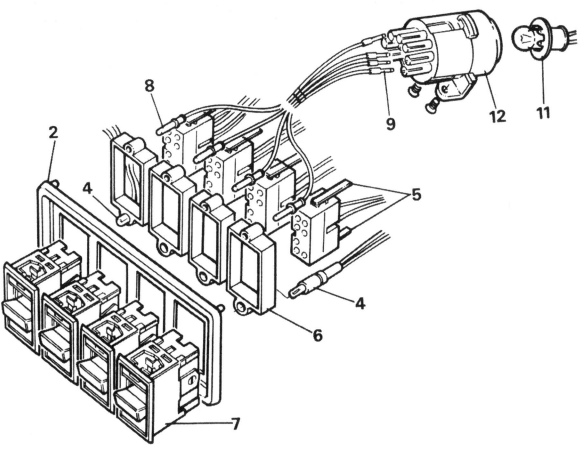

RR446M

INSTRUMENT ILLUMINATION
ELECTRONIC DIMMING CONTROL — MID 1984 TO 1985

The electronic dimming control switch is located on the lower fascia panel adjacent to the steering column. Rotate the control upwards to fully illuminate the instruments and downwards to reduce intensity.

The dimming control unit also controls the clock, heater and cigar lighter illumination.

— Remove and refit

Removing
1. Disconnect the battery.
2. Remove the lower fascia panel by releasing the six securing screws.
3. Disconnect the dimming control multi-plug.
4. Remove the two screws securing the dimmer control switch to the underside of the lower fascia panel.

Refitting
Reverse operations 1 to 4.

IGNITION/STARTER SWITCH

– Remove and refit

Removing
1. Disconnect the battery.
2. Remove the steering wheel centre cover.
3. Remove the lower fascia panel.
4. Remove the four screws securing bottom shroud to top shroud.
5. Remove the single screw securing top shroud to switch housing bracket.
6. Remove top shroud and lower the bottom shroud.
7. Disconnect the ignition switch cables at the multi-plug.
8. Remove the rubber cover protecting the switch.
9. Remove the single screw securing the ignition/starter switch to the housing.
10. Withdraw the switch.

Refitting
11. Reverse the removal procedure.
12. Locate the lugs on the sides of the switch with the grooves on the inside of the housing.

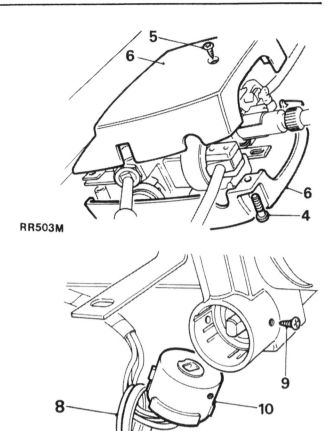

RR503M

The steering column switch layout has been standardised for left and right hand drive vehicles and is as follows:–

LEFT HAND CONTROLS

Lower switch – Main lighting switch
Upper switch – Main and dipped beam, direction indicators and horn.

RIGHT HAND CONTROLS

Lower switch – Rear screen programmed wash/wipe
Upper switch – Windscreen programmed wash/wipe.

RR504M

MAIN LIGHTING SWITCH

– Remove and refit

Removing
1. Disconnect the battery.
2. Remove the steering wheel centre cover.
3. Remove the lower fascia panel.
4. Remove the four screws securing bottom shroud to top shroud.
5. Remove the single screw securing top shroud to switch housing bracket.
6. Remove top shroud, and lower the bottom shroud.
7. Disconnect cables at snap connectors.
8. Loosen the switch retaining lock-nut.
9. Slide switch unit away from its bracket.

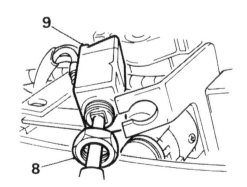

Refitting
10. Reverse the removal procedure.

REVERSE LIGHT SWITCH – Automatic gearbox

– Remove and refit

Removing

The reverse light switch is an integral part of the start inhibitor switch and is located on the left hand side of the gearbox, accessible from inside the vehicle through the gearbox tunnel side aperture.

1. Disconnect the battery.
2. Fold back the LH front footwell rubber mat.
3. Fold back the gearbox tunnel carpet to expose the cover plate.
4. Remove the twelve retaining bolts and remove the cover.
5. Disconnect the plug from the inhibitor switch.
6. Unscrew the switch and remove complete with 'O' ring.
7. Remove the 'O' ring.

Refitting

8. Reverse the removal procedure, fitting a new 'O' ring.

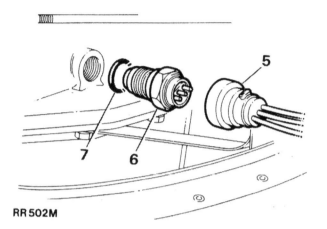

RR 502M

WINDSCREEN PROGRAMMED WASH/WIPE SWITCH

MAIN AND DIPPED BEAM, DIRECTION INDICATORS AND HORN SWITCH

– Remove and refit

Removing

1. Disconnect the battery.
2. Remove the steering wheel centre cover.
3. Remove the lower fascia panel.
4. Remove the four screws securing the bottom shroud to the top shroud.
5. Remove the single screw securing the top shroud to the switch housing bracket.
6. Remove the top shroud and lower the bottom shroud.
7. Disconnect the electrical leads at the multi-plugs.
8. Remove the two screws securing the windscreen wash/wipe switch to the column switch bracket.
9. Remove the switch to give access to the screws securing the main and dipped beam switch.
10. Release the two screws securing the upper switch to the switch bracket.
11. Slide the switch and bracket off the steering column.

Refitting

12. Reverse the removal procedure.

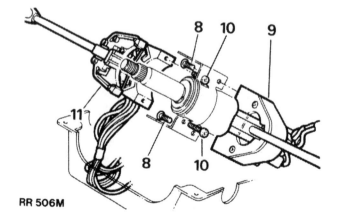

RR 506M

REAR SCREEN PROGRAMMED WASH/WIPE SWITCH

– Remove and refit

Removing
1. Disconnect the battery.
2. Remove the steering wheel centre cover.
3. Remove the lower fascia panel.
4. Remove the top and bottom shroud (refer to main and dipped beam switch remove and refit – items 4 to 7).
5. Disconnect the electrical leads at the multi-plug.
6. Remove the four small bolts securing the switch to the switch mounting bracket.
7. Remove the switch from the bracket.

Refitting
8. Reverse the removal procedure.

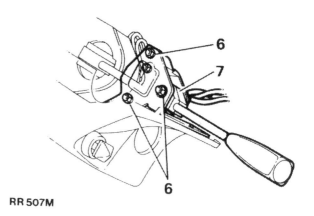

RR 507M

HANDBRAKE WARNING SWITCH

– Remove and refit

Removing
1. Disconnect the battery.
2. Apply the handbrake.
3. Fold back the gearbox tunnel carpet and sound deadening pads to reveal the screws securing the rubber gaiter.
4. Remove the screws securing the bottom of the rubber gaiter.
5. Peel back the gaiter to give access to the handbrake switch.
6. Release the locknut securing the switch to the mounting bracket and remove the switch from the bracket.
7. Pull the switch through the handbrake aperture to reveal the electrical connections.
8. Disconnect the two leads and withdraw the switch.

Refitting
9. Reverse the removal procedure.
   NOTE: On Left Hand Drive vehicle the switch is removed from underneath the vehicle.

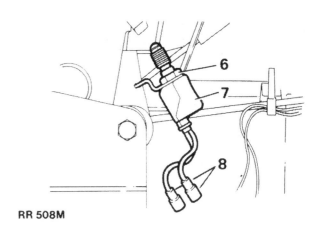

RR 508M

STOP LIGHT SWITCH

– Remove and refit

Removing
1. Disconnect the battery.
2. Remove the lower fascia panel.
3. Depress the foot brake.
4. Remove the rubber protector from switch (where fitted).
5. Remove the hexagon nut.
6. Withdraw the switch.
7. Disconnect the electrical leads.

Refitting
8. Reverse the removal procedure.

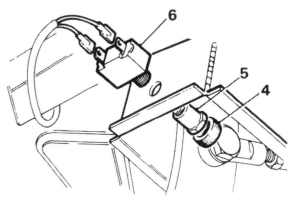

RR509M

CHOKE WARNING LIGHT SWITCH

— Remove and refit

Removing
1. Disconnect the battery.
2. Remove the six screws securing the lower fascia panel.
3. Remove the lower fascia panel to give access to the rear of the upper fascia.
4. Remove the two electrical leads from the switch.
5. Remove the screw and clip securing the switch to the choke cable and slide the clip off the switch.
6. Remove the switch.

Refitting
7. Reverse the removal operations.
8. Ensure that the three pegs on the switch locate in the corresponding hole on the choke outer cable.

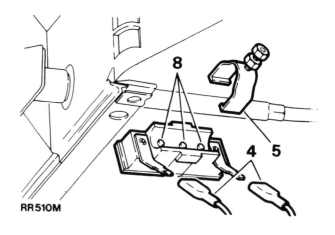

RR 510M

CIGAR LIGHTER

— Remove and refit

Removing
1. Disconnect the battery.
2. Carefully prise out the auxiliary switch panel from the centre console to gain access to the rear of the cigar lighter.
3. Pull the switch panel forward as far as the electrical leads will permit.
4. Remove the electrical leads at the rear of the cigar lighter.
5. Remove the push in switch from the lighter outer body.
6. Supporting the front of the console, and applying pressure to the plastic surround (shown by the two arrows) push the outer body through the plastic surround.
7. Manoeuvre the body and outer plastic surround to enable it to be withdrawn from the console complete with bulb holder.
8. Pull the bulb holder out of the plastic surround.
9. Pull the 1.2 watt 'wedge' type bulb from the bulb holder.

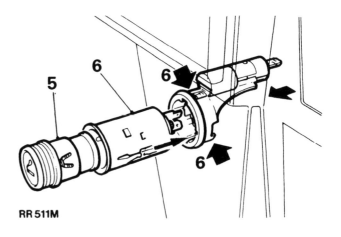

RR 511M

Refitting
10. Reverse the removal procedure.
11. Fit the bulb holder to the surround and reconnect the electrical lead.
12. Push the plastic surround into the console and pull the remaining two leads through the surround.
13. Push the lighter outer body into the surround ensuring that the raised positioning tab is located in either of the two small keyways in the plastic surround.
14. Push the outer body into the surround until fully secured.

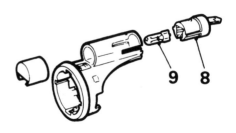

RR 512M

WINDOW LIFT SWITCHES

— Remove and refit

Removing
1. Disconnect the battery.
2. Carefully prise the switch(es) out of the centre floor mounted cover.
3. Disconnect the multi-plug at the rear of the switch(es).
4. Remove the switch(es).
   NOTE: If the multi-plugs are difficult to remove from the rear of the switches it may be necessary to remove the cubby box to disconnect the plugs from behind the floor mounted cover.

Refitting
5. Reverse the removal procedure items 1 to 4.

HEADLAMP ASSEMBLY

— Remove and refit

Removing
1. Disconnect the battery.
2. Remove screws and washers securing the headlamp frame to the body.
3. Ease the headlamp assembly forward and disconnect.
4. Remove the two adjusting screws and one clamp to separate lamp unit from the frame.
5. Remove the rubber seal.
6. Separate the lamp unit from rim by loosening the three retaining screws.

Refitting
7. Reverse removal procedure.

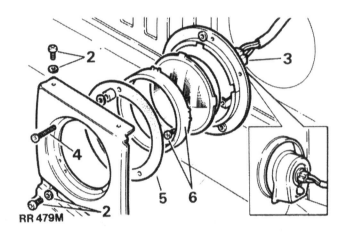

RR 479M

HEADLAMP BULB REPLACEMENT

— Remove and refit (Halogen)

Removing
1. Prop open the bonnet. Two large clearance holes are provided, one on each side of the front valance, to give access to the respective bulb holders in the headlamp reflectors.
   NOTE: To obtain access to the right-hand clearance hole it will be necessary to remove the battery from the vehicle.
2. Disconnect the multi-plug lead.
3. Remove the rubber dust cover.
4. Release the bulb retaining spring clip.
5. Remove the faulty bulb.

Refitting
6. Fit the correct 'Halogen' type. The bulb holder is keyed to facilitate fitting.
   IMPORTANT: Do not touch the quartz envelope of the bulb with the fingers. If contact is accidentally made wipe gently with methylated spirits.
7. Refit the bulb retaining spring clip rubber dust cover and multi-plug lead.
8. In the case of right-hand bulb replacement, refit the battery.
9. Reverse remaining removal procedure.

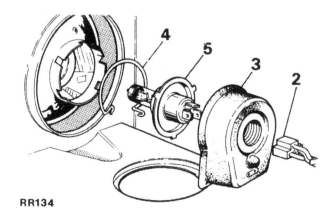

RR134

SIDE LIGHT AND FLASHER LAMP ASSEMBLY
RH AND LH AND BULB REPLACEMENT

– Remove and refit

Removing
1. Disconnect the battery.
2. Remove the four screws securing the lamp lens.
3. Release the lamp lens from the lamp body.
4. Remove the foam rubber seal.
5. Remove the two bayonet fitting bulbs.
6. Remove the two screws securing the lamp body.
7. Ease the lamp body forward to reveal the electrical connection.
8. Disconnect the electrical plug at the rear of the lamp beds.
9. Remove the lamp body.

Refitting
10. Reverse the removal procedure.

RR 480M

REFLECTORS

– Remove and refit

Removing
1. Remove the four screws securing reflector.
2. Remove reflector.
3. Remove rubber seal.

Refitting
4. Reverse the removal procedure.

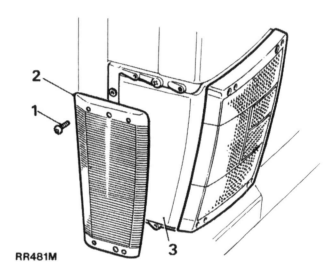

RR481M

TAIL, STOP, REVERSE, FOG GUARD AND
FLASHER LAMP ASSEMBLY RH AND LH

– Remove and refit

Removing
1. Disconnect the battery.
2. Remove the four lens retaining screws.
3. Remove lens.
4. Remove sealing rubber.
5. Remove the bulbs.
6. Remove the four screws securing the lamp unit to the body.
7. Remove the two through-screws from the reflector side, which also secure the lamp unit to the body.
8. Ease the lamp unit forward and disconnect leads at moulded connections.

Refitting
9. Reverse the removal procedure.

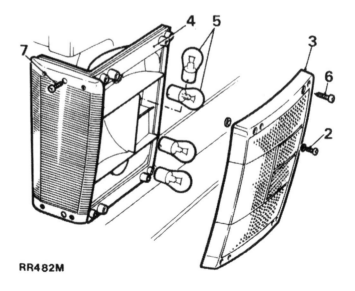

RR482M

ENGINE COMPARTMENT LAMP ASSEMBLY

— Remove and refit

Removing

1. Disconnect the battery.
2. Remove the two securing screws.
3. Remove the lamp glass.
4. Pull the 5 watt 'wedge' type bulb from the bulb holder.
5. Disconnect the electrical leads located below the bonnet lamp switch attached to the inner wing.
6. Pull the rubber grommet off the leads and pull the lamp and leads up through the bonnet stiffener channel.

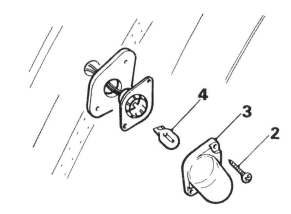

RR483M

Refitting

7. Reverse operations 1 to 6.
   NOTE: A piece of bent wire will be needed to pull the electrical leads out of the channel exit hole when fitting a new lamp assembly.

SIDE REPEATER LAMPS

— Remove and refit

Removing

1. Disconnect the battery.
2. Remove the single screw retaining the lamp lens.
3. Remove the lamp lens and rubber seal.
4. Remove the 4 watt bayonet fitting bulb.
5. Remove the two nuts and spring washers from the rear of the lamp body accessible from behind the front wing.
6. Remove the earth wire from the rear of the lamp.
7. Disconnect the twin snap connector from within the engine compartment located directly behind the lamp.
8. Remove the lamp body from the outer wing.

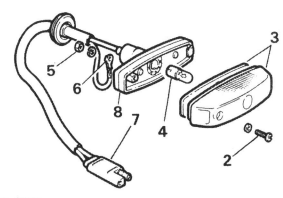

Refitting

9. Reverse the removal procedure.

RR457M

NUMBER PLATE LAMP ASSEMBLY
AND BULB REPLACEMENT

– Remove and refit

Removing
1. Disconnect the battery.
2. Remove the two self-tapping screws and fibre washers.
3. Detach the lens surround and lamp lens.
4. Remove the bulb.
   NOTE: Carefully pull the electrical leads out of the bottom of the lower tailgate panel to reveal the snap connectors.
5. Disconnect the electrical connections located at the bottom of the lower tailgate.
6. Remove the bulb holder surround.
7. Carefully pull the electrical leads up through the inside of the lower tailgate panels.

Refitting
8. Reverse the removal procedure.

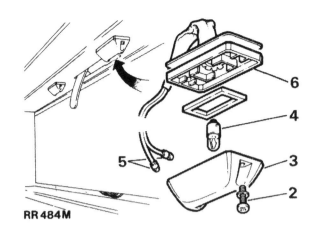

RR 484M

INTERIOR ROOF LAMPS

– Remove and refit

The interior roof lamps are operated automatically via the side door and tailgate courtesy switches or by independent switch located on the auxiliary switch panel.

Removing
1. Disconnect the battery.
2. Remove the lens from the courtesy lamp by pressing upward and turning it anti-clockwise.
3. Withdraw bulb from spring clip holder.
4. Remove screws securing lamp base to roof panel.
5. Lower the lamp to reveal the cable snap connections.
6. Disconnect the electrical connections.

Refitting
7. Reverse the removal procedure.

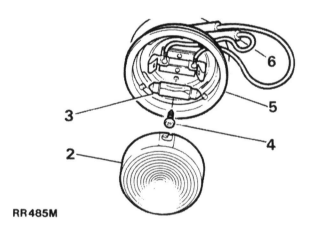

RR 485M

INTERIOR ROOF LAMPS CIRCUIT DELAY

– Remove and refit

The roof lamp circuit incorporates a delay function which is designed to allow the lamps to remain on for 12 to 18 seconds after either of the front doors are closed.
NOTE: Switching on the ignition (with both doors closed) will immediately override this feature, switching the interior lamps off.

continued

Removing

1. Disconnect the battery.
2. Remove the six screws securing the lower fascia panel.
3. Lower the fascia panel to gain access to the red delay unit attached to the steering column support bracket.
4. Remove the delay unit by pushing the unit up off its retaining bracket, to clear the steering column support bracket.
5. Pull the red multi-plug off the delay unit.

Refitting

6. Reverse the removal operations.

RELAYS

Incorporated into the vehicle electrical circuits are several relays some of which are located on the front and rear of the engine compartment closure panel, on the opposite side of the vehicle to the steering column. The remaining relays are located behind the lower fascia panel attached to the steering column support bracket.

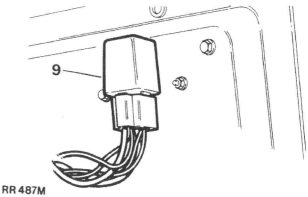

RR 487M

Closure panel viewed from the engine compartment.

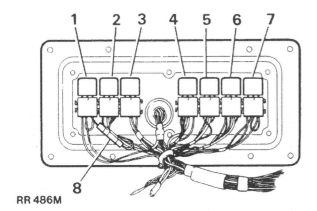

RR 486M

Closure panel viewed from inside the vehicle.

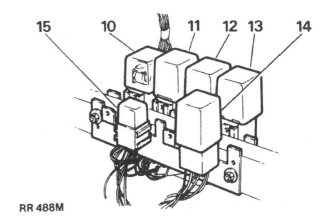

RR 488M

Viewed from inside the vehicle.

1.	Brake failure warning check relay	(79 on main circuit diagram)	(Metal case)
2.	Heated rear window relay	(95 on main circuit diagram)	(Metal case)
3.	Starter solenoid relay	(93 on main circuit diagram)	(Metal case)
4.	Fan relay	(8 on optional Air con circuit diagram)	(Metal case)
5.	Compressor clutch relay	(9 on optional Air con circuit diagram)	(Metal case)
6.	Air conditioning relay (ignition controlled)	(7 on optional Air con circuit diagram)	(Metal case)
7.	Air conditioning controlled fan relay	(10 on optional Air con circuit diagram)	(Metal case)
8	Diode	(96 on main circuit diagram)	
9.	Headlamp wash timer relay	(61 on main circuit diagram)	(Black case)
10.	Hazard and flasher relay	(81 on main circuit diagram)	(Transparent blue case)
11.	Rear wiper delay	(116 on main circuit diagram)	(Black case)
12.	Interior light delay	(30 on main circuit diagram)	(Red case)
13.	Voltage sensitive switch (option – fitted to vehicles which will split charge or air conditioning	(5 on split charge circuit diagram) (14 on air conditioning circuit diagram) (97 on main circuit diagram)	(Yellow case)
14.	Overspeed relay (Saudi only)		(Green case)
15.	Rear wiper relay	(100 on main circuit diagram)	(Metal case)

continued

RELAYS – (mounted on the steering column support bracket)

– Remove and refit

Removing
1. Disconnect the battery.
2. Remove the lower fascia panel to give access to the relays attached to the steering column support bracket.
3. Pull the appropriate relay multi-plug off the retaining bracket.
4. Pull the relay off the multi-plug.

Refitting
5. Reverse the removal procedure.
   NOTE: The windscreen wiper delay unit is fitted to the inside of the bulkhead on the drivers side of the vehicle. To gain access to the black control box remove the lower fascia panel. To remove the control box, disconnect the electrical leads and remove the two screws securing the unit to the bulkhead.

RELAYS – (mounted on engine compartment closure plate)

– Remove and refit

Removing
6. Disconnect the battery.
7. Remove the washer bottle reservoir.
8. Remove the six bolts securing the closure plate.
9. Detach the closure plate from the surround to give access to the seven relays located on the inside of the plate.
10. Pull the appropriate relay from the multi-plug.

Refitting
11. Reverse the removal procedure.

INSTRUMENT BINNACLE — MID 1984 TO 1985

Description
The electro-mechanical instrument pack contains four gauges, a tachometer, temperature indicator, fuel indicator and a speedometer with odometer and trip mileage recorder. It also includes a control warning light panel.

Tachometer
The tachometer trigger lead is connected to the alternator 'phase tap' terminal. The multi-function unit receives the pulses and sends them on to the tachometer.

Temperature Gauge
The engine temperature thermistor (sensor) is located in the front of the inlet manifold and provides a resistance, varying with the engine temperature, to operate the gauge. The multi-function unit contains the stabilised 10 volt supply system.

Fuel Gauge
The variable resistance type fuel level sensor, fitted inside the fuel tank, is controlled by a float. The gauge senses the level of fuel in the tank by the resistance in the circuit. The multi-function unit contains the stabilised 10V supply system and controls the circuit of the low fuel warning light in the warning light panel.

Speedometer
The speedometer with odometer and trip mileage recording is a purely mechanical instrument driven by a jointed (two-piece) cable. The bottom of the lower part of the cable is fitted to the transfer gearbox speedometer housing on the rear output shaft housing.

Warning Light Panel
The central panel incorporates fifteen warning light symbols, as listed in the following chart, including two alternatives to meet Australian brake symbol requirements and an additional warning light symbol for high gearbox oil temperature in automatic models.

WARNING LIGHT SYMBOLS

 Direction indicator – left turn (green)

 Direction indicator – right turn (green)

 Park brake on – Australia only (red)

 Headlamp main beam on (blue)

 Trailer connected – flashes with direction indicators (green)

 Rear fog guard lamps on (amber)

 Ignition on (red)

 Automatic gearbox oil temperature – high (red)

 Engine oil pressure, low (red)

 Cold start, engaged (amber)

 Fuel indicator, low (amber)

Differential lock engaged (amber)

Transmission handbrake on – except Australia (red)

Brake fluid pressure, failure – Australia only (red)

Brake fluid pressure, failure – except Australia (red)

NOTE: The ignition and engine oil pressure symbols will be automatically illuminated when the ignition is switched on and extinguished wh... the engine is running. The brake fluid pressure symbol will also be illuminated while the ignition key is being held over to actuate the starter, confirming that the warning circuit is functioning correctly.

Direction indicator arrow warning lights

 Connected to the direction indicator switch. The appropriate green arrow flashes in conjunction with the selected set of indicator and side repeater lights. In addition the flasher unit is audible while the lights are flashing.

Should an indicator bulb fail, the warning lights will not function and the flasher unit will not be heard.

NOTE: The above warning lights will all flash together when the hazard warning switch on the auxiliary switch panel is operating.

Park (differential) brake warning light (Australia only)

 Connected to a switch mounted on the park brake operating linkage the warning light is illuminated when the park brake is applied with the ignition switched on.

Main beam warning light

 This warning light is connected to the dip switch and will be illuminated when the main beams are switched on.

The warning light will also be illuminated when the headlamp flasher switch is used.

Trailer warning light

 When a trailer is connected to the vehicle via a seven-pin socket – option.

It will flash in conjunction with the vehicle indicator warning lights, thus ensuring that the trailer indicator lamps are functioning correctly. In the event of an indicator bulb failure on the trailer, the trailer warning light will NOT be illuminated.

Rear fog guard lamps warning light

 Operated by the rear fog guard switch in the auxiliary switch panel. This is connected to the dip beam circuit.

Ignition/No charge warning light

 This warning light is connected to the alternator field winding and will be illuminated when the ignition is switched on and extinguished when the engine is running.

High oil temperature warning light – automatic gearbox only

 This warning light operates when the thermal switch in the oil cooler reaches a pre-set temperature.

Low engine oil pressure warning light

 Connected to the oil pressure switch in the cylinder block, this warning light will be illuminated when the ignition is switched on and extinguished when the engine is running.

continued

Cold start warning light

 The warning light is connected to a switch on the cold start control and will be illuminated while the control remains out in the operating position.

Low fuel indicator warning light

 The warning light is connected to the fuel gauge circuit and will be illuminated when there are approximately 9 litres (2 gallons) of fuel remaining in the tank.

Differential lock warning light

 Positive engagement of the differential lock actuates a switch in the transfer gearbox which illuminates the warning light.

Transmission hand brake (except Australia)

 Connected to a switch mounted on the hand-brake linkage the warning light is illuminated when the handbrake is applied with the ignition switched on.

Brake failure warning light (Australia only)

 The warning light is illuminated when the pressure differential warning actuator switch in the brake reservoir body operates, due to loss of pressure in one of the two brake systems.

Brake failure warning light (except Australia)

 The warning light is illuminated when the P.D.W.A. switch in the brake reservoir body operates, due to loss of pressure in one of the two brake systems.

RENEWAL OF PANEL AND WARNING LIGHT BULBS

1. Disconnect the battery.
2. Unclip the back of the cowl from the instrument binnacle to give access to the panel and warning light bulbs in the back of the instrument case.
3. Remove the appropriate bulb holder unit by rotating it anti-clockwise and withdrawing it.
   NOTE: The No charge ignition warning light, identified by its red coloured bulb holder, is of a higher wattage and is the only bulb which can be pulled from its holder and replaced independently.
4. Fit a new bulb holder unit and rotate clockwise to lock in position. The correct bulb type is a 1.2 watt bulb/holder unit, except the ignition bulb which is 2 watt wedge base type.
5. Refit the cowl and reconnect the battery.
   NOTE: If difficulty is experienced in changing bulbs, due to the limited space available the instrument binnacle fixings should be removed to enable the binnacle to be raised above the fascia as far as other connections permit. See 'Instrument Binnacle removal' for details of binnacle mounting bracket fixing.

INSTRUMENT BINNACLE – MID 1984 TO 1985

Removing from vehicle

1. Disconnect the battery.
2. Remove the lower fascia by releasing the six retaining screws.
3. Remove the four nuts (with spring and plain washers) from under the top fascia rail which secure the instrument binnacle to the vehicle.
4. Unclip the binnacle cowl, from the rear, to provide access to the two-part speedometer cable.
5. Disconnect the two-part speedometer cable from the speedometer drive on the back of the instrument case. Alternatively, from under the top fascia rail, release the cable connector ring at the intermediate clamped connection. This is located some 470mm (18.5in) from the speedometer drive. This connection is provided to facilitate seperate renewal of either the upper or lower part of the cable in service.
6. Disconnect the two multi-plugs from the printed circuit connectors.
7. Lift the instrument binnacle from the top fascia rail and transfer it to the workbench.

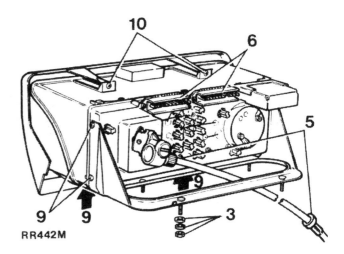

RR442M

Refitting to vehicle

8. Reverse the removal instructions 1 to 7.
   NOTE: On LHD vehicles, where an over-speed buzzer is fitted, the intermediate speedometer cable connections are threaded and retain a sensor unit between the two parts of the cable.
   A lead from the sensor unit is connected to a black two-pin socket (black and white leads) below the binnacle, above the steering column area. The adjacent buzzer will be audible at approximately 120 kph (75 mph).

Removing Instrument Pack

9. Having removed the instrument binnacle from the vehicle, detach the binnacle mounting bracket. This is secured to the instrument case by two screws and to the bottom of the binnacle bezel by two similar screws.

continued

10. Remove the two screws retaining the top of the bezel to the front housing and detach the bezel.
11. Separate the instrument case from the binnacle housing by releasing the two wire clips.
12. Detach the curved lens from the binnacle housing by releasing the wire clip at the top.

Refitting Instrument Pack to Binnacle

13. Reverse removal instructions 9 to 12.

continued

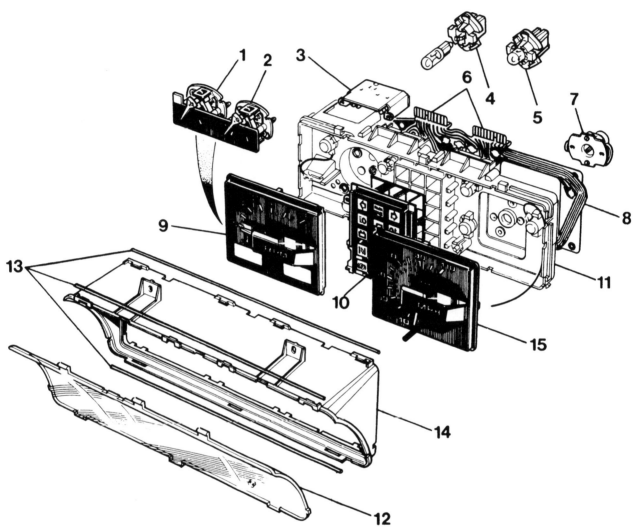

RR443M

Instrument Pack
1. Fuel gauge
2. Temperature gauge
3. Voltage stabiliser
4. Ignition warning bulb (with separate red holder unit)
5. Panel/warning lights bulb/holder
6. Printed circuit input tags (for harness connection)
7. Speedometer drive unit
8. Printed circuit
9. Tachometer
10. Warning lights panel
11. Instrument case (front)
12. Curved lens
13. Wire connecting clips
14. Binnacle housing
15. Speedometer

Renewing panel and warning lamp bulbs

14. Remove the appropriate bulb holder unit from the back of the instrument case by rotating the bulb holder anti-clockwise and withdrawing it.

    NOTE: The No charge ignition warning light, identified by its red coloured bulb holder is of a higher wattage and is the only bulb which can be separated from its holder and replaced independently.

15. Fit a new bulb holder unit to the printed circuit and rotate clockwise to lock in position.

    The correct bulb type is a 1.2 watt bulb/holder unit, except the ignition bulb which is 2 watt wedge base type.

continued

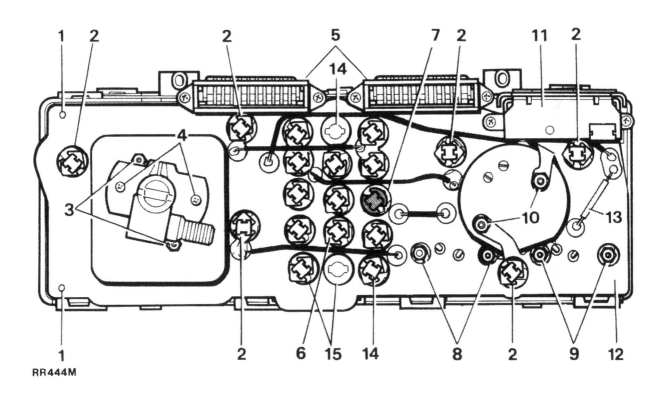

RR444M

Instrument case (back)

1. Locating pegs
2. Panel light bulbs
3. Speedometer securing screws
4. Speedometer drive securing screws
5. Harness connectors
6. Warning light bulbs
7. No charge warning light bulb (red holder)
8. Temperature gauge securing nuts
9. Fuel gauge securing nuts
10. Tachometer securing nuts
11. Multi-function unit
12. Printed circuit
13. Pull-up resistor — high temperature gearbox oil
14. Alternative symbols — park/hand brake
15. Alternative symbols — brake failure

Removing printed circuit

16. Remove the two tachometer nuts (with washers) to release the printed circuit connecting tags.
17. Remove the four nuts (with washers) securing the fuel and temperature gauges to release the printed circuit from the fixing studs.
18. Release the two screws retaining the multi-function unit and lift off to release the printed circuit connecting tag.
19. Remove the two harness connectors, retained by four screws, to release the printed circuit tags.
20. Carefully ease the printed circuit from its four locating pegs.

Refitting the Printed Circuit

21. Reverse the removal procedure, items 16 to 20.
22. Ensure that the fuel and temperature gauge mounting studs are correctly located before pressing the printed circuit on to its four locating pegs.

Removing Tachometer

23. Carefully prise the needle shroud from the tachometer and disconnect the fibre optic element underneath the shroud.
24. Remove the two nuts (with washers) at the back of the instrument case which retains the tachometer and release the printed circuit tags.
25. Slacken the four nuts retaining the fuel and temperature gauges and carefully manoeuvre the tachometer from the front of the instrument case.

Refitting the Tachometer

26. Reverse the removal procedure, items 23 to 25.

Removing Fuel and Temperature Gauge Unit

27. Carefully prise the needle shroud from the tachometer and disconnect the fibre optic element underneath the shroud.
28. Remove the two nuts (with washers) retaining the tachometer and release the printed circuit tags.
29. Remove the four nuts (with washers) retaining the fuel and temperature gauges and carefully manoeuvre the tachometer, fuel and temperature gauge unit from the front of the instrument case.

Refitting the Fuel and Temperature Gauges

30. Locate the fuel and temperature gauge unit in the instrument panel but DO NOT fit the washers and nuts at this stage.
31. Feed the fibre optic element through the aperture in the tachometer then locate the tachometer in the instrument panel.
32. Position the printed circuit tags over the two tachometer studs, fit the washers and fit and tighten the retaining nuts.
33. Fit the washers to the four fuel and temperature gauge studs and fit and tighten the retaining nuts.

Removing Fuel Gauge Tank Unit

34. Disconnect the battery.

35. Chock the front wheels, raise the rear wheels clear of the ground and support the vehicle on stands.
36. Remove the left side rear wheel to provide easy access to the gauge unit which is fitted in the side of the fuel tank.
37. Disconnect the electrical leads from the gauge unit.
38. Release the fuel feed pipe from the gauge unit by unscrewing the hexagon nut.
39. Using tool 18G 1001 release the tank unit locking ring.
40. Remove the gauge unit and sealing washer.

Refitting the Fuel Gauge Tank Unit

41. Locate the fuel gauge unit in the tank, with a new seal.
42. Using tool 18G 1001 secure the locking ring.
43. Connect the green/black lead to the white terminal and the black earth lead to the centre terminal on the gauge unit.
    NOTE: The red terminal is not used.
44. Connect the fuel feed pipe to the gauge unit.
45. Refit the rear wheel and lower the vehicle.
46. Connect the battery.
47. Fill the fuel tank.

Removing the Speedometer and Speedometer Drive Unit

48. Carefully prise the needle shroud from the speedometer and disconnect the fibre optic element underneath the shroud.
49. Remove the two hexagonal headed screws (with washers) at the back of the instrument case which retain the speedometer.
50. Carefully remove the speedometer from the front of the instrument case.
51. To release the speedometer drive unit, remove the two self-tapping screws securing it to the back of the instrument case.

Refitting the Speedometer and Speedometer Drive Unit

52. Reverse the removal procedure items 48 to 51 ensuring that the rubber gasket is fitted behind the speedometer drive unit.

PRINTED CIRCUIT HARNESS CONNECTIONS

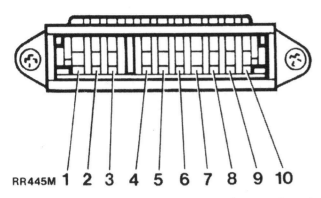
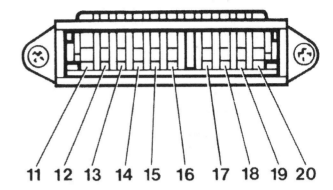

RR445M

Sequence of connections viewed from back of instrument case)

CIRCUIT SERVED

Tacho signal 1
Ignition switch 12V+ 2
Ignition warning light 3
Trailer warning light 4
Main beam warning light 5
Earth 6
Direction indicators left hand 7
Rear fog warning light 8
Direction indicators right hand 9
Oil pressure warning light 10
High oil temperature warning light (Auto gearbox) . . . 11
Cold start warning light 12
Differential lock warning light 13
Brake failure warning light 14
Brake failure warning light (Australia only) 15
Panel illumination warning light 16
Brake failure warning light (Australia only) 17
Park brake warning light 18
Fuel tank gauge 19
Coolant temperature gauge 20

Additional wired circuit on RHD vehicles for
alternative Australian park brake warning light
symbol. Connected to a black two-pin socket 21
(white & black/pink leads) located under the
binnacle, above the steering column area

MULTI-FUNCTION UNIT CONNECTIONS

A . . 12V+ supply
B . . Input to high oil temp warning light circuit
C . . Tacho drive
D . . Tachometer
E . . Spare
F . . 10V+ stabilised
G . . Input to low fuel warning light circuit
H . . Tacho signal
I . . Low fuel warning light
J . . 12V+ protected
K . . High oil temperature warning light
L . . Earth

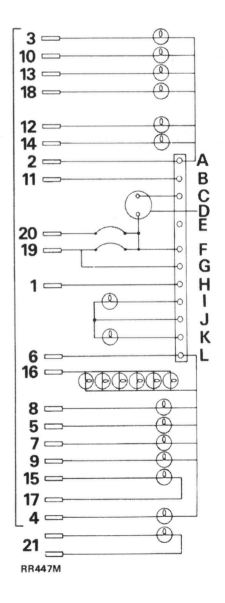

RR447M

CLOCK

– Remove and refit

Removing
1. Disconnect the battery.
2. Carefully prise the clock out of the fascia panel to reveal the electrical connections.
3. Disconnect the two electrical leads.
4. Remove the illumination lead complete with holder and bulb.
   NOTE: The clock is illuminated by a 2 watt bayonet type bulb.

Refitting
5. Reverse the removal procedure.

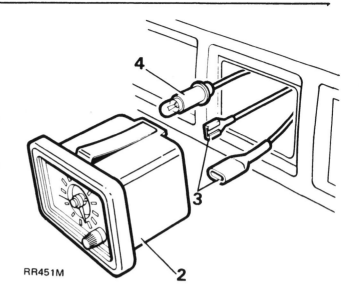

RR451M

SPEEDOMETER CABLE ASSEMBLY

The speedometer cable is a two part assembly, consisting of an upper cable, connected to the rear of the binnacle and a lower cable connected to the speedometer drive housing at the rear of the transfer gearbox. The two cables are joined by a connector ring behind the lower fascia, this connection is provided to facilitate separate renewal of either the upper or lower part of the cable in service.
To remove the upper cable refer to instrument binnacle removal.

LOWER SPEEDOMETER CABLE

– Remove and refit

Removing
1. Disconnect the battery.
2. Remove the lower fascia panel by releasing the six retaining screws.
3. Release the cable connector ring between the upper and lower cables and withdraw the cable and grommet from the bulkhead.
4. Remove the single nyloc nut and clamp securing the cable to the speedometer drive housing at the rear of the transfer gearbox.
5. Release the cable from the two retaining clips.
   NOTE: On left hand drive vehicles with automatic gearbox the speedometer cable is secured by a further two clips located above the cross-member attached to the chassis side-member.
6. Withdraw the cable from the speed drive housing.

Refitting
7. Reverse the removal procedure.

TRAILER SOCKET — OPTION

Incorporated in the vehicle electrical circuit is a facility for fitting a seven pin trailer lighting socket.

The pick up point is located behind the left hand rear tail light cluster and is accessible by removing the tail light assembly.

The pick up point consists of a seven pin pre-wired plug, a separate auxiliary fused live feed and reverse light lead.

1. Disconnect the battery.
2. Remove the rear tail light assembly and disconnect the electrical plug.
3. Remove the protective cap from the trailer pick up point plug.
4. Feed a seven core cable (fitted with a pre-wired plug to one end — suitable for connection to pick up point) down between the inner and outer body panels through the rear light aperture.
5. Feed the cable alongside the existing rear lighting harness.
6. Pull the cable through the aperture between the chassis side member and fuel tank.
7. Fit two retaining clips to the cable and secure it to the rear end cross member.
8. Connect the electrical leads to the vehicle trailer socket (refer to current trailer wiring regulations).
9. Secure trailer socket to the tow bar.
10. If it is necessary to provide a live feed and reverse light feed, provision is made for this by the presence of two extra leads in the rear light aperture. Means of identification are as follows.
    Fused auxiliary live feed — PINK LEAD
    Reverse light feed — GREEN/BROWN LEAD
11. Refit rear tail light.
12. Reconnect the battery.

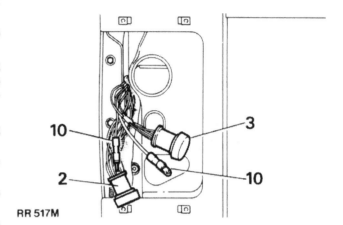

RR 517M

EXTERIOR DRIVING MIRRORS — 4 DOOR ONLY — MID 1984 TO 1985

1. The mirror housing is hinged vertically and should be set in one of the two fixed angle positions provided to suit the respective left or right side mirror location.
2. Additionally, for safety and convenience, the mirror housing is designed to fold completely forwards or rearwards against the vehicle body.

NOTE: Flat mirrors are fitted to Australian vehicles.

Setting the Mirror — Manual Version
3. The glass angle is finely adjusted by moving it vertically or horizontally as required.

continued

Setting the Mirror – Electrical-operated

4. Fine adjustment is controlled by an electric motor inside the mirror housing. This is operated by individual finger-tip operated controls fitted on either side of the steering column lower cover. To adjust, move the head of the appropriate control to the left, right, up or down as required. The mirror selected will respond accordingly.

5. The mirror also incorporates a demist facility, activated by operation of the rear window demist switch.

Renewing the Mirror Glass – Manual and Electric Versions

6. Press the inner (wider) end of the glass inwards to its full extent.

7. Insert the fingers under the outer (narrower) end of the glass, and pull outwards until the glass is released from its four retaining clips.

8. On electrical versions disconnect the two demister leads attached to back of the glass unit.

9. To replace the glass, locate the inner (wider) end of the glass in the mirror housing first.

10. Carefully press the outer (narrower) end of the glass inwards until it is safely held by its four retaining clips.

11. Reset the fine adjustment as required.

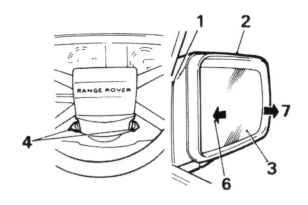

RR521M

EXTERIOR DRIVING MIRRORS – MID 1984 TO 1985

ELECTRIC MOTORS

– Remove and refit

Removing

12. Disconnect the battery.
13. Remove the mirror glass, as described in items 6 to 8.
14. Remove the four self-tapping screws securing the motor assembly to the mirror body.
15. Manoeuvre the motor assembly to reveal the electrical connections at the rear of the motor.
16. Pull the leads from the rear of the motor assembly.

Refitting

17. Reverse operations 12 to 16, ensuring that the electrical leads are correctly refitted (see electric mirrors, circuit diagram).

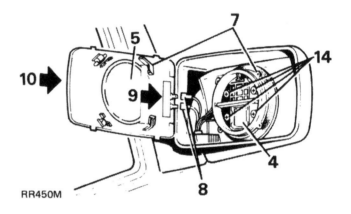

RR450M

EXTERIOR DRIVING MIRRORS – FINGER TIP CONTROLLED SWITCHES

– Remove and refit

Service Tool: 18G 1014 Extractor for Steering Wheel
18G 1014-2 Adaptor pins

Removing

18. Disconnect the battery.

continued

19. Release the screw retaining the centre cover to the steering wheel and remove the cover.
20. Remove the retaining nut and washer securing the steering wheel.
21. Remove the steering wheel using the correct service tool.
22. Remove the lower fascia panel by releasing the six securing screws.
23. Remove the four screws securing the bottom shroud to the top shroud.
24. Remove the single screw securing the top shroud to the switch housing bracket.
25. Remove the top shroud.
26. Release the light switch locknut and remove the switch from the mounting bracket.
27. Manoeuvre the bottom shroud to clear the ignition switch/steering lock assembly.
28. Retrieve the small spacing collar located on the forward left hand side of the bottom shroud.
    NOTE: It is important that the spacing collar is refitted on assembly.
29. Pull the multi-plug from the rear of the finger tip controlled mirror switch.
30. Carefully prise off the finger tip button at the operating end of the switch.
31. Unscrew the black plastic retaining collar securing the switch to the bottom shroud.
32. Remove the switch from the shroud.

Refitting

33. Reverse operations 18 to 32, ensuring that the black spacing collar is refitted.
    NOTE: To prevent damage occurring to the electrical wiring within the top and bottom shrouds, the leads should be arranged carefully to avoid contact with mating faces on re-assembly.

continued

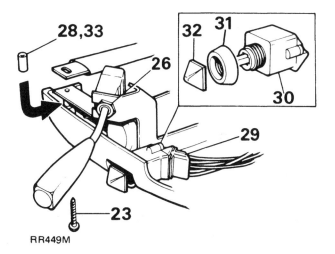

RR449M

687

EXTERIOR DRIVING MIRRORS

COMPLETE ASSEMBLY

— Remove and refit

Removing
34. Disconnect the battery.
35. Carefully prise off the interior finisher plate to reveal the three securing screws and electrical wiring.
36. Disconnect the two electrical plugs (one two pin, one three pin).
37. Supporting the exterior mirror assembly remove the three securing screws (with plain and spring washers).
38. Pull the inner mounting plate away from the inner door frame complete with the two retaining clips.
39. Detach the mirror assembly from the outer door frame.
40. Remove the sealing rubber.

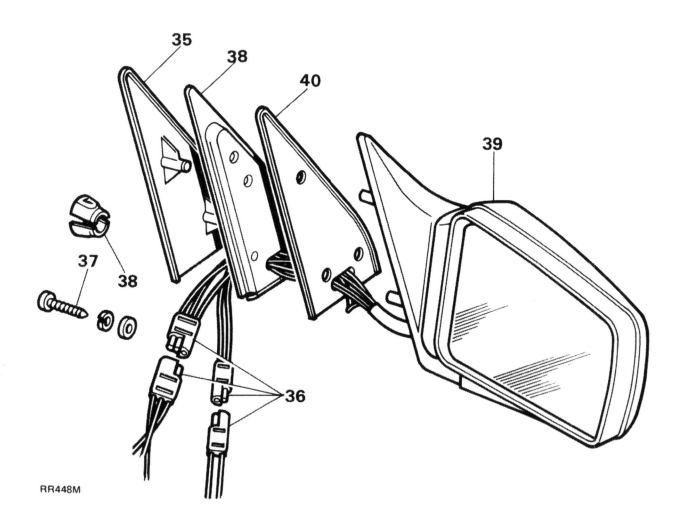

RR448M

Refitting
41. Reverse operations 34 to 40.
    NOTE: To prevent damage occurring to the electrical wiring, do not push the leads down inside the door casing.

ELECTRICALLY OPERATED SIDE DOOR WINDOWS – MID 1984 TO 1985

All side windows are operated from either front seat by four rocker switches fitted in the centre floor-mounted cover which control individual electric motors in each door. Additionally, for the convenience of rear seat passengers, both rear side doors are also fitted with a rocker switch, integral with the door pull handles, which provides independent rear side window control.

If required, these additional switch controls can be rendered inoperative, or restored by operation of an isolating rocker switch, also fitted in the centre floor-mounted cover.

The electrical circuit is directed through the fuse box and is protected by a 25 amp fuse (No. 20 in fuse box).

NOTE: Windows are only operative whilst the ignition is switched on.

1. Left side, front window ⎤
2. Right side, front window ⎥— operating switches
3. Left side, rear window ⎥
4. Right side, rear window ⎦

5. Isolating switch – door fitted switches, rear windows.

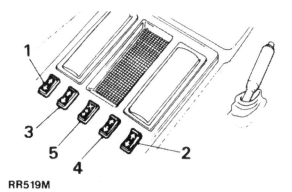

RR519M

Operating switches (1 to 4)
To lower glass depress rear of switch, to raise glass depress front of switch.
Release switch as soon as window is fully open or closed.

Isolating switch (5)
To isolate door fitted switches on rear windows depress front of the central switch in the floor mounted cover.
To restore independent rear passenger control depress rear of switch.

Rear side door switches (6)
Depress switch as indicated to lower or raise the glass.

CAUTION: Do not attempt to raise or lower a window when it is jammed by ice. Should a window be obstructed, when being raised or lowered, a thermal cut-out will render the window inoperative. In this event, release the switch and remove the obstruction. The window can be re-operated after two seconds. In the event of a motor failure, the window involved will become inoperative necessitating the renewal of the motor unit.

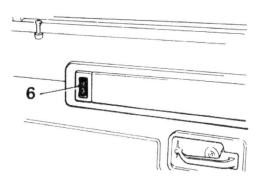

WARNING: Particular care should be taken that children are kept away from windows when raising or lowering is in progress.

WINDOW LIFT MOTOR – FRONT DOORS

– Remove and refit

Removing

1. Ensure the window is in its fully closed position and secure it with adhesive tape to prevent the window dropping down.
2. Disconnect the battery.
3. Detach the armrest/door pull finisher to reveal the two securing screws.
4. Remove the two securing screws (with plain washers) to enable the armrest/door pull to be detached from the inner door panel.
5. Remove the interior door handle finisher button to reveal the screw retaining the handle surround.
6. Remove the screw and detach the handle surround from the inner door panel.
7. Detach the inner door trim pad by inserting a screw-driver between the trim pad and inner door panel gently prising out the nine plastic securing clips from their respective holes in the inner door panel.
8. Disconnect the two speaker connections inside the door and remove the door trim pad complete with speaker.
   NOTE: At this stage the speaker can be removed by releasing the four nuts (with plain washers) located on the back of the trim pad.
9. Peel back the front top corner of the plastic weather sheet to reveal the window lift motor
10. Release the window lift motor harness from the three retaining clips to allow the harness to be pulled out of the aperture at the front of the inner door panel.
11. Disconnect the window lift motor multi-plug from the main door harness.
12. Supporting the motor, remove the three securing bolts.
13. Withdraw the motor through the top front aperture of the door.

Refitting

14. Reverse operations 1 to 13.
    NOTE: Ensure the drive gear is engaged and correctly aligned with the window lift linkage before proceeding.

continued

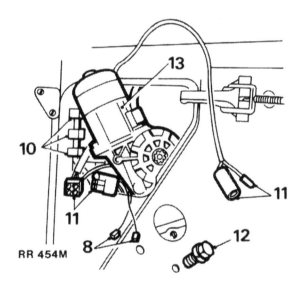

RR 454M

WINDOW LIFT MOTOR — REAR

— Remove and refit

Removing

15. Ensure the window is in its fully closed position, and secure it with adhesive tape over the top of the door to prevent the window dropping down.
16. Disconnect the battery.
17. Remove the armrest/door pull finisher to reveal the two securing screws.
18. Remove the two securing screws (with plain washers) detach the armrest/door pull from the inner door panel. Disconnect the multi-plug at the rear of the window lift switch.

    NOTE: At this stage the window lift switch can be removed by applying pressure to the rear of the switch, pushing it through the door pull moulding.
19. Remove the interior handle finisher button to reveal the screw retaining the handle surround.
20. Remove the screw and detach the handle surround from the door trim pad.
21. Detach the door trim pad by inserting a screwdriver between the trim pad and inner door panel, gently prising out the six plastic securing clips from their respective holes in the inner door panel.
22. Displace the bottom half of the plastic weather sheet to reveal the window lift motor.
23. Release the lift motor harness from the retaining clips.
24. Disconnect the lift motor harness snap connections from the main door harness.
25. Supporting the lift motor release the three bolts securing the motor to the inner door panel.
26. Withdraw the lift motor from the lower aperture in the inner door panel.

Refitting

27. Reverse operations 15 to 26.
28. Ensure the lift motor drive gear is engaged and correctly aligned with the window lift linkage before proceeding.

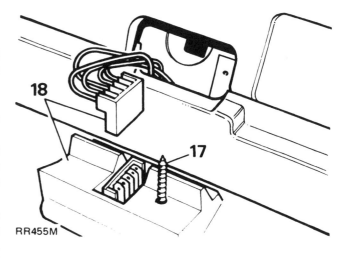

RR455M

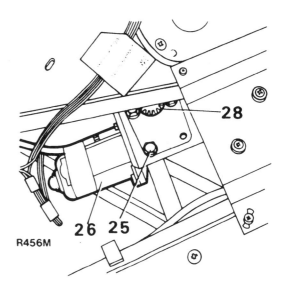

R456M

ELECTRICALLY OPERATED CENTRAL DOOR LOCKING SYSTEM – MID 1984 TO 1985

An electrically operated central door locking system is fitted as an option on four door models.

Locking or unlocking the drivers door from outside by key operation, or from inside by sill knob automatically locks or unlocks all four doors.

Rear doors can be independently locked or unlocked from inside by sill knob operation but can be overridden by further operation of the drivers door locking control.

On earlier models the central door locking system is controlled by either front door as shown in circuit diagram Number 1. This system incorporated a control box, located under the fascia adjacent to the steering column.

On later models the locking system is controlled by the drivers door lock actuator as detailed in circuit diagram Number 2.

The circuit is directed through the fuse box and is protected by a 10 amp fuse (No. 19 in fuse box).

On rear doors only a childrens safety lock is provided which can be mechanically preset to render the interior door release handles inoperative.

NOTE: Circuit diagrams have been previously released in Workshop Bulletin 01/85.

Failure of an actuator will not affect the electric locking of the remaining three doors. The door with the inoperative actuator can still be unlocked or locked manually.

NOTE: The door lock actuator units contain non-serviceable parts. If a fault should occur replace the unit concerned with a new one. Before carrying out any maintenance work disconnect the battery.

DOOR LOCK ELECTRICAL CONTROL UNIT

– Fitted to vehicles with two-door locking control only

– Remove and refit

Removing
1. Release the lower fascia panel by removing the six self-tapping screws.
2. Disconnect the electrical leads by releasing the multi-plug from the bottom end of the control unit.
3. Remove the two self-tapping screws securing the control unit to the outer side of the steering column support bracket and remove the control unit.

Refitting
4. Reverse instructions 1 to 3.

FRONT DOOR ACTUATOR UNITS

– Remove and refit

Removing
1. Ensure the window is in its fully closed position.
2. Remove the armrest/door pull finisher to reveal the two retaining screws securing the armrest door pull to the inner door panel.
3. Remove the interior door handle finisher button to reveal the screw retaining the handle surround.
4. Release the screw and remove the handle surround from the interior door trim pad.
   NOTE: On earlier models with manually operated side windows it is necessary to remove the window regulator handle to enable the trim pad to be removed.
5. Release the door trim pad by inserting a screwdriver between the trim pad and the inner door panel, gently prising out the nine plastic clips from their respective holes around the edges of the trim pad.
6. Disconnect the two speaker connections inside the door and remove the door trim pad complete with speaker.
7. Peel back the top of the plastic weather sheet at the rear of the inner door panel to expose the lock actuator unit.
8. Remove the four screws and plain washers securing the lock actuator mounting plate to the inner door panel.
9. Release the clip retaining the electrical cable.
10. Manoeuvre the actuator assembly to detach the operating rod 'eye' from the hooked end of the actuator link on the door lock.
11. Withdraw the actuator assembly from the door until the electrical cable is pulled out of its channel sufficiently to expose the connectors which can then be detached.
12. Remove the actuator assembly from the door.
13. The actuator unit may be changed if necessary by removing the two rubber mounted screws which secure it to the mounting plate.

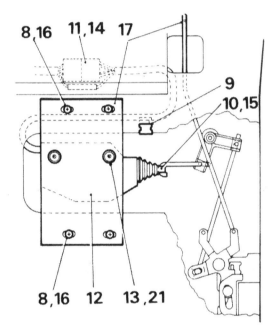

Refitting
14. Locate the actuator assembly in the inner door panel and fit the electrical cable connectors. The cable, and connectors, are pulled back into the channel from the front end and the cable clip refitted.
15. Manoeuvre the actuator assembly to engage the operating rod 'eye' on the hooked actuator link.
16. Loosely fit the actuator mounting plate to the inner door panel with the four screws, setting the mounting plate in the centre of the slotted holes.
17. Ensure that manual operation of the sill locking control is not restricted by the operation of the actuator operating rod and vice versa, resetting the mounting plate as necessary.
18. Reconnect the vehicle battery.

continued

19. Check that electrical operation of the door lock occurs when the sill locking control is moved through half of its total movement. Reset the mounting plate if necessary and tighten the four screws.
    NOTE: The above adjustment ensures that the full tolerance on the switching operation is utilised.

REAR DOOR ACTUATOR UNITS

Remove and refit

Instructions as for front doors with the following exceptions:

20. No radio speaker is involved.
21. The electrical cable and plug is retained to the inner door panel by two spring clips and is immediately accessible through the large aperture in the door.
22. Instruction 19 does not apply to rear actuator units which are not fitted with switches.
    NOTE: If necessary the lock actuator may be detached from its mounting plate to facilitate the removal of the lock actuator from the connector rod inside the door panel.

SPLIT CHARGING FACILITY – OPTION

The circuit provides an additional source of electrical supply allowing separate charging and discharging of an additional battery for auxiliary equipment without affecting the charge state of the vehicle's main battery.

A terminal bracket, heavy duty relay and cables are fitted on the left hand front wing valance.

The additional battery, leads and fixing clamps are NOT included in the option.

The split charging system is controlled by a voltage sensitive switch which energises the relay when the ignition voltage exceeds a pre-set level, thus supplying current to the positive terminal on the terminal bracket. Conversely, if the ignition voltage falls below the pre-set level the split charging circuit will cut-out.

To operate split charging system

1. Install the additional battery.
2. Remove the fixing bolt securing the terminal bracket cover.
3. Ensure that the positive and negative leads are correctly connected to the terminals and the battery in accordance with the markings on the terminal bracket cover.
4. Start the engine. As soon as the alternator is charging the vehicle battery and the voltage exceeds the pre-set level the split charge function will operate.
5. After charging replace the terminal bracket cover.

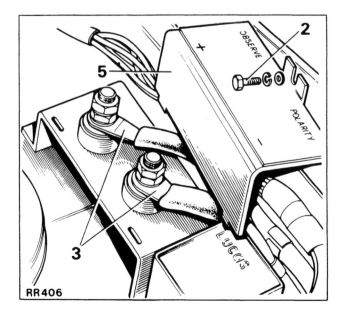

RR406

General service information

1. Introduction

Before any component of the air conditioning system is removed the system must be depressurised. When the component is replaced the system must be evacuated to remove all traces of old refrigerant and moisture. Then the system must be recharged with new refrigerant.

Any service work that requires loosening of a refrigerant line connection should be performed only by qualified service personnel. Refrigerant and/or oil will escape whenever a hose or pipe is disconnected.

All work involving the handling of refrigerant requires special equipment, a knowledge of its proper use and attention to safety measures.

2. Servicing equipment

The following equipment is required for full servicing of air conditioning.

Charging trolley.
Service valve adaptors.
Valve core removers.
Leak detector.
Tachometer.
Lock ring spanner.
Valve key.
Safety goggles.
Refrigerant charging line gaskets.
Compressor dip stick.
5/8in. UNC bolt or Union nut (Part Number 534127) for extraction of the compressor pulley.
Thermometer $-20°C$ to $-60°C$ ($0°F$ to $-120°F$).

3. Servicing materials

Refrigerant: Freon R12.
Nominal charge weight:
    RHD vehicles — 1.25 kg (44 oz)
    LHD vehicles — 1.08 kg (38 oz)

CAUTION: Methychloride refrigerants must not be used. Compressor oil: See Recommended Lubricants, Section 09, alternatives.

4. Precautions in handling refrigerant

Refrigerant 12 is transparent and colourless in both the gaseous and liquid state. It has a boiling point of $-30°C$ ($-22°F$) and at all normal pressures and temperatures it is a vapour. The vapour is heavier than air, non-flammable and non-explosive. It is non-poisonous except when in contact with an open flame, and non-corrosive until it comes into contact with water.

The following precautions in handling refrigerant 12 should be observed at all times:

a. Do not leave a drum of refrigerant without its heavy metal cap fitted.

b. Do not carry a drum in the passenger compartment of a car.

c. Do not subject drums to a high temperature.

d. Do not weld or steam clean near an air conditioning system.

e. Do not discharge refrigerant vapour into an area with an exposed flame, or into the engine air intake. Heavy concentrations of refrigerant in contact with a live flame will produce a toxic gas that will also attack metal.

f. Do not expose the eyes to liquid refrigerant. ALWAYS wear safety goggles.

5. Precautions in handling refrigerant lines

WARNING: Always wear safety goggles when opening refrigerant connections.

a. When disconnecting any pipe or flexible connection the system must be discharged of all pressure. Proceed cautiously, regardless of gauge readings. Open connections slowly, keeping hands and face well clear, so that no injury occurs if there is liquid in the line. If pressure is noticed allow it to bleed off slowly.

b. Lines, flexible end connections and components must be capped immediately they are opened to prevent the entrance of moisture and dirt.

c. Any dirt or grease on fittings must be wiped off with a clean alcohol dampened cloth. Do not use chlorinated solvents such as trichloroethylene. If dirt, grease or moisture cannot be removed from inside pipes, they must be replaced with new.

d. All replacement components and flexible end connections are sealed, and should only be opened immediately prior to making the connection. (They must be at room temperature before uncapping to prevent condensation of moisture from the air that enters.)

e. Components must not remain uncapped longer than 15 minutes. In the event of delay the caps must be replaced.

f. Receiver driers must never be left uncapped as they contain Silica Gel which will absorb moisture from the atmosphere. A receiver drier left uncapped must be replaced, and not used.

g. A new compressor contains an initial charge of 11 UK fluid ozs (312.5 ml) of oil when received, part of which is distributed throughout the system when it has been run. The compressor contains a holding charge of gas when received which should be retained until the hoses are connected.

h. The compressor shaft must not be rotated until the system is entirely assembled and contains a charge of refrigerant.

CONDENSER FANS AND MOTORS (TWIN FAN SYSTEM)

– Remove and refit   82.15.01

Removing fans and motors
1. Open the bonnet and disconnect the battery.
2. Remove the six screws and withdraw the grille panel.
3. Remove the insulation tape from the wiring harness to expose snap connectors. Make a note of the wiring colours to facilitate reconnection and disconnect snap connector.
4. Disconnect earth wiring retaining bolt.
5. Remove wiring securing clip.
6. Slacken the two upper bolts securing the left hand and right hand bonnet striker support stays.
7. Remove the lower bolts securing the lower ends of the stays and pivot both stays forward.
8. Remove the two upper bolts securing the fans (one for each fan).
9. Remove the four lower bolts securing the fans (two for each fan).
10. Turn each fan and motor assembly in an anti-clockwise direction and carefully remove from the vehicle.

NOTE: Later models have fan motor assemblies mounted on two bars across the front of the condenser. Follow instructions 1 to 7 above, then remove two nuts and washers securing each fan motor and withdraw the assembly.

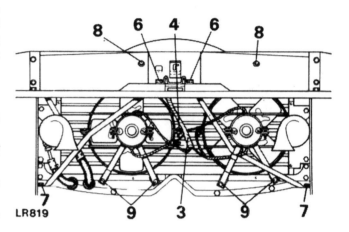

LR819

To dismantle fan motor and cowl assembly
NOTE: The fan cowl is deleted on later models.
11. Slacken the fan blade grub screw and withdraw the fan blades from the motor drive shaft. Make a note of the exact location of the fan blades on the shaft to facilitate reassembly.
12. Remove the wiring securing clip.
13. Slacken the fan cowl clamp screws.
14. Slacken the stay bracket clamp screws and remove the fan cowl and stay bracket.

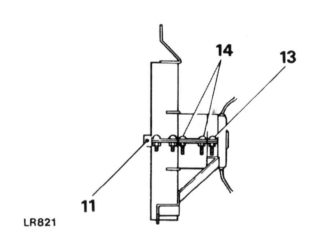

LR821

To reassemble fan motor and cowl assembly
15. Fit stay bracket to fan motor.
16. Fit the fan cowl.
17. Correctly position fan motor and tighten clamp screws.
18. Locate fan in correct position on drive shaft and secure with grub screw. Check that the fan blade rotates freely.
19. Fit cable securing clip.

To fit fans and motors to vehicle
20. Offer up each assembly into its mounting position.
21. Fit the upper securing bolts (one for each assembly) but leave slack for the time being.
22. Fit the four lower securing bolts (two for each assembly).
23. Tighten all the securing bolts.
24. Secure earth wire.
25. Fit and tighten the bonnet striker support stays.
26. Fasten snap connector and apply insulation tape to the harness.
27. Fit the front grille and secure with the six screws.

CONDENSER (TWIN FAN SYSTEM)

– Remove and refit   82.15.07

Removing

NOTE: On later models it is not possible to withdraw the condenser through the grille aperture.

Ignore instructions 4 to 9 below. Remove radiator 26.40.04. Remove six condenser mounting bolts and withdraw condenser complete with fan motor assemblies.

1. Open the bonnet and disconnect the battery.
2. Depressurise the air condition system – refer to operation 82.30.05.
3. Remove six screws and withdraw the front grille panel.
4. Disconnect the left-hand and right-hand horn electrical leads.
5. Remove the horn bracket securing bolts and remove both horns.
6. Mark the position of the bonnet striker plate.
7. Remove the two bolts securing the striker plate and striker plate diagonal support stays.
8. Move the support stays aside to gain access to the condenser.
9. Remove the fan and motor assemblies, see operation 82.15.01 instructions 8 to 10.

CAUTION: Before carrying out instruction 10 protect the eyes with safety goggles and wear protective gloves.

10. Using two spanners on each union, carefully disconnect the pipes at the condenser end. Fit blanks to the exposed ends of the pipes.
11. Remove the six bolts retaining the condenser and withdraw the condenser, from the vehicle, through the grille aperture.

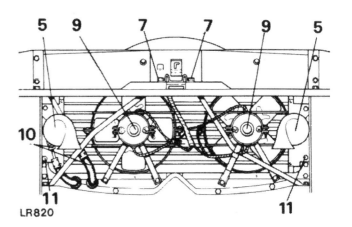

LR820

Refitting

12. Place the condenser into position and secure with the six bolts.
13. Remove the blanks from the pipes and fit new 'O' rings to the pipes.
14. Apply refrigerant oil to the pipe threads to aid sealing.
15. Connect the pipes to the condenser and tighten to the correct torque:–
    Compressor hose 3.4 to 3.9 kgf.m (24 to 29 lbf.ft)
    Receiver drier hose 1.4 to 21. kgf.m (10 to 15 lbf.ft)
16. Fit the fan and motor assemblies.
17. Locate the bonnet striker support stays and align striker to previously made marks and tighten the retaining bolts.
18. Tighten the striker plate diagonal support stay lower bolts.
19. Refit the two horns, and connect the electrical leads.
20. Connect the battery and test the horns.
21. To compensate the oil loss, add 2 UK fluid ozs (56.8 ml) of the correct oil to the compressor.
22. Evacuate the system, operation 82.30.06.
23. Charge the system, operation 82.30.08.
24. Carry out a leak test on the disturbed joints, see operation 82.30.09.
25. Check the complete system as described in operation 82.30.16.
26. Fit the grille panel.

AIR CONDITIONING SYSTEM

– Charge   **82.30.08**

CAUTION: Do not charge liquid refrigerant into the compressor. Liquid cannot be compressed, and if liquid refrigerant enters the compressor inlet valve severe damage is possible. In addition, the oil charge may be absorbed, with consequent damage when the compressor is operated.

NOTE: Nominal charge weight:
    RHD vehicles 1.25 kg (44 oz)
    LHD vehicles 1.08 kg (38 oz)

Charging

1. Fit the charging and testing equipment, 82.30.01, as described for evacuating.
2. Evacuate the air conditioning system, 82.30.06, allowing 1.37 kg (3 lb) of refrigerant to enter the charging cylinder.
3. Put on safety goggles.
4. Close the low pressure valve (No. 1).
5. Open the refrigerant control valve (No. 4) and release liquid refrigerant into the system through the compressor discharge valve. The pressure in the system will eventually balance.
6. If the full charge of liquid refrigerant will not enter the system, proceed with items 7 to 12.
7. Reconnect the charging and testing equipment as described for charging with gaseous refrigerant, 82.30.01.
8. Open the low pressure valve (No. 1).
9. Open valve No. 3.
10. Close the high pressure valve (No. 2).
11. Start and run the engine at 1000 to 1500 rev/min and allow refrigerant to be drawn through the low pressure valve (No. 1) until the full charge has been drawn into the system.
12. Close valve number 1 and 3.
13. Close valve No. 4.
14. Check that the air conditioning system is operating satisfactorily by carrying out a pressure test 82.30.10.

CAUTION: Do not overcharge the air conditioning system as this will cause excessive head pressure.

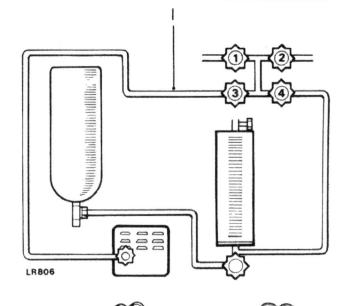

LR806

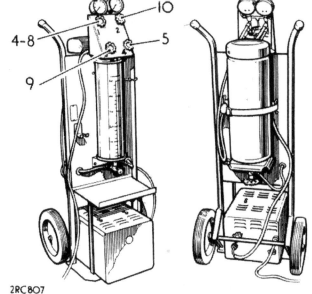

2RC807

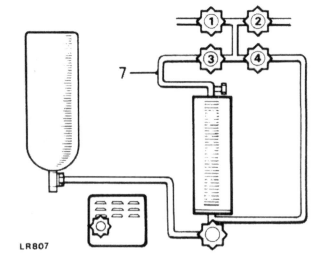

LR807

Brooklands Books Ltd., PO Box 146, Cobham,
Surrey KT11 1LG, England Phone: (44) 1932 865051
E-mail: sales@brooklands-books.com   www.brooklandsbooks.com

ISBN: 9781855201224   Part No. AKM3630   Ref: RR10WH   5T5/2363

LAND ROVER OFFICIAL FACTORY PUBLICATIONS

Land Rover Series 1 Workshop Manual	4291
Land Rover Series 1 1948-53 Parts Catalogue	4051
Land Rover Series 1 1954-58 Parts Catalogue	4107
Land Rover Series 1 Instruction Manual	4277
Land Rover Series 1 and II Diesel Instruction Manual	4343
Land Rover Series II and IIA Workshop Manual	AKM8159
Land Rover Series II and Early IIA Bonneted Control Parts Catalogue	605957
Land Rover Series IIA Bonneted Control Parts Catalogue	RTC9840CC
Land Rover Series IIA, III and 109 V8 Optional Equipment Parts Catalogue	RTC9842CE
Land Rover Series IIA/IIB Instruction Manual	LSM64IM
Land Rover Series 2A and 3 88 Parts Catalogue Supplement (USA Spec)	606494
Land Rover Series III Workshop Manual	AKM3648
Land Rover Series III Workshop Manual V8 Supplement (edn. 2)	AKM8022
Land Rover Series III 88, 109 and 109 V8 Parts Catalogue	RTC9841CE
Land Rover Series III Owners Manual 1971-1978	607324B
Land Rover Series III Owners Manual 1979-1985	AKM8155
Military Land Rover (Lightweight) Series III Parts Catalogue	61278
Military Land Rover Series III (L.W.B.) User Handbook	608179
Military Land Rover (Lightweight) Series III User Manual	608180
Land Rover 90/110 and Defender Workshop Manual 1983-1992	SLR621ENWM
Land Rover Defender Workshop Manual 1993-1995	LDAWMEN93
Land Rover Defender 300 Tdi and Supplements Workshop Manual 1996-1998	LRL0007ENGBB
Land Rover Defender Td5 Workshop Manual and Supplements 1999-2006	LRL0410BB
Land Rover Defender Electrical Manual Td5 1999-06 and 300Tdi 2002-2006	LRD5EHBB
Land Rover 110 Parts Catalogue 1983-1986	RTC9863CE
Land Rover Defender Parts Catalogue 1987-2006	STC9021CC
Land Rover 90 • 110 Handbook 1983-1990 MY	LSM0054
Land Rover Defender 90 • 110 • 130 Handbook 1991 MY - Feb. 1994	LHAHBEN93
Land Rover Defender 90 • 110 • 130 Handbook Mar. 1994 - 1998 MY	LRL0087ENG/2
Military Land Rover 90/110 All Variants (Excluding APV and SAS) User Manual	2320-D-122-201
Military Land Rover 90 and 110 2.5 Diesel Engine Versions User Handbook	SLR989WDHB
Military Land Rover Defender XD - Wolf Workshop Manual - 2320D128 -	302 522 523 524
Military Land Rover Defender XD - Wolf Parts Catalogue	2320D128711
Discovery Workshop Manual 1990-1994 (petrol 3.5, 3.9, Mpi and diesel 200 Tdi)	SJR900ENWM
Discovery Workshop Manual 1995-1998 (petrol 2.0 Mpi, 3.9, 4.0 V8 and diesel 300 Tdi)	LRL0079BB
Discovery Series II Workshop Manual 1999-2003 (petrol 4.0 V8 and diesel Td5 2.5)	VDR100090/6
Discovery Parts Catalogue 1989-1998 (2.0 Mpi, 3.5, 3.9 V8 and 200 Tdi and 300 Tdi)	RTC9947CF
Discovery Parts Catalogue 1999-2003 (petrol 4.0 V8 and diesel Td5 2.5)	STC9049CA
Discovery Owners Handbook 1990-1991 (petrol 3.5 V8 and diesel 200 Tdi)	SJR820ENHB90
Discovery Series II Handbook 1999-2004 MY (petrol 4.0 V8 and Td5 diesel)	LRL0459BB
Freelander Workshop Manual 1998-2000 (petrol 1.8 and diesel 2.0)	LRL0144
Freelander Workshop Manual 2001-2003 ON (petrol 1.8L, 2.5L and diesel Td4 2.0)	LRL0350ENG/4
Land Rover 101 1 Tonne Forward Control Workshop Manual	RTC9120
Land Rover 101 1 Tonne Forward Control Parts Catalogue	608294B
Land Rover 101 1 Tonne Forward Control User Manual	608239
Range Rover Workshop Manual 1970-1985 (petrol 3.5)	AKM3630
Range Rover Workshop Manual 1986-1989	SRR660ENWM &
(petrol 3.5 and diesel 2.4 Turbo VM)	LSM180WS4/2
Range Rover Workshop Manual 1990-1994	
(petrol 3.9, 4.2 and diesel 2.5 Turbo VM, 200 Tdi)	LHAWMENA02
Range Rover Workshop Manual 1995-2001 (petrol 4.0, 4.6 and BMW 2.5 diesel)	LRL0326ENGBB
Range Rover Workshop Manual 2002-2005 (BMW petrol 4.4 and BMW 3.0 diesel)	LRL0477
Range Rover Electrical Manual 2002-2005 UK version (petrol 4.4 and 3.0 diesel)	RR02KEMBB
Range Rover Electrical Manual 2002-2005 USA version (BMW petrol 4.4)	RR02AEMBB
Range Rover Parts Catalogue 1970-1985 (petrol 3.5)	RTC9846CH
Range Rover Parts Catalogue 1986-1991 (petrol 3.5, 3.9 and diesel 2.4 and 2.5 Turbo VM)	RTC9908CB
Range Rover Parts Catalogue 1992-1994 MY and 95 MY Classic	
(petrol 3.9, 4.2 and diesel 2.5 Turbo VM, 200 Tdi and 300 Tdi)	RTC9961CB
Range Rover Parts Catalogue 1995-2001 MY (petrol 4.0, 4.6 and BMW 2.5 diesel)	RTC9970CE
Range Rover Owners Handbook 1970-1980 (petrol 3.5)	606917
Range Rover Owners Handbook 1981-1982 (petrol 3.5)	AKM8139
Range Rover Owners Handbook 1983-1985 (petrol 3.5)	LSM0001HB
Range Rover Owners Handbook 1986-1987 (petrol 3.5 and diesel 2.4 Turbo VM)	LSM129HB

Engine Overhaul Manuals for Land Rover and Range Rover

300 Tdi Engine, R380 Manual Gearbox and LT230T Transfer Gearbox Overhaul Manuals	LRL003, 070 & 081
Petrol Engine V8 3.5, 3.9, 4.0, 4.2 and 4.6 Overhaul Manuals	LRL004 & 164
Land Rover/Range Rover Driving Techniques	LR369
Working in the Wild - Manual for Africa	SMR684MI
Winching in Safety - Complete guide to winching Land Rovers and Range Rovers	SMR699MI

Workshop Manual Owners Edition
Land Rover 2 / 2A / 3 Owners Workshop Manual 1959-1983
Land Rover 90, 110 and Defender Workshop Manual Owners Edition 1983-1995
Land Rover Discovery Workshop Manual Owners Edition 1990-1998

All titles available from Amazon or Land Rover specialists
Brooklands Books Ltd., P.O. Box 146, Cobham, Surrey, KT11 1LG, England, UK
Phone: +44 (0) 1932 865051   info@brooklands-books.com   www.brooklands-books.com

www.brooklandsbooks.com

Printed in Great Britain
by Amazon

57269520R00388